Nonlinear Dynamics of Continuous Elastic Systems

Springer

Berlin
Heidelberg
New York
Hong Kong
London
Milan
Paris
Tokyo

J. Awrejcewicz · V. Krys'ko · A.F. Vakakis

Nonlinear Dynamics of Continuous Elastic Systems

With 125 Figures and 27 Tables

Springer

Authors

Prof.Dr. Jan Awrejcewicz
Technical University Lodz
Dept. of Automatics and Biomechanics
ul. Stefanowskiego 1/15
90-924 Lodz, Poland
E-mail: awrejcew@p.lodz.pl

Prof. Vadim A. Krys'ko
Saratov State Technical University
Dept. of Mathematics/Mechanics
Polyteshnycheskaya ul. 77
410054 Saratov, Russia
E-mail: kva@agma.sstu.saratov.su

Prof. Alexander F. Vakakis
National Technical
University of Athens
Dept. of Mechanical Engineering
Heroon Polytechnioupoli 9
157 10 Athens, Greece
E-mail: vakakis@central.ntua.gr

ISBN 3-540-20515-2   Springer-Verlag Berlin Heidelberg New York

Cataloging-in-Publication Data applied for
A catalog record for this book is available from the Library of Congress.
Bibliographic information published by Die Deutsche Bibliothek
Die Deutsche Bibliothek lists this publication in the Deutsche Nationalbibliografie; detailed bibliographic data is available in the Internet at <http://dnb.ddb.de>.

This work is subject to copyright. All rights are reserved, whether the whole or part of the material is concerned, specifically the rights of translation, reprinting, reuse of illustrations, recitation, broadcasting, reproduction on microfilm or in other ways, and storage in data banks. Duplication of this publication or parts thereof is permitted only under the provisions of the German Copyright Law of September 9, 1965, in its current version, and permission for use must always be obtained from Springer-Verlag. Violations are liable for prosecution under German Copyright Law.

Springer-Verlag is a part of Springer Science+Business Media

springeronline.com

© Springer-Verlag Berlin Heidelberg 2004
Printed in Germany

The use of general descriptive names, registered names, trademarks, etc. in this publication does not imply, even in the absence of a specific statement, that such names are exempt from the relevant protective laws and regulations and therefore free for general use.

Typesetting: Data conversion by the authors
Final processing by PTP-Berlin Protago-TeX-Production GmbH, Berlin
Cover-Design: deblik, Berlin
Printed on acid-free paper   62/3020Yu - 5 4 3 2 1 0

Preface

There are many monographs in the existing literature devoted to the static and dynamic behavior of plates and shells. Plates and shells are encountered often in engineering applications being integral parts of a wide range of constructions, such as machines, vehicles, airplanes, rockets, ships, bridges, buildings, and containers, to name a few. In addition to the usual requirements posed by engineers related to lightweightness, sufficient rigidity or flexibility, and robust stability properties, there is an additional class of *a priori* dynamical properties required by modern engineering applications where non-homogeneity and non-uniformity of structural components is often the norm in certain applications. In addition, strict operational requirements in modern engineering applications towards higher speeds, lighter construction, robust and reliable performance, dictates smaller margins of error or deviations from prescribed performances in adverse or uncertain forcing environments. This, in turn, requires the development of new analytical and computational tools capable of addressing challenging and not very well developed topics, such as, nonlinearities affecting the system performance, the effects of unmodeled dynamics on the stability of operation, and the role of uncertainties in the system parameters on the structural response. As a result, there is an ongoing effort to address such issues, leading to the development of new analytical and computational tools, some of which are discussed in this monograph.

The monograph follows an approach based on an integrated treatment of analysis and computation. Such a hybrid approach, coupled with computer algebra, can lead to results that cannot be obtained by other standard theories in the field. We show, that in a wide class of problems only a carefully prepared numerical experiment followed by purely mathematical considerations can finally lead to the sought results. The numerous analytical constructions are illustrated by examples of application and computational results. The figures, mathematical derivations and described algorithms are balanced so to attract potential readers, even those without a rigorous mathematical background.

Certain of the topics presented in this work can be considered as novel for a monograph on continuous elastic structures of this type. These include techniques and bibliography from the Russian literature not available until now in the Western Scientific and Technical Community; one of the first expositions in a monograph of the Proper Orthogonal (or Karhunen-Loeve) Decomposition (POD) methodology as applied to the dynamics of elastic continuous structures; application of a new technique based on non-smooth transformations to the analytic/numerical study of discretely supported or forced continuous systems; and certain techniques dealing with clearance and vibro-impact localization phenomena in coupled continuous oscillators. Many of these results were published dispersedly in the archival literature during the last five years, and, as a result, have not attracted yet the attention of the vast part of the engineering community.

Chapter 1 of the monograph is devoted to the analysis of plates and shells with added elements (discrete additives). First a literature and research review is given implying the necessity of developing more accurate techniques for studying the dynamics of plates and shells in the framework of both non-classical and three dimensional (3D) elasticity theories. Special emphasis is given to the analysis of the full stress-strain distribution along the thickness of the plate (which is three dimensional in the neighborhoods of the joints between the continuous system and the discrete added masses). Furthermore, the obtained results within the context of 3D-theory provide the limits of validity of classical lower dimensional theories. Numerical algorithms together with numerical applications are included.

Chapter 2 deals with the so-called rational design of plates and shells. After a brief introduction, frequency spectra properties of orthotropic shells with varying thickness are described. In addition, the finite dimensional approximation of the problem of free vibration of a Timoshenko-like shell is discussed. Special attention is paid to two different methods of numerical analysis of plates and shells with variable thickness, namely the method of Bubnov-Galerkin (BG), and the method of variational iterations (MVI). These methods are widely recognized as being the most effective for solving various statics and dynamics problems in the theory of plates and shells. The main problem of the BG method for dynamic problems relates to the lack of a relatively simple estimate for the errors involved in the discrete approximation, and to the lack of a strict mathematical formulation for the proper choice of basis functions. Both aspects are addressed in this monograph. Then, the issue of mass optimization of plates and shells with free vibration frequency constraints is formulated and rigorously discussed. Finite dimensional approximate techniques solutions for solving the optimization problem are outlined and numerically implemented. In addition, the optimization of the surfaces of plates and shells with constraints related to the distribution of their spectral lines is considered, together with optimization studies related to vibration isolation of plates subject to applied harmonic loads.

In the following three Chapters we present some new techniques for order reduction, vibro-impact analysis and discrete effects estimation of elastic structures. In Chapter 3 we introduce the method of proper orthogonal decomposition (POD). Whereas this method has been successfully applied in the past to many engineering fields including, fluid mechanics, acoustics, and optics engineering, its applications to structural dynamics are rather limited and very recent. This method provides an optimal orthogonal base in functional space for decomposing the dynamics of discrete or continuous oscillators. The POD modes (e.g., the elements of the basis) are axes of inertia of the 'cloud' of data in time and space that is obtained by measuring time-series of the structural response at different 'sensors' along the structure. We show that by using POD analysis one is able to produce low order models of structures with high modal densities and local nonlinearities, where traditional modal analysis or system identification methods cannot be applied. The method is applied to the analysis of the dynamics of a light weight truss and of systems with clearance nonlinearities. A special characteristic of the method is that it can be applied to linear as well as nonlinear problems, thus eliminating a drawback of many of the existing inverse techniques that are based on the assumption of linearity.

In Chapter 4 we provide some special mathematical techniques for analyzing discreteness effects in the dynamics of elastic structures. We introduce a non-smooth transformation of variables that permits the complete (exact) analytical elimination of singularities from the equations of periodically discretely supported or discretely forced continuous structures. Examples of application of this technique are given with the study of the static buckling of a discretely forced ring, and of the dynamics of a free or forced string resting on a discrete elastic foundation. This non-smooth technique paves the way for the analytic treatment of linear and nonlinear problems with singularities in their equations of motion due to forcing or boundary conditions. Indeed, we are unaware of any other existing analytical technique that can account exactly for nonlinear effects in elastic structures arising due to spatially periodic boundary supports.

In Chapter 5 we examine the dynamics of continuous systems with backlash nonlinearities, with motion-limiting constraints that introduce vibro-impact nonlinearities in the response, and with essential local nonlinearities. These types of nonlinearities are rather common in engineering applications. Due to the discontinuous nature of the backlash and vibro-impact nonlinearities, the associated dynamic analysis is quite challenging. We show that backlash and vibro-impact nonlinearities can be used to induce localization and motion confinement properties in symmetric, spatially extended structures, such as trusses. In other applications, we demonstrate that even small clearances can introduce highly unstable and sudden dynamic phenomena that, if unaccounted for, can result in early and sudden failure of the structurec. Hence, the examples provided demonstrate clearly the beneficial or destructive influence that clearance or vibro-impact nonlinearities can have on the dynamics of continuous systems, and urge for careful consideration of the dynamic effects of this type of nonlinearities in practical structures.

We would like to acknowledge the significant help of our graduate students and other collaborators who contributed significantly in the results reported in this monograph. In addition, we would like to thank the following sponsoring Agencies and Institutions for providing support, through travel or research grants for some of the topics included in this work: The US National Science Foundation (NSF), the Kosciuszko Foundation, the US Office of Naval Research (ONR), the Electric Power Research Institute (EPRI), and the Department of Mechanical and Industrial Engineering of the University of Illinois at Urbana - Champaign. Last but not least we would like to thank our family members for their continuous and unconditional support and understanding. Without their support and understanding the research presented in this book would never have been undertaken, and this book would never have been written.

Lodz, Saratov and Athens
August 2003

Jan Awrejcewicz
Vadim A. Krys'ko
Alexander F. Vakakis

Table of Contents

Preface V

Table of Contents IX

1 **Vibration of Plates and Shells with Added Masses** 1
  1.1 Introduction 1
  1.2 3D-Theory of Orthotropic Shallow Shells with Added Masses 4
    1.2.1 Curvilinear Orthogonal Coordinates 4
    1.2.2 Fundamental Relations and Hypotheses 9
    1.2.3 Variational Equations 10
    1.2.4 Differential Equations, Boundary and Initial Conditions 20
  1.3 Analysis of Orthotropic Shallow Shells with Added Masses... 23
    1.3.1 Two Dimensional Theory 23
    1.3.2 Simplified Model 29
    1.3.3 Added Masses Stiffness 33
    1.3.4 Determination of Dynamical Characteristics 38
    1.3.5 Algorithms and Numerical Results 51
  1.4 Dynamical Characteristics of Plates with Added Masses 56
    1.4.1 Timoshenko's Kinematic Model 56
    1.4.2 Spatial Theory of Elasticity 59
    1.4.3 Three Dimensional Problem 65
    1.4.4 Algorithms 69
    1.4.5 Frequency Spectra Comparison of Three Dimensional
        and Approximate Theories 73
    1.4.6 Influence of the Added Masses - Numerical Investigation 82
    1.4.7 Fundamental Conclusions 91

2 **Rational Design of Plates and Shells** 93
  2.1 Introduction 93
  2.2 Orthotropic Plates and Shells with Variable Thickness and
    Low Stiffness 94
    2.2.1 Frequency Spectra Properties of Orthotropic Shells
        with Variable Thickness 94
    2.2.2 Finite Dimensional Approximation of Free Vibrations
        of the Timoshenko Like Shell 101

- 2.2.3 Algorithms for the MB and MVI Methods 107
- 2.2.4 Analysis of Free Vibrations of Transversal-Isotropic Plates 122
- 2.3 Rational Design of Plates with Finite Transversal Stiffness ... 125
  - 2.3.1 Plates and Shells Mass Optimization with Free Vibration Frequencies constraints 126
  - 2.3.2 Finite Dimensional Approximations of the Rational Design of Plates and Shells 129
  - 2.3.3 Timoshenko-Like Rectangular Plates with Stiff and Rolling Supports 134
- 2.4 Optimization of Plates and Shells Surfaces with Constraints . 144
  - 2.4.1 Shells Optimization - Formulation of the Problem 144
  - 2.4.2 Algorithm for Optimal Surface Search 147
  - 2.4.3 Design of Plates and Shells with Constant Thickness and Rolling Supports 151
- 2.5 Vibroisolation of a Construction with Shells Elements 159
  - 2.5.1 Nonlinear Forced Vibrations of the Timoshenko Like Shells 160
  - 2.5.2 Optimal Vibroisolation-Formulation of the Problem and Vibroisolation Harmonic Excitation 170
  - 2.5.3 Algorithm of the Optimal Vibroisolation of Timoshenko-Like Shells 175
- 2.6 Generalisations 175

3 Order Reduction by Proper Orthogonal Decomposition (POD) Analysis 177
- 3.1 The POD Method 177
  - 3.1.1 Introduction 177
  - 3.1.2 Theoretical Basis 179
- 3.2 Application of POD to Vibro-impact Continuous Oscillators . 183
  - 3.2.1 Vibro-impacting Beam 183
  - 3.2.2 Overhung Rotor 191
- 3.3 Coherent Spatial Structures in Extended Systems 200
  - 3.3.1 Computation of the Transient Dynamics 201
  - 3.3.2 POD Analysis of the Truss Dynamics 210
  - 3.3.3 POD Based Reduced-Order Models 213
  - 3.3.4 Experimental Results 222
- 3.4 POD Study of the Interaction of Slow and Fast Dynamics ... 228
  - 3.4.1 Formulation of the Problem 228
  - 3.4.2 Regular Motions 231
  - 3.4.3 Chaotic Motions 233

4 Analytic Modelling of Discreteness Effects in Continuous Systems ... 239
4.1 The Method of Non-smooth Transformations ... 239
4.2 Elastic Continuum on Discrete Elastic Nonlinear Foundation . 245
4.2.1 Asymptotic Analysis ... 245
4.3 Static Buckling of a Circular Ring Compressive Loads ... 256

5 Continuous Systems with Non-smooth Nonlinearities ... 269
5.1 Transient Localization Due to Backlash Nonlinearities ... 269
5.1.1 Transient Localization and Beat Phenomena Due to Clearance ... 270
5.1.2 Order Reduction and Response Reconstruction ... 274
5.1.3 Nonlinear Analysis ... 279
5.1.4 Additional Examples ... 282
5.2 Nonlinear Localization in Systems of Vibro-impacting Beams ... 286
5.2.1 Numerical Results - Nonlinear Localization Due to Vibro-impacts ... 290
5.2.2 Experimental Results - Nonlinear Localization Due to Vibro-impacts ... 298
5.3 Vibro-impact Motions of a Rotary System ... 303

References ... 315

Index ... 329

1 Vibration of Plates and Shells with Added Masses

1.1 Introduction

This chapter is devoted to analysis of plates and shells with added elements (discrete additives). In the analysis a complex system will be simplified allowing for an efficient investigation of the influence of added elements' on the dynamic characteristics of plates' and shells'. "Added elements" are defined either as concentrated masses [53, 129] or as continuous masses (lying on a small surface) and absolutely stiffly attached to the plates and shells. In addition, they can be lumped oscillators or lumped (continuous) joint elements.

One of the first works focused on investigation of a plate with added masses vibration, belongs to Gershgorin [82]. The problem has been reduced to the consideration of forced vibration of plate jointly supported with the masses. The equations of plates and masses have been integrated separately (similar approaches have been often applied in many dynamical problems of constructions with added masses). The added element is considered to be "unjoined" during analysis and therefore its' and plate's vibrations are analysed separately. The dynamic influence on the plate has been reduced to analyzing harmonic vibrations. Using physical model of "plate (shell) - added mass" one finds many interesting dynamical phenomena.

This classical approach has been step by step expanded by including many different properties such as:

a) Geometry of plates and shells (circle plates and shells [75, 76], ring plates and shells, trapezoidal plates and shells, triangle plates and shells [200]);
b) Real properties and material structure (orthotropy, anisotropy [28], elasticity and damping [33], composites [96], elastic support [220], ribs [124]), and so on.

Now a brief review of the research oriented on investigation of different lumped-continuous constructions with an emphasis on the results of free plates and shells vibration with discrete masses in the frame of the linear theory will be given. This review includes only the works that are relevant to this chapter subject rather than the totality of them.

It should be pointed out that during analysis of the added masses' influence on constructions with plates and shells, a few simple and fundamental methods have been used, leading to analysis of "shell - mass" or "plate - mass" models.

The applied calculation models can be divided into the following groups.

1. *Continuous and absolutely stiff models.*

   It has been assumed that mass joints on plates or shells occupy small surfaces, which is typical for technical situations. It has also been assumed that a contact surface is either square or rectangular (in a case of a shell a curvature is negligible). This model has been used by Christienko [54, 53] during the vibration analysis of shells with discrete added elements and by Palamarčuk [179] during the vibration analysis of a system consisting of a rib cylindrical shell and an absolutely stiff body. Chomčenko [50] has considered vibrations of orthotropic shallow shells with two curvatures and the added masses jointed stiffly with a shell. An increase of the frequencies of free vibrations has been observed. Due to the author's explanation, the jointed mass has increased the construction's stiffness and also the potential energy. If this effect dominates, according to Rayleigh's principle the free vibration frequencies will increase.

   In reference [178], vibrations of a system consisting of a ribbed cylindrical shell with a beam have been analysed. It has been shown that frequencies have increased (with an increase of the beam's stiffness) and asymptotically approached the frequencies corresponding to an absolutely stiff beam model. To conclude, models with added masses always have increased frequencies in comparison to the corresponding systems with no added masses.

   The above mentioned researchers have applied the so called partition method and have assumed that transversal inertiality has played the role of pressure acting on a construction.

   In [5, 180] the added mass and a load capacity construction have been considered separately, and a force reaction of the mass has been substituted by discrete loads and moments. After fulfilment of the same translations and rotation angles conditions in the centres of the joint surfaces, a mathematical model has been established.

   Free vibrations of a circular isotropic plate with a circular and absolutely stiff added mass have been analysed by Okamoto and Sekiya [171]. A rectangular plate with a few symmetrically situated cylindrical masses have been analysed by Stokey and Zorowski [225] using the Lagrange method.

2. *Added concentrated masses stiffly jointed.*

   Both notions of "concentrated mass" and "force" notions are abstract. In the theory of plates and shells the notion of concentrated force is interpreted as a sequence of loads acting on an elementary surface approaching zero. A similar method is applied while solving the problems with added masses. Using a limited case describing the vibrating system "plate (shell) - attached stiff mass", when stick surfaces approach zero, the differential equations characterising a point of mass - shell interaction have been derived [54].

One of the most effective methods revealing the dynamical characteristics of plates and shells with added masses is supported by generalised functions theory. The concentrated masses' effects are introduced to the input equations using δ -(Dirac) functions, and the mass density is added to the plate density; therefore, an inertia is included [2, 124, 146]. For instance, in reference [207] Lagrange's principle and generalized function theory have been applied to solving differential equations.

The generalised function method has been also successfully applied to a wide class of different shells: shallow with two curvatures [50]; closed cylindrical [115, 150]; spherical [126] and others.

In the above mentioned papers only transversal inertia of the added masses have been taken into account. In the investigations of Christienko and Kalko [51], it has been shown that the inertia of the concentrated masses may lead to a sufficient decrease of higher eigenfrequencies of plates and shells.

In addition, there are works devoted to the following aspects of vibration analysis of plates and shells with: a) joined elasticly with discrete masses (oscillators) [6]; b) elastic and continuous joints [148, 149]; c) elastic supports [122]; d) point clamping [200]. In [89, 130, 149] inertial effects have been analysed.

The obtained differential equations have been solved using different methods. In references [33, 47, 76, 101, 122, 164], a method of decomposition for eigenfunctions of the homogeneous problem has been used. In [50] an integral (Fourier) transform has been applied. Using the variational Ritz method many problems connected with dynamics of rotational shells with discrete masses [129, 127], or orthotropic and isotropic closed cylindrical shells with added masses have been solved [52].

In practice, especially on a design stage, simple methods are very profitable for defining the eigenfrequencies. In the literature [128] it is possible to find a description of approximate methods applied to shells with small and large added masses values. In reference [7] approximate formulas of fundamental eigenfrequency estimation have been given. In the general case, eigenfrequencies and corresponding modes of plates and shells with the added masses have been computed using complex algorithms.

It should be pointed out that the application of classical theories to analyse continuous systems with added masses omits many essential and fundamental problems. In addition, improving the computational accuracy in the frame of these theories is often not possible.

A qualitatively new stage of finding a solution to the described problem has appeared thanks to dimensional reduction. More subtle theories have been developed in references [86, 184]. Many practical results on statics and dynamics of plates and shells were obtained using non-classical theories, as described in the monographs [3, 4, 88, 105, 113, 117, 185, 197, 232, 231].

Eigenfrequencies estimation of plates and shells on the basis of non-classical theories [86] has led to the conclusion that a low construction stiffness can lead to different results compared to classical theories. This indicates that advanced theories should be applied for analyzing practical problems, especially those associated with composite materials.

Only few papers have been devoted to vibration analysis of plates and shells with local singularities in a frame of non-classical theory. In reference [90] free vibrations of a thin isotropic plate circumfurencly compressed with a constant thickness with transversal offsets and rotation inertia has been analysed on a basis of the Timoshenko's model. It has been shown, that the error in the analysis increases with increasing of the order of the harmonic and the ratio of the added mass to the plate mass.

However, an investigation of dynamical characteristics of the shells with discrete elements on the basis of the non-classical theory has not been carried out widely.

The aforementioned literature review implies the necessity of developing more accurate methods, based on the definition of dynamics of plates and shells in the framework of both non-classical as well as three dimensional (3D) elasticity theories (see [16, 17, 18, 21]).

Application of three dimensional theory allows for a detailed analysis of physical effects, that are not detected using classical approaches (for instance, higher frequencies). In addition, it gives a full picture of stress and displacement distribution along plate thickness, which is essentially three dimensional in the neighbourhood of the joints between the continuous system and a discrete mass. The obtained results (using 3D theory) enable a validity estimation of the classical theories and reveals the range of their applications. Because this approach is connected with direct integration of 3D elasticity equations, and it is associated with many formal mathematical difficulties, this approach is not widely followed in the literature.

1.2 3D-Theory of Orthotropic Shallow Shells with Added Masses

1.2.1 Curvilinear Orthogonal Coordinates

The following assumptions and hypotheses of the linear theory of an elastic anisotropic body will be considered for the following analysis:

1. A shell is shallow and also becomes shallow after deformations. Tension moments are negligible, and the tension tensor is symmetric.
2. Small deformations are considered, therefore relations between deformation components and their derivatives along coordinates are linear.

1.2 3D-Theory of Orthotropic Shallow Shells with Added Masses

3. Hook's principle is valid. It means that relations between tension and deformation components are linear with constant coefficients.
4. The initial (possible) deformations of the continua considered are not taken into account.

Let us consider an orthotropic shell with thickness $2h$, attached to curvilinear and orthogonal coordinates α, β (α, β cover the main curvature lines of the mean shell surface); γ is normal to α and β and it describes the distance along the normal from the point (α, β) to the point (α, β, γ).

For the taken coordinate system we get the following Lamé coefficients [243]
$$H_1 = A(1 + k_1\gamma), \quad H_2 = A(1 + k_2\gamma), \quad H_3 = 1, \qquad (1.1)$$
where, $A = A(\alpha, \beta)$ and $B = B(\alpha, \beta)$ are the coefficients of the second power form of the mean surface; $k_1 = k_1(\alpha, \beta)$, $k_2 = k_2(\alpha, \beta)$ are the main curvatures of a shell surface along lines $\alpha = const.$, $\beta = const.$

The following formulas [245] define the Lamé coefficients
$$H_1^2 = \left(\frac{\partial x}{\partial \alpha}\right)^2 + \left(\frac{\partial y}{\partial \alpha}\right)^2 + \left(\frac{\partial z}{\partial \alpha}\right)^2, \quad \overleftrightarrow{(1,2,3)}, \quad \overleftrightarrow{(\alpha,\beta,\gamma)}. \qquad (1.2)$$

Notation $\overleftrightarrow{(\alpha,\beta,\gamma)}$, $\overleftrightarrow{(x,y,z)}$ denotes that other formulas are obtained using a simultaneous circular shift of the symbols.

The Lamé coefficients are mutually independent and satisfy the following differential equations [196]

$$\frac{\partial}{\partial \alpha}\left(\frac{1}{H_1}\frac{\partial H_2}{\partial \alpha}\right) + \frac{\partial}{\partial \beta}\left(\frac{1}{H_2}\frac{\partial H_1}{\partial \beta}\right) + \frac{1}{H_3^2}\frac{\partial H_1}{\partial \gamma}\frac{\partial H_2}{\partial \gamma} = 0,$$

$$\frac{\partial}{\partial \beta}\left(\frac{1}{H_2}\frac{\partial H_3}{\partial \beta}\right) + \frac{\partial}{\partial \gamma}\left(\frac{1}{H_3}\frac{\partial H_2}{\partial \gamma}\right) + \frac{1}{H_1^2}\frac{\partial H_2}{\partial \alpha}\frac{\partial H_3}{\partial \alpha} = 0,$$

$$\frac{\partial}{\partial \gamma}\left(\frac{1}{H_3}\frac{\partial H_1}{\partial \gamma}\right) + \frac{\partial}{\partial \alpha}\left(\frac{1}{H_1}\frac{\partial H_3}{\partial \alpha}\right) + \frac{1}{H_2^2}\frac{\partial H_3}{\partial \beta}\frac{\partial H_1}{\partial \beta} = 0,$$

$$\frac{\partial^2 H_1}{\partial \beta \partial \gamma} - \frac{1}{H_2}\frac{\partial H_2}{\partial \gamma}\frac{\partial H_1}{\partial \beta} - \frac{1}{H_3}\frac{\partial H_3}{\partial \beta}\frac{\partial H_1}{\partial \gamma} = 0,$$

$$\frac{\partial^2 H_2}{\partial \gamma \partial \alpha} - \frac{1}{H_3}\frac{\partial H_3}{\partial \alpha}\frac{\partial H_2}{\partial \gamma} - \frac{1}{H_1}\frac{\partial H_1}{\partial \gamma}\frac{\partial H_2}{\partial \alpha} = 0,$$

$$\frac{\partial^2 H_3}{\partial \alpha \partial \beta} - \frac{1}{H_1}\frac{\partial H_1}{\partial \beta}\frac{\partial H_3}{\partial \alpha} - \frac{1}{H_2}\frac{\partial H_2}{\partial \alpha}\frac{\partial H_3}{\partial \beta} = 0. \qquad (1.3)$$

Substituting H_1, H_2, H_3 from (1.1) to (1.3) we get the Gauss formulas for a shell surface defined by the equation $\gamma = 0$.

$$\frac{\partial}{\partial \alpha}\left(\frac{1}{A}\frac{\partial B}{\partial \alpha}\right) + \frac{\partial}{\partial \beta}\left(\frac{1}{B}\frac{\partial A}{\partial \beta}\right) = -k_1 k_2,$$

$$\frac{\partial}{\partial \beta}(Ak_1) = k_2 \frac{\partial A}{\partial \beta}, \quad \frac{\partial}{\partial \alpha}(Bk_2) = k_1 \frac{\partial B}{\partial \alpha}. \tag{1.4}$$

Some formulas necessary for future considerations of the elasticity theory are given below.

The coordinates u_1, u_2 and u_3 denote the shell displacement vector projections for an tangent directions to the coordinates α, β, γ.

The deformable state of a three dimensional shallow shell is characterized by six deformations $e_{11}, ..., e_{22}, ..., e_{23}$, connected with the displacement vector components due to the formulas [169]:

$$e_{11} = \frac{1}{H_1}\frac{\partial u_1}{\partial \alpha} + \frac{1}{H_1 H_2}\frac{\partial H_1}{\partial \beta}u_2 + \frac{\partial H_1}{\partial \gamma}u_3,$$

$$e_{22} = \frac{1}{H_2}\frac{\partial u_2}{\partial \beta} + \frac{1}{H_2}\frac{\partial H_2}{\partial \gamma}u_3 + \frac{1}{H_1 H_2}\frac{\partial H_2}{\partial \alpha}u_1,$$

$$e_{33} = \frac{\partial u_3}{\partial \gamma},$$

$$e_{12} = \frac{H_1}{H_2}\frac{\partial}{\partial \beta}\left(\frac{1}{H_1}u_1\right) + \frac{H_2}{H_1}\frac{\partial}{\partial \alpha}\left(\frac{1}{H_2}u_2\right),$$

$$e_{13} = \frac{1}{H_1}\frac{\partial u_3}{\partial \alpha} + H_1 \frac{\partial}{\partial \gamma}\left(\frac{1}{H_1}u_1\right),$$

$$e_{23} = H_2 \frac{\partial}{\partial \gamma}\left(\frac{1}{H_2}u_2\right) + \frac{1}{H_2}\frac{\partial u_3}{\partial \beta}. \tag{1.5}$$

The rotation vector $\vec{\omega}$ has the following coordinates:

$$\omega_1 = \frac{1}{2H_2}\left(\frac{\partial u_3}{\partial \beta} - \frac{\partial H_2 u_2}{\partial \gamma}\right) = \frac{1}{2}(\mathrm{rot}\bar{u})_\alpha,$$

$$\omega_2 = \frac{1}{2H_1}\left(\frac{\partial H_1 u_1}{\partial \gamma} - \frac{\partial u_3}{\partial \alpha}\right) = \frac{1}{2}(\mathrm{rot}\bar{u})_\beta,$$

$$\omega_3 = \frac{1}{2H_1}\frac{1}{H_2}\left(\frac{\partial H_2 u_2}{\partial \alpha} - \frac{\partial H_1 u_1}{\partial \beta}\right) = \frac{1}{2}(\mathrm{rot}\bar{u})_\gamma. \tag{1.6}$$

A motion of the cut shell element $d\alpha\, d\beta\, d\gamma$ is governed by the following equations [169]

$$\frac{\partial}{\partial \alpha}(H_2 \sigma_{11}) - \sigma_{22}\frac{\partial H_2}{\partial \alpha} + \frac{1}{H_1}\frac{\partial}{\partial \beta}(H_1^2 \sigma_{12}) + \frac{1}{H_1}\frac{\partial}{\partial \gamma}(H_1^2 H_2 \sigma_{13}) =$$

$$\rho H_1 H_2 \frac{\partial^2 u_1}{\partial t^2},$$

$$\frac{\partial}{\partial \beta}(H_1 \sigma_{22}) - \sigma_{11}\frac{\partial H_1}{\partial \beta} + \frac{1}{H_2}\frac{\partial}{\partial \alpha}(H_2^2 \sigma_{21}) + \frac{1}{H_2}\frac{\partial}{\partial \gamma}(H_1 H_2^2 \sigma_{23}) =$$

1.2 3D-Theory of Orthotropic Shallow Shells with Added Masses

$$\rho H_1 H_2 \frac{\partial^2 u_2}{\partial t^2},$$

$$\frac{\partial}{\partial \gamma}(H_1 H_2 \sigma_{33}) - \sigma_{11} H_2 \frac{\partial H_1}{\partial \gamma} - \sigma_{22} H_1 \frac{\partial H_2}{\partial \gamma} + \frac{\partial}{\partial \alpha}(H_2 \sigma_{13}) + \frac{\partial}{\partial \beta}(H_1 \sigma_{23}) =$$

$$\rho H_1 H_2 \frac{\partial^2 u_3}{\partial t^2}, \qquad (1.7)$$

$$\sigma_{21} = \sigma_{12}, \quad \sigma_{31} = \sigma_{13}, \quad \sigma_{32} = \sigma_{23},$$

where, $\sigma_{11}, \sigma_{22}, ..., \sigma_{23}$ are the tension coordinates related to the deformation coordinates $e_{11}, e_{22}, ..., e_{23}$ using the general Hook's principle.

Assume that at every point of a body meet three perpendicular planes of an elastic symmetry. Assuming that at every point of an anisotropic body the planes are perpendicular to the corresponding coordinates α, β, γ the general Hook's principle equations have the following form [232]

$$e_{11} = a_{1111}\sigma_{11} + a_{1122}\sigma_{22} + a_{1133}\sigma_{33}, \quad \overleftrightarrow{(1,2,3)}$$

$$e_{12} = a_{1212}\sigma_{12}, \quad e_{13} = a_{1313}\sigma_{13}, \quad e_{23} = a_{2323}\sigma_{23}. \qquad (1.8)$$

In that case there are nine constant and independent elasticity coefficients a_{ijkl} defined by the following relations

$$a_{1111} = \frac{1}{E_1}, \quad a_{1122} = -\frac{\nu_{12}}{E_2}, \quad a_{1133} = -\frac{\nu_{13}}{E_3}, \quad a_{2211} = -\frac{\nu_{21}}{E_1},$$

$$a_{2222} = \frac{1}{E_2}, \quad a_{2233} = -\frac{\nu_{23}}{E_3}, \quad a_{3311} = -\frac{\nu_{31}}{E_1}, \quad a_{3322} = -\frac{\nu_{32}}{E_2},$$

$$a_{3333} = \frac{1}{E_3}, \quad a_{1313} = \frac{1}{G_{13}}, \quad a_{2323} = \frac{1}{G_{23}}, \quad a_{1212} = \frac{1}{G_{12}}. \qquad (1.9)$$

Because the above equations are symmetric it holds that,

$$E_2 \nu_{21} = E_1 \nu_{12}, \quad E_3 \nu_{32} = E_2 \nu_{23}, \quad E_1 \nu_{13} = E_3 \nu_{31}. \qquad (1.10)$$

We define a body as 'orthotropic' if at each of its points three mutually perpendicular planes of an elastic symmetry meet. We define a plane as 'isotropic', when all directions of its points are equivalent because of the elasticity properties. If the isotropic plane is attached to each body's point, then the body's material is defined as the transversal and isotropic one [232], and the number of independent constant elasticity coefficients a_{ijkl} is reduced to the following ones

$$a_{1111} = \frac{1}{E}, \quad a_{1122} = -\frac{\nu}{E}, \quad a_{1133} = -\frac{\nu'}{E'}, \quad a_{2222} = a_{1111},$$

$$a_{2233} = a_{1133}, \quad a_{3333} = \frac{1}{E'}, \quad a_{3322} = a_{2233},$$

$$a_{3311} = a_{1133}, \quad a_{1313} = \frac{1}{G'}, \quad a_{2323} = \frac{1}{G'}, \quad a_{1212} = \frac{2(1+\nu)}{E}. \quad (1.11)$$

E denotes Young's modulus for the isotropic plane directions, and E' the Young's modulus for the perpendicular to isotropic plane directions; ν the Poisson's coefficient characterising shorting in the isotropic plane due to extension at the same plane; ν' an analogous Poisson's coefficient in the perpendicular direction to that plane; G' the shear modulus for normal planes to the isotropic plane; $G = E/2(1+\nu)$ the shear modulus for the parallel planes to the isotropic plane.

Solving equations of the generalised Hook's law according to the tension components $\sigma_{11}, \sigma_{22}, ..., \sigma_{23}$ we get the reversed formulas:

$$\sigma_{11} = A_{1111}e_{11} + A_{1122}e_{22} + A_{1133}e_{33}, \quad \overleftrightarrow{(1,2,3)}$$

$$\sigma_{12} = A_{1212}e_{12}, \quad \sigma_{13} = A_{1313}e_{13},$$

$$\sigma_{23} = A_{2323}e_{23}, \quad (1.12)$$

where the stiffness coefficients have the following form

$$A_{1111} = (a_{2222}a_{3333} - a_{2233}a_{3322})/\Delta, \quad \overleftrightarrow{(1,2,3)}$$

$$A_{1122} = (a_{1133}a_{3322} - a_{1122}a_{3333})/\Delta, \quad \overleftrightarrow{(1,2,3)}$$

$$A_{1212} = \frac{1}{a_{1212}}, \quad A_{1313} = \frac{1}{a_{1313}}, \quad A_{2323} = \frac{1}{a_{2323}},$$

$$A_{2211} = A_{1122}, \quad \overleftrightarrow{(1,2,3)}, \quad \Delta = \det[a_{iijj}]_{i,j\,=\,\overline{1,3}}. \quad (1.13)$$

For a transversal and isotropic material the above coefficients are as follows:

$$A_{1111} = E(1 - \nu'\nu'')/\Omega, \quad A_{1122} = E(\nu'\nu'' + \nu)/\Omega,$$

$$A_{1133} = E\nu'(1+\nu)/\Omega, \quad A_{3333} = E'(1-\nu^2)/\Omega,$$

$$A_{2222} = A_{1111}, \quad A_{2233} = A_{1133}, \quad A_{1313} = G',$$

$$A_{2323} = G', \quad A_{1212} = E/2(1+\nu), \quad \nu''E' = \nu'E,$$

$$\Omega = (1+\nu)(1 - 2\nu'\nu'' - \nu). \quad (1.14)$$

1.2 3D-Theory of Orthotropic Shallow Shells with Added Masses

1.2.2 Fundamental Relations and Hypotheses

Here the fundamental relationships further utilised during the dynamic model "shallow shell - added mass" will be given. The necessary relations are obtained from the general equations of elasticity theory taking into account many additional conditions.

Let us consider a shallow orthotropic shell which has a projected rectangular shape with the sides a and b. Suppose that x and y are the Cartesian coordinates of the shell surface, then the second power of the linear element in the plane $x0y$ is given by the relation

$$ds^2 = dx^2 + dy^2, \tag{1.15}$$

which defines the coefficients of the first second power form

$$A = B = 1. \tag{1.16}$$

For the considered shallow shell with the curvilinear orthogonal coordinate system α, β, γ we have

$$ds^2 = d\alpha^2 + d\beta^2, \tag{1.17}$$

which means that $A \approx 1$, $B \approx 1$.

Suppose that a mean surface is defined by $z = f(x,y)$. Thus, taking a mesh defined by $x = const.$, $y = const.$, from (1.2) we get

$$A = \sqrt{1 + \left(\frac{\partial f}{\partial x}\right)^2}, \quad B = \sqrt{1 + \left(\frac{\partial f}{\partial y}\right)^2}. \tag{1.18}$$

A shell will be enough shallow [243], if in its each point of the mean surface we have:

$$\left(\frac{\partial f}{\partial x}\right)^2 \ll 1; \quad \left(\frac{\partial f}{\partial y}\right)^2 \ll 1. \tag{1.19}$$

Therefore, in all relations of the previous section we can take $\alpha = x$, $\beta = y$, and the first second power form coefficients are equal to one. From the differential geometry it is known [196], that for the shallow shells we have

$$k_1 = -\frac{\partial^2 f}{\partial x^2}, \quad k_2 = -\frac{\partial^2 f}{\partial y^2}, \quad k_{12} = \frac{\partial^2 f}{\partial x \partial y}, \tag{1.20}$$

where $k_{12} = 0$ if the coordinate axes cover with the main curvature lines. In the latter case we have:

$$k_1 = -\frac{\partial^2 f}{\partial x^2}, \quad k_2 = -\frac{\partial^2 f}{\partial y^2}, \quad k_{12} = 0. \tag{1.21}$$

The following additional assumptions are also made:

a) In the first two equations the terms $k_1\sigma_{11}$ and $k_2\sigma_{12}$ are negligible;
b) In the relations between shear deformations and displacements the terms with curvature coefficients are negligible.

Finally, we get the following relationships and equations for a shallow shell:

A. Motion equations

$$\frac{\partial \sigma_{11}}{\partial x} + \frac{\partial \sigma_{12}}{\partial y} + \frac{\partial \sigma_{13}}{\partial z} = \rho \frac{\partial^2 u}{\partial t^2}, \quad (\overleftrightarrow{1,2}) \quad (\overleftrightarrow{u,v}),$$

$$\frac{\partial \sigma_{13}}{\partial x} + \frac{\partial \sigma_{23}}{\partial y} + \frac{\partial \sigma_{33}}{\partial z} - k_1\sigma_{11} - k_2\sigma_{22} = \rho \frac{\partial^2 w}{\partial t^2}. \tag{1.22}$$

B. Geometrical relationships

$$e_{11} = \frac{\partial u}{\partial x} + k_1 w, \quad (\overleftrightarrow{1,2}), \quad (\overleftrightarrow{x,y}),$$

$$e_{12} = \frac{\partial u}{\partial y} + \frac{\partial v}{\partial x}, \quad (\overleftrightarrow{1,2,3}), \quad (\overleftrightarrow{x,y,z}),$$

$$e_{33} = \frac{\partial w}{\partial z}. \tag{1.23}$$

C. Rotational vector components

$$\omega_1 = \frac{1}{2}\left(\frac{\partial w}{\partial y} + \frac{\partial v}{\partial z}\right), \quad (\overleftrightarrow{1,2,3}), \quad (\overleftrightarrow{x,y,z}). \tag{1.24}$$

The shell tensions are defined by relations (1.12). For $k_1 = k_2 = 0$ equations (1.22)–(1.24) govern the plates' behaviour.

Suppose that a sloped shell is loaded by an arbitrary number of added masses, situated on the rectangular elements ΔS ($i = \overline{1,N}$) on a top shells' surface, which are bounded by two line segment pairs $x = x_i - \tilde{c}_1^i$, $x = x_i - \tilde{\tilde{c}}_1^i$ and $y = y_i - \tilde{c}_2^i$, $y = y_i - \tilde{\tilde{c}}_2^i$, where x_i, y_i are the first two coordinates of the added mass centre $O^i(x_i, y_i, z_i)$ (Fig. 1.1). Denoting $\tau_1^i = \tilde{\tilde{c}}_1^i/\tilde{c}_1^i$, $\tau_2^i = \tilde{\tilde{c}}_2^i/\tilde{c}_2^i$ (where τ_1^i, τ_2^i are characterized by a degree of deviation from its geometrical centre), for homogeneous material we have $\tilde{\tilde{c}}_1^i = \tilde{c}_1^i$, $\tilde{\tilde{c}}_2^i = \tilde{c}_2^i$ (because of the symmetry) which means that $\tau_1^i = \tau_2^i = 1$. Denoting i-th added mass height by h_i we do not introduce any additional constraints. It is assumed that the contact surfaces $\tilde{c}_1^i(1 + \tau_1^i) \cdot \tilde{c}_2^i(1 + \tau_2^i)$ are small in comparison to the shell surface.

1.2.3 Variational Equations

The variational Hamilton principle will be used for deriving the differential equations governing the shell's dynamics, together with boundary and initial conditions.

1.2 3D-Theory of Orthotropic Shallow Shells with Added Masses

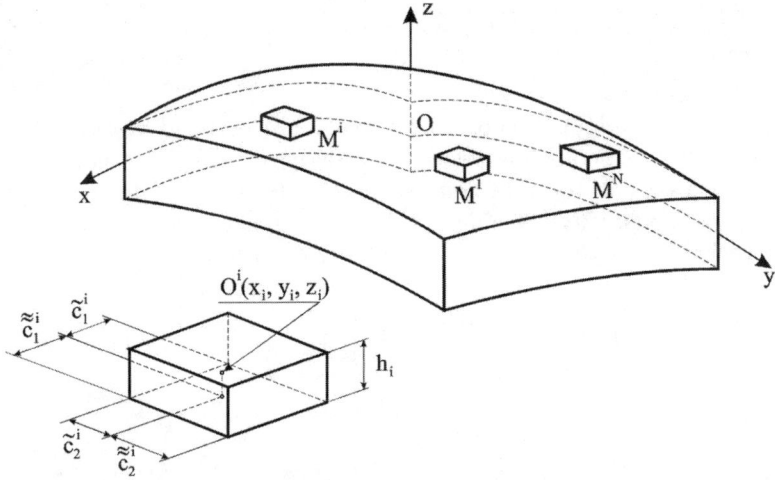

Figure 1.1. A shell loaded by the added masses M^i.

Let us consider a motion process between two time instants t_0 and t. We compare the different trajectories between these points. Physically realiable trajectories are defined by the following condition [244]

$$\int_{t_0}^{t} (\delta K + \delta A - \delta \Pi)\, dt = 0, \qquad (1.25)$$

where δK denotes the kinetic energy variation; δA the external forces work variation; and $\delta \Pi$ the potential energy deformation variation.

In the next considerations we focus in one the eigenfrequencies of either the "shell - mass" or "plate - mass" system. The free vibrations of the thin-walled structures considered will now be analyzed, assuming no external forces. Assuming lack of external forces, we take $\delta A = 0$. Taking into account the above assumptions we get the following Hamilton principle

$$\int_{t_0}^{t} \delta(K - \Pi)\, dt = 0, \qquad (1.26)$$

where $L = K - \Pi$ is the Lagrange function. This function in the case of "shell - mass" vibrating system, possesses the following form

$$L = \frac{1}{2} \iiint_V \rho \left[\left(\frac{\partial u}{\partial t} \right)^2 + \left(\frac{\partial v}{\partial t} \right)^2 + \left(\frac{\partial w}{\partial t} \right)^2 \right] dx\, dy\, dz -$$

$$\frac{1}{2}\iiint\limits_V (\sigma_{11}e_{11} + \sigma_{22}e_{22} + \sigma_{33}e_{33} + \sigma_{12}e_{12} + \sigma_{13}e_{13} + \sigma_{23}e_{23})\,dxdydz.$$

(1.27)

V denotes total volume occupied by a shell and the added masses. The original discrete-continuous construction, consisting of a shallow shell with constant thickness and the attached elements, is transformed to an equivalent continuous model with variable thickness. A similar approach has been applied earlier in the theory of ribbed shells [259]. The L function is given by

$$L = \frac{1}{2}\int_{-h}^{z^*}\int_0^a\int_0^b \rho\left[\left(\frac{\partial u}{\partial t}\right)^2 + \left(\frac{\partial v}{\partial t}\right)^2 + \left(\frac{\partial w}{\partial t}\right)^2\right] dxdydz -$$

$$\frac{1}{2}\int_{-h}^{z^*}\int_0^a\int_0^b (\sigma_{11}e_{11} + \sigma_{22}e_{22} + \sigma_{33}e_{33} + \sigma_{12}e_{12} + \sigma_{13}e_{13} + \sigma_{23}e_{23})\,dxdydz,$$

(1.28)

where

$$z^* = h + \sum_{i=1}^N h_i \Theta^*\left(x, x_i, \tilde{c}_1^i, \tilde{\tilde{c}}_1^i\right) \cdot \Theta^{**}\left(y, y_i, \tilde{c}_2^i, \tilde{\tilde{c}}_2^i,\right) \quad (1.29)$$

$$\Theta^*\left(x, x_i, \tilde{c}_1^i, \tilde{\tilde{c}}_1^i\right) = \Theta_1\left[x - (x_i - \tilde{c}_1^i)\right] - \Theta_2\left[x - (x_i + \tilde{\tilde{c}}_1^i)\right],$$

$$\Theta^{**}\left(y, y_i, \tilde{c}_2^i, \tilde{\tilde{c}}_2^i\right) = \Theta_1\left[y - (y_i - \tilde{c}_2^i)\right] - \Theta_2\left[y - (y_i + \tilde{\tilde{c}}_2^i)\right], \quad (1.30)$$

and Θ_1, Θ_2 are the characteristic functions [104] with the properties

$$\Theta_1 = \begin{cases} 0, x < x_i - \tilde{c}_1^i \\ 1, x \geq x_i - \tilde{c}_1^i \end{cases}, \Theta_2 = \begin{cases} 0, x < x_i - \tilde{\tilde{c}}_1^i \\ 1, x \geq x_i - \tilde{\tilde{c}}_1^i \end{cases}. \quad (1.31)$$

Integrals occurring in (1.28) can be transformed to the following forms

$$L_1 = \frac{1}{2}\int_{-h}^h\int_0^a\int_0^b \rho\left[\left(\frac{\partial u}{\partial t}\right)^2 + \left(\frac{\partial v}{\partial t}\right)^2 + \left(\frac{\partial w}{\partial t}\right)^2\right] dxdydz +$$

$$\frac{1}{2}\sum_{i=1}^N \int_h^{h+h_i}\int_0^a\int_0^b \rho\left[\left(\frac{\partial u}{\partial t}\right)^2 + \left(\frac{\partial v}{\partial t}\right)^2 + \left(\frac{\partial w}{\partial t}\right)^2\right]\Theta^*\Theta^{**} dxdydz, \quad (1.32)$$

$$L_2 = \frac{1}{2}\int_{-h}^h\int_0^a\int_0^b (\sigma_{11}e_{11} + \sigma_{22}e_{22} + \sigma_{33}e_{33} + \sigma_{12}e_{12} + \sigma_{13}e_{13} +$$

$$+ \sigma_{23}e_{23})\,dxdydz + \frac{1}{2}\sum_{i=1}^N \int_h^{h+h_i}\int_0^a\int_0^b (\sigma_{11}e_{11} + \sigma_{22}e_{22} + \sigma_{33}e_{33} +$$

1.2 3D-Theory of Orthotropic Shallow Shells with Added Masses

$$\sigma_{12}e_{12} + \sigma_{13}e_{13} + \sigma_{23}e_{23})\,\Theta^*\Theta^{**}dxdydz. \tag{1.33}$$

As an example we consider the second integral of (1.28) in the form

$$\frac{1}{2}\int_{-h}^{z*}\int_0^a\int_0^b (\ldots)dxdydz = \frac{1}{2}\int_{-h}^{h}\int_0^a\int_0^b (\ldots)dxdydz + \frac{1}{2}\int_h^{z*}\int_0^a\int_0^b (\ldots)dxdydz, \tag{1.34}$$

where $(\ldots) = (\sigma_{11}e_{11} + \sigma_{22}e_{22} + \sigma_{33}e_{33} + \sigma_{12}e_{12} + \sigma_{13}e_{13} + \sigma_{23}e_{23})$. The second term of (1.34) on the basis of (1.29) and using characteristic function properties, is expressed

$$\frac{1}{2}\int_h^{z*}\int_0^a\int_0^b (\ldots)dxdydz = \frac{1}{2}\sum_{i=1}^N \int_h^{h+h_i}\int_0^a\int_0^b (\ldots)\Theta^*\Theta^{**}dxdydz. \tag{1.35}$$

From (1.35) and (1.34) we get (1.32) and (1.33).

We have to emphasise that during derivation of relationships (1.32) and (1.33) no additional conditions are introduced, which proves that our considerations are general. In addition, alternative shell, plate configurations can be considered in the framework of the introduced theory. And finally, this method can serve as a tool for investigation of newly developed models and for accuracy investigations of existing ones.

To continue the further investigations and to build a "shell - mass" model, it is necessary to introduce certain physical and geometrical simplifications. As the added element we take an absolutely stiff mass concentrated on a small surface. We are going to get high accuracy results by taking additional forced terms caused by added masses interaction. Up to now, this has been negligible by the earlier cited researchers. Of course, that direction of investigations is not the only one possible and the other proposed models will also be discussed.

The assumptions about the added masses joints which lead to the omission of the internal deformation of the added masses, allow for cancellation of the second integral in equation (1.33) (which characterizes deformation energy of the added masses). We are also going to transform the second term of (1.32), which characterizes the kinetic energy of the added masses. At first, we investigate a problem dealing with velocity distributions of the points inside a small volume covered by the added mass in relation to a certain point $O^i(x_i, y_i, z_i)$ being the mass centre. It is assumed, that the velocity field is continuous and posseses first order derivatives.

Suppose that the velocity of point $O^i(x_i, y_i, z_i)$ is equal to $\bar{\xi}_0^i$, and that velocity of the point of the added mass is equal to $\bar{\xi}_{1i}^i(x, y, z)$. Let us develop $\bar{\xi}_1^i$ in Taylor series in a neighbourhood of O^i taking only linear terms of ρ_i, where $\rho_i = |\bar{u} - \bar{u}_i|$, $\bar{u}_i = u(x_i, y_i, z_i)$ (curvature of the joint surface is negligible):

$$\xi_{1x}^i = \xi_{0x}^i + \frac{\partial \xi_{1x}^i}{\partial x}(x - x_i) + \frac{\partial \xi_{1x}^i}{\partial y}(y - y_i) + \frac{\partial \xi_{1x}^i}{\partial z}(z - z_i), \quad \overleftrightarrow{(1,2,3)}, \quad (1.36)$$

$$i = \overline{1, N}.$$

Thus, the following relationships are obtained from (1.36) for a displacement of an arbitrarily taken added mass point

$$u_1^i = u_0^i + \frac{\partial \xi_{1x}^i}{\partial x}(x - x_i)\Delta t + \frac{\partial \xi_{1x}^i}{\partial y}(y - y_i)\Delta t + \frac{\partial \xi_{1x}^i}{\partial z}(z - z_i)\Delta t, \quad (1.37)$$

$$\overleftrightarrow{(1,2,3)}, \quad i = \overline{1, N}.$$

Defining an arbitrarily taken point velocity by the mass centre velocity O^i from (1.36) we get

$$\xi_{1x}^i = \xi_{0x}^i + \frac{\partial \xi_{1x}^i}{\partial x}(x - x_i) + \frac{1}{2}\left(\frac{\partial \xi_{1x}^i}{\partial y} + \frac{\partial \xi_{1y}^i}{\partial x}\right)(y - y_i) +$$

$$\frac{1}{2}\left(\frac{\partial \xi_{1x}^i}{\partial y} - \frac{\partial \xi_{1y}^i}{\partial x}\right)(y - y_i) + \frac{1}{2}\left(\frac{\partial \xi_{1x}^i}{\partial z} + \frac{\partial \xi_{1z}^i}{\partial x}\right)(z - z_i) +$$

$$\frac{1}{2}\left(\frac{\partial \xi_{1x}^i}{\partial z} - \frac{\partial \xi_{1z}^i}{\partial x}\right)(z - z_i), \quad (1.38)$$

$$\overleftrightarrow{(1,2,3)}, \quad i = \overline{1, N}.$$

In the above relationships both symmetric and antisymmetric tensors are used, which are defined in the three dimensional vector by

$$\dot{\omega}_1^i = \frac{1}{2}\left(\frac{\partial \xi_{1z}^i}{\partial y} - \frac{\partial \xi_{1y}^i}{\partial z}\right), \quad \overleftrightarrow{(1,2,3)}. \quad (1.39)$$

Introducing the following notation

$$\hat{\xi}_x^i = \frac{\partial \xi_{1x}^i}{\partial x}(x - x_i) + \frac{1}{2}\left(\frac{\partial \xi_{1x}^i}{\partial y} + \frac{\partial \xi_{1y}^i}{\partial x}\right)(y - y_i) +$$

$$\frac{1}{2}\left(\frac{\partial \xi_{1x}^i}{\partial z} + \frac{\partial \xi_{1z}^i}{\partial x}\right)(z - z_i), \quad \overleftrightarrow{(1,2,3)}, \quad (1.40)$$

and taking into account (1.39) and (1.40) in (1.38) we get:

$$\xi_{1x}^i = \xi_{0x}^i + \hat{\xi}_x^i + \dot{\omega}_2^i(z - z_i) - \dot{\omega}_3^i(y - y_i), \quad \overleftrightarrow{(1,2,3)}. \quad (1.41)$$

1.2 3D-Theory of Orthotropic Shallow Shells with Added Masses

Using the symbolic notation

$$\bar{\omega} = \omega_1 \bar{i} + \omega_2 \bar{j} + \omega_3 \bar{k} = \frac{1}{2} \begin{vmatrix} \bar{i} & \bar{j} & \bar{k} \\ \frac{\partial}{\partial x} & \frac{\partial}{\partial y} & \frac{\partial}{\partial z} \\ \xi^i_{1x} & \xi^i_{1y} & \xi^i_{1z} \end{vmatrix}, \qquad (1.42)$$

the equation (1.41) is expressed in the following form

$$\xi^i_{1x} = \xi^i_{0x} + \hat{\xi}^i_x + \overleftarrow{[\omega^i \times (\bar{u} - \bar{u}_i)]}_x, \quad (\overline{1,2,3}). \qquad (1.43)$$

To conclude, velocities of the added mass points can be defined in a form of three components' sum. The first one, $\bar{\xi}^i_0$ does not depend on the coordinates x, y, z and the translatory motion velocity of a whole body is equal to the mass centre velocity. The second component is related to the relative extension and shear deformation velocities between the element and the added mass. Assuming, that there are no deformations inside the added mass we get $\hat{\xi}^i_x = \hat{\xi}^i_y = \hat{\xi}^i_z = 0$ (velocity deformation tensor components are equal zero). The third component in the relationship (1.43) defines the components of the vector $\dot{\bar{\omega}}^i$ describing an instantaneous angular velocity vector of the body treated as absolutely stiff. Taking into account the latter observation the relationships (1.43) get the form

$$\xi^i_{1x} = \xi^i_{0x} + \overleftarrow{[\omega^i \times (\bar{u} - \bar{u}_i)]}_x, \quad (\overline{1,2,3}). \qquad (1.44)$$

Now, taking into account (1.44), we get the following approximation for the kinetic energy

$$L'' = \frac{1}{2} \sum_{i=1}^{N} \int_{h}^{h+h_i} \int_{0}^{a} \int_{0}^{b} \rho \left[\left(\frac{\partial u}{\partial t}\right)^2 + \left(\frac{\partial v}{\partial t}\right)^2 + \left(\frac{\partial w}{\partial t}\right)^2 \right] \Theta^* \Theta^{**} dx\, dy\, dz =$$

$$\frac{1}{2} \sum_{i=1}^{N} \int_{h}^{h+h_i} \int_{0}^{a} \int_{0}^{b} \rho \Big\{ (\xi^i_{0x})^2 + (\xi^i_{0y})^2 + (\xi^i_{0z})^2 + (\dot{\omega}^i_2)^2 \left[(z-z_i)^2 + \right.$$

$$(x-x_i)^2 \Big] + (\dot{\omega}^i_1)^2 \left[(z-z_i)^2 + (y-y_i)^2 \right] + (\dot{\omega}^i_3)^2 \left[(y-y_i)^2 + \right.$$

$$(x-x_i)^2 \Big] - 2\dot{\omega}^i_2 \dot{\omega}^i_3 (z-z_i)(y-y_i) - 2\dot{\omega}^i_3 \dot{\omega}^i_1 (x-x_i)(y-y_i) -$$

$$2\dot{\omega}^i_1 \dot{\omega}^i_2 (y-y_i)(x-x_i) \Big\} \Theta^* \Theta^{**} dx\, dy\, dz. \qquad (1.45)$$

The dimension of the i-th added mass with coordinates x_i, y_i, z_i is given by $\tilde{c}^i_1 (1 + \tau^i_1) \tilde{c}^i_2 (1 + \tau^i_2)$. Multiplying and dividing (1.45) by that quantity and taking into account that $dV = \tilde{c}^i_1 (1 + \tau^i_1) \tilde{c}^i_2 (1 + \tau^i_2) dz$, after integration of (1.45) one gets

16   1 Vibration of Plates and Shells with Added Masses

$$L_1'' = \frac{1}{2} \sum_{i=1}^{N} \int_0^a \int_0^b \left\{ \tilde{M}^i \left[\left(\xi_{0x}^i\right)^2 + \left(\xi_{0y}^i\right)^2 + \left(\xi_{0z}^i\right)^2 \right] + \tilde{J}_{xx}^i \left(\dot{\omega}_1^i\right)^2 + \tilde{J}_{yy}^i \left(\dot{\omega}_2^i\right)^2 + \right.$$

$$\left. \tilde{J}_{zz}^i \left(\dot{\omega}_3^i\right)^2 - 2\tilde{J}_{xy}^i \dot{\omega}_1^i \dot{\omega}_2^i - 2\tilde{J}_{xz}^i \dot{\omega}_1^i \dot{\omega}_3^i - 2\tilde{J}_{yx}^i \dot{\omega}_2^i \dot{\omega}_3^i \right\} \Theta^* \Theta^{**} dxdy. \qquad (1.46)$$

$\tilde{M}^i, \tilde{J}_{xx}^i, \tilde{J}_{yy}^i, \tilde{J}_{zz}^i, \tilde{J}_{xy}^i, \tilde{J}_{yz}^i, \tilde{J}_{xz}^i$ denote masses and mass inertial moments of the added mass related to a unit joint contact surface uniquely distributed on it.

In a limited case, when the concentrated masses are located on a shell we use δ functions. It means that in (1.46) we take $\tilde{c}_1^i \to 0$, $\tilde{x}_2^i \to 0$, and in addition we obtain

$$\lim_{\tilde{c}_1^i \to 0} \frac{\Theta^*\left(x, x_i, \tilde{c}_1^i, \tilde{c}_1^i\right)}{\tilde{c}_1^i \left(1 + \tau_1^i\right)} = \delta(x - x_i), \quad \lim_{\tilde{c}_2^i \to 0} \frac{\Theta^{**}\left(y, y_i, \tilde{c}_2^i, \tilde{c}_2^i\right)}{\tilde{c}_2^i \left(1 + \tau_2^i\right)} = \delta(y - y_i). \qquad (1.47)$$

Then, L_1'' is transformed to the following form

$$L_1'' = \frac{1}{2} \sum_{i=1}^{N} \int_0^a \int_0^b \left\{ M^i \left[\left(\xi_{0x}^i\right)^2 + \left(\xi_{0y}^i\right)^2 + \left(\xi_{0z}^i\right)^2 \right] + J_{xx}^i \left(\dot{\omega}_1^i\right)^2 + \right.$$

$$J_{yy}^i \left(\dot{\omega}_2^i\right)^2 + J_{zz}^i \left(\dot{\omega}_3^i\right)^2 - 2J_{xy}^i \dot{\omega}_1^i \dot{\omega}_2^i - 2J_{xz}^i \dot{\omega}_1^i \dot{\omega}_3^i -$$

$$\left. 2J_{yx}^i \dot{\omega}_2^i \dot{\omega}_3^i \right\} \delta(x - x_i)\delta(y - y_i) \, dxdy, \qquad (1.48)$$

where ω_j^i are defined by (1.39).

Introducing (1.46) to (1.32) and (1.33) to (1.28) we get the sought Lagrange function of the "shallow shell - concentrated mass" system

$$L = \frac{1}{2} \int_{-h}^{h} \int_0^a \int_0^b \rho \left[\left(\frac{\partial u}{\partial t}\right)^2 + \left(\frac{\partial v}{\partial t}\right)^2 + \left(\frac{\partial w}{\partial t}\right)^2 \right] dxdydz +$$

$$\frac{1}{2} \sum_{i=1}^{N} \int_0^a \int_0^b \left\{ M^i \left[\left(\xi_{0x}^i\right)^2 + \left(\xi_{0y}^i\right)^2 + \left(\xi_{0z}^i\right)^2 \right] + J_{xx}^i \left(\dot{\omega}_1^i\right)^2 + \right.$$

$$J_{yy}^i \left(\dot{\omega}_2^i\right)^2 + J_{zz}^i \left(\dot{\omega}_3^i\right)^2 - 2J_{xy}^i \dot{\omega}_1^i \dot{\omega}_2^i - 2J_{xz}^i \dot{\omega}_1^i \dot{\omega}_3^i -$$

$$\left. 2J_{yx}^i \dot{\omega}_2^i \dot{\omega}_3^i \right\} \delta(x - x_i)\delta(y - y_i)dxdy -$$

$$\frac{1}{2} \int_{-h}^{h} \int_0^a \int_0^b (\sigma_{11}e_{11} + \sigma_{22}e_{22} + \sigma_{33}e_{33} + \sigma_{12}e_{12} + \sigma_{13}e_{13} + \sigma_{23}e_{23}) \, dxdydz.$$

$$(1.49)$$

1.2 3D-Theory of Orthotropic Shallow Shells with Added Masses

A variational principle similar to (1.25) related to (1.49) should be formulated taking into account the generalised Hook's law (1.12) and the geometrical relationships (1.33). In addition, we need expressions for the mass centre velocity vector components $\bar{\xi}_0^i$ and the angular velocity vector components $\dot{\bar{\omega}}^i$.

To conclude, we have got the important relationship (1.49) fully described by energetic characteristics of the investigated system in the three dimensional space of displacements. Now we are going to derive the equations of motion of the shallow shell with the added concentrated masses.

First we define the potential energy variation $\delta\Pi$ due to the shell's deformation. Taking into account the relationships

$$\delta\left(\frac{\partial u}{\partial x}\right) = \frac{\partial(\delta u)}{\partial x}$$

and integrating by a parts one gets

$$\iiint_V \left[\frac{\partial(\delta u)}{\partial x}\sigma_{11} + \left(\frac{\partial(\delta u)}{\partial y} + \frac{\partial(\delta v)}{\partial x}\right)\sigma_{12}+ \right.$$

$$\left. \left(\frac{\partial(\delta u)}{\partial z} + \frac{\partial(\delta w)}{\partial x}\right)\sigma_{13} + ...\right] dxdydz =$$

$$\iint_S [(\sigma_{11}l + \sigma_{12}m + \sigma_{13}n)\delta u + ...] dxdy -$$

$$\iiint_V \left[\left(\frac{\partial\sigma_{11}}{\partial x} + \frac{\partial\sigma_{12}}{\partial y} + \frac{\partial\sigma_{13}}{\partial z}\right)\delta u + ...\right] dxdydz.$$

Grouping the terms multiplied by $\delta u, \delta v$ and δw one obtains

$$\delta\Pi = -\int_{-h}^{h}\int_0^a\int_0^b \left[\left(\frac{\partial\sigma_{11}}{\partial x} + \frac{\partial\sigma_{12}}{\partial y} + \frac{\partial\sigma_{13}}{\partial z}\right)\delta u + \left(\frac{\partial\sigma_{12}}{\partial x} + \frac{\partial\sigma_{22}}{\partial y} + \frac{\partial\sigma_{23}}{\partial z}\right)\delta v + \right.$$

$$\left. + \left(\frac{\partial\sigma_{13}}{\partial x} + \frac{\partial\sigma_{23}}{\partial y} + \frac{\partial\sigma_{33}}{\partial z} - k_1\sigma_{11} - k_2\sigma_{22}\right)\delta w\right] + \int_{-h}^{h}\int_0^b [\sigma_{11}\delta u +$$

$$+ \sigma_{12}\delta v + \sigma_{13}\delta w]|_0^a\, dydz + \int_{-h}^{h}\int_0^a [\sigma_{22}\delta v + \sigma_{12}\delta u + \sigma_{23}\delta w]|_0^b\, dxdz +$$

$$+ \int_0^a\int_0^b [\sigma_{33}\delta w + \sigma_{13}\delta u + \sigma_{23}\delta v]|_{-h}^{h}\, dxdy. \quad (1.50)$$

18   1 Vibration of Plates and Shells with Added Masses

Let us consider the variation of the kinetic energy of the system "shallow shell - mass". From (1.45) and (1.32) one obtains the following value of the kinetic energy

$$K = \frac{1}{2}\int_{-h}^{h}\int_0^a\int_0^b \rho\left[\left(\frac{\partial u}{\partial t}\right)^2 + \left(\frac{\partial v}{\partial t}\right)^2 + \left(\frac{\partial w}{\partial t}\right)^2\right]dxdydz+$$

$$\sum_{i=1}^N \int_0^a \int_0^b \{M^i\left[\left(\xi_{0x}^i\right)^2 + \left(\xi_{0y}^i\right)^2 + \left(\xi_{0z}^i\right)^2\right] + J_{xx}^i\left(\dot\omega_1^i\right)^2 +$$

$$J_{yy}^i\left(\dot\omega_2^i\right)^2 + J_{zz}^i\left(\dot\omega_3^i\right)^2 - 2J_{xy}^i\dot\omega_1^i\dot\omega_3^i - 2J_{xz}^i\dot\omega_1^i\dot\omega_3^i -$$

$$2J_{yx}^i\dot\omega_2^i\dot\omega_3^i\}\delta(x-x_i)\delta(y-y_i)dxdy. \qquad (1.51)$$

We are going to characterize the values $\xi_{0x}^i, \xi_{0y}^i, \xi_{0z}^i, \dot\omega_1^i, \dot\omega_2^i, \dot\omega_3^i$ now. The following displacements changes with respect to i-th mass thickness are introduced

$$u_i = u^h - (z_i - h)\frac{\partial w^h}{\partial x}, \quad v_i = v^h - (z_i - h)\frac{\partial w^h}{\partial y}, \quad w_i = w^h, \qquad (1.52)$$

where: u^h, v^h, w^h are displacements of the top shell surface in contact with the i-th added mass. Taking into account (1.39) we get the following relationships

$$\xi_{0x}^i = \frac{\partial u^h}{\partial t} - (z_i - h)\frac{\partial^2 w^h}{\partial x \partial t},$$

$$\xi_{0y}^i = \frac{\partial v^h}{\partial t} - (z_i - h)\frac{\partial^2 w^h}{\partial y \partial t},$$

$$\xi_{0z}^i = \frac{\partial w^h}{\partial t}, \quad \dot\omega_1^i = \frac{\partial^2 w^h}{\partial y \partial t},$$

$$\dot\omega_2^i = \frac{\partial^2 w^h}{\partial x \partial t}, \quad \dot\omega_3^i = \frac{1}{2}\left[\frac{\partial^2 v^h}{\partial x \partial t} - \frac{\partial^2 u^h}{\partial y \partial t}\right]. \qquad (1.53)$$

Coming back to integral of the kinetic energy variation of a shell with added masses described by (1.51) and taking into account (1.53) one gets

$$\int_{t_0}^t \delta K dt = \int_{t_0}^t \sum_{i=1}^N \int_0^a \int_0^b \{M^i\left[\frac{\partial u}{\partial t}\frac{\partial(\delta u)}{\partial t} + (z_i-h)^2\frac{\partial^2 w}{\partial x \partial t}\frac{\partial^2(\delta w)}{\partial x \partial t} - \right.$$

$$\frac{\partial(\delta u)}{\partial t}(z_i-h)\frac{\partial^2 w}{\partial x \partial t} - \frac{\partial u}{\partial t}(z_i-h)\frac{\partial^2(\delta w)}{\partial x \partial t} + \frac{\partial v}{\partial t}\frac{\partial(\delta v)}{\partial t} + (z_i-h)^2\frac{\partial^2 w}{\partial y \partial t}\times$$

1.2 3D-Theory of Orthotropic Shallow Shells with Added Masses

$$\frac{\partial^2(\delta w)}{\partial y \partial t} - \frac{\partial(\delta w)}{\partial t}(z_i - h)\frac{\partial^2 w}{\partial y \partial t} - \frac{\partial^2 v}{\partial t^2}(z_i - h)\frac{\partial^2(\delta w)}{\partial y \partial t} + \frac{\partial w}{\partial t}\frac{\partial(\delta w)}{\partial t}\bigg] +$$

$$J^i_{xx}\frac{\partial^2 w}{\partial y \partial t}\frac{\partial^2(\delta w)}{\partial y \partial t} + J^i_{yy}\frac{\partial^2 w}{\partial x \partial t}\frac{\partial^2(\delta w)}{\partial x \partial t} + \frac{1}{4}J^i_{zz}\left[\frac{\partial^2 v}{\partial x \partial t} - \frac{\partial^2 u}{\partial y \partial t}\right]\left(\frac{\partial^2(\delta v)}{\partial x \partial t} -\right.$$

$$\left.\frac{\partial^2(\delta u)}{\partial y \partial t}\right) + J^i_{xy}\frac{\partial^2(\delta w)}{\partial y \partial t}\frac{\partial^2 w}{\partial x \partial t} + J^i_{xy}\frac{\partial^2 w}{\partial y \partial t}\frac{\partial^2(\delta w)}{\partial x \partial t} - J^i_{xz}\frac{\partial^2(\delta w)}{\partial y \partial t}\left[\frac{\partial^2 v}{\partial x \partial t} -\right.$$

$$\left.\frac{\partial^2 u}{\partial y \partial t}\right] - J^i_{xz}\frac{\partial^2 w}{\partial y \partial t}\left[\frac{\partial^2(\delta v)}{\partial x \partial t} - \frac{\partial^2(\delta u)}{\partial y \partial t}\right] + J^i_{yz}\frac{\partial^2(\delta w)}{\partial x \partial t}\left[\frac{\partial^2 v}{\partial x \partial t} - \frac{\partial^2 u}{\partial y \partial t}\right] +$$

$$\left.J^i_{yz}\frac{\partial^2 w}{\partial x \partial t}\left[\frac{\partial^2(\delta v)}{\partial x \partial t} - \frac{\partial^2(\delta u)}{\partial y \partial t}\right]\right\}\bigg|_{z=h}\delta(x - x_i)\delta(y - y_i)dxdydt+$$

$$\int_{t_0}^{t}\int_{-h}^{h}\int_{0}^{a}\int_{0}^{b}\rho\left[\frac{\partial u}{\partial t}\frac{\partial(\delta u)}{\partial t} + \frac{\partial v}{\partial t}\frac{\partial(\delta v)}{\partial t} + \frac{\partial w}{\partial t}\frac{\partial(\delta w)}{\partial t}\right]\bigg|_{z=h}dxdydzdt. \quad (1.54)$$

After transformating the components using integration by parts and substituting the obtained expression together with (1.50) to (1.26) we obtain

$$\int_{t_0}^{t}\int_{0}^{a}\int_{0}^{b}\int_{-h}^{h}\left[\left(\frac{\partial\sigma_{11}}{\partial x} + \frac{\partial\sigma_{12}}{\partial y} + \frac{\partial\sigma_{13}}{\partial z} - \rho\frac{\partial^2 u}{\partial t^2}\right)\delta u + \left(\frac{\partial\sigma_{12}}{\partial x} + \frac{\partial\sigma_{22}}{\partial y} + \frac{\partial\sigma_{23}}{\partial z} - \rho\frac{\partial^2 v}{\partial t^2}\right)\delta v +\right.$$

$$\left.\left(\frac{\partial\sigma_{13}}{\partial x} + \frac{\partial\sigma_{23}}{\partial y} + \frac{\partial\sigma_{33}}{\partial z} - k_1\sigma_{11} - k_2\sigma_{22}\rho\frac{\partial^2 w}{\partial t^2}\right)\delta w\right] - \int_{t_0}^{t}\int_{0}^{b}\int_{-h}^{h}(\sigma_{11}\delta u+$$

$$\sigma_{12}\delta v + \sigma_{13}\delta w)|_0^a\, dydzdt - \int_{t_0}^{t}\int_{0}^{a}\int_{-h}^{h}(\sigma_{22}\delta v + \sigma_{12}\delta u + \sigma_{23}\delta w)|_0^b\, dxdzdt-$$

$$\int_{t_0}^{t}\int_{0}^{a}\int_{0}^{b}\left\{\sigma_{13} - \sum_{i=1}^{N}\left\{M^i\left[-\frac{\partial^2 u}{\partial t^2} + (z_i - h)\frac{\partial^3 w}{\partial x \partial t^2}\right]\delta(x - x_i)\delta(y - y_i)-\right.\right.$$

$$\frac{1}{4}J^i_{zz}\frac{\partial}{\partial y}\left[\delta(y - y_i)\left(\frac{\partial^3 v}{\partial x \partial t^2} - \frac{\partial^3 u}{\partial y \partial t^2}\right)\right] + \frac{1}{2}J^i_{xz}\frac{\partial}{\partial y}\left[\delta(y - y_i)\frac{\partial^3 w}{\partial y \partial t^2}\right]\delta(x - x_i)-$$

$$\left.\frac{1}{2}J^i_{yz}\frac{\partial}{\partial y}\left[\delta(y-y_i)\frac{\partial^3 w}{\partial x \partial t^2}\right]\delta(x - x_i)\right\}\delta u\bigg|_{z=h} - \sigma_{13}\delta u|_{z=-h} + \left\{\sigma_{23} - \sum_{i=1}^{N}M^i\times\right.$$

$$\left[-\frac{\partial^2 v}{\partial t^2} + (z_i - h)\frac{\partial^3 w}{\partial y \partial t^2}\right]\delta(x - x_i)\delta(y - y_i) + \frac{1}{4}J^i_{zz}\frac{\partial}{\partial x}\left[\delta(x - x_i)\left(\frac{\partial^3 v}{\partial x \partial t^2}-\right.\right.$$

$$\left.\left.\frac{\partial^3 u}{\partial y \partial t^2}\right)\right]\delta(y-y_i) - \frac{1}{2}J^i_{xz}\frac{\partial}{\partial x}\left[\delta(x - x_i)\frac{\partial^3 w}{\partial y \partial t^2}\right]\delta(y-y_i) + \frac{1}{2}J^i_{yz}\frac{\partial}{\partial x}\left[\delta(x - x_i)\times\right.$$

$$\left.\frac{\partial^3 w}{\partial x \partial t^2}\right]\delta(y-y_i)\bigg\}\delta v\bigg|_{z=h} - \sigma_{23}\delta v\big|_{z=-h} + \bigg\{\sigma_{33} - \sum_{i=1}^{N} M^i\bigg[-\frac{\partial^2 w}{\partial t^2} + (z_i-h)^2 \times$$

$$\frac{\partial}{\partial x}\left[\delta(x-x_i)\frac{\partial^3 w}{\partial x \partial t^2}\right]\delta(y-y_i) - (z_i-h)\frac{\partial}{\partial x}\left[\delta(x-x_i)\frac{\partial^2 u}{\partial t^2}\right]\delta(y-y_i) +$$

$$(z_i-h)^2 \frac{\partial}{\partial y}\left[\delta(y-y_i)\frac{\partial^3 w}{\partial y \partial t^2}\right]\delta(x-x_i) - (z_i-h)\frac{\partial}{\partial y}\left[\delta(y-y_i)\frac{\partial^2 v}{\partial t^2}\right]\delta(x-x_i) +$$

$$J^i_{xx}\frac{\partial}{\partial y}\left[\delta(y-y_i)\frac{\partial^3 w}{\partial y \partial t^2}\right]\delta(x-x_i) + J^i_{yy}\frac{\partial}{\partial x}\left[\delta(x-x_i)\frac{\partial^3 w}{\partial x \partial t^2}\right]\delta(y-y_i) +$$

$$J^i_{xy}\frac{\partial}{\partial y}\left[\delta(y-y_i)\frac{\partial^3 w}{\partial x \partial t^2}\right]\delta(x-x_i) + J^i_{xy}\frac{\partial}{\partial x}\left[\delta(x-x_i)\frac{\partial^3 w}{\partial y \partial t^2}\right]\delta(y-y_i) -$$

$$\frac{1}{2}J^i_{xz}\frac{\partial}{\partial y}\left[\delta(y-y_i)\left(\frac{\partial^3 v}{\partial x \partial t^2} - \frac{\partial^3 u}{\partial y \partial t^2}\right)\right]\delta(x-x_i) +$$

$$\frac{1}{2}J^i_{yz}\frac{\partial}{\partial x}\left[\delta(x-x_i)\left(\frac{\partial^3 v}{\partial x \partial t^2} - \frac{\partial^3 u}{\partial y \partial t^2}\right)\right]\delta(y-y_i)\bigg\}\delta w\bigg|_{z=h} -$$

$$\sigma_{33}\delta w\big|_{z=-h}\bigg\}dxdydt + \int_{-h}^{h}\int_{0}^{a}\int_{0}^{b}\left[\rho\left(\frac{\partial u}{\partial t}\delta u + \frac{\partial v}{\partial t}\delta v + \frac{\partial w}{\partial t}\delta w\right)\right]\bigg|_{t=0}^{t}dxdydz =$$

$$\int_0^t \delta K dt. \qquad (1.55)$$

1.2.4 Differential Equations, Boundary and Initial Conditions

Comparing expressions standing by displacement variations $\delta u, \delta v, \delta w$, the following differential equations governing a stress-deformation dynamical state of the sloped shell are obtained

$$\frac{\partial \sigma_{11}}{\partial x} + \frac{\partial \sigma_{12}}{\partial y} + \frac{\partial \sigma_{13}}{\partial z} = \rho \frac{\partial^2 u}{\partial t^2}, \quad (\overleftrightarrow{1,2}),$$

$$\frac{\partial \sigma_{13}}{\partial x} + \frac{\partial \sigma_{23}}{\partial y} + \frac{\partial \sigma_{33}}{\partial z} - k_1\sigma_{11} - k_2\sigma_{22} = \rho\frac{\partial^2 w}{\partial t^2}. \qquad (1.56)$$

So, we have got the same equations as (1.32), which govern an elastic medium motion in curvilinear coordinates. Equations (1.56) can be substituted by an equivalent system. Let us focus in the tension components using (1.36) and (1.33). Then the vibration problem of a shallow orthotropic shell with the added elements is reduced to the determination of the displacements components u, v, w satisfying the following equations

1.2 3D-Theory of Orthotropic Shallow Shells with Added Masses

$$A_{1111}\frac{\partial^2 u}{\partial x^2} + A_{1212}\frac{\partial^2 u}{\partial y^2} + A_{1313}\frac{\partial^2 u}{\partial z^2} + (A_{1122} + A_{1212})\frac{\partial^2 v}{\partial x \partial y} +$$

$$(A_{1133} + A_{1313})\frac{\partial^2 w}{\partial x \partial z} - (A_{1111}k_1 + A_{1122}k_2)\frac{\partial w}{\partial x} = \rho\frac{\partial^2 u}{\partial t^2}, \quad \overleftrightarrow{(1,2)},$$

$$A_{3333}\frac{\partial^2 w}{\partial z^2} + A_{1313}\frac{\partial^2 w}{\partial x^2} + A_{2323}\frac{\partial^2 w}{\partial y^2} + (A_{1133} + A_{1313})\frac{\partial^2 u}{\partial x \partial z} +$$

$$(A_{2323} + A_{2233})\frac{\partial^2 v}{\partial y \partial z} - 2(A_{1133}k_1 + A_{2233}k_2)\frac{\partial w}{\partial z} - (A_{1111}k_1 +$$

$$(A_{2323} + A_{2233})\frac{\partial^2 v}{\partial y \partial z} - 2(A_{1133}k_1 + A_{2233}k_2)\frac{\partial w}{\partial z} - (A_{1111}k_1 +$$

$$A_{2211}k_2)\frac{\partial u}{\partial x} - (A_{1122}k_1 + A_{2222}k_2)\frac{\partial v}{\partial y} + w\left(k_1^2 + k_2^2\right) = \rho\frac{\partial^2 w}{\partial t^2}. \quad (1.57)$$

After introducing the dimensionless parameters

$$x = \bar{x}a, \quad y = \bar{y}b, \quad z = 2h\bar{z}, \quad u = 2h\bar{u}, \quad v = 2h\bar{v}, \quad w = 2h\bar{w},$$

$$\lambda_1 = a/2h, \quad \lambda_2 = b/2h, \quad k_1 = 2h/a^2\bar{k}_1, \quad k_2 = 2h/b^2\bar{k}_2, \quad \lambda = a/b,$$

$$A_{1111} = A_{1111} \cdot \overline{A}_{1111}, \quad A_{1122} = \overline{A}_{1122} \cdot A_{1111}, \quad A_{1133} = A_{1111} \cdot \overline{A}_{1133},$$

$$A_{2222} = \overline{A}_{2222} \cdot A_{1111}, \quad A_{2233} = \overline{A}_{2233} \cdot A_{1111}, \quad A_{3333} = \overline{A}_{3333} \cdot A_{1111},$$

$$A_{1212} = A_{1111} \cdot \overline{A}_{1212}, \quad A_{1313} = A_{1111} \cdot \overline{A}_{1313}, \quad A_{2323} = A_{1111} \cdot \overline{A}_{2323},$$

$$t = \frac{ab}{2h}\sqrt{\rho/A_{1111}} \cdot \bar{t}, \quad \omega = \frac{2h}{ab}\sqrt{A_{1111}/\rho} \cdot \bar{\omega}, \quad \bar{M}^i = M^i/M_0;$$

$$J_{xx}^i = ab^3 2h\} \bar{J}_{xx}^i, \quad J_{yy}^i = a^3 b 2h\} \bar{J}_{yy}^i, \quad J_{xy}^i = a^2 b^2 2h\} \bar{J}_{xy}^i,$$

$$J_{xz}^i = a^2 b(2h)^3 \rho \bar{J}_{zz}^i, \quad J_{yz}^i = ab^2(2h)^2 \rho \bar{J}_{yz}^i, \quad J_{zz}^i = ab(2h)^3 \rho \bar{J}_{zz}^i, \quad (1.58)$$

the following dimensionless equations are obtained from (1.57) (bars are omitted):

$$A_{1111}\lambda_2^2\frac{\partial^2 u}{\partial x^2} + A_{1212}\lambda_1^2\frac{\partial^2 u}{\partial y^2} + A_{1313}\lambda_1^2\lambda_2^2\frac{\partial^2 u}{\partial z^2} + (A_{1122} + A_{1212})\lambda_1\lambda_2\frac{\partial^2 v}{\partial x \partial y} +$$

$$(A_{1133} + A_{1313})\lambda_1\lambda_2\frac{\partial^2 w}{\partial x \partial z} - \left(A_{1111}\lambda_2^2 k_1 + A_{1122}\lambda_1^2 k_2\right)\frac{1}{\lambda_1}\frac{\partial w}{\partial x} = \frac{\partial^2 u}{\partial t^2}, \quad \overleftrightarrow{(1,2)},$$

$$A_{3333}\lambda_1^2\lambda_2^2\frac{\partial^2 w}{\partial z^2} + A_{1313}\lambda_2^2\frac{\partial^2 w}{\partial x^2} + A_{2323}\lambda_1^2\frac{\partial^2 w}{\partial y^2} + (A_{1133} + A_{1313})\lambda_1\lambda_2\frac{\partial^2 u}{\partial x \partial z} +$$

$$(A_{2323} + A_{2233})\lambda_1^2\lambda_2\frac{\partial^2 v}{\partial y \partial z} - 2\left(A_{1133}\lambda_2^2 k_1 + A_{2233}\lambda_1^2 k_2\right)\frac{\partial w}{\partial z} - \left(A_{1111}\lambda_2^2 k_1 +\right.$$

$$A_{2211}\lambda_1^2 k_2)\frac{1}{\lambda_1}\frac{\partial u}{\partial x} - \left(\Lambda_{1122}\lambda_2^2 k_1 + A_{2222}\lambda_1^2 k_2\right)\frac{\partial v}{\partial y} + w\left(\frac{k_1^2}{\lambda^2} + k_2^2\lambda^2\right) = \frac{\partial^2 w}{\partial t^2}.$$
(1.59)

In order to integrate the above equations the boundary conditions are needed. Equation (1.55) gives the possibility of prescribing three boundary conditions for each shell edge, remembering that the lateral edges are statically unloaded. For the stressed components we have homogeneous boundary conditions; for instance, for the edges $x = 0, 1$ we have

$$\sigma_{11} = 0, \quad \sigma_{12} = 0, \quad \sigma_{13} = 0. \tag{1.60}$$

For the contents of the displacement vector's in a general case, we have

$$\delta u = 0, \quad \delta v = 0, \quad \delta w = 0. \tag{1.61}$$

Let us define the boundary conditions for the edge $x = 0, 1$ using the standard clamping conditions.

1. Rigid clamping of the whole edge

$$u = 0, \quad v = 0, \quad w = 0. \tag{1.62}$$

2. Sliding cover

$$\sigma_{11} = 0, \quad v = 0, \quad w = 0. \tag{1.63}$$

3. Free edge

$$\sigma_{11} = 0, \quad \sigma_{12} = 0, \quad \sigma_{13} = 0. \tag{1.64}$$

In a similar way the boundary conditions are defined for $y = 0, 1$ edge.

Let us formulate conditions for a surface bounded shell with $z = 0.5; -0.5$. For free vibration we have

$$\sigma_{13} = 0, \quad \sigma_{23} = 0, \sigma_{33} = 0. \tag{1.65}$$

Suppose that the top surface will be loaded by N concentrated masses, and the bottom one will be free. According to (1.55) we have the following stresses conditions for $z = 0.5$

$$\sigma_{i3} = f_i(x, y, t), \quad i = 1, 2, 3, \tag{1.66}$$

where surface load functions include added mass influence on a shell. According to (1.55) we have

$$f_1(x,y,t) = \sum_{i=1}^{N}\left\{M^i\left[-\frac{\partial^2 u}{\partial t^2} + (z_i - h)\frac{\partial^3 w}{\partial x\partial t^2}\right]\delta(x-x_i)\delta(y-y_i) - \frac{1}{4}J_{zz}^i \times\right.$$

$$\frac{\partial}{\partial y}\left[\delta(y-y_i)\left(\frac{\partial^3 w}{\partial x\partial t^2} - \frac{\partial^3 u}{\partial y\partial t^2}\right)\right]\delta(x-x_i) + \frac{1}{2}J_{xz}^i\frac{\partial}{\partial y}\left[\delta(y-y_i)\frac{\partial^3 w}{\partial y\partial t^2}\right]\times$$

$$\delta(x-x_i) - \frac{1}{2}J^i_{yz}\frac{\partial}{\partial y}\left[\delta(y-y_i)\frac{\partial^3 w}{\partial x \partial t^2}\right]\delta(x-x_i)\bigg\}, \quad \overleftrightarrow{(1,2)}, \quad \overleftrightarrow{(x,y)},$$

$$f_3(x,y,t) = \sum_{i=1}^{N} M^i \bigg\{ -\frac{\partial^2 w}{\partial t^2} + (z_i-h)\frac{\partial}{\partial x}\left[\delta(x-x_i)\frac{\partial^3 w}{\partial x \partial t^2}\right]\delta(y-y_i) -$$

$$(z_i-h)\frac{\partial}{\partial x}\left[\delta(x-x_i)\frac{\partial^2 u}{\partial t^2}\right]\delta(y-y_i) + (z_i-h)^2\frac{\partial}{\partial y}\left[\delta(y-y_i)\frac{\partial^3 w}{\partial y \partial t^2}\right]\delta(x-x_i) -$$

$$(z_i-h)\frac{\partial}{\partial y}\left[\delta(y-y_i)\frac{\partial^2 v}{\partial t^2}\right]\delta(x-x_i) + J^i_{xx}\frac{\partial}{\partial y}\left[\delta(y-y_i)\frac{\partial^3 w}{\partial y \partial t^2}\right]\delta(x-x_i) +$$

$$J^i_{xy}\frac{\partial}{\partial x}\left[\delta(x-x_i)\frac{\partial^3 w}{\partial x \partial t^2}\right]\delta(y-y_i) + J^i_{xy}\frac{\partial}{\partial y}\left[\delta(y-y_i)\frac{\partial^3 w}{\partial x \partial t^2}\right]\delta(x-x_i) +$$

$$J^i_{xy}\frac{\partial}{\partial x}\left[\delta(x-x_i)\frac{\partial^3 w}{\partial y \partial t^2}\right]\delta(y-y_i) - \frac{1}{2}J^i_{xy}\frac{\partial}{\partial y}\left[\delta(y-y_i)\right]\left(\frac{\partial^3 v}{\partial x \partial t^2} - \right.$$

$$\left. \frac{\partial^3 u}{\partial y \partial t^2}\right)\bigg]\delta(x-x_i) + \frac{1}{2}J^i_{yz}\frac{\partial}{\partial x}\left[\delta(x-x_i)\left(\frac{\partial^3 v}{\partial x \partial t^2} - \frac{\partial^3 u}{\partial y \partial t^2}\right)\right]\delta(y-y_i)\bigg\}. \tag{1.67}$$

In order to get the initial conditions the equation (1.55) will be used. For arbitrarily taken variations $\delta u, \delta v, \delta w$ and t_0 we get the following initial conditions

$$u = u_0, \quad v = v_0, \quad w = w_0,$$
$$\frac{\partial u}{\partial t} = u'_0, \quad \frac{\partial v}{\partial t} = v'_0, \quad \frac{\partial w}{\partial t} = w'_0. \tag{1.68}$$

1.3 Analysis of Orthotropic Shallow Shells with Added Masses

1.3.1 Two Dimensional Theory

The considered alternative theory is based on Timoshenko's kinematic model [233]. It is assumed that shell fibres in an initial state are normal to a mean shell surface and become linear after rotation caused by deformation, but do not remain perpendicular to the deformed surface.

A displacement field along the shell thickness is defined by the following relationships [185]

$$u^z = u + z\gamma_x, \quad \overleftrightarrow{(x,y)}, \quad w^z = w. \tag{1.69}$$

Deformations are described by the equations

$$e_{11} = \varepsilon_{11} + z\aleph_{11}, \quad \overleftrightarrow{(1,2)}, \quad e_{12} = \varepsilon_{12} + z\aleph_{12},$$

$$e_{13} = \gamma_x + \frac{\partial w}{\partial x}, \quad (\overleftrightarrow{1,2}), \quad (\overleftrightarrow{x,y}), \tag{1.70}$$

where tangent deformation components $\varepsilon_{11}, \varepsilon_{22}, \varepsilon_{12}$ and bending components $\aleph_{11}, \aleph_{22}, \aleph_{12}$ of the mean surface are the following

$$\varepsilon_{11} = \frac{\partial u}{\partial x} + k_1 w, \quad (\overleftrightarrow{1,2}), \quad \varepsilon_{12} = \frac{\partial u}{\partial y} + \frac{\partial v}{\partial x},$$

$$\aleph_{11} = \frac{\partial \gamma_x}{\partial x}, \quad \aleph_{22} = \frac{\partial \gamma_y}{\partial y}, \quad \aleph_{12} = \frac{\partial \gamma_x}{\partial y} + \frac{\partial \gamma_y}{\partial x}. \tag{1.71}$$

Stress-displacement relationships are as follows:

$$\sigma_{11} = A_{1111} e_{11} + A_{1122} e_{22}, \quad (\overleftrightarrow{1,2}),$$

$$\sigma_{13} = A_{1313} e_{13}, \quad \sigma_{23} = A_{2323} e_{23}, \quad \sigma_{12} = A_{1212} e_{12}. \tag{1.72}$$

For an orthotropic material elasticity coefficients A_{ijkl} are defined by

$$A_{1111} = \frac{E_1}{1 - \nu_1 \nu_2}, \quad A_{1122} = \frac{E_1 \nu_2}{1 - \nu_1 \nu_2}, \quad (\overleftrightarrow{1,2}),$$

$$A_{1212} = G_{12}, \quad A_{1313} = G_{13}, \quad A_{2323} = G_{23}. \tag{1.73}$$

Integrating (1.72) along thickness one obtains

$$T_1 = \int_{-h}^{h} \sigma_{11} dz - 2h \left(A_{1111} \varepsilon_{11} + A_{1122} \varepsilon_{22} \right), \quad (\overleftrightarrow{1,2}),$$

$$S = \int_{-h}^{h} \sigma_{12} dz = 2h A_{1212} \varepsilon_{12},$$

$$Q_1 = \int_{-h}^{h} \sigma_{13} dz = 2h A_{1313} \left(\gamma_x + \frac{\partial w}{\partial x} \right), \quad (\overleftrightarrow{1,2}). \tag{1.74}$$

Multiplying $\sigma_{11}, \sigma_{22}, \sigma_{12}$ from (1.72) by 'z' and integrating along thickness we have the following force moments

$$M_1 = \int_{-h}^{h} \sigma_{11} z dz = \frac{2h^3}{3} \left(A_{1111} \frac{\partial \gamma_x}{\partial x} + A_{1122} \frac{\partial \gamma_y}{\partial y} \right), \quad (\overleftrightarrow{1,2}),$$

$$M_{12} = \int_{-h}^{h} \sigma_{12} z dz = \frac{2h^3}{3} A_{1212} \left(\frac{\partial \gamma_x}{\partial y} + \frac{\partial \gamma_y}{\partial x} \right). \tag{1.75}$$

1.3 Analysis of Orthotropic Shallow Shells with Added Masses

In order to get the differential equations of motion and the boundary as well as the initial conditions, again the Hamilton rule is used.

The final expressions defining the kinetic and potential energies of a shell with added masses kinetic energy is obtained from (1.49) after integration along thickness and by introducing the integral characteristics defined by (1.74) and (1.75)

$$\Pi = \frac{1}{2} \int_0^a \int_0^b (T_1 \varepsilon_{11} + T_2 \varepsilon_{22} + S \varepsilon_{12} + M_1 \aleph_{11} + M_2 \aleph_{22} + M_{12} \aleph_{12} +$$

$$Q_1 \varepsilon_{13} + Q_2 \varepsilon_{23}) \, dxdy,$$

$$K_1 = \frac{1}{2} \int_0^a \int_0^b \rho \left[2h \left(\frac{\partial u}{\partial t} \right)^2 + 2h \left(\frac{\partial v}{\partial t} \right)^2 + 2h \left(\frac{\partial w}{\partial t} \right)^2 + \frac{2h^3}{3} \left(\frac{\partial \gamma_x}{\partial x} \right)^2 + \right.$$

$$\left. \frac{2h^3}{3} \left(\frac{\partial \gamma_y}{\partial y} \right)^2 \right] dxdy,$$

$$K_2 = \frac{1}{2} \sum_{i=1}^{N} \int_0^a \int_0^b \{M^i \left[(\xi_1^i)^2 + (\xi_2^i)^2 + (\xi_3^i)^2 \right] + J_{xx}^i (\dot{\omega}_1^i)^2 + J_{yy}^i (\dot{\omega}_2^i)^2 +$$

$$J_{zz}^i (\dot{\omega}_3^i)^2 - 2J_{xy}^i \dot{\omega}_1^i \dot{\omega}_2^i - 2J_{xz}^i \dot{\omega}_1^i \dot{\omega}_3^i - 2J_{yz}^i \dot{\omega}_2^i \dot{\omega}_3^i \} \times \delta(x - x_i) \delta(y - y_i) dxdy. \quad (1.76)$$

The velocity vector components $\bar{\xi}^i$ of the added element mass centre and the angular velocity components $\dot{\bar{\omega}}$ are determined from (1.53) using (1.69)

$$\xi_1^i = \frac{\partial u}{\partial t} + h \frac{\partial \gamma_x}{\partial t} - (z_i - h) \frac{\partial^2 w}{\partial x \partial t}, \quad \overrightarrow{(1,2)}, \quad \overrightarrow{(x,y)},$$

$$\xi_3^i = \frac{\partial w}{\partial t}, \quad \dot{\omega}_1^i = \frac{\partial^2 w}{\partial y \partial t}, \quad \dot{\omega}_2^i = -\frac{\partial^2 w}{\partial x \partial t},$$

$$\dot{\omega}_3^i = \frac{1}{2} \left[\frac{\partial^2 v}{\partial x \partial t} - \frac{\partial^2 u}{\partial y \partial t} + h \left(\frac{\partial^2 \gamma_y}{\partial x \partial t} - \frac{\partial^2 \gamma_x}{\partial y \partial t} \right) \right]. \quad (1.77)$$

Variation of the independent functions u, v, w, γ_x, γ_y in the expression

$$\int_{t_0}^{t} \delta L dt = \delta \int_{t_0}^{t} (\Pi - K) \, dt = 0 \quad (1.78)$$

lead to the following variational equation

$$\int_{t_0}^{t} \int_0^a \int_0^b \rho \left[2h \frac{\partial u}{\partial t} \frac{\partial (\delta u)}{\partial t} + 2h \frac{\partial v}{\partial t} \frac{\partial (\delta v)}{\partial t} + 2h \frac{\partial w}{\partial t} \frac{\partial (\delta w)}{\partial t} + \frac{2h^3}{3} \frac{\partial \gamma_x}{\partial t} \frac{\partial (\delta \gamma_x)}{\partial t} + \right.$$

$$\left.\frac{2h}{3}\frac{\partial\gamma_y}{\partial t}\frac{\partial(\delta\gamma_y)}{\partial t}\right]dxdy\,dt+\int\limits_{t_0}^{t}\int\limits_{0}^{a}\int\limits_{0}^{b}\sum_{i=1}^{N}\left\{M^i\left[\frac{\partial u}{\partial t}\frac{\partial(\delta u)}{\partial t}+h^2\frac{\partial\gamma_x}{\partial t}\frac{\partial(\delta\gamma_x)}{\partial t}+\right.\right.$$

$$(z_i-h)^2\frac{\partial^2 w}{\partial x\partial t}\frac{\partial(\delta w)}{\partial t}+h\frac{\partial(\delta u)}{\partial t}\frac{\partial\gamma_x}{\partial t}+h\frac{\partial u}{\partial t}\frac{\partial(\delta\gamma_x)}{\partial t}-(z_i-h)\frac{\partial(\delta u)}{\partial t}\frac{\partial^2 w}{\partial x\partial t}-$$

$$(z_i-h)\frac{\partial u}{\partial t}\frac{\partial^2(\delta w)}{\partial x\partial t}-h(z_i-h)\frac{\partial(\delta\gamma_x)}{\partial t}\frac{\partial^2 w}{\partial x\partial t}-h(z_i-h)\frac{\partial\gamma_x}{\partial t}\frac{\partial^2(\delta w)}{\partial x\partial t}+$$

$$\frac{\partial v}{\partial t}\frac{\partial(\delta v)}{\partial t}+h^2\frac{\partial\gamma_y}{\partial t}\frac{\partial(\delta\gamma_y)}{\partial t}+(z_i-h)^2\frac{\partial^2(\delta w)}{\partial y\partial t}\frac{\partial^2 w}{\partial y\partial t}+h\frac{\partial(\delta v)}{\partial t}\frac{\partial\gamma_y}{\partial t}+$$

$$h\frac{\partial v}{\partial t}\frac{\partial(\delta\gamma_y)}{\partial t}-(z_i-h)\frac{\partial^2 w}{\partial y\partial t}\frac{\partial(\delta v)}{\partial t}-(z_i-h)\frac{\partial v}{\partial t}\frac{\partial^2(\delta w)}{\partial y\partial t}-h(z_i-h)\times$$

$$\frac{\partial(\delta\gamma_y)}{\partial t}\frac{\partial^2 w}{\partial y\partial t}-h(z_i-h)\frac{\partial\gamma_y}{\partial t}\frac{\partial^2(\delta w)}{\partial t^2}+\frac{\partial(\delta w)}{\partial t}\frac{\partial w}{\partial t}\right]+J^i_{xx}\frac{\partial^2 w}{\partial y\partial t}\frac{\partial^2(\delta w)}{\partial y\partial t}+$$

$$J^i_{yy}\frac{\partial^2 w}{\partial x\partial t}\frac{\partial^2(\delta w)}{\partial x\partial t}+J^i_{zz}\frac{1}{4}\left[\frac{\partial^2 v}{\partial x\partial t}\frac{\partial^2(\delta v)}{\partial x\partial t}+\frac{\partial^2 u}{\partial y\partial t}\frac{\partial^2(\delta u)}{\partial t\partial y}+h^2\frac{\partial^2\gamma_y}{\partial x\partial t}\frac{\partial^2(\delta\gamma_y)}{\partial x\partial t}+\right.$$

$$h^2\frac{\partial^2\gamma_x}{\partial y\partial t}\frac{\partial^2(\delta\gamma_x)}{\partial y\partial t}-\frac{\partial^2(\delta v)}{\partial x\partial t}\frac{\partial^2 u}{\partial y\partial t}-\frac{\partial^2 v}{\partial x\partial t}\frac{\partial^2(\delta u)}{\partial y\partial t}+h\frac{\partial^2(\delta v)}{\partial x\partial t}\frac{\partial^2\gamma_y}{\partial x\partial t}+$$

$$h\frac{\partial^2 v}{\partial x\partial t}\frac{\partial^2(\delta\gamma_y)}{\partial x\partial t}-h\frac{\partial^2(\delta v)}{\partial x\partial t}\frac{\partial^2\gamma_x}{\partial y\partial t}-h\frac{\partial^2 v}{\partial x\partial t}\frac{\partial^2(\delta\gamma_x)}{\partial y\partial t}-h\frac{\partial^2(\delta u)}{\partial y\partial t}\frac{\partial^2\gamma_y}{\partial x\partial t}-$$

$$h\frac{\partial^2 u}{\partial y\partial t}\frac{\partial^2(\delta\gamma_y)}{\partial x\partial t}+h\frac{\partial^2(\delta u)}{\partial y\partial t}\frac{\partial^2\gamma_x}{\partial y\partial t}+h\frac{\partial^2 u}{\partial y\partial t}\frac{\partial^2(\delta\gamma_x)}{\partial y\partial t}-h^2\frac{\partial^2(\delta\gamma_y)}{\partial x\partial t}\frac{\partial^2\gamma_x}{\partial y\partial t}-$$

$$\left.h^2\frac{\partial^2\gamma_y}{\partial x\partial t}\frac{\partial^2(\delta\gamma_x)}{\partial y\partial t}\right]+J^i_{xy}\frac{\partial^2(\delta w)}{\partial y\partial t}\frac{\partial^2 w}{\partial x\partial t}+J^i_{xy}\frac{\partial^2 w}{\partial y\partial t}\frac{\partial^2(\delta w)}{\partial x\partial t}-\frac{1}{2}J^i_{xz}\frac{\partial^2(\delta w)}{\partial y\partial t}\times$$

$$\left[\frac{\partial^2(\delta v)}{\partial x\partial t}-\frac{\partial^2(\delta u)}{\partial y\partial t}+h\left(\frac{\partial^2\gamma_y}{\partial x\partial t}-\frac{\partial^2\gamma_x}{\partial y\partial t}\right)\right]-\frac{1}{2}J^i_{xz}\frac{\partial^2 w}{\partial y\partial t}\left[\frac{\partial^2(\delta v)}{\partial x\partial t}-\frac{\partial^2(\delta u)}{\partial y\partial t}+\right.$$

$$\left.h\left(\frac{\partial^2(\delta\gamma_y)}{\partial x\partial t}-\frac{\partial^2(\delta\gamma_x)}{\partial y\partial t}\right)\right]+\frac{1}{2}J^i_{yz}\frac{\partial^2(\delta w)}{\partial x\partial t}\left[\frac{\partial^2 v}{\partial x\partial t}-\frac{\partial^2 u}{\partial y\partial t}+h\left(\frac{\partial^2\gamma_y}{\partial x\partial t}-\frac{\partial^2\gamma_x}{\partial y\partial t}\right)\right]+$$

$$\left.\left.\frac{1}{2}J^i_{yz}\frac{\partial^2 w}{\partial x\partial t}\left[\frac{\partial^2(\delta v)}{\partial x\partial t}-\frac{\partial^2(\delta u)}{\partial y\partial t}+h\left(\frac{\partial^2(\delta\gamma_y)}{\partial x\partial t}-\frac{\partial^2(\delta\gamma_x)}{\partial y\partial t}\right)\right]\right\}\delta\left(x-x_i\right)\times$$

$$\delta\left(x-x_i\right)dxdydt-\int\limits_{t_0}^{t}\int\limits_{0}^{a}\int\limits_{0}^{b}(T_1\delta\varepsilon_{11}+T_2\delta\varepsilon_{22}+S\delta\varepsilon_{12}+M_1\delta\aleph_{11}+M_2\delta\aleph_{22}+$$

$$M_{12}\delta\aleph_{12}+Q_1\delta\varepsilon_{13}+Q_2\delta\varepsilon_{23})\,dxdydt=0. \tag{1.79}$$

During derivation of the above equation some technical aspects of a shell with finite rotation stiffness are used [185].

1.3 Analysis of Orthotropic Shallow Shells with Added Masses

Assuming $\gamma_x = -\partial w/\partial x$, $\gamma_y = -\partial w/\partial y$ the classical shallow shell theory equations are obtained from (1.79). Then, in the frame of the classical theory and using the Kirchhoff-Love hypothesis, some of the terms of (1.79) are omitted. From (1.79) we get the equation considered in [244] for $M^i = J^i_{xx} = J^i_{yy} = J^i_{zz} = J^i_{xy} = J^i_{xz} = J^i_{yz} = 0$. This assumption describes the system with no added masses. The differential equations fulfil the above mentioned assumptions. In addition, the boundary and initial conditions are obtained.

Taking into account γ_x, γ_y, u, v, w variations, using integration by parts and equalling to zero the components multiplied by $\delta\gamma_x$, $\delta\gamma_y$, δw, δu, δv variations the following equations of Timoshenko's kinematic model are obtained

$$\frac{\partial T_1}{\partial x} + \frac{\partial S}{\partial y} = 2h\rho \frac{\partial^2 u}{\partial t^2} + f_1(x,y,t), \quad \overleftrightarrow{(1,2)}, \quad \overleftrightarrow{(x,y)},$$

$$\frac{\partial Q_1}{\partial x} + \frac{\partial Q_2}{\partial y} - k_1 T_1 - k_2 T_2 = 2h\rho \frac{\partial^2 w}{\partial t^2} + f_3(x,y,t),$$

$$\frac{\partial M_1}{\partial x} + \frac{\partial M_{12}}{\partial y} - Q_1 = \frac{2h^3}{3}\rho \frac{\partial^2 \gamma_x}{\partial t^2} + \varphi_1(x,y,t), \quad \overleftrightarrow{(1,2)}, \quad \overleftrightarrow{(x,y)}, \quad (1.80)$$

where f_1, f_2, f_3, φ_1 and φ_2 the inertial components that depend on the physical and geometrical added masses characteristics

$$f_1(x,y,t) = \sum_{i=1}^{N} \left\{ M^i \left[\frac{\partial^2 u}{\partial t^2}\delta(x-x_i)\delta(y-y_i) + h\frac{\partial^2 \gamma_x}{\partial t^2}\delta(x-x_i)\delta(y-y_i) - \right.\right.$$

$$(z_i - h)\frac{\partial^3 w}{\partial x \partial t^2}\delta(x-x_i)\delta(y-y_i)\bigg] - \frac{1}{4}J^i_{zz}\frac{\partial}{\partial y}\left[\frac{\partial^3 u}{\partial y \partial t^2}\delta(y-y_i)\right]\delta(x-x_i)+$$

$$\frac{1}{4}J^i_{zz}\frac{\partial}{\partial y}\left[\frac{\partial^3 v}{\partial x \partial t^2}\delta(y-y_i)\right]\delta(x-x_i) + \frac{1}{4}J^i_{zz}h\frac{\partial}{\partial y}\left[\frac{\partial^3 \gamma_y}{\partial x \partial t^2}\delta(y-y_i)\right]\delta(x-x_i)-$$

$$\frac{1}{4}J^i_{zz}h\frac{\partial}{\partial y}\left[\frac{\partial^3 \gamma_x}{\partial y \partial t^2}\delta(y-y_i)\right]\delta(x-x_i) - \frac{1}{2}J^i_{xz}\frac{\partial}{\partial y}\left[\frac{\partial^3 w}{\partial y \partial t^2}\delta(y-y_i)\right]\delta(x-x_i)+$$

$$\frac{1}{2}J^i_{yz}\frac{\partial}{\partial y}\left[\frac{\partial^3 w}{\partial x \partial t^2}\delta(y-y_i)\right]\delta(x-x_i)\bigg\}, \quad \overleftrightarrow{(1,2)},$$

$$f_3(x,y,t) = \sum_{i=1}^{N} M^i \left\{ -(z_i-h)^2 \frac{\partial}{\partial x}\left[\frac{\partial^3 w}{\partial x \partial t^2}\delta(x-x_i)\right]\delta(y-y_i) - \right.$$

$$(z_i-h)\frac{\partial}{\partial x}\left[\frac{\partial^2 u}{\partial t^2}\delta(x-x_i)\right]\delta(y-y_i) + h(z_i-h)\frac{\partial}{\partial x}\left[\frac{\partial^2 \gamma_x}{\partial t^2}\delta(x-x_i)\right]\delta(y-y_i) -$$

$$(z_i-h)^2 \frac{\partial}{\partial y}\left[\frac{\partial^3 w}{\partial y \partial t^2}\delta(y-y_i)\right]\delta(x-x_i) + (z_i-h)\frac{\partial}{\partial y}\left[\frac{\partial^2 v}{\partial t^2}\delta(y-y_i)\right]\delta(x-x_i) +$$

$$h(z_i-h)\frac{\partial}{\partial y}\left[\frac{\partial^2 \gamma_y}{\partial t^2}\delta(y-y_i)\right]\delta(x-x_i) + \frac{\partial^2 w}{\partial t^2}\delta(x-x_i)\delta(y-y_i)\bigg\} -$$

$$\sum_{i=1}^{N}\left\{J_{xx}^{i}\frac{\partial}{\partial y}\left[\frac{\partial^{3}w}{\partial y\partial t^{2}}\delta(y-y_{i})\right]\delta(x-x_{i})-J_{yy}^{i}\frac{\partial}{\partial x}\left[\frac{\partial^{3}w}{\partial x\partial t^{2}}\delta(x-x_{i})\right]\delta(y-y_{i})-\right.$$
$$J_{xy}^{i}\frac{\partial}{\partial y}\left[\frac{\partial^{3}w}{\partial x\partial t^{2}}\delta(y-y_{i})\right]\delta(x-x_{i})-J_{xy}^{i}\frac{\partial}{\partial x}\left[\frac{\partial^{3}w}{\partial y\partial t^{2}}\delta(x-x_{i})\right]\delta(y-y_{i})+$$
$$J_{xz}^{i}\frac{\partial}{\partial y}\left\{\left[\frac{\partial^{3}v}{\partial x\partial t^{2}}-\frac{\partial^{3}u}{\partial y\partial t^{2}}+h\left(\frac{\partial^{3}\gamma_{y}}{\partial x\partial t^{2}}-\frac{\partial^{3}\gamma_{x}}{\partial y\partial t^{2}}\right)\right]\delta(y-y_{i})\right\}\delta(x-x_{i})-$$
$$\left.J_{yz}^{i}\frac{\partial}{\partial x}\left\{\left[\frac{\partial^{3}v}{\partial x\partial t^{2}}-\frac{\partial^{3}u}{\partial y\partial t^{2}}+h\left(\frac{\partial^{3}\gamma_{y}}{\partial x\partial t^{2}}-\frac{\partial^{3}\gamma_{x}}{\partial y\partial t^{2}}\right)\right]\delta(x-x_{i})\right\}\delta(y-y_{i})\right\},\ \overleftrightarrow{(1,2)},$$

$$\varphi_{1}(x,y,t)=\sum_{i=1}^{N}M^{i}\left\{h^{2}\frac{\partial^{2}\gamma_{x}}{\partial t^{2}}\delta(x-x_{i})\delta(y-y_{i})+h\frac{\partial^{2}u}{\partial t^{2}}\delta(x-x_{i})\delta(y-y_{i})-\right.$$
$$\left.h(z_{i}-h)\frac{\partial^{3}w}{\partial x\partial t^{2}}\delta(x-x_{i})\delta(y-y_{i})\right\}+\sum_{i=1}^{N}\left\{J_{zz}^{i}\frac{1}{4}\left[-h^{2}\frac{\partial}{\partial y}\left(\frac{\partial^{3}\gamma_{x}}{\partial y\partial t^{2}}\delta(y-y_{i})\right)\right]\times\right.$$
$$\delta(x-x_{i})+h\frac{\partial}{\partial y}\left[\frac{\partial^{3}v}{\partial x\partial t^{2}}\delta(y-y_{i})\right]\delta(x-x_{i})-h\frac{\partial}{\partial y}\left[\frac{\partial^{3}v}{\partial x\partial t^{2}}\delta(y-y_{i})\right]\delta(x-x_{i})-$$
$$h\frac{\partial}{\partial y}\left[\frac{\partial^{3}u}{\partial y\partial t^{2}}\delta(y-y_{i})\right]\delta(x-x_{i})+h^{2}\frac{\partial}{\partial y}\left[\frac{\partial^{3}\gamma_{y}}{\partial x\partial t^{2}}\delta(y-y_{i})\right]\delta(x-x_{i})-\frac{1}{2}J_{xz}^{i}h\times$$
$$\left.\frac{\partial}{\partial y}\left[\frac{\partial^{3}w}{\partial y\partial t^{2}}\delta(y-y_{i})\right]\delta(x-x_{i})+\frac{1}{2}h\frac{\partial}{\partial y}\left[\frac{\partial^{3}w}{\partial x\partial t^{2}}\delta(y-y_{i})\right]\delta(x-x_{i})\right\}. \quad (1.81)$$

The boundary conditions are additionally needed for the integration of the kinematic model's (1.80).

Using the Kirchhoff-Love hypothesis, a normal element position is defined by four degrees of freedom (four boundary conditions are formulated on each part of that element). Using Timoshenko's model the normal element's degrees of freedom is equal to five. Using variational equation for each part of that element, five boundary conditions are needed. Equation (1.79) allows for the formulation of five boundary conditions for each of the shell's edges. Kinematic and conjugated static homogeneous boundary conditions for the edge $x = 0$, $x = a$ are given below:

$$\gamma_x = \gamma_x^0, \quad M_1 = 0, \quad \gamma_y = \gamma_y^0, \quad M_{12} = 0,$$
$$w = w_0, \quad Q_1 = 0, \quad u = u_0, \quad T_1 = 0, \quad v = v_0, \quad S = 0, \quad (1.82)$$

where: $u_0, v_0, w_0, \gamma_x^0, \gamma_y^0$ define displacement and rotation of the shell's edge.

In real constructions with shell elements different supports are possible and therefore different mathematical models can be formulated.

Let us consider certain boundary conditions for the shell's sides $x = 0$, $x = a$.

1. Roller support
$$w = \gamma_y = M_1 = T_1 = v = 0. \qquad (1.83)$$

2. Free support
$$w = \gamma_y = M_1 = T_1 = S = 0. \qquad (1.84)$$

3. Roller clamping
$$u = v = w = M_1 = \gamma_y = 0. \qquad (1.85)$$

The above mentioned boundary conditions do not exhaust all possibilities [117, 185].

The initial conditions defining the kinematic characteristics and the velocity of their changes in the time moment t_0 are as follows

$$u = \tilde{u}_0, \quad \frac{\partial u}{\partial t} = \tilde{u}'_0; \quad v = \tilde{v}_0, \quad \frac{\partial v}{\partial t} = \tilde{v}'_0;$$

$$w = \tilde{w}_0, \quad \frac{\partial w}{\partial t} = \tilde{w}'_0; \quad \gamma_x = \tilde{\gamma}^0_x, \quad \frac{\partial \gamma_x}{\partial t} = \tilde{\gamma}'_x; \quad (\overleftrightarrow{x,y}). \qquad (1.86)$$

1.3.2 Simplified Model

The earlier described model of a shallow orthotropic shell with concentrated masses including tangential rotations is defined by five differential equations, whose right-hand side terms include the inertial effects introduced by the added concentrated masses. These terms are rather complicated and the search for a real solution would generally cause many difficulties. Therefore, we need to make some simplifications and variants, which include typical problem properties, and can be applied in practise.

Consider a case, when the added bodies are homogeneous. Taking into account the symmetry properties of (1.81), one takes $J^i_{xy} = 0$, $J^i_{xz} = 0$, $J^i_{yz} = 0$. An inertial moment J_{zz} for the mass centre is equal to

$$\int\limits_{(m)} \left(x^2 + y^2\right) dm.$$

If the added masses possess small clamping diameters then $J^i_{zz} \ll J^i_{xx}$, $J^i_{zz} \ll J^i_{yy}$.

The quantity ω^i_3, which characterizes an average rotation of a top plate surface is small. Therefore all terms including the multiplier J^i_{zz} may be neglected.

We get the following relationships for f_1, f_2, f_3, φ_1 and φ_2:

$$f_1(x,y,t) = \sum_{i=1}^{N} M^i \left[\frac{\partial^2 u}{\partial t^2} \delta(x - x_i)\delta(y - y_i) + h\frac{\partial^2 \gamma_x}{\partial t^2}\delta(x - x_i)\delta(y - y_i) - \right.$$

$$\left.(z_i - h)\frac{\partial^3 w}{\partial x \partial t^2}\delta(x - x_i)\delta(y - y_i)\right], \quad \overleftrightarrow{(1,2)}, \quad \overleftrightarrow{(x,y)}, \qquad (1.87)$$

$$f_3(x,y,t) = \sum_{i=1}^{N}\left\{M^i\left[(z_i - h)^2\frac{\partial}{\partial x}\left[\frac{\partial^3 w}{\partial x \partial t^2}\delta(x - x_i)\right]\delta(y - y_i) - (z_i - h)\times\right.\right.$$

$$\frac{\partial}{\partial x}\left[\frac{\partial^2 u}{\partial t^2}\delta(x - x_i)\right]\delta(y - y_i) + h(z_i - h)\frac{\partial}{\partial x}\left[\frac{\partial^2 \gamma_x}{\partial t^2}\delta(x - x_i)\right]\delta(y - y_i) -$$

$$(z_i - h)^2\frac{\partial}{\partial y}\left[\frac{\partial^3 w}{\partial y \partial t^2}\delta(y - y_i)\right]\delta(x - x_i) + (z_i - h)\frac{\partial}{\partial y}\left[\frac{\partial^2 v}{\partial t^2}\delta(y - y_i)\right]\times$$

$$\delta(x - x_i) + h(z_i - h)\frac{\partial}{\partial y}\left[\frac{\partial^2 \gamma_y}{\partial t^2}\delta(y - y_i)\right]\delta(x - x_i) + \frac{\partial^2 w}{\partial t^2}\delta(x - x_i)\delta(y - y_i)\right] -$$

$$\left.J_{xx}^i\frac{\partial}{\partial y}\left[\frac{\partial^3 w}{\partial y \partial t^2}\delta(y - y_i)\right]\delta(x - x_i) - J_{yy}^i\frac{\partial}{\partial x}\left[\frac{\partial^3 w}{\partial x \partial t^2}\delta(x - x_i)\right]\delta(y - y_i)\right\},$$

$$\varphi_1(x,y,t) = \sum_{i=1}^{N}M^i\left[h^2\frac{\partial^2 \gamma_x}{\partial t^2}\delta(x - x_i)\delta(y - y_i) + h\frac{\partial^2 u}{\partial t^2}\delta(x - x_i)\delta(y - y_i) -\right.$$

$$\left.h(z_i - h)\frac{\partial^3 w}{\partial x \partial t^2}\delta(x - x_i)\delta(y - y_i)\right], \quad \overleftrightarrow{(1,2)}, \quad \overleftrightarrow{(x,y)}.$$

Taking into account (1.80) and (1.87), one comes to the conclusion, that the occurrence of the concentrated masses of the top plate surface is equivalent to a certain load surface $\bar{q}$, reduced to an average plate surface and having the following components $q_1 = -f(x,y,t)$, $q_2 = -f_2(x,y,t)$, $q_3 = -f_3(x,y,t)$. External load moment components are equal to $m_1 = -\varphi(x,y,t)$ and $m_2 = -\varphi_2(x,y,t)$.

This analogy generalizes approach commonly used in the investigation of classical plates and shells (only the q_3 component is considered, since tangential components as well as the moments m_1 and m_2 are negligible).

The different load forms and shapes acting on real constructions cause various kinds of vibrations. If ones considers the problems when the first bending modes are of fundamental significance, then further simplification of (1.87) is possible by assuming the tangential inertial components are negligible. Then, we put $f_1 = f_2 = 0$ in (1.87) and the other terms are defined below

$$f_3(x,y,t) = \sum_{i=1}^{N}\left\{M^i\left[\frac{\partial^2 w}{\partial t^2}\delta(x - x_i)\delta(y - y_i) - (z_i - h)^2\frac{\partial}{\partial x}\left[\frac{\partial^3 w}{\partial x \partial t^2}\delta(x - x_i)\right]\times\right.\right.$$

$$\delta(y - y_i) + h(z_i - h)\frac{\partial}{\partial x}\left[\frac{\partial^2 \gamma_x}{\partial t^2}\delta(x - x_i)\right]\delta(y - y_i) - (z_i - h)^2\times$$

$$\frac{\partial}{\partial y}\left[\frac{\partial^3 w}{\partial y \partial t^2}\delta(y - y_i)\right]\delta(x - x_i) + h(z_i - h)\frac{\partial}{\partial y}\left[\frac{\partial^2 \gamma_y}{\partial t^2}\delta(y - y_i)\right]\delta(x - x_i)\right] -$$

1.3 Analysis of Orthotropic Shallow Shells with Added Masses

$$J_{xx}^i \frac{\partial}{\partial y}\left[\frac{\partial^3 w}{\partial y \partial t^2}\delta(y-y_i)\right]\delta(x-x_i) - J_{yy}^i \frac{\partial}{\partial x}\left[\frac{\partial^3 w}{\partial x \partial t^2}\delta(x-x_i)\right]\delta(y-y_i)\right\},$$

$$\varphi_1(x,y,t) = \sum_{i=1}^{N} M^i \left[h^2 \frac{\partial^2 \gamma_x}{\partial t^2}\delta(x-x_i)\delta(y-y_i) - h(z_i - h)\frac{\partial^3 w}{\partial x \partial t^2}\delta(x-x_i)\delta(y-y_i)\right]. \quad (1.88)$$

Let us introduce F defined as [243]

$$T_1 = \frac{\partial^2 F}{\partial y^2}, \quad T_2 = \frac{\partial^2 F}{\partial x^2}, \quad S = \frac{\partial^2 F}{\partial x \partial y}. \quad (1.89)$$

Substituting (1.89) to the first two equations of (1.80) leads to an identity. Substituting forces and moments by displacements and the Airy's function owing to (1.74), (1.75) and (1.89), the rest of the equations of (1.74) are as follows

$$-\nabla_k^2 F + 2hA_{1313}\left(\frac{\partial \gamma_x}{\partial x} + \frac{\partial^2 w}{\partial x^2}\right) + 2hA_{2323}\left(\frac{\partial \gamma_y}{\partial y} + \frac{\partial^2 w}{\partial y^2}\right) = 2h\rho\frac{\partial^2 w}{\partial t^2} + f_3(x,y,t),$$

$$h^2\left[A_{1111}\frac{\partial^2 \gamma_x}{\partial x^2} + (A_{1122} + A_{1212})\frac{\partial^2 \gamma_y}{\partial x \partial y} + A_{1212}\frac{\partial^2 \gamma_x}{\partial y^2}\right] -$$

$$-3A_{1313}\left(\gamma_x + \frac{\partial w}{\partial x}\right) = \rho h^2 \frac{\partial^2 \gamma_x}{\partial t^2} + \varphi_1(x,y,t), \quad \overleftrightarrow{(1,2)}, \quad \overleftrightarrow{(x,y)}. \quad (1.90)$$

The above equations define four independent functions w, γ_x, γ_y and F. An additional equation is obtained using the continuous deformation equation

$$\frac{\partial^2 \varepsilon_{11}}{\partial y^2} + \frac{\partial^2 \varepsilon_{22}}{\partial x^2} - \frac{\partial^2 \varepsilon_{12}}{\partial x \partial y} - k_1 \frac{\partial^2 w}{\partial y^2} - k_2 \frac{\partial^2 w}{\partial x^2} = 0. \quad (1.91)$$

Taking the solutions of (1.74) and (1.89), and substituting them into (1.91) we get the following continuity deformation of equation

$$\frac{1}{2h}\nabla^4 F - \nabla_k^2 w = 0, \quad (1.92)$$

with

$$\nabla_k^2(\bullet) = k_1 \frac{\partial^2(\bullet)}{\partial y^2} + k_2 \frac{\partial^2(\bullet)}{\partial x^2},$$

$$\nabla^4(\bullet) = a_{2222}\frac{\partial^4(\bullet)}{\partial x^4} + (a_{1122} + a_{2211} - a_{1212})\frac{\partial^4(\bullet)}{\partial x^2 \partial y^2} + a_{1111}\frac{\partial^4(\bullet)}{\partial y^4}. \quad (1.93)$$

The boundary conditions are defined by (1.83)–(1.85).

Finally, we consider a second problem related to the vibrations of a shallow shell with added masses in the frame of the Kirchhoff-Love theory. According to the Kirchhoff-Love assumptions the following definitions are included in (1.79)

$$\gamma_x = -\frac{\partial w}{\partial x}, \quad \gamma_y = -\frac{\partial w}{\partial y}, \tag{1.94}$$

which means that $\varepsilon_{13} = \varepsilon_{23} = 0$.
In addition, we assume that,

$$u_i = u - z_i \frac{\partial w}{\partial x}, \quad v_i = v - z_i \frac{\partial w}{\partial y},$$

$$w_i = w, \quad w_3^i = 0, \quad i = \overline{1, N}. \tag{1.95}$$

Applying a variational procedure we get (after some transformations) the following system of differential equations, governing the motions of the shallow shell with added masses:

$$\frac{\partial T_1}{\partial x} + \frac{\partial S}{\partial y} = 2h\rho \frac{\partial^2 u}{\partial t^2} + f_1(x, y, t), \quad \overleftrightarrow{(1,2)}, \quad \overleftrightarrow{(x,y)},$$

$$\frac{\partial^2 M_1}{\partial x^2} + 2\frac{\partial^2 M_{12}}{\partial x \partial y} + \frac{\partial^2 M_2}{\partial y^2} - T_1 k_1 - T_2 k_2 = 2h\rho \frac{\partial^2 w}{\partial t^2} + f_3(x, y, t), \tag{1.96}$$

where:

$$f_1(x, y, t) = \sum_{i=1}^{N} M^i \left[\frac{\partial^2 u}{\partial t^2} \delta(x - x_i)\delta(y - y_i) - 2z_i \frac{\partial^3 w}{\partial x \partial t^2} \delta(x - x_i)\delta(y - y_i) \right],$$

$$\overleftrightarrow{(1,2)}, \quad \overleftrightarrow{(x,y)}, \tag{1.97}$$

$$f_3(x,y,t) = \sum_{i=1}^{N} \left\{ M^i \left[z_i \frac{\partial}{\partial x} \left[\frac{\partial^2 u}{\partial t^2} \delta(x-x_i) \right] \delta(y-y_i) - z_i^2 \frac{\partial}{\partial x} \left[\frac{\partial^3 w}{\partial x \partial t^2} \delta(x-x_i) \right] \times \right. \right.$$

$$\delta(y-y_i) + z_i \frac{\partial}{\partial y} \left[\frac{\partial^2 v}{\partial t^2} \delta(y-y_i) \right] \delta(x-x_i) - z_i^2 \frac{\partial}{\partial y} \left[\frac{\partial^3 w}{\partial y \partial t^2} \delta(y-y_i) \right] \times$$

$$\delta(x-x_i) + \frac{\partial^2 w}{\partial t^2} \delta(x-x_i)\delta(y-y_i) \right] - J_{xx}^i \frac{\partial}{\partial y} \left[\frac{\partial^3 w}{\partial y \partial t^2} \delta(y-y_i) \right] \delta(x-x_i) -$$

$$J_{yy}^i \frac{\partial}{\partial x} \left[\frac{\partial^3 w}{\partial x \partial t^2} \delta(x-x_i) \right] \delta(y-y_i) - J_{xy}^i \frac{\partial}{\partial y} \left[\frac{\partial^3 w}{\partial x \partial t^2} \delta(y-y_i) \right] \delta(x-x_i) -$$

$$J_{xy}^i \frac{\partial}{\partial x} \left[\frac{\partial^3 w}{\partial y \partial t^2} \delta(x-x_i) \right] \delta(y-y_i) \right\}.$$

The differential equations (1.96), (1.97) generalize the equations used in [51] taking into account the added masses eccentricity (a distance measured from the masses centre to the mean shell surface), and the inertia tangent components. For instance, the equations used in reference [51], can be obtained

1.3 Analysis of Orthotropic Shallow Shells with Added Masses

from (1.96) and (1.97) by taking $z_i = 0$ (it is assumed that the added masses centre lies on the mean shell surface), and by neglecting tangential components of the inertial shell and masses forces. Taking $J_{xx}^i = J_{yy}^i = J_{xy}^i = 0$, we get the equations governing dynamic interaction between the added masses and the shell in a normal direction to the mean shell's surface.

Some of the boundary conditions for $x = 0$, a will be given for the Kirchhoff-Love model:

1. Roller support
$$w = M_1 = T_1 = v = 0. \tag{1.98}$$

2. Roller clamping
$$u = v = w = M_1 = 0. \tag{1.99}$$

3. Free support
$$w = M_1 = T_1 = S = 0. \tag{1.100}$$

The other boundary conditions for the Kirchhoff-Love model are given in the monograph [244].

1.3.3 Added Masses Stiffness

The considerations described so far, are idealized, and can be used if the added masses stiffness is higher than the shell's stiffness. Otherwise, the added masses stiffness should also be taken into account. Below we consider one possible variant of the discussed problem, where the shell carrying masses have variable thickness and stiffness characteristics. The application of the theory of generalized functions allows for effective solutions to the corresponding vibration problem.

Suppose that a shell consists of N added masses attached on small parts of the top shell's surface. Suppose additionally, that the stiffness modulus E_k^i, the shear modulus D_{kl}^i and the Poisson's coefficient ν_{kl}^i of the homogeneous and orthotropic concentrated masses are similar to that of the shell's material.

Using Kirchhoff-Love hypothesis and using (1.33) the potential energy is defined by

$$\Pi = \frac{1}{2} \int_0^a \int_0^b (T_1^* \varepsilon_{11} + T_2^* \varepsilon_{22} + S^* \varepsilon_{12} + M_1^* \aleph_{11} + M_2^* \aleph_{22} + M_{12}^* \aleph_{12}) \, dx dy, \tag{1.101}$$

where the tangential deformation components ε_{11}, ε_{22}, ε_{12} and the bending components $\aleph_{11}$, $\aleph_{22}$, $\aleph_{12}$ of the mean surface are as follows

$$\varepsilon_{11} = \frac{\partial u}{\partial x} + k_1 w, \quad \overleftrightarrow{(1,2)}, \quad \varepsilon_{12} = \frac{\partial u}{\partial y} + \frac{\partial v}{\partial x},$$

$$\aleph_{11} = -\frac{\partial^2 w}{\partial x^2}, \quad \overleftrightarrow{(1,2)}, \quad \aleph_{12} = -2\frac{\partial^2 w}{\partial x \partial y}, \tag{1.102}$$

whereas the forces and moments are given by

$$T_1^* = \int_{-h}^{z*} \sigma_{11} dz, \quad \overleftrightarrow{(1,2)}, \quad S^* = \int_{-h}^{z*} \sigma_{12} dz,$$

$$M_1^* = \int_{-h}^{z*} \sigma_{11}(z-\tilde{z}) dz, \quad \overleftrightarrow{(1,2)}, \quad M_{12}^* = \int_{-h}^{z*} \sigma_{12}(z-\tilde{z}) dz, \qquad (1.103)$$

where: z^* is defined by (1.29).
Let us transform the relationships (1.103). For T_1^* one obtains

$$T_1^* = \int_{-h}^{h} \sigma_{11} dz + \int_{-h}^{z*} \sigma_{11} dz. \qquad (1.104)$$

Taking into account (1.29) and the properties of the characteristic function, the last term can be expressed by

$$T'' = \sum_{i=1}^{N} \int_{h}^{h+h_i} \sigma_{11} \Theta^* \Theta^{**} dz. \qquad (1.105)$$

The i-th added mass dimension in the x, y directions are given by $w\tilde{x}_1^i$ and $2\tilde{c}_2^i$. Dividing and multiplying (1.105) by these quantities and taking into account that $dV_i(z) = 4\tilde{c}_1^i \tilde{c}_2^i dz$ is the elementary volume inside which the whole added mass is contained, one obtains (the changes in $\partial u/\partial x$, $\partial v/\partial y$ along the volume of the added masses are neglected):

$$T_1'' = \sum_{i=1}^{N} T_{1i} \Theta^* \Theta^{**} \left(4\tilde{c}_1^i \tilde{c}_2^i\right)^{-1}, \qquad (1.106)$$

where

$$T_{1i} = \int_{h}^{h+h_i} \sigma_{11} dV_i(z). \qquad (1.107)$$

If the dimensions of the added mass are small in comparison to the shell dimensions a and b, then in the formula (1.107) we can reach a limit as $\tilde{c}_1^i \to 0$, $\tilde{c}_2^i \to 0$; using (1.47) we obtain

$$T_1'' = \sum_{i=1}^{N} T_{1i} \delta(x-x_i) \delta(y-y_i). \qquad (1.108)$$

Then (1.104) takes the following form

1.3 Analysis of Orthotropic Shallow Shells with Added Masses

$$T_1^* = T_1 + \sum_{i=1}^{N} T_{1i}\delta(x - x_i)\delta(y - y_i), \quad (1.109)$$

where

$$T_1 = \int_{-h}^{h} \sigma_{11} dz. \quad (1.110)$$

Working similarly we get

$$T_2^* = T_2 + \sum_{i=1}^{N} T_{2i}\delta(x - x_i)\delta(y - y_i),$$

$$S^* = S + \sum_{i=1}^{N} S_i \delta(x - x_i)\delta(y - y_i),$$

$$M_1^* = M_1 + \sum_{i=1}^{N} M_{1i}\delta(x - x_i)\delta(y - y_i),$$

$$M_{12}^* = M_{12} + \sum_{i=1}^{N} M_{12i}\delta(x - x_i)\delta(y - y_i), \quad (1.111)$$

where

$$T_2 = \int_{-h}^{h} \sigma_{22} dz, \quad S = \int_{-h}^{h} \sigma_{12} dz,$$

$$M_1 = \int_{-h}^{h} \sigma_{11} z\, dz, \quad \overleftrightarrow{(1,2)}, \quad M_{12} = \int_{-h}^{h} \sigma_{12} z\, dz,$$

$$T_{2i} = \int_{h}^{h+h_i} \sigma_{22} dV_i(z), \quad S_i = \int_{h}^{h+h_i} \sigma_{12} dV_i(z),$$

$$M_{1i} = \int_{h}^{h+h_i} \sigma_{11}(z - z_i) dV_i(z), \quad \overleftrightarrow{(1,2)},$$

$$M_{12i} = \int_{h}^{h+h_i} \sigma_{12}(z - z_i) dV_i(z), \quad i = \overline{1, N}. \quad (1.112)$$

Substituting (1.111) in (1.102) with (1.101) we get the following formula for the potential energy of an orthotropic shell with added masses

$$\Pi = \frac{1}{2} \int_0^a \int_0^b \left\{ T_1 \left(\frac{\partial u}{\partial x} + k_1 w \right) + T_2 \left(\frac{\partial v}{\partial y} + k_2 w \right) + S \left(\frac{\partial u}{\partial y} + \frac{\partial v}{\partial x} \right) - \right.$$

$$M_1 \frac{\partial^2 w}{\partial x^2} - M_2 \frac{\partial^2 w}{\partial y^2} - 2 M_{12} \frac{\partial^2 w}{\partial x \partial y} + \sum_{i=1}^{N} \left[T_{1i} \left(\frac{\partial u_i}{\partial x} + k_1 w \right) + \right.$$

$$T_{2i} \left(\frac{\partial v_i}{\partial y} + k_2 w \right) + S_i \left(\frac{\partial u_i}{\partial y} + \frac{\partial v_i}{\partial x} \right) - M_{1i} \frac{\partial^2 w}{\partial x^2} -$$

$$\left. M_{2i} \frac{\partial^2 w}{\partial y^2} - 2 M_{12i} \frac{\partial^2 w}{\partial x \partial y} \right] \delta(x - x_i) \delta(y - y_i) \bigg\} dx dy, \tag{1.113}$$

where

$$T_1 = 2h \left(A_{1111} \varepsilon_{11} + A_{1122} \varepsilon_{22} \right), \quad \overleftrightarrow{(1,2)},$$

$$S = 2h A_{1212} \varepsilon_{12},$$

$$M_1 = \frac{2h^3}{3} \left(A_{1111} \aleph_{11} + A_{1122} \aleph_{22} \right), \quad \overleftrightarrow{(1,2)},$$

$$M_{12} = \frac{2h^3}{3} A_{1212} \aleph_{12},$$

$$T_{1i} = V_i^{(0)} \left(A_{1111}^i \varepsilon_{11}^i + A_{1122}^i \varepsilon_{22}^i \right), \quad \overleftrightarrow{(1,2)},$$

$$S_i = V_i^{(0)} A_{1212} \varepsilon_{12}^i,$$

$$M_{1i} = V_i^{(2)} \left(A_{1111}^i \aleph_{11} + A_{1122}^i \aleph_{22} \right), \quad \overleftrightarrow{(1,2)},$$

$$M_{2i} = V_i^{(2)} A_{1212} \aleph_{12}. \tag{1.114}$$

The material stiffness of the added masses is defined by the coefficients A_{jmkl}^i. The following notation is introduced in (1.114):

$$\varepsilon_{11}^i = \frac{\partial u_i}{\partial x} + k_1 w, \quad \overleftrightarrow{(1,2)}, \quad \varepsilon_{12}^i = \frac{\partial v_i}{\partial x} + \frac{\partial u_i}{\partial y},$$

$$u_i = u - z_i \frac{\partial w}{\partial x}, \quad \overleftrightarrow{(1,2)}, \quad V_i^{(k)} = \int_h^{h+h_i} (z - z_i)^k \, dV_i(z), \quad k = 0, 2. \tag{1.115}$$

Let us define the potential energy variation $\delta \Pi$ taking into account (1.114), (1.115) and (1.113)

$$\delta \Pi = \int_0^a \int_0^b \left\{ T_1 \left[\frac{\partial (\delta u)}{\partial x} + k_1 \delta w \right] + T_2 \left[\frac{\partial (\delta v)}{\partial y} + k_2 \delta w \right] + S \left[\frac{\partial (\delta u)}{\partial y} + \frac{\partial (\delta v)}{\partial x} \right] - \right.$$

1.3 Analysis of Orthotropic Shallow Shells with Added Masses

$$M_1 \frac{\partial^2(\delta w)}{\partial x^2} - M_2 \frac{\partial^2(\delta w)}{\partial y^2} - 2M_{12} \frac{\partial^2(\delta w)}{\partial x \partial y} + \sum_{i=1}^{N} \left[T_{1i} \left(\frac{\partial(\delta u)}{\partial x} - z_i \frac{\partial^2(\delta w)}{\partial x^2} + k_1 \delta w \right) + T_{2i} \left(\frac{\partial(\delta v)}{\partial y} - z_i \frac{\partial^2(\delta w)}{\partial y^2} + k_2 \delta w \right) + S_i \left(\frac{\partial(\delta u)}{\partial y} + \frac{\partial(\delta v)}{\partial x} - 2z_i \frac{\partial^2(\delta w)}{\partial x \partial y} \right) - M_{1i} \frac{\partial^2(\delta w)}{\partial x^2} - M_{2i} \frac{\partial^2(\delta w)}{\partial y^2} - 2M_{12i} \frac{\partial^2(\delta w)}{\partial x \partial y} \right] \delta(x - x_i) \delta(y - y_i) \right\}. \quad (1.116)$$

For the kinetic energy of the system "shell - added masses" the formula (1.32) is valid.

Finding the kinetic energy variation and substituting it together with (1.116) in the Hamilton equation, we finally obtain the equations governing the system's vibrations

$$\frac{\partial T_1}{\partial x} + \frac{\partial S}{\partial y} + \sum_{i=1}^{N} \left\{ \frac{\partial}{\partial x} [T_{1i} \delta(x - x_i)] \delta(y - y_i) + \frac{\partial}{\partial y} [S_i \delta(y - y_i)] \delta(x - x_i) \right\} = 0, \quad (\overleftrightarrow{1,2}), \quad (\overleftrightarrow{x,y}),$$

$$\frac{\partial^2 M_1}{\partial x^2} + 2 \frac{\partial^2 M_{12}}{\partial x \partial y} + \frac{\partial^2 M_2}{\partial y^2} - T_1 k_1 - T_2 k_2 + \sum_{i=1}^{N} \left\{ \frac{\partial^2}{\partial x^2} [M_{1i} \delta(x - x_i)] \delta(y - y_i) + \frac{\partial^2}{\partial y^2} [M_{2i} \delta(y - y_i)] \delta(x - x_i) + 2 \frac{\partial^2}{\partial x \partial y} [M_{12i} \delta(x - x_i) \delta(y - y_i)] - T_{1i} k_1 \delta(x - x_i) \delta(y - y_i) - T_{2i} k_2 \delta(x - x_i) \delta(y - y_i) + z_i \frac{\partial^2}{\partial x^2} [T_{1i} \delta(x - x_i)] \times \delta(y - y_i) + z_i \frac{\partial^2}{\partial y^2} [T_{2i} \delta(y - y_i)] \delta(x - x_i) + 2 z_i \frac{\partial^2}{\partial x \partial y} [S_i \delta(x - x_i) \delta(y - y_i)] \right\} =$$

$$2 h \rho \frac{\partial^2 w}{\partial t^2} + \sum_{i=1}^{N} \left\{ \left[M^i \frac{\partial^2 w}{\partial t^2} \delta(x - x_i) \delta(y - y_i) \right] + J^i_{yoz} \frac{\partial}{\partial x} \left[\frac{\partial^3 w}{\partial x \partial t^2} \delta(x - x_i) \right] \times \delta(y - y_i) - J^i_{xoz} \frac{\partial}{\partial y} \left[\frac{\partial^3 w}{\partial y \partial t^2} \delta(x - x_i) \delta(y - y_i) \right] \right\}. \quad (1.117)$$

During the kinetic energy calculations the assumption about neglecting the tangential components of the inertial forces of the shell and the added masses is used. In (1.117) the loading forces, the moments as well as the integral characteristics of the elastic-deformable state in the places of the attached masses are defined by (1.114).

In many cases the equations may be simplified. For instance, it can be assumed, that only some of the added masses have its dynamic characteristics comparable to that of the shell. The equations are simplified also when $z_i = 0$, which means that the $i - th$ mass centre is located very close to the inertial axis. In addition, it is sometimes allowed to neglect some terms responsible for the added masses reaction.

1.3.4 Determination of Dynamical Characteristics

As it has been mentioned earlier, the dynamics of shallow orthotropic shell with the concentrated masses is reduced to finding solutions to the differential equations (1.80), with the functions $f_1, f_2, f_3, \varphi_1, \varphi_2$ depending on the added element structure. The integration of (1.80) is carried out using the boundary conditions (1.83)–(1.85).

Substituting (1.74) and (1.75) and taking into account (1.70), (1.71), from (1.80) the following system of dimensionless equations is obtained with the unknowns $u, v, w, \gamma_x, \gamma_y$:

$$A_{1111}\lambda_2^2 \frac{\partial^2 u}{\partial x^2} + A_{1212}\lambda_1^2 \frac{\partial^2 u}{\partial y^2} + (A_{1212} + A_{1122})\lambda_1\lambda_2 \frac{\partial^2 v}{\partial x \partial y} +$$

$$(A_{1111}\lambda_2^2 k_1 + A_{1122}\lambda_1^2 k_2) \frac{1}{\lambda_1} \frac{\partial w}{\partial x} = \frac{\partial^2 u}{\partial t^2} + f_1(x, y, t), \quad \overleftrightarrow{(1,2)},$$

$$A_{1313}\lambda_2^2 \left(\frac{\partial^2 w}{\partial x^2} + \frac{\partial \gamma_x}{\partial x} \right) + A_{2323}\lambda_1^2 \left(\frac{\partial^2 w}{\partial y^2} + \frac{\partial \gamma_y}{\partial y} \right) - (A_{1111}\lambda_2^2 k_1 + A_{2211}\lambda_1^2 k_2) \times$$

$$\left(\frac{1}{\lambda_1} \frac{\partial u}{\partial x} + \frac{k_1}{\lambda_1^2} w \right) - (A_{1122}\lambda_2^2 k_1 + A_{2222}\lambda_1^2 k_2) \left(\frac{1}{\lambda_2} \frac{\partial v}{\partial y} + \frac{k_2}{\lambda_2^2} w \right) =$$

$$\frac{\partial^2 w}{\partial t^2} + f_3(x, y, t),$$

$$\frac{1}{12} \left[A_{1111}\lambda_2^2 \frac{\partial^2 \gamma_x}{\partial x^2} + A_{1122}\lambda_1^2 \frac{\partial^2 \gamma_y}{\partial x \partial y} + A_{1212}\lambda_1^2 \left(\frac{\partial^2 \gamma_x}{\partial y^2} + \frac{\partial^2 \gamma_y}{\partial x \partial y} \right) \right] -$$

$$A_{1313}\lambda_1^2\lambda_2^2 \left(\gamma_x + \frac{\partial w}{\partial x} \right) = \frac{1}{12} \frac{\partial^2 \gamma_x}{\partial t^2} + \varphi_1(x, y, t), \quad \overleftrightarrow{(1,2)}. \quad (1.118)$$

Assume that $f_1, f_2, f_3, \varphi_1, \varphi_2$ are defined by (1.87). Physically, it means that the added masses are homogeneous and their inertial moments J_{zz}^i do not vary. From (1.87) we get in the nondimensional form

$$f_1(x, y, t) = \sum_{i=1}^{N} M^i \left[\frac{\partial^2 u}{\partial t^2} + \frac{1}{2\lambda_1} \frac{\partial^2 \gamma_x}{\partial t^2} - (z_i - 0.5) \frac{\partial^3 w}{\partial x \partial t^2} \right] \times$$

$$\delta(x - x_i)\delta(y - y_i), \quad \overleftrightarrow{(1,2)},$$

1.3 Analysis of Orthotropic Shallow Shells with Added Masses

$$f_3(x,y,t) = \sum_{i=1}^{N} M^i \left\{ (z_i - 0.5)^2 \frac{1}{\lambda_1^2} \frac{\partial}{\partial x} \left[\frac{\partial^3 w}{\partial x \partial t^2} \delta(x - x_i) \right] \delta(y - y_i) - \right.$$

$$(z_i - 0.5) \frac{1}{\lambda_1} \frac{\partial}{\partial x} \left[\frac{\partial^2 u}{\partial t^2} \delta(x - x_i) \right] \delta(y - y_i) + \frac{1}{2\lambda_1^2}(z_i - 0.5) \times$$

$$\frac{\partial}{\partial x}\left[\frac{\partial^2 \gamma_x}{\partial t^2}\delta(x-x_i)\right]\delta(y-y_i) - (z_i-0.5)^2\frac{1}{\lambda_2^2}\frac{\partial}{\partial y}\left[\frac{\partial^3 w}{\partial y \partial t^2}\delta(y-y_i)\right] \times$$

$$\delta(x-x_i) + (z_i - 0.5)\frac{1}{\lambda_2}\frac{\partial}{\partial y}\left[\frac{\partial^2 v}{\partial t^2}\delta(y-y_i)\right]\delta(x-x_i) +$$

$$\left. \frac{1}{2\lambda_2^2}(z_i - 0.5)\frac{\partial}{\partial y}\left[\frac{\partial^2 \gamma_y}{\partial t^2}\delta(y-y_i)\right]\delta(x-x_i) + \frac{\partial^2 w}{\partial t^2}\delta(x-x_i)\delta(y-y_i)\right\} +$$

$$\sum_{i=1}^{N}\left\{-J_{xx}^i\frac{\partial}{\partial y}\left[\frac{\partial^3 w}{\partial y \partial t^2}\delta(y-y_i)\right]\delta(x-x_i) - J_{yy}^i\frac{\partial}{\partial x}\left[\frac{\partial^3 w}{\partial x \partial t^2}\delta(x-x_i)\right]\delta(y-y_i)\right\},$$

$$\varphi_1(x,y,t) = \sum_{i=1}^{N} M^i \left[\frac{1}{2}\frac{\partial^2 \gamma_x}{\partial t^2} + \frac{\lambda_1}{2}\frac{\partial^2 u}{\partial t^2} - \frac{1}{2}(z_i - 0.5)\frac{\partial^3 w}{\partial x \partial t^2} \right] \times$$

$$\delta(x-x_i)\delta(y-y_i), \quad \overleftrightarrow{(1,2)}. \tag{1.119}$$

Now we consider the vibration analysis of an orthotropic shell after rejecting the terms $f_1, f_2, f_3, \varphi_1, \varphi_2$. We are looking for the solution of the following form of the homogeneous system (1.118) satisfying the boundary conditions (1.83):

$$u = A\cos\alpha_m x \sin\beta_n y \sin\omega_{mn}t, \quad v = B\sin\alpha_m x \cos\beta_n y \sin\omega_{mn}t,$$

$$w = C\sin\alpha_m x \sin\beta_n y \sin\omega_{mn}t, \quad \gamma_x = D\cos\alpha_m x \sin\beta_n y \sin\omega_{mn}t,$$

$$\gamma_y = E\sin\alpha_m x \cos\beta_n y \sin\omega_{mn}t, \tag{1.120}$$

where A, B, C, D, E are constans, $\alpha_m = m\pi$, $\beta_n = n\pi$, and m, n are integers.

Substituting (1.120) into (1.118) and after elementary transformations we get a linear algebraic system of equations in terms of the unknowns A, B, C, D, E. Its determinant is defined as

$$\det[e_{ik}]_{5\times 5} = 0, \tag{1.121}$$

where

$$e_{11} = -A_{1111}\lambda_2^2\alpha_m^2 - A_{1212}\lambda_1^2\beta_n^2 + \omega_{mn}^2, \quad e_{12} = -(A_{1212} + A_{1122})\lambda_1\lambda_2\alpha_m\beta_n,$$

$$e_{13} = \left(A_{1111}\lambda_2^2 k_1 + A_{1122}\lambda_1^2 k_2\right)\alpha_m/\lambda_1, \quad e_{14} = 0, \quad e_{15} = 0,$$

$$e_{21} = -(A_{1212} + A_{2211})\lambda_1\lambda_2\alpha_m\beta_n, \quad e_{22} = -A_{2222}\lambda_1^2\beta_n^2 - A_{1212}\lambda_2^2\alpha_m^2 + \omega_{mn}^2,$$

$$e_{23} = \left(A_{2211}\lambda_2^2 k_1 + A_{2222}\lambda_1^2 k_2\right)\beta_n/\lambda_2, \quad e_{24} = 0, \quad e_{25} = 0,$$

$$e_{31} = \left(A_{1111}\lambda_2^2 k_1 + A_{2211}\lambda_1^2 k_2\right)\alpha_m/\lambda_1, \quad e_{32} = \left(A_{1122}\lambda_2^2 k_1 + A_{2222}\lambda_1^2 k_2\right)\beta_n/\lambda_2,$$

$$e_{33} = -A_{1313}\lambda_2^2\alpha_m^2 - A_{2323}\lambda_1^2\beta_n^2 - \left(A_{1111}\lambda_2^2 k_1 + A_{2211}\lambda_1^2 k_2\right)k_1/\lambda_1^2 -$$
$$\left(A_{1122}\lambda_2^2 k_1 + A_{2222}\lambda_1^2 k_2\right)k_2/\lambda_2^2 + \omega_{mn}^2,$$

$$e_{34} = -A_{1313}\lambda_2^2\alpha_m^2, \quad e_{35} = -A_{2323}\lambda_1^2\beta_n^2,$$

$$e_{41} = 0, \quad e_{42} = 0, \quad e_{43} = -A_{1313}\lambda_1^2\lambda_2^2\alpha_m,$$

$$e_{44} = \frac{1}{12}\left[-A_{1111}\lambda_2^2\alpha_m^2 - A_{1212}\lambda_1^2\beta_n^2\right] - A_{1313}\lambda_1^2\lambda_2^2\alpha_m + \frac{\omega_{mn}^2}{12},$$

$$e_{45} = \frac{1}{12}\left[-A_{1122}\lambda_1^2\alpha_m\beta_n + A_{1212}\lambda_1^2\alpha_m\beta_n\right],$$

$$e_{51} = 0, \quad e_{52} = 0, \quad e_{53} = -A_{2323}\lambda_1^2\lambda_2^2\beta_n,$$

$$e_{54} = \frac{1}{12}\left[-A_{2211}\lambda_2^2\alpha_m\beta_n + A_{1212}\lambda_2^2\alpha_m\beta_n\right],$$

$$e_{55} = \frac{1}{12}\left[A_{2222}\lambda_1^2\beta_n^2 - A_{1212}\lambda_2^2\alpha_m^2\right] - A_{2323}\lambda_1^2\lambda_2^2\beta_n\frac{\omega_{mn}^2}{12}. \qquad (1.122)$$

Finally, the following characteristic equation is obtained (ω_{mn} denotes the frequency of free vibrations):

$$r_1\omega_{mn}^{10} + r_2\omega_{mn}^8 + r_3\omega_{mn}^6 + r_4\omega_{mn}^4 + r_5\omega_{mn}^2 + r_6 = 0. \qquad (1.123)$$

Newton's method is used for the determination of the roots (1.123) [59]. For each pair (m,n), we get five frequency spectra. In reference [215] it is pointed out, that the spectra number defined by two dimensional theory is equal to the internal degrees-of-freedom of the two dimensional model of the medium. If, for instance, we consider the i-th spectrum, then the internal degrees-of-freedom number should be not less than i. The two-dimensional theory used in this book, including shear deformations and rotation inertia, allows for a model definition with five degrees-of-freedom, which means that five frequency backbone curves can be obtained.

Let us determine the frequencies of free vibration a shallow shell with added masses. The solution to (1.118) is sought in the form

$$u = \sin\omega t \sum_{m,n}^{\infty} A_{mn}\cos\alpha_m x \sin\beta_n y, \quad v = \sin\omega t \sum_{m,n}^{\infty} B_{mn}\sin\alpha_m x \cos\beta_n y,$$

$$w = \sin\omega t \sum_{m,n}^{\infty} C_{mn}\sin\alpha_m x \sin\beta_n y, \quad \gamma_x = \sin\omega t \sum_{m,n}^{\infty} D_{mn}\cos\alpha_m x \sin\beta_n y,$$

$$\gamma_y = \sin\omega t \sum_{m,n}^{\infty} E_{mn}\sin\alpha_m x \cos\beta_n y. \qquad (1.124)$$

1.3 Analysis of Orthotropic Shallow Shells with Added Masses

For the functions on the right-hand sides of (1.118) possesing the singular coefficients of the δ-type the following distribution is valid (with respect to the basis function W)

$$W\delta(x-x_i)\delta(y-y_i) = 4\sum_{m,n}^{\infty} W_i \sin\alpha_m x_i \sin\beta_n y_i \sin\alpha_m x \sin\beta_n y,$$

$$\frac{\partial W}{\partial x}\delta(x-x_i)\delta(y-y_i) = 4\sum_{m,n}^{\infty} \frac{\partial W_i}{\partial x}\cos\alpha_m x_i \sin\beta_n y_i \cos\alpha_m x \times \sin\beta_n y,$$

$$\frac{\partial}{\partial x}\left[\frac{\partial W}{\partial x}\delta(x-x_i)\right]\delta(y-y_i) =$$

$$-4\sum_{m,n}^{\infty} \frac{\partial W_i}{\partial x}\alpha_m \cos\alpha_m x_i \sin\beta_n y_i \times \sin\alpha_m x \sin\beta_n y. \qquad (1.125)$$

In the above, the following notation has been assumed

$$u = U\sin\omega t, \quad v = V\sin\omega t, \quad w = W\sin\omega t,$$

$$\gamma_x = \Gamma_x \sin\omega t, \quad \gamma_y = \Gamma_y \sin\omega t,$$

$$U_i = U(x_i, y_i), \quad V_i = V(x_i, y_i), \quad W_i = W(x_i, y_i),$$

$$\Gamma_{xi} = \Gamma_x(x_i, y_i), \quad \Gamma_{yi} = \Gamma_y(x_i, y_i), \quad (i = \overline{1, N}). \qquad (1.126)$$

Substituting (1.124) and (1.125) with (1.118) the following nonhomogeneous algebraic equation is obtained, which can be used for determining A_{mn}, B_{mn}, C_{mn}, D_{mn}, E_{mn}

$$LA = Q, \qquad (1.127)$$

where

$$L = \begin{bmatrix} e_{11} & e_{12} & \cdots & e_{15} \\ e_{21} & e_{22} & \cdots & e_{25} \\ \cdots & \cdots & \cdots & \cdots \\ e_{51} & e_{52} & \cdots & e_6 \end{bmatrix}, \quad A = \begin{bmatrix} A_{mn} \\ B_{mn} \\ C_{mn} \\ D_{mn} \\ E_{mn} \end{bmatrix}, \quad Q = \begin{bmatrix} g_1 \\ g_2 \\ g_3 \\ g_4 \\ g_5 \end{bmatrix},$$

$$g_1 = \sum_{i=1}^{N}\left(d_{11}^i U_i + d_{12}^i \Gamma_{xi} + d_{13}^i \frac{\partial W_i}{\partial x}\right), \quad g_2 = \sum_{i=1}^{N}\left(d_{21}^i V_i + d_{22}^i \Gamma_{yi} + d_{23}^i \frac{\partial W_i}{\partial y}\right),$$

$$g_3 = \sum_{i=1}^{N}\left(d_{31}^i U_i + d_{32}^i V_i + d_{33}^i W_i + d_{34}^i \Gamma_{xi} + d_{35}^i \Gamma_{yi} + d_{36}^i \frac{\partial W_i}{\partial x} + d_{37}^i \frac{\partial W_i}{\partial y}\right),$$

$$g_4 = \sum_{i=1}^{N}\left(d_{41}^i U_i + d_{42}^i \Gamma_{xi} + d_{43}^i \frac{\partial W_i}{\partial x}\right), \quad g_5 = \sum_{i=1}^{N}\left(d_{51}^i V_i + d_{52}^i \Gamma_{yi} + d_{53}^i \frac{\partial W_i}{\partial y}\right).$$

The L-matrix elements depend on ω, the physical as well as the geometrical parameters of the shell, and they are found from (1.122). The parameters are defined below

$$d_{11}^i = -4\omega^2 M^i \cos\alpha_m x_i \sin\beta_n y_i,$$

$$d_{12}^i = -2\omega^2 \frac{M^i}{\lambda_1} \cos\alpha_m x_i \sin\beta_n y_i,$$

$$d_{13}^i = 4\omega^2 M^i \frac{(z_i - 0.5)}{\lambda_1} \cos\alpha_m x_i \sin\beta_n y_i,$$

$$d_{21}^i = -4\omega^2 M^i \sin\alpha_m x_i \cos\beta_n y_i,$$

$$d_{22}^i = -2\omega^2 \frac{M^i}{\lambda_2} \sin\alpha_m x_i \cos\beta_n y_i,$$

$$d_{23}^i = 4\omega^2 \frac{M^i}{\lambda_2} (z_i - 0.5) \sin\alpha_m x_i \cos\beta_n y_i,$$

$$d_{31}^i = -4\omega^2 \frac{M^i}{\lambda_1} (z_i - 0.5)\alpha_m \cos\alpha_m x_i \sin\beta_n y_i,$$

$$d_{32}^i = 4\omega^2 M^i \frac{(z_i - 0.5)}{\lambda_2} \beta_n \sin\alpha_m x_i \cos\beta_n y_i,$$

$$d_{33}^i = -4\omega^2 M^i \sin\alpha_m x_i \sin\beta_n y_i,$$

$$d_{34}^i = 2\omega^2 \frac{M^i}{\lambda_1^2} (z_i - 0.5)\alpha_m \cos\alpha_m x_i \sin\beta_n y_i,$$

$$d_{35}^i = 2\omega^2 M^i \left(\frac{z_i - 0.5}{\lambda_2^2}\right) \beta_n \sin\alpha_m x_i \cos\beta_n y_i,$$

$$d_{36}^i = 4\omega^2 M^i \left(\frac{z_i - 0.5}{\lambda_1^2}\right) \alpha_m \cos\alpha_m x_i \sin\beta_n y_i -$$
$$4\omega^2 J_{xx}^i \beta_n \sin\alpha_m x_i \cos\beta_n y_i,$$

$$d_{37}^i = -4\omega^2 \frac{M^i}{\lambda_2^2} (z_i - 0.5)^2 \beta_n \sin\alpha_m x_i \cos\beta_n y_i -$$

$$4\omega^2 J_{yy}^i \alpha_m \cos\alpha_m x_i \sin\beta_n y_i, \quad d_{41}^i = -2\omega^2 M^i \lambda_1 \cos\alpha_m x_i \sin\beta_n y_i,$$

$$d_{42}^i = -M^i \omega^2 \cos\alpha_m x_i \sin\beta_n y_i, \quad d_{43}^i = 2\omega^2 M^i (z_i - 0.5) \cos\alpha_m x_i \sin\beta_n y_i,$$

$$d_{51}^i = -2\omega^2 M^i \lambda_2 \sin\alpha_m x_i \cos\beta_n y_i, \quad d_{52}^i = -\omega^2 M^i \sin\alpha_m x_i \cos\beta_n y_i,$$

$$d_{53}^i = 2\omega^2 M^i (z_i - 0.5) \sin\alpha_m x_i \cos\beta_n y_i, \quad (i = \overline{1, N}). \tag{1.128}$$

The solution of (1.127) leads to the determination of A_{mn}, B_{mn}, C_{mn}, D_{mn}, E_{mn}; therefore the displacement vector components and the rotation angles of the average shell's surface with the added masses are given by

$$U = \sum_{i=1}^{N} \sum_{m,n}^{\infty} \frac{1}{\Delta(m,n)} \left(U_i \mathrm{p}_{11}^i + V_i \mathrm{p}_{12}^i + W_i \mathrm{p}_{13}^i + \Gamma_{xi} \mathrm{p}_{14}^i + \Gamma_{yi} \mathrm{p}_{15}^i + \right.$$

1.3 Analysis of Orthotropic Shallow Shells with Added Masses

$$\left.\frac{\partial W_i}{\partial x} \mathrm{p}_{16}^i + \frac{\partial W_i}{\partial y} \mathrm{p}_{17}^i\right) \cos \alpha_m x \sin \beta_n y,$$

$$V = \sum_{i=1}^{N} \sum_{m,n}^{\infty} \frac{1}{\Delta(m,n)} \left(U_i \mathrm{p}_{21}^i + V_i \mathrm{p}_{22}^i + W_i \mathrm{p}_{23}^i + \Gamma_{xi} \mathrm{p}_{24}^i + \Gamma_{yi} \mathrm{p}_{25}^i + \right.$$

$$\left.\frac{\partial W_i}{\partial x} \mathrm{p}_{26}^i + \frac{\partial W_i}{\partial y} \mathrm{p}_{27}^i\right) \sin \alpha_m x \cos \beta_n y,$$

$$W = \sum_{i=1}^{N} \sum_{m,n}^{\infty} \frac{1}{\Delta(m,n)} \left(U_i \mathrm{p}_{31}^i + V_i \mathrm{p}_{32}^i + W_i \mathrm{p}_{33}^i + \Gamma_{xi} \mathrm{p}_{34}^i + \Gamma_{yi} \mathrm{p}_{35}^i + \right.$$

$$\left.\frac{\partial W_i}{\partial x} \mathrm{p}_{36}^i + \frac{\partial W_i}{\partial y} \mathrm{p}_{37}^i\right) \sin \alpha_m x \sin \beta_n y,$$

$$\Gamma_x = \sum_{i=1}^{N} \sum_{m,n}^{\infty} \frac{1}{\Delta(m,n)} \left(U_i \mathrm{p}_{41}^i + V_i \mathrm{p}_{42}^i + W_i \mathrm{p}_{43}^i + \Gamma_{xi} \mathrm{p}_{44}^i + \Gamma_{yi} \mathrm{p}_{45}^i + \right.$$

$$\left.\frac{\partial W_i}{\partial x} \mathrm{p}_{46}^i + \frac{\partial W_i}{\partial y} \mathrm{p}_{47}^i\right) \cos \alpha_m x \sin \beta_n y,$$

$$\Gamma_y = \sum_{i=1}^{N} \sum_{m,n}^{\infty} \frac{1}{\Delta(m,n)} \left(U_i \mathrm{p}_{51}^i + V_i \mathrm{p}_{52}^i + W_i \mathrm{p}_{53}^i + \Gamma_{xi} \mathrm{p}_{54}^i + \Gamma_{yi} \mathrm{p}_{55}^i + \right.$$

$$\left.\frac{\partial W_i}{\partial x} \mathrm{p}_{56}^i + \frac{\partial W_i}{\partial y} \mathrm{p}_{57}^i\right) \sin \alpha_m x \cos \beta_n y,$$

$$\frac{\partial W}{\partial x} = \sum_{i=1}^{N} \sum_{m,n}^{\infty} \frac{1}{\Delta(m,n)} \left(U_i \mathrm{p}_{31}^i + V_i \mathrm{p}_{32}^i + W_i \mathrm{p}_{33}^i + \Gamma_{xi} \mathrm{p}_{34}^i + \Gamma_{yi} \mathrm{p}_{35}^i + \right.$$

$$\left.\frac{\partial W_i}{\partial x} \mathrm{p}_{36}^i + \frac{\partial W_i}{\partial y} \mathrm{p}_{37}^i\right) \alpha_m \cos \alpha_m x \sin \beta_n y,$$

$$\frac{\partial W}{\partial y} = \sum_{i=1}^{N} \sum_{m,n}^{\infty} \frac{1}{\Delta(m,n)} \left(U_i \mathrm{p}_{31}^i + V_i \mathrm{p}_{32}^i + W_i \mathrm{p}_{33}^i + \Gamma_{xi} \mathrm{p}_{34}^i + \Gamma_{yi} \mathrm{p}_{35}^i + \right.$$

$$\left.\frac{\partial W_i}{\partial x} \mathrm{p}_{36}^i + \frac{\partial W_i}{\partial y} \mathrm{p}_{37}^i\right) \beta_n \sin \alpha_m x \cos \beta_n y. \quad (1.129)$$

The terms, which are being sought, include the components of the displacement vector, the rotation angles, and the first derivatives of the normal bending functions at the joining points between the masses and the shell.

Substituting $x = x_i$, $y = y_i$ with (1.129) ($i = 1, N$) and fulfilling the conditions of continuity at the joint points (the first derivatives of N), we get the following homogeneous equations defining the amplitudes u_k, V_k, W_k, Γ_{xk}, Γ_{yk}, $\partial W_k/\partial x$, $\partial W_k/\partial y$:

$$U_k = \sum_{i=1}^{N} \sum_{m,n}^{\infty} R_{i1}^{mn}\left(U_i, V_i, W_i, \Gamma_{xi}, \Gamma_{yi}, \frac{\partial W_i}{\partial x}, \frac{\partial W_i}{\partial y}\right) \cos\alpha_m x_k \sin\beta_n y_k,$$

$$V_k = \sum_{i=1}^{N} \sum_{m,n}^{\infty} R_{i2}^{mn}\left(U_i, V_i, W_i, \Gamma_{xi}, \Gamma_{yi}, \frac{\partial W_i}{\partial x}, \frac{\partial W_i}{\partial y}\right) \sin\alpha_m x_k \cos\beta_n y_k,$$

$$W_k = \sum_{i=1}^{N} \sum_{m,n}^{\infty} R_{i3}^{mn}\left(U_i, V_i, W_i, \Gamma_{xi}, \Gamma_{yi}, \frac{\partial W_i}{\partial x}, \frac{\partial W_i}{\partial y}\right) \sin\alpha_m x_k \sin\beta_n y_k,$$

$$\Gamma_{xk} = \sum_{i=1}^{N} \sum_{m,n}^{\infty} R_{i4}^{mn}\left(U_i, V_i, W_i, \Gamma_{xi}, \Gamma_{yi}, \frac{\partial W_i}{\partial x}, \frac{\partial W_i}{\partial y}\right) \cos\alpha_m x_k \sin\beta_n y_k,$$

$$\Gamma_{yk} = \sum_{i=1}^{N} \sum_{m,n}^{\infty} R_{i5}^{mn}\left(U_i, V_i, W_i, \Gamma_{xi}, \Gamma_{yi}, \frac{\partial W_i}{\partial x}, \frac{\partial W_i}{\partial y}\right) \sin\alpha_m x_k \cos\beta_n y_k,$$

$$\frac{\partial W_k}{\partial x} = \sum_{i=1}^{N} \sum_{m,n}^{\infty} R_{i3}^{mn}\left(U_i, V_i, W_i, \Gamma_{xi}, \Gamma_{yi}, \frac{\partial W_i}{\partial x}, \frac{\partial W_i}{\partial y}\right) \alpha_m \cos\alpha_m x_k \sin\beta_n y_k,$$

$$\frac{\partial W_k}{\partial y} = \sum_{i=1}^{N} \sum_{m,n}^{\infty} R_{i3}^{mn}\left(U_i, V_i, W_i, \Gamma_{xi}, \Gamma_{yi}, \frac{\partial W_i}{\partial x}, \frac{\partial W_i}{\partial y}\right) \beta_n \sin\alpha_m x_k \cos\beta_n y_k,$$

(1.130)

$$(k = \overline{1,N}),$$

where: $R_{ij}^{mn}(...)$ denote the coefficients of the corresponding function distributions of (1.129) that depend linearly on the variables U_k, V_k, W_k, Γ_{xk}, Γ_{yk}, $\partial W_k/\partial x$, $\partial W_k/\partial y$.

After equalling to zero the determinant of system (1.130) we obtain the following characteristic equation

$$\det[a_{ik}]_{7N \times 7N} = 0. \quad (1.131)$$

The determinant elements a_{ij} are given in Table 1.1, and the parameters p_{ij}^k have the following form

$$p_{j1}^k = d_{11}^k \Delta_{1j} + d_{31}^k \Delta_{3j} + d_{41}^k \Delta_{4j}, \quad p_{j2}^k = d_{21}^k \Delta_{2j} + d_{51}^k \Delta_{5j} + d_{32}^k \Delta_{3j},$$

$$p_{j3}^k = d_{33}^k \Delta_{3j}, \quad p_{j4}^k = d_{12}^k \Delta_{1j} + d_{42}^k \Delta_{4j} + d_{34}^k \Delta_{3j},$$

$$p_{j5}^k = d_{22}^k \Delta_{2j} + d_{52}^k \Delta_{5j} + d_{35}^k \Delta_{3j}, \quad p_{j6}^k = d_{13}^k \Delta_{1j} + d_{43}^k \Delta_{4j} + d_{36}^k \Delta_{3j},$$

$$p_{j7}^k = d_{23}^k \Delta_{2j} + d_{53}^k \Delta_{5j} + d_{37}^k \Delta_{3j}, \quad (1.132)$$

$$\left(k = \overline{1,N}, \quad j = \overline{1,5}\right).$$

In the above δ denotes the determinant of (1.130), and Δ_{ij} the complements of the determinant elements (1.130) lying on the i-th row and the j-th column.

1.3 Analysis of Orthotropic Shallow Shells with Added Masses

The roots of characteristic equation (1.131) define free vibrations frequency spectra of an orthotropic shallow shell with two curvatures and added masses (see also [15]).

As it has been mentioned earlier, in many practical cases it is suitable to use a simplified variant of Timoshenko's shell theory, i.e. including only tangential shell's inertial forces.

The governing system of equations of motion in the nondimensional form, satisfied by w, γ_x, γ_y and F, has the following form

$$-\nabla_k^2 F + A_{1313}\lambda_2^2 \left(\frac{\partial^2 w}{\partial x^2} + \frac{\partial \gamma_x}{\partial x}\right) + A_{2323}\lambda_1^2 \left(\frac{\partial^2 w}{\partial y^2} + \frac{\partial \gamma_y}{\partial y}\right) = \frac{\partial^2 w}{\partial t^2} + f_3(x,y,t),$$

$$\frac{1}{12}\left[A_{1111}\lambda_2^2 \frac{\partial^2 \gamma_x}{\partial x^2} + A_{1122}\lambda_1^2 \frac{\partial^2 \gamma_y}{\partial x \partial y} + A_{1212}\lambda_1^2\left(\frac{\partial^2 \gamma_x}{\partial y^2} + \frac{\partial^2 \gamma_y}{\partial x \partial y}\right)\right] -$$

$$- A_{1313}\lambda_1^2\lambda_2^2 \left(\gamma_x + \frac{\partial w}{\partial x}\right) = \frac{1}{12}\frac{\partial^2 \gamma_x}{\partial t^2} + \varphi_1(x,y,t), \quad \overleftrightarrow{(1,2)}, \quad \overleftrightarrow{(x,y)},$$

$$a_{1111}\lambda^2 \frac{\partial^4 F}{\partial y^4} + \frac{a_{2222}}{\lambda^2}\frac{\partial^4 F}{\partial x^4} + (a_{1122} + a_{2211} - a_{1212})\frac{\partial^4 F}{\partial x^2 \partial y^2} -$$

$$k_1\frac{\partial^2 w}{\partial y^2} - k_2\frac{\partial^2 w}{\partial x^2} = 0. \qquad (1.133)$$

Also the f_3 surface load components and components of the moment vector φ_1, φ_2 can be simplified as following

$$f_3(x,y,t) = \sum_{i=1}^{N} M^i \left\{\frac{(z_i - 0.5)}{\lambda_1^2}\frac{\partial}{\partial x}\left[\frac{\partial^3 w}{\partial x \partial t^2}\delta(x - x_i)\right]\delta(y - y_i) + \frac{1}{2\lambda_1^2}(z_i - 0.5) \times\right.$$

$$\frac{\partial}{\partial x}\left[\frac{\partial^2 \gamma_x}{\partial t^2}\delta(x - x_i)\right]\delta(y - y_i) - (z_i - 0.5)\frac{1}{\lambda_2^2}\frac{\partial}{\partial y}\left[\frac{\partial^3 w}{\partial y \partial t^2}\delta(y - y_i)\right]\delta(x - x_i) +$$

$$\left.\frac{1}{2\lambda_2^2}(z_i - 0.5)\frac{\partial}{\partial y}\left[\frac{\partial^2 \gamma_y}{\partial t^2}\delta(y - y_i)\right]\delta(x - x_i) + \frac{\partial^2 w}{\partial t^2}\delta(x - x_i)\delta(y - y_i)\right\} -$$

$$\sum_{i=1}^{N}\left\{J_{xx}^i\frac{\partial}{\partial y}\left[\frac{\partial^3 w}{\partial y \partial t^2}\delta(y - y_i)\right]\delta(x - x_i) - J_{yy}^i\frac{\partial}{\partial x}\left[\frac{\partial^3 w}{\partial x \partial t^2}\delta(x - x_i)\right]\delta(y - y_i)\right\},$$

$$\varphi_1(x,y,t) = \sum_{i=1}^{N} M^i \left[\frac{1}{2}\frac{\partial^2 \gamma_x}{\partial t^2} - \frac{1}{2}(z_i - 0.5)\frac{\partial^3 w}{\partial x \partial t^2}\right]\delta(x - x_i)\delta(y - y_i),$$

$$(1.134)$$

$$\overleftrightarrow{(1,2)}, \quad \overleftrightarrow{(x,y)}.$$

Considering free vibration case (without concentrated masses) the following w, γ_x, γ_y, F functions are assumed

Table 1.1. Elements a_{ij} of the determinant in the characteristic equation (1.131).

$\sum_{m,n}^{\infty} \dfrac{p_{11}^1 \lambda_1^1}{\Delta(m,n)} - 1$	$\sum_{m,n}^{\infty} \dfrac{p_{12}^1 \lambda_1^1}{\Delta(m,n)}$	$\sum_{m,n}^{\infty} \dfrac{p_{17}^1 \lambda_1^1}{\Delta(m,n)}$	$\sum_{m,n}^{\infty} \dfrac{p_{11}^2 \lambda_1^1}{\Delta(m,n)}$	$\sum_{m,n}^{\infty} \dfrac{p_{17}^2 \lambda_1^1}{\Delta(m,n)}$	$\cdots$	$\sum_{m,n}^{\infty} \dfrac{p_{11}^N \lambda_1^1}{\Delta(m,n)}$	$\sum_{m,n}^{\infty} \dfrac{p_{17}^N \lambda_1^1}{\Delta(m,n)}$
$\sum_{m,n}^{\infty} \dfrac{p_{21}^1 \lambda_1^2}{\Delta(m,n)}$	$\sum_{m,n}^{\infty} \dfrac{p_{22}^1 \lambda_1^2}{\Delta(m,n)} - 1$	$\sum_{m,n}^{\infty} \dfrac{p_{27}^1 \lambda_1^2}{\Delta(m,n)}$	$\sum_{m,n}^{\infty} \dfrac{p_{21}^2 \lambda_1^2}{\Delta(m,n)}$	$\sum_{m,n}^{\infty} \dfrac{p_{27}^2 \lambda_1^2}{\Delta(m,n)}$	$\cdots$	$\sum_{m,n}^{\infty} \dfrac{p_{21}^N \lambda_1^2}{\Delta(m,n)}$	$\sum_{m,n}^{\infty} \dfrac{p_{27}^N \lambda_1^2}{\Delta(m,n)}$
$\sum_{m,n}^{\infty} \dfrac{p_{31}^1 \lambda_1^7}{\Delta(m,n)}$	$\sum_{m,n}^{\infty} \dfrac{p_{32}^1 \lambda_1^7}{\Delta(m,n)}$	$\sum_{m,n}^{\infty} \dfrac{p_{37}^1 \lambda_1^1}{\Delta(m,n)} - 1$	$\sum_{m,n}^{\infty} \dfrac{p_{31}^2 \lambda_1^7}{\Delta(m,n)}$	$\sum_{m,n}^{\infty} \dfrac{p_{37}^2 \lambda_1^7}{\Delta(m,n)}$	$\cdots$	$\sum_{m,n}^{\infty} \dfrac{p_{31}^N \lambda_1^7}{\Delta(m,n)}$	$\sum_{m,n}^{\infty} \dfrac{p_{37}^N \lambda_1^7}{\Delta(m,n)}$
$\sum_{m,n}^{\infty} \dfrac{p_{11}^1 \lambda_2^1}{\Delta(m,n)}$	$\sum_{m,n}^{\infty} \dfrac{p_{12}^1 \lambda_2^1}{\Delta(m,n)}$	$\sum_{m,n}^{\infty} \dfrac{p_{17}^1 \lambda_2^1}{\Delta(m,n)}$	$\sum_{m,n}^{\infty} \dfrac{p_{11}^2 \lambda_2^1}{\Delta(m,n)} - 1$	$\sum_{m,n}^{\infty} \dfrac{p_{17}^2 \lambda_2^1}{\Delta(m,n)}$	$\cdots$	$\sum_{m,n}^{\infty} \dfrac{p_{11}^N \lambda_2^1}{\Delta(m,n)}$	$\sum_{m,n}^{\infty} \dfrac{p_{17}^N \lambda_2^1}{\Delta(m,n)}$
$\sum_{m,n}^{\infty} \dfrac{p_{31}^1 \lambda_2^7}{\Delta(m,n)}$	$\sum_{m,n}^{\infty} \dfrac{p_{32}^1 \lambda_2^7}{\Delta(m,n)}$	$\sum_{m,n}^{\infty} \dfrac{p_{37}^1 \lambda_2^7}{\Delta(m,n)}$	$\sum_{m,n}^{\infty} \dfrac{p_{31}^2 \lambda_2^7}{\Delta(m,n)}$	$\sum_{m,n}^{\infty} \dfrac{p_{37}^2 \lambda_2^7}{\Delta(m,n)} - 1$	$\cdots$	$\sum_{m,n}^{\infty} \dfrac{p_{31}^N \lambda_2^7}{\Delta(m,n)}$	$\sum_{m,n}^{\infty} \dfrac{p_{37}^N \lambda_2^7}{\Delta(m,n)}$
$\sum_{m,n}^{\infty} \dfrac{p_{11}^1 \lambda_N^1}{\Delta(m,n)}$	$\sum_{m,n}^{\infty} \dfrac{p_{12}^1 \lambda_N^1}{\Delta(m,n)}$	$\sum_{m,n}^{\infty} \dfrac{p_{17}^1 \lambda_N^1}{\Delta(m,n)}$	$\sum_{m,n}^{\infty} \dfrac{p_{11}^2 \lambda_N^1}{\Delta(m,n)}$	$\sum_{m,n}^{\infty} \dfrac{p_{17}^2 \lambda_N^1}{\Delta(m,n)}$	$\cdots$	$\sum_{m,n}^{\infty} \dfrac{p_{11}^N \lambda_N^1}{\Delta(m,n)} - 1$	$\sum_{m,n}^{\infty} \dfrac{p_{17}^N \lambda_N^1}{\Delta(m,n)}$
$\sum_{m,n}^{\infty} \dfrac{p_{31}^1 \lambda_N^7}{\Delta(m,n)}$	$\sum_{m,n}^{\infty} \dfrac{p_{32}^1 \lambda_N^7}{\Delta(m,n)}$	$\sum_{m,n}^{\infty} \dfrac{p_{37}^1 \lambda_N^7}{\Delta(m,n)}$	$\sum_{m,n}^{\infty} \dfrac{p_{31}^2 \lambda_N^7}{\Delta(m,n)}$	$\sum_{m,n}^{\infty} \dfrac{p_{37}^2 \lambda_N^7}{\Delta(m,n)}$	$\cdots$	$\sum_{m,n}^{\infty} \dfrac{p_{31}^N \lambda_N^7}{\Delta(m,n)}$	$\sum_{m,n}^{\infty} \dfrac{p_{37}^N \lambda_N^7}{\Delta(m,n)} - 1$

$\lambda_i^1 = \cos\alpha_m x_i \sin\beta_n y_i, \quad \lambda_i^2 = \sin\alpha_m x \cos\alpha_m y_i, \quad \lambda_i^3 = \sin\alpha_m x_i \sin\beta_n y_i,$

$\lambda_i^4 = \lambda_i^1, \quad \lambda_i^5 = \lambda_i^2, \quad \lambda_i^6 = \alpha_m \lambda_i^1, \quad \lambda_i^7 = \beta_n \lambda_i^2, \quad i = \overline{1, N}$

1.3 Analysis of Orthotropic Shallow Shells with Added Masses

$$w = A \sin \alpha_m x \sin \beta_n y \sin \omega_{mn} t,$$

$$\gamma_x = B \cos \alpha_m x \sin \beta_n y \sin \omega_{mn} t,$$

$$\gamma_y = C \sin \alpha_m x \cos \beta_n y \sin \omega_{mn} t,$$

$$F = D \sin \alpha_m x \sin \beta_n y \sin \omega_{mn} t. \tag{1.135}$$

The assumed solution satisfies (1.133) and boundary conditions (1.83).

Substituting (1.135) into (1.133) an algebraic system of linear equations with the unknowns A, B, C, D is obtained. Equalling its determinant to zero, we get the following characteristic equation

$$r_1 \omega_{mn}^6 + r_2 \omega_{mn}^4 + r_3 \omega_{mn}^2 + r_4 = 0. \tag{1.136}$$

The polynomial r_i coefficients are functions of the determinant elements e_{ij}:

$$e_{11} = -A_{1313}\lambda_2^2 \alpha_m^2 - A_{2323}\lambda_1^2 \beta_n^2 + \omega_{mn}^2,$$

$$e_{12} = -A_{1313}\lambda_2^2 \alpha_m, \quad e_{13} = -A_{2323}\lambda_1^2 \beta_n,$$

$$e_{14} = k_1 \beta_n^2 + k_2 \alpha_m^2, \quad e_{21} = -A_{1313}\lambda_1^2 \lambda_2^2 \alpha_m,$$

$$e_{22} = \frac{1}{12}\left[-A_{1111}\lambda_2^2 \alpha_m^2 - A_{1212}\lambda_1^2 \beta_n^2 + \frac{1}{12}\omega_{mn}^2\right] - A_{1313}\lambda_1^2 \lambda_2^2,$$

$$e_{23} = \frac{1}{12}\left[-A_{1122}\lambda_1^2 \alpha_m \beta_n - A_{1212}\lambda_1^2 \alpha_m \beta_n\right], \quad e_{24} = 0,$$

$$e_{31} = -A_{2323}\lambda_1^2 \lambda_2^2 \beta_n, \quad e_{32} = \frac{1}{12}\left[-A_{2211}\lambda_2^2 \alpha_m \beta_n - A_{1212}\lambda_1^2 \alpha_m \beta_n\right],$$

$$e_{33} = \frac{1}{12}\left[-A_{2222}\lambda_1^2 \beta_n^2 - A_{1212}\lambda_2^2 \alpha_m^2\right] - A_{2323}\lambda_1^2 \lambda_2^2 + \frac{1}{12}\omega_{mn}^2,$$

$$e_{34} = 0, \quad e_{41} = k_1 \beta_n^2 + k_2 \alpha_m^2,$$

$$e_{42} = 0, \quad e_{43} = 0,$$

$$e_{44} = a_{1111}\lambda^2 \beta_n^4 + \frac{a_{2222}}{\lambda^2}\alpha_m^4 + (a_{1122} + a_{2211} - a_{1212})\alpha_m^2 \beta_n^2. \tag{1.137}$$

For each pair (m,n) the solutions of (1.137) give three frequency spectra of free vibrations.

For example, the relationship leading to the determination of free vibration frequencies of an orthotropic and rollerly supported shallow shell is as follows

$$\omega_{mn}^2 = \frac{1}{12}\left[\frac{A_{1111}}{\lambda^2}\alpha_m^4 + 2(A_{1122} + 2A_{1212})\alpha_m^2 \beta_n^2 + A_{2222}\beta_n^4 \lambda^2 + (k_1 \alpha_m^2 + k_2 \beta_n^2)\right] \Big/ \left(\frac{a_{2222}}{\lambda^2}\alpha_m^4 + (a_{1122} + a_{2211} - a_{1212})\alpha_m^2 \beta_n^2 + a_{1111}\beta_n^4 \lambda^2\right). \tag{1.138}$$

1 Vibration of Plates and Shells with Added Masses

Comparing (1.137) and (1.138) it is concluded, that the correction brought by taking into consideration the tangential displacements depends on the shell dimensions, its stiffness characteristics and on have number of vibrationsin x and y directions. In the formula of the dimensionless frequency the dimensionless λ_1, λ_2 thicknesses and the stiffness constants A_{1313}, A_{2323} characterizing the transversal deformation elements do not appear. However, many researchers (for example [3, 4, 117]) established the validity of using the transversal displacements in many practically important problems.

We consider a procedure for the determination of the dynamic characteristics of "shell - added masses" on the basis of equations (1.133). In this problem the influence of added masses has been expressed in the governing equations by additional inertial terms that include coefficients with the δ-functions singularities. The transversal components of the concentrated masses inertial forces and the vector moment components due to the transversal inertial load are taken into account.

For a shell with the boundary conditions defined by (1.83) the following solution is being sought

$$w = \sin\omega t \sum_{m,n}^{\infty} A_{mn} \sin\alpha_m x \sin\beta_n y,$$

$$\gamma_x = \sin\omega t \sum_{m,n}^{\infty} B_{mn} \cos\alpha_m x \sin\beta_n y,$$

$$\gamma_y = \sin\omega t \sum_{m,n}^{\infty} C_{mn} \sin\alpha_m x \cos\beta_n y,$$

$$F = \sin\omega t \sum_{m,n}^{\infty} D_{mn} \sin\alpha_m x \sin\beta_n y. \qquad (1.139)$$

Using (1.139) in (1.133) are obtains

$$A_{mn} = \sum_{i=1}^{N} \left(W_i \frac{p_{11}^i}{\Delta(m,n)} + \Gamma_{xi} \frac{p_{12}^i}{\Delta(m,n)} + \Gamma_{yi} \frac{p_{13}^i}{\Delta(m,n)} + \frac{\partial W_i}{\partial x} \frac{p_{14}^i}{\Delta(m,n)} + \frac{\partial W_i}{\partial y} \frac{p_{15}^i}{\Delta(m,n)} \right),$$

$$B_{mn} = \sum_{i=1}^{N} \left(W_i \frac{p_{21}^i}{\Delta(m,n)} + \Gamma_{xi} \frac{p_{22}^i}{\Delta(m,n)} + \Gamma_{yi} \frac{p_{23}^i}{\Delta(m,n)} + \frac{\partial W_i}{\partial x} \frac{p_{24}^i}{\Delta(m,n)} + \frac{\partial W_i}{\partial y} \frac{p_{25}^i}{\Delta(m,n)} \right),$$

1.3 Analysis of Orthotropic Shallow Shells with Added Masses 49

$$C_{mn} = \sum_{i=1}^{N} \left(W_i \frac{p_{31}^i}{\Delta(m,n)} + \Gamma_{xi} \frac{p_{32}^i}{\Delta(m,n)} + \Gamma_{yi} \frac{p_{33}^i}{\Delta(m,n)} + \frac{\partial W_i}{\partial x} \frac{p_{34}^i}{\Delta(m,n)} + \frac{\partial W_i}{\partial y} \frac{p_{35}^i}{\Delta(m,n)} \right). \quad (1.140)$$

For the function derivatives we have

$$\frac{\partial w}{\partial x} = \sum_{m,n}^{\infty} A_{mn}\alpha_m \cos\alpha_m x \sin\beta_n y, \quad \frac{\partial w}{\partial y} = \sum_{m,n}^{\infty} A_{mn}\beta_n \sin\alpha_m x \cos\beta_n y. \quad (1.141)$$

We take $x = x_i$, $y = y_i$ ($l = \overline{1,N}$) in (1.139), (1.141) and using (1.140) on obtains a system of homogeneous equations in terms of w_i, Γ_{xi}, Γ_{yi}, $\partial w_i/\partial x$, $\partial w_i/\partial y$. It possesses the nontrivial solutions when

$$\det [a_{ik}]_{5N \times 5N} = 0. \quad (1.142)$$

The determinant elements are given in Table 1.2, where the following notation is used

$$p_{j1}^i = d_{11}^i \Delta_{1j}, \quad p_{j2}^i = d_{21}^i \Delta_{2j} + d_{12}^i \Delta_{1j},$$

$$p_{j3}^i = d_{31}^i \Delta_{3j} + d_{13}^i \Delta_{1j}, \quad p_{j4}^i = d_{22}^i \Delta_{2j} + d_{14}^i \Delta_{1j},$$

$$p_{j5}^i = \Delta_{3j} d_{32}^i + d_{15}^i \Delta_{1j}, \quad (j = \overline{1,3}) ;$$

$$d_{11}^i = -4\omega^2 M^i \sin\alpha_m x_i \sin\beta_n y_i,$$

$$d_{12}^i = 4\omega^2 \frac{M^i}{2\lambda_1^2}(z_i - 0.5)\cos\alpha_m x_i \sin\beta_n y_i,$$

$$d_{13}^i = 4\omega^2 M^i \frac{(z_i - 0.5)}{2\lambda_2^2} \beta_n \sin\alpha_m x_i \cos\beta_n y_i,$$

$$d_{14}^i = 4\omega^2 M^i \frac{(z_i - 0.5)^2}{\lambda_1^2}\alpha_m \cos\alpha_m x_i \sin\beta_n y_i - 4\omega^2 J_{xx}^i \beta_n \sin\alpha_m x_i \cos\beta_n y_i,$$

$$d_{15}^i = -4\omega^2 M^i \frac{(z_i - 0.5)^2}{\lambda_1^2}\beta_n \sin\alpha_m x_i \cos\beta_n y_i - 4\omega^2 J_{yy}^i \alpha_m \cos\alpha_m x_i \sin\beta_n y_i,$$

$$d_{21}^i = -\omega^2 M^i \cos\alpha_m x_i \sin\beta_n y_i, \quad d_{22}^i = 2\omega^2 M^i(z_i - 0.5)\cos\alpha_m x_i \sin\beta_n y_i,$$

$$d_{31}^i = -\omega^2 M^i \sin\alpha_m x_i \cos\beta_n y_i,$$

$$d_{32}^i = 2\omega^2 M^i(z_i - 0.5)\sin\alpha_m x_i \cos\beta_n y_i, \quad (i = \overline{1,N}). \quad (1.143)$$

To conclude, the determination of the free vibration frequencies of a shallow orthotropic shell with attached discrete masses is reduced to finding the solution to the characteristic equations (1.131) and (1.142) obtained from the $7N$ dimensional and $5N$ dimensional determinants, correspondingly. The elements of the determinant are infinite series of function dependent on all

Table 1.2. Elements a_{ik} of the determinant in the characteristic equation (1.142).

$\sum\limits_{m,n}^{\infty} \dfrac{p_{11}^1 \lambda_1^1}{\Delta(m,n)} - 1$	$\sum\limits_{m,n}^{\infty} \dfrac{p_{12}^1 \lambda_1^1}{\Delta(m,n)}$	$\sum\limits_{m,n}^{\infty} \dfrac{p_{15}^1 \lambda_1^1}{\Delta(m,n)}$	$\sum\limits_{m,n}^{\infty} \dfrac{p_{11}^2 \lambda_1^1}{\Delta(m,n)}$	$\sum\limits_{m,n}^{\infty} \dfrac{p_{15}^2 \lambda_1^1}{\Delta(m,n)}$	...	$\sum\limits_{m,n}^{\infty} \dfrac{p_{11}^N \lambda_1^1}{\Delta(m,n)}$	$\sum\limits_{m,n}^{\infty} \dfrac{p_{15}^N \lambda_1^1}{\Delta(m,n)}$
$\sum\limits_{m,n}^{\infty} \dfrac{p_{21}^1 \lambda_1^2}{\Delta(m,n)}$	$\sum\limits_{m,n}^{\infty} \dfrac{p_{22}^1 \lambda_1^2}{\Delta(m,n)} - 1$	$\sum\limits_{m,n}^{\infty} \dfrac{p_{25}^1 \lambda_1^1}{\Delta(m,n)}$	$\sum\limits_{m,n}^{\infty} \dfrac{p_{21}^2 \lambda_1^1}{\Delta(m,n)}$	$\sum\limits_{m,n}^{\infty} \dfrac{p_{25}^2 \lambda_1^1}{\Delta(m,n)}$	...	$\sum\limits_{m,n}^{\infty} \dfrac{p_{21}^N \lambda_1^1}{\Delta(m,n)}$	$\sum\limits_{m,n}^{\infty} \dfrac{p_{25}^N \lambda_1^1}{\Delta(m,n)}$
$\sum\limits_{m,n}^{\infty} \dfrac{p_{11}^1 \lambda_1^5}{\Delta(m,n)}$	$\sum\limits_{m,n}^{\infty} \dfrac{p_{12}^1 \lambda_1^5}{\Delta(m,n)}$	$\sum\limits_{m,n}^{\infty} \dfrac{p_{15}^1 \lambda_1^5}{\Delta(m,n)} - 1$	$\sum\limits_{m,n}^{\infty} \dfrac{p_{11}^2 \lambda_1^5}{\Delta(m,n)}$	$\sum\limits_{m,n}^{\infty} \dfrac{p_{15}^2 \lambda_1^5}{\Delta(m,n)}$	...	$\sum\limits_{m,n}^{\infty} \dfrac{p_{11}^N \lambda_1^5}{\Delta(m,n)}$	$\sum\limits_{m,n}^{\infty} \dfrac{p_{15}^N \lambda_1^5}{\Delta(m,n)}$
$\sum\limits_{m,n}^{\infty} \dfrac{p_{11}^1 \lambda_2^1}{\Delta(m,n)}$	$\sum\limits_{m,n}^{\infty} \dfrac{p_{12}^1 \lambda_2^1}{\Delta(m,n)}$	$\sum\limits_{m,n}^{\infty} \dfrac{p_{15}^1 \lambda_2^1}{\Delta(m,n)}$	$\sum\limits_{m,n}^{\infty} \dfrac{p_{11}^2 \lambda_2^1}{\Delta(m,n)} - 1$	$\sum\limits_{m,n}^{\infty} \dfrac{p_{15}^2 \lambda_2^1}{\Delta(m,n)}$	...	$\sum\limits_{m,n}^{\infty} \dfrac{p_{11}^N \lambda_2^1}{\Delta(m,n)}$	$\sum\limits_{m,n}^{\infty} \dfrac{p_{15}^N \lambda_2^1}{\Delta(m,n)}$
$\sum\limits_{m,n}^{\infty} \dfrac{p_{11}^1 \lambda_2^5}{\Delta(m,n)}$	$\sum\limits_{m,n}^{\infty} \dfrac{p_{12}^1 \lambda_2^5}{\Delta(m,n)}$	$\sum\limits_{m,n}^{\infty} \dfrac{p_{15}^1 \lambda_2^5}{\Delta(m,n)}$	$\sum\limits_{m,n}^{\infty} \dfrac{p_{11}^2 \lambda_2^5}{\Delta(m,n)}$	$\sum\limits_{m,n}^{\infty} \dfrac{p_{15}^2 \lambda_2^5}{\Delta(m,n)} - 1$	...	$\sum\limits_{m,n}^{\infty} \dfrac{p_{11}^N \lambda_2^5}{\Delta(m,n)}$	$\sum\limits_{m,n}^{\infty} \dfrac{p_{15}^N \lambda_2^5}{\Delta(m,n)}$
$\sum\limits_{m,n}^{\infty} \dfrac{p_{11}^1 \lambda_N^1}{\Delta(m,n)}$	$\sum\limits_{m,n}^{\infty} \dfrac{p_{12}^1 \lambda_N^1}{\Delta(m,n)}$	$\sum\limits_{m,n}^{\infty} \dfrac{p_{15}^1 \lambda_N^1}{\Delta(m,n)}$	$\sum\limits_{m,n}^{\infty} \dfrac{p_{11}^2 \lambda_N^1}{\Delta(m,n)}$	$\sum\limits_{m,n}^{\infty} \dfrac{p_{15}^2 \lambda_N^1}{\Delta(m,n)}$	...	$\sum\limits_{m,n}^{\infty} \dfrac{p_{11}^N \lambda_N^1}{\Delta(m,n)} - 1$	$\sum\limits_{m,n}^{\infty} \dfrac{p_{15}^N \lambda_N^1}{\Delta(m,n)}$
$\sum\limits_{m,n}^{\infty} \dfrac{p_{11}^1 \lambda_N^5}{\Delta(m,n)}$	$\sum\limits_{m,n}^{\infty} \dfrac{p_{12}^1 \lambda_N^5}{\Delta(m,n)}$	$\sum\limits_{m,n}^{\infty} \dfrac{p_{15}^1 \lambda_N^5}{\Delta(m,n)}$	$\sum\limits_{m,n}^{\infty} \dfrac{p_{11}^2 \lambda_N^5}{\Delta(m,n)}$	$\sum\limits_{m,n}^{\infty} \dfrac{p_{15}^2 \lambda_N^5}{\Delta(m,n)}$	...	$\sum\limits_{m,n}^{\infty} \dfrac{p_{11}^N \lambda_N^5}{\Delta(m,n)}$	$\sum\limits_{m,n}^{\infty} \dfrac{p_{15}^N \lambda_N^5}{\Delta(m,n)} - 1$

$$\lambda_i^1 = \sin\alpha_m x_i \sin\beta_n y_i, \quad \lambda_i^2 = \cos\alpha_m x_i \sin\beta_n y_i, \quad \lambda_i^3 = \sin\alpha_m x_i \cos\beta_n y_i,$$

$$\lambda_i^4 = \alpha_m \lambda_i^2, \quad \lambda_i^5 = \beta_n \lambda_i^3, \quad i = \overline{1, N}.$$

1.3 Analysis of Orthotropic Shallow Shells with Added Masses

input parameters and the frequency ω. Each series term possesses a fractional-polynomial structure, i.e. occurs as a division of two polynomials of different degrees.

Because the solution of the characteristic equations (1.131), (1.142) can not be found in the closed form, approximate methods will be used in what follows.

1.3.5 Algorithms and Numerical Results

In the previous chapters a general approach, using generalized functions to analyse the vibrations of plates and shallow shells, has been presented. The different methods of finding solutions to the obtained differential equations in the frame of the subtle Timoshenko's model and the three dimensional problem have been discussed and illustrated. The effect on the dynamics added masses is reflected in the forms of the frequency equations. The complex calculation algorithms have to be used to find the solutions of these equations.

A traditional analysis causes many calculation difficulties in finding solutions to the frequency equations of the "shell - mass" system. As it has been shown in [72], the free vibration frequencies of elastic systems with discrete added elements may be found as the eigenvalues of a certain class matrix with trigonometric series as its elements. Thus, the problem becomes reduced to solving the following equation

$$\det \left[\sum_{m,n}^{\infty} a_{mn}^{ik} \right] = 0, \qquad (1.144)$$

where

$$a_{mn}^{ik} = \frac{A_{mn}^{ik} \omega^2}{\omega_{mn}^2 - \omega^2}.$$

In the relations above, ω_{mn} is the frequency of free vibration and A_{mn}^{ik} are coefficients depending on the magnitude and location of the added masses.

Instead of using (1.144) the equivalent formula, suitable for developing of the algorithm, is used

$$\det \left[f^{ik} \left(\omega^2 \right) \right] = 0, \qquad (1.145)$$

where $f^{ik}(\omega^2)$ are the polynomials of degree mn depending on ω^2.

It should be pointed out, that the eigenvalue problem based on the above mentioned matrices is not easy, even when the elements of the matrices are numbers. The obtained matrices have polynomial structure, and special methods leading to their solutions should be developed. Also, during numerical calculations a simplified problem and a suitable economic algorithm should be proposed.

Using a "traditional" approach we take $J_{xx}^i = J_{yy}^i = 0$ and we assume that the added mass centre is situated on the top surface of a shell or a plate. The first assumption is justified because the inertial effects caused by the mass rotation does not change significantly the lowest free vibration frequency [51]. The second assumption results in neglecting of the added mass height in comparison to the plate or the shell height, i.e. it is assumed that the added masses the sufficeintly "flat". Although these assumptions introduce inaccracies, yet they can serve as the first step in a systematic numerical strategy towards the qualitative solutions of the described types of problems.

The above discussed assumptions lead to the following characteristic equation, which serves for determining the frequencies of the system "shell (plate) - mass"

$$\det \left[\sum_{m,n}^{\infty} a_{mn}^{ik} \right] = 0, \qquad (1.146)$$

where

$$a_{mn}^{ik} = \begin{cases} \dfrac{\Delta_1}{\Delta} c_{mn}^{ik}, & i \neq k \\ \dfrac{\Delta_1}{\Delta} c_{mn}^{ik} - \dfrac{1}{4M^k}, & i = k \end{cases}. \qquad (1.147)$$

The above determinant is of fourth order, whose elements are obtained from (1.137), and Δ_1 is the co-factor of the determinant's element situated on the crossing point of the first row and the first column, and

$$c_{mn}^{ik} = \left[\sin \alpha_m x_j \sin \alpha_m \tilde{c}_1^j \sin \beta_n y_i \sin \beta_n \tilde{c}_2^j \times \right.$$

$$\left. \times \sin \alpha_m x_i \sin \beta_n y_i \right] / \left[\pi^2 m n \tilde{c}_1^i \tilde{c}_2^i \right], \quad (i,j = \overline{1, N}). \qquad (1.148)$$

Numerical experiments have shown that a substitution of the matrix (1.146) by an equivalent matrix with the polynomial elements, and then its reduction to an algebraic equation is not an effective procedure due to of accuracy requirements. It leads to the result, that for $S > 5$, where S denotes the number terms remaining in (1.146), there are imaginary roots of the characteristic equation, which is in contradiction to physical interpretation. However, we can give the following explanation. If only 'S' terms in the determinants (1.144), (1.146) are used then we develop the determinants and we get a characteristic equation of $(N \times S)$ and $(3 \times N \times S)$ degrees, respectively. Thus, for the development of the second determinant more arithmetic transformations are needed, which leads to inacurate calculations. In addition, in this case the equation degree is higher than in the first case, which brings in limitations of the terms number S remaining in the term series of the determinant (1.146). It is expected that the described algorithm does not have any applications in finding solutions to the more complicated characteristic equations (1.131), (1.142), leading to the algebraic equations of the $(7 \times N \times S)$ and $(S \times N \times S)$ order, respectively.

A step by step method jointed with an iteration procedure of finding function roots in certain intervals is a heart of the algorithm developed for finding solutions to the characteristic equations (1.146).

1.3 Analysis of Orthotropic Shallow Shells with Added Masses

Some examples of the numerical analysis of the free vibration shell (with the added masses) frequencies will be given below.

In Fig. 1.2 the dependence of nondimensional frequency parameter $\omega^* = a\sqrt{p/A_{1111}}\omega$ (corresponding to the lowest frequency) on the relative thickness $\lambda_1 = a/2h = \lambda_2 = b/2h$ of a cylindrical shell ($k_1 = 6$, $k_2 = 0$) for different materials with $G_{13}/E = G_{23}/E = G'/E$ is presented. The solid curves 1 and 2 correspond to the free shell's vibrations for $G'/E = 0.2; 0.01$. If we limit the series (1.139) to only one term we get an expression which is not accurate enough for frequency determination. The calculation results for a shell with an attached mass at its centre of mass $M = 0.5 M_0$ for different G'/E values are labeled by 3 and 5 (dashed curves). The dashed lines 4 and 6 are related to more accurate frequency values by taking into account nine terms in the sum (1.139).

In Fig. 1.3, the fundamental frequency dependence on a mass ($\tilde{M} = M_0/4M$), situated at the shell's centre is shown ($\lambda_1 = 10$: for $G'/E = 0.2$ - curve 1, for $G'/E = 0.01$ - curve 2; and for $\lambda_1 = 50$: $G'/E = 0.2$ for - curve 3, for $G'/E = 0.01$ - curve 4).

Low frequency values of a spherical shell with the mass $M = 0.5 M_0$ versus $k_1 = k_2 = K$ are reported in Fig. 1.4. The curves 1, 2 (7, 8) depict the free vibration frequency against the parameter $\lambda_1 = 10$ ($\lambda_1 = 50$) for $G'/E = 0.2; 0.01$, whereas the curves 3, 5 (9, 10) depict the same but using only the first approximation. The method accounting nine terms of the series (1.139) for $G'/E = 0.2; 0.01$, has led to the values lying closer to the dashed curves 4 and 6 (for $\lambda_1 = 10$).

The transversal deformation leads to a decrease of the free vibration frequencies obtained on the basis of the classical theory. For a material ($G'/E = 0.2$) with similar properties to that of the orthotropic one, both theories results converge. A decrease of transversal stiffness leads to an essential decrease of the fundamental free vibration frequency in comparison with the classical theory results. For instance, for $\lambda_1 = 5$ the difference is 63% (Fig. 1.2). An increase of λ_1 causes both curves 1 and 2 to draw close. It shows that the influence of transversal deformations decreases as λ_1 decreases and can be neglected for $\lambda_1 < 50$. It means that in the latter case the calculations can be conducted according to the classical theory.

The correction introduced by the influence of transversal deformations on the fundamental frequency depends on the shell's slope. The most important influence is observable for shells with small curvature parameters. The transversal displacements' influence is of less importance. For $\lambda_1 = 10$ the fundamental vibration frequency of the plate for $G'/E = 0.2$ is of amount of 1.75 times of the vibration plate frequency with $G'/E = 0.01$. For a spherical shell ($K = 6$) this ratio achieves 1.2.

The masses added to a shell cause an essential change of the free vibration frequencies and the corresponding modes. They depend on the added mass value, its position, as well as the geometric (λ_1, λ_2, k_1, k_2) and physical

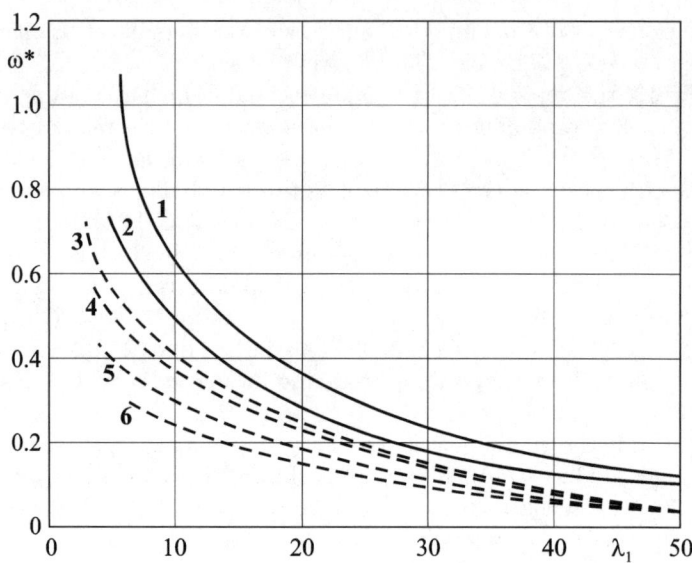

Figure 1.2. Non-dimensional frequency ω^* versus relative shell thickness λ_1.

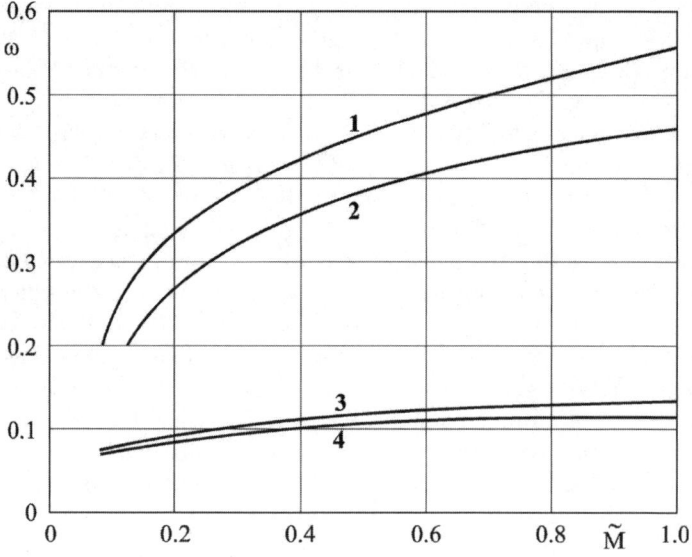

Figure 1.3. Fundamental frequency w versus mass $\tilde{M}$ attached in the shell centre.

(G'/E) shell's parameters. It causes a decrease of the fundamental frequency, which is more evident for thick shells. Besides, with the shell's thickness increase, the error in the determination of the first fundamental frequency

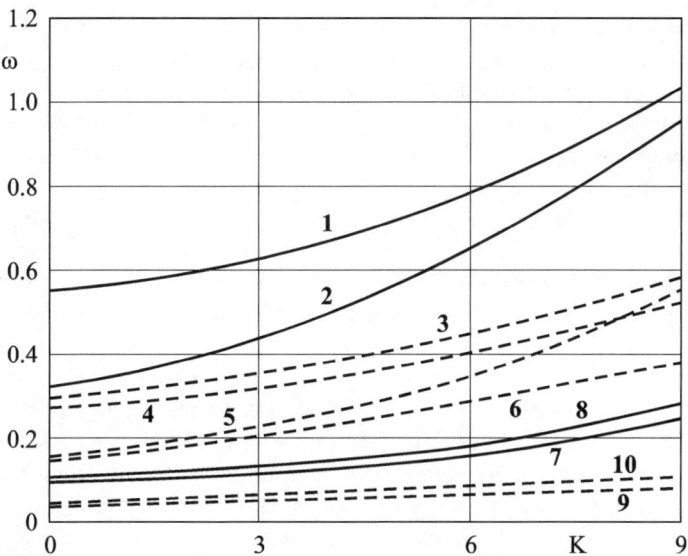

Figure 1.4. Low frequency of shallow vibrations ($M = 0.5M_0$) versus curvature K.

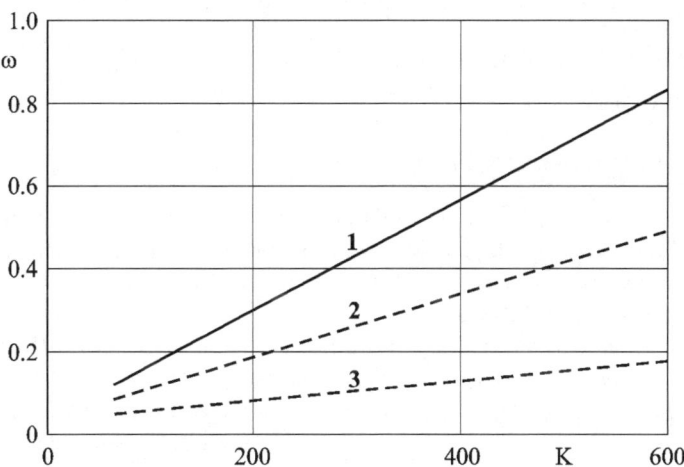

Figure 1.5. Fundamental frequency of spherical shell ($\lambda_1 = 700$, $G/E = 0.2$; $M = 0.5M_0$) against curvature K.

increases. In the case of a cylindrical shell with $G/E = 0.01$ for $\lambda_1 = 10$ the free vibration frequency is 2.14 times higher than a corresponding frequency with added masses (the first approximation error is equal to 23%), whereas for $\lambda_1 = 50$ that ratio is equal to 1.82 (4,9%). The error increases on the shell's height increases; it is 14% in the case of a plate and 68% in the case of a spheric shell ($k_1 = k_2 = 9$, $G/E = 0.01$, $\lambda_1 = 10$).

For an increase of the added mass the fundamental frequency decreases. The shell vibrations extend to high frequencies. With $\tilde{M} = M_0/4M \to \infty$ the frequencies approach the corresponding values of the free shell vibration frequencies.

In the previous cases the influence of the added mass on vibrations of the shallow shells with small relative thickness values ($\lambda_1 \leq 50$) has been considered. However, in practical constructions with shells and added masses the parameter λ_1 can be bigger than 500. Therefore, this kind of constructions is very important.

In Fig 1.5 the spherical shell fundamental frequency variations are depicted for parameters $\lambda_1 = 700$, $G/E = 0.2$. In the shell centre the mass $M = 0.5M_0$ is attached. From that figure it is clearly seen that for such shells the added mass has bigger influence in comparison to the earlier analysed shells. For instance, for $k_1 = 500$ a ratio of free vibration frequency (curve 1) to vibration frequency of the loaded shell (curve 3) is equal to 4.3; using the first order approximation (curve 2) this ratio is equal to 1.73. Therefore, in order to achieve enough accuracy of the fundamental frequency determination at least nine terms of (1.139) are needed.

1.4 Dynamical Characteristics of Plates with Added Masses

1.4.1 Timoshenko's Kinematic Model

Setting $k_1 = k_2 = 0$, in equations (1.80) the equations governing orthotropic plates vibrations with the transversal transformations and rotation inertia as well as all external surface load components (1.80) resulting from the added masses iterations are obtained.

Consider now a special case, of an orthotropic shell with attached concentrated masses with zero inertial moments J_{xy}^i, J_{xz}^i, J_{yz}^i, J_{zz}^i. The system of equation (1.80) governing vibrational behaviour of a plate with the added masses and with the load defined by (1.87) does not allow are to get two separate subsystems governing the symmetric and antisymmetric vibration modes. The procedure leading to the solution is completely reduced to the previously considered case described by (1.118), (1.119) for $k_1 = k_2 = 0$. The frequency vibration spectra of the "plate - mass" system are determined from the characteristic equation (1.131).

Now we formulate and solve the problem of shell vibrations with the added masses in the frame of the second variant, taking into account the equations (1.118) and the condition (1.119). In this variant we do not need to omit tangent inertial components. It is enough to neglect the terms with γ_x, γ_y, w in the first two expressions for f_1 and f_2, and also the terms u, v in the

1.4 Dynamical Characteristics of Plates with Added Masses

expressions of f_3, φ_1, φ_2. Substituting $k_1 = k_2 = 0$ to the equations (1.118) we get two equation groups. The first one includes displacements components u, v and describes the transversal vibrations, whereas the second one governs the dynamics in the normal direction w. In the latter case we have the following equations

$$A_{1313}\lambda_2^2 \left(\frac{\partial^2 w}{\partial x^2} + \frac{\partial \gamma_x}{\partial x}\right) + A_{2323}\lambda_1^2 \left(\frac{\partial^2 w}{\partial y^2} + \frac{\partial \gamma_y}{\partial y}\right) = \frac{\partial^2 w}{\partial t^2} + f_3(x,y,t),$$

$$\frac{1}{12}\left\{A_{1111}\lambda_2^2 \frac{\partial^2 \gamma_x}{\partial x^2} + A_{1122}\lambda_1^2 \frac{\partial^2 \gamma_y}{\partial x \partial y} + A_{1212}\lambda_1^2 \left(\frac{\partial^2 \gamma_x}{\partial y^2} + \frac{\partial^2 \gamma_y}{\partial x \partial y}\right)\right\} -$$

$$A_{1313}\lambda_1^2\lambda_2^2 \left(\gamma_x + \frac{\partial w}{\partial x}\right) = \frac{1}{12}\frac{\partial^2 \gamma_x}{\partial t^2} + \varphi_1(x,y,t), \quad (\overleftrightarrow{1,2}), \quad (\overleftrightarrow{x,y}), \quad (1.149)$$

where f_3, φ_1, φ_2 are defined by (1.119).

The system of equations (1.149) governs free vibrations of the orthotropic plate with added masses including the transversal deformations and inertial effects caused by rotation.

The roller support boundary conditions are added in the following form

$$\text{for} \quad x = 0, 1 \quad w = 0, \quad \gamma_y = 0, \quad M_{11} = 0, \quad (\overleftrightarrow{1,2}). \quad (1.150)$$

In the case of free vibrations of an unloaded orthotropic plate the following solution is sought

$$w = A \sin \alpha_m x \sin \beta_n y \sin \omega_{mn} t, \quad \gamma_x = B \cos \alpha_m x \sin \beta_n y \sin \omega_{mn} t,$$

$$\gamma_y = C \sin \alpha_m x \cos \beta_n y \sin \omega_{mn} t, \quad (1.151)$$

which fulfils the boundary conditions on each edge. In (1.151) A, B, C are constants, and $\alpha_m = m\pi$, $\beta_n = n\pi$, where m, n are integers.

Substituting (1.151) to (1.149) (with $f_3 = \varphi_1 = \varphi_2 = 0$) we get, after some transformations, an algebraic equation system with unknowns A, B, C. Nontrivial solution exist if

$$\det [e_{ik}]_{3\times 3} = 0, \quad (1.152)$$

where

$$e_{11} = -A_{1313}\lambda_2^2\alpha_m^2 - A_{2323}\lambda_1^2\beta_n^2 + \omega_{mn}^2,$$

$$e_{12} = -A_{1313}\lambda_2^2\alpha_m, \quad e_{13} = -A_{2323}\lambda_1^2\beta_n, \quad e_{21} = -A_{1313}\lambda_1^2\lambda_2^2\alpha_m,$$

$$e_{22} = \frac{1}{12}\left[-A_{1111}\lambda_2^2\alpha_m^2 - A_{1212}\lambda_1^2\beta_n^2 + \omega_{mn}^2\right] - A_{1313}\lambda_1^2\lambda_2^2,$$

$$e_{23} = \frac{1}{12}\left(-A_{1122}\lambda_1^2\alpha_m\beta_n - A_{1212}\lambda_1^2\alpha_m\beta_n\right),$$

$$e_{31} = -A_{2323}\lambda_1^2\lambda_2^2\beta_n, \quad e_{32} = \frac{1}{12}\left(-A_{2211}\lambda_2^2\alpha_m\beta_n - A_{1212}\lambda_1^2\alpha_m\beta_n\right),$$

$$e_{33} = \frac{1}{12}\left(-A_{2222}\lambda_1^2\beta_n^2 - A_{1212}\lambda_2^2\alpha_m^2 + \omega_{mn}^2\right) - A_{2323}\lambda_1^2\lambda_2^2, \quad (1.153)$$

and ω_{mn} is the free vibration frequency.

From terms of ω_{mn}^2 is obtained

$$r_1\omega_{mn}^6 + r_2\omega_{mn}^4 + r_3\omega_{mn}^2 + r_4 = 0. \quad (1.154)$$

The relationship for determining the free vibration frequency of the orthotropic plate roller supported on all edges is given below (using a classical approach):

$$\omega_{mn}^2 = \frac{1}{12}\left(\frac{A_{1111}}{\lambda^2}\alpha_m^4 + 2\left(A_{1122} + 2A_{1212}\right)\alpha_m^2\beta_n^2 + A_{2222}\beta_n^4\lambda^2\right). \quad (1.155)$$

Substituting (1.155) to (1.154), one finds that the correction introduced by transversal displacements depends on the plate dimension, its stiffness characteristics and the vibration wavenumbers in the x and y directions. The expression obtained using classical theory and defining the dimensionless frequency does not contain the parameters describing the dimensionless thickness and the transversal stiffness coefficients, which shows the classical theory.

Consider now a problem of the dynamical orthotropic shells and a way of how to determine the characteristics of the added masses.

The functions being sought have the following form

$$w = \sin\omega t \sum_{m,n}^{\infty} A_{mn}\sin\alpha_m x \sin\beta_n y,$$

$$\gamma_x = \sin\omega t \sum_{m,n}^{\infty} B_{mn}\cos\alpha_m x \sin\beta_n y,$$

$$\gamma_y = \sin\omega t \sum_{m,n}^{\infty} C_{mn}\sin\alpha_m x \cos\beta_n y. \quad (1.156)$$

Substituting (1.156) to (1.149) and taking into account (1.125) the following nonhomogeneous algebraic equations system in matrix form, is obtained

$$LA = Q, \quad (1.157)$$

where

$$L = \begin{bmatrix} e_{11} & e_{12} & e_{13} \\ e_{21} & e_{22} & e_{23} \\ e_{31} & e_{32} & e_{33} \end{bmatrix}, \quad A = \begin{bmatrix} A_{mn} \\ B_{mn} \\ C_{mn} \end{bmatrix}, \quad Q = \begin{bmatrix} g_1 \\ g_2 \\ g_3 \end{bmatrix}, \quad (1.158)$$

and

$$g_1 = \sum_{i=1}^{N}\left(d_{11}^i W_i + d_{12}^i \Gamma_{xi} + d_{13}^i \Gamma_{yi} + d_{14}^i \frac{\partial W_i}{\partial x} + d_{15}^i \frac{\partial W_i}{\partial y}\right),$$

$$g_2 = \sum_{i=1}^{N} \left(d_{21}^i \Gamma_{xi} + d_{22}^i \frac{\partial W_i}{\partial x} \right),$$

$$g_3 = \sum_{i=1}^{N} \left(d_{31}^i \Gamma_{yi} + d_{32}^i \frac{\partial W_i}{\partial y} \right), \qquad (1.159)$$

with A_{mn}, B_{nm}, C_{mn} being unknown coefficients.

The elements of matrix L are obtained from (1.153), whereas the parameters D_{jk}^i are found from (1.143). From (1.157) the expressions for the coefficients A_{mn}, B_{mn}, C_{mn} are obtained (see (1.140)) which are substituted to (1.156). Next, fulfilment of the added masses contact conditions due to the displacements (1.156) and due to their first derivatives (1.141) is needed. Finally, we get a characteristic equation expressed by $5N$ determinant's order, whose elements are given in Table 1.2. In the relationships (1.142) Δ denotes the L matrix determinant - defined by (1.158), and Δ_{ij} denotes a minor of the determinant's element situated in the crossing point of i-th row and j-th column.

1.4.2 Spatial Theory of Elasticity

A large amount of literature is devoted to the topic of reducing the three dimensional problem to a two dimensional one. Kilčevskij [162] has already pointed out, that a general method for achieving that reduction is thought analytical constructions characterizing the stress-deformation state by introducing quantities defined in the $x0y$ coordinate system.

In reference [86] it has been shown, that because of the lack of universal calculation models and of one-valued solutions, there are many different approximate methods applied.

An application range of the approximate theories is defined by the full three dimensional theory. Therefore, the next logical step will include a comparison of the two dimensional theory results with the three dimensional one. This approach can be used for the qualitative estimation of the different approximate, practically oriented theories. It allows for direct comparisons of different variables such as displacements, frequencies and modes of vibrations. It also seems that the errors of estimation are introduced by two dimensional theories in comparison to the three dimensional theory are valid especially for certain classes of problems.

Now such a comparison will be outlined on the basis of the following considerations:

A) Free vibrations of an unloaded orthotropic (isotropic) plate;
B) Free vibrations of an elastic system "orthotropic (isotropic) plate - concentrated masses".

In order to analyse free vibration frequencies and the orthotropic plate vibration modes, a solution of the differential equations (1.59) will be constructed using a condition allowing for slip on the plate edges and assuming the lack of the additional masses on its external surface.

The displacement functions which fulfil the above mentioned boundary conditions are as follows

$$u = \sum_{m,n}^{\infty} \phi(z) \cos \alpha_m x \sin \beta_n y \sin \omega_{mn} t,$$

$$v = \sum_{m,n}^{\infty} \psi(z) \sin \alpha_m x \cos \beta_n y \sin \omega_{mn} t,$$

$$w = \sum_{m,n}^{\infty} \chi(z) \sin \alpha_m x \sin \beta_n y \sin \omega_{mn} t. \tag{1.160}$$

Substituting (1.160) in (1.59) with $k_1 = k_2 = 0$ we get the following uncoupled three ordinary differential equations for different m, n

$$\xi_{11} \frac{d^2 \phi}{dz^2} = \xi_{12} \frac{d\chi}{dz} + \xi_{13} \phi + \xi_{14} \psi,$$

$$\xi_{21} \frac{d^2 \psi}{dz^2} = \xi_{22} \frac{d\chi}{dz} + \xi_{23} \phi + \xi_{24} \psi,$$

$$\xi_{31} \frac{d^2 \chi}{dz^2} = \xi_{32} \frac{d\phi}{dz} + \xi_{33} \frac{d\psi}{dz} + \xi_{34} \chi, \tag{1.161}$$

where ξ_{ij} are functions of m, n and of physical and geometrical parameters, and are defined as:

$$\xi_{11} = A_{1313} \lambda_1^2 \lambda_2^2, \quad \xi_{12} = -(A_{1133} + A_{1313}) \lambda_1 \lambda_2^2 \alpha_m,$$

$$\xi_{13} = A_{1111} \lambda_2^2 \alpha_m^2 + A_{1212} \lambda_1^2 \beta_n^2 - \omega_{mn}^2,$$

$$\xi_{14} = (A_{1122} + A_{1212}) \lambda_1 \lambda_2 \alpha_m \beta_n, \quad \xi_{21} = A_{2323} \lambda_1^2 \lambda_2^2,$$

$$\xi_{22} = -(A_{2233} + A_{2323}) \lambda_1^2 \lambda_2 \beta_n, \quad \xi_{23} = (A_{1122} + A_{1212}) \lambda_1 \lambda_2 \alpha_m \beta_n,$$

$$\xi_{24} = A_{2222} \lambda_1^2 \beta_n^2 + A_{1212} \lambda_2^2 \alpha_m^2 - \omega_{mn}^2, \quad \xi_{31} = A_{3333} \lambda_1^2 \lambda_2^2,$$

$$\xi_{32} = (A_{1313} + A_{1133}) \lambda_1 \lambda_2^2 \alpha_m, \quad \xi_{33} = (A_{2233} + A_{2323}) \lambda_1^2 \lambda_2 \beta_n,$$

$$\xi_{34} = A_{1313} \lambda_2^2 \alpha_m^2 + A_{2323} \lambda_1^2 \beta_n^2 - \omega_{mn}^2. \tag{1.162}$$

The solutions of (1.161) are sought in the form

$$\{\phi, \psi, \chi\} = \{A, B, C\} e^{\tau z}, \tag{1.163}$$

where τ is a parameter obtained from the equation:

1.4 Dynamical Characteristics of Plates with Added Masses

$$r_1\tau^6 + r_2\tau^4 + r_3\tau^2 + r_4 = 0. \tag{1.164}$$

With each of the six roots of (1.164) corresponds a particular solution denoted as η_i. Therefore, the general solution has the following form

$$[\eta] = \sum_{i=1}^{6} [\eta_i]. \tag{1.165}$$

A procedure for solving in terms of (1.165) posses difficulties because it is impossible to find the solutions of (1.164) for each orthotropic material. Therefore, at this stage we need to assume the existence of different combinations of roots (without the conjugated roots). Theoretically, the following roots combinations are possible: 1) simple; 2) one simple and a multiple pair 3) all equal. Now we consider the solution construction for each of the mentioned cases [21, 222].

1. Simple roots of (1.164). Substituting the solution (1.163) to the differential equation system (1.161), and dividing by an exponential multiplier the following algebraic equations are obtained:

$$LR = 0, \tag{1.166}$$

where

$$L = \begin{bmatrix} \xi_{13} - \xi_{11}\tau^2 & \xi_{14} & \xi_{12}\tau \\ \xi_{23} & \xi_{14} - \xi_{21}\tau^2 & \xi_{22}\tau \\ \xi_{32}\tau & \xi_{33}\tau & \xi_{34} - \xi_{31}\tau^2 \end{bmatrix}, \quad R = \begin{bmatrix} A \\ B \\ C \end{bmatrix}. \tag{1.167}$$

The determinant of coefficients multiplying the unknowns A, B, C serves to find the roots. The rank of coefficients matrix has to be equal to two and then there exists at least one minor of the second order. From the three linearly dependent equations we take those two, which possess the coefficients dependent on the minor. From the obtained two linear equations we define two constants as a function of the third one. For instance, if the second order minor constructed by the coefficients standing by A and B in the last two equations of the analysed system is different from zero, then the following vector of the fundamental solutions is defined

$$[\eta'] = C \begin{bmatrix} \dfrac{-\xi_{22}\xi_{33}\tau^2 - (\xi_{34} - \xi_{31}\tau^2)(\xi_{31}\tau^2 - \xi_{24})}{\xi_{23}\xi_{33}\tau + \xi_{32}\tau(\xi_{21}\tau^2 - \xi_{34})} \\ \dfrac{\xi_{23}(\xi_{31}\tau^2 - \xi_{34}) + \xi_{32}\xi_{22}\tau}{\xi_{23}\xi_{33}\tau + \xi_{32}\tau(\xi_{21}\tau^2 - \xi_{34})} \\ 1 \end{bmatrix} \cdot e^{\tau z}. \tag{1.168}$$

2. Multiple roots of (1.164). When $\tau = \tau^x$ is the multiple root then the corresponding solution (1.163) is substituted to equation (1.161). Transforming the algebraic system (1.166) with the unknowns A, B, C we get a rank which is equal or greater than 1. Suppose, that the rank of the

system coefficients matrix is equal to one. In that case, the analysed system is equivalent to an arbitrary taken one from two equations, in which at least one coefficient is different from zero. It leads to the conclusion that to define three unknown constants we have one relationship. Therefore, two constants can be chosen arbitrarily and the fundamental system possesses two vectors. For example, using the third equation of (1.167) we have

$$-A\xi_{32}\tau - B\xi_{33}\tau + C\left(\xi_{31}\tau^2 - \xi_{34}\right) = 0. \qquad (1.169)$$

Thus, the eigenvector may be defined by

$$[\eta'] = \begin{bmatrix} 1 & 0 \\ 0 & 1 \\ \frac{(\xi_{31}\tau^2 - \xi_{34})}{\xi_{22}\tau} & \frac{(\xi_{31}\tau^2 - \xi_{34})}{\xi_{34}\tau} \end{bmatrix} \begin{bmatrix} C_1 \\ C_2 \end{bmatrix} \cdot e^{\tau^x z}. \qquad (1.170)$$

Suppose that τ^x is doubled. Then the vector components have the form

$$[\eta'] = \begin{bmatrix} A_1 + B_1 z \\ A_2 + B_2 z \\ A_3 + B_3 z \end{bmatrix} \cdot e^{\tau^x} z. \qquad (1.171)$$

Substituting the relationship (1.171) to the differential equations and comparing the coefficients standing by the same degrees we get six linear equations. The rank of the obtained [6 × 6] matrix is equal to four and therefore we have four equations to define six unknown. Taking two unknowns arbitrary we get the fundamental solution system which consists of two vectors with six components.

To conclude, it is possible in the analysed case, to construct the solutions to the investigated differential equations (1.161).

3. Triple root. We take the solution (1.171) and substitute it to the system (1.161). Comparing the coefficients multiplied by the same 'z' powers, we get six linear equations. The matrix rank is greater or equal to three. Proceeding similarly to the above case, we can also construct a general solution. We do not focus on a detailed analysis here but we have to emphasise that this case is practically realised during the analysis of a three dimensional isotropic plate deflection substituted to the external surface load action [223].

The detailed calculations have shown, that for an isotropic plate material we have real and different roots of the characteristic equation (1.164). A matrix rank of the linear system (1.166) is equal to two. The minors constructed of coefficients standing by the unknowns in the last two algebraic equations are different from 0. Suppose that (1.164) has the roots τ_1, τ_2, τ_3. A general solution of (1.161) may be obtained as a linear combination of the particular solutions (1.168),

$$\phi(z) = A d_{11}^{(1)} ch\tau_1 z + B d_{11}^{(1)} sh\tau_1 z + C d_{11}^{(2)} ch\tau_2 z +$$

1.4 Dynamical Characteristics of Plates with Added Masses

$$Dd_{11}^{(2)} sh\tau_2 z + Ed_{11}^{(3)} ch\tau_3 z + Fd_{11}^{(3)} sh\tau_3 z,$$

$$\psi(z) = Ad_{12}^{(1)} ch\tau_1 z + Bd_{12}^{(1)} sh\tau_1 z + Cd_{12}^{(2)} ch\tau_2 z +$$

$$Dd_{12}^{(2)} sh\tau_2 z + Ed_{12}^{(3)} ch\tau_3 z + Fd_{12}^{(3)} sh\tau_3 z,$$

$$\chi(z) = Ad_{13}^{(1)} sh\tau_1 z + Bd_{13}^{(1)} ch\tau_1 z + Cd_{13}^{(2)} sh\tau_2 z +$$

$$Dd_{13}^{(2)} ch\tau_2 z + Ed_{13}^{(3)} sh\tau_3 z + Fd_{13}^{(3)} ch\tau_3 z, \qquad (1.172)$$

where

$$d_{11}^{(i)} = \left(\xi_{34} + \tau_i^2 \xi_{21}\right)\left(\xi_{34} + \xi_{31}\tau_i^2\right) + \xi_{22}\xi_{33}\tau_i^2,$$

$$d_{12}^{(i)} = -\xi_{32}\xi_{22}\tau_i^2 - \xi_{23}\left(\xi_{34} - \xi_{31}\tau_i^2\right),$$

$$d_{13}^{(i)} = \tau_i \left[\xi_{23}\xi_{33} - \xi_{32}\left(\xi_{24} + \tau_i^2 \xi_{21}\right)\right], \quad i = \overline{1,3}. \qquad (1.173)$$

In the free vibration case, the external bounded conditions are formulated using (1.65). They have the following form (because of ϕ, ψ, χ and for $z = -0.5$, and 0.5):

$$-A_{1133}\tilde{\alpha}_m \phi - A_{2233}\tilde{\beta}_n \psi + A_{3333}\chi' = 0,$$

$$\psi' + \tilde{\beta}_n \chi = 0, \quad \phi' + \tilde{\alpha}_m \chi = 0,$$

$$\tilde{\alpha}_m = \alpha_m/\lambda_1, \tilde{\beta}_n = \beta_n/\lambda_2. \qquad (1.174)$$

Substituting the expressions for ϕ, ψ, χ from (1.172) to (1.174) we get the following system of the homogeneous algebraic equations for each pair of (m,n):

$$\begin{bmatrix} [L(0.5)] \\ [L(-0.5)] \end{bmatrix} \begin{bmatrix} A \\ B \\ C \\ D \\ E \\ F \end{bmatrix} = \begin{bmatrix} 0 \\ 0 \\ 0 \\ 0 \\ 0 \\ 0 \end{bmatrix}, \qquad (1.175)$$

where:

$$L(z) = \begin{bmatrix} e_1(\tau_1)ch\tau_1 z & e_1(\tau_1)sh\tau_1 z & \ldots & e_1(\tau_3)ch\tau_3 z & e_1(\tau_3)sh\tau_3 z \\ e_2(\tau_1)sh\tau_1 z & e_2(\tau_1)ch\tau_1 z & \ldots & e_2(\tau_3)sh\tau_3 z & e_2(\tau_3)ch\tau_3 z \\ e_3(\tau_1)sh\tau_1 z & e_3(\tau_1)ch\tau_1 z & \ldots & e_3(\tau_3)sh\tau_3 z & e_3(\tau_3)ch\tau_3 z \end{bmatrix},$$

$$e_1(\tau_i) = A_{1133}\tilde{\alpha}_m d_{11}^{(i)}(\tau_i) + A_{2233}\tilde{\beta}_n d_{12}^{(i)}(\tau_i) + A_{3333}d_{13}^{(i)}(\tau_i)\tau_i,$$

$$e_2(\tau_i) = d_{11}^{(i)}(\tau_i)\tau_i + d_{13}^{(i)}(\tau_i)\tilde{\alpha}_m,$$

$$d_{11}^{(i)} = \left(\xi_{24} - \tau_i^2 \xi_{21}\right)\left(\xi_{34} - \xi_{31}\tau_i^2\right) - \xi_{22}\xi_{33}\tau_i^2,$$

$$d_{12}^{(i)} = \xi_{32}\xi_{22}\tau_i^2 - \xi_{23}\left(\xi_{34} - \xi_{31}\tau_i^2\right),$$

$$d_{13}^{(i)} = \tau_i \left[\xi_{23}\xi_{33} - \xi_{32} \left(\xi_{24} - \tau_i^2 \xi_{21} \right) \right],$$

$$e_3(\tau_i) = d_{11}^{(i)}(\tau_i)\tau_i + d_{13}^{(i)}(\tau_i)\tilde{\alpha}_m, \qquad (1.176)$$

$$(i = \overline{1,3}).$$

In order to get the nontrivial solutions we have

$$\det \begin{bmatrix} [L(0.5)] \\ [L(-0.5)] \end{bmatrix} = 0. \qquad (1.177)$$

As a result, we derive a transcendental equation whose roots correspond to the transversally isotropic plate vibration frequencies. For each pair (m,n), we get an infinite set of eigenvalues.

The free vibrations analysis of a cubicoid made from the isotropic material is of particular importance. In that case the general considerations are simplified and lead to relatively simple characteristic equations. It is necessary to investigate the particular solutions (modes) of isotropic plates vibrations. They are needed for both free and forced vibrations as well as for investigations of nonstationary processes. For that reason, these solutions (modes) have been considered by many researchers and have been used mainly for the analysis of nonstationary processes. In the case of a harmonically vibrating isotropic plate with frequency ω the fundamental solutions (normal modes) are governed by the following equations:

$$\frac{1}{\lambda_1^2}\frac{\partial^2 u}{\partial x^2} + \frac{1}{\lambda_2^2}\frac{\partial^2 u}{\partial y^2} + \frac{\partial^2 u}{\partial z^2} + \frac{1}{(1-2\nu)}\left(\frac{1}{\lambda_1^2}\frac{\partial^2 u}{\partial x^2} + \frac{1}{\lambda_1\lambda_2}\frac{\partial^2 v}{\partial x \partial y} + \right.$$

$$\left. \frac{1}{\lambda_1}\frac{\partial^2 w}{\partial x \partial z} \right) + u\omega^2 = 0, \quad (\overleftrightarrow{1,2}), \quad (\overleftrightarrow{x,y}),$$

$$\frac{1}{\lambda_1^2}\frac{\partial^2 w}{\partial x^2} + \frac{1}{\lambda_2^2}\frac{\partial^2 w}{\partial y^2} + \frac{\partial^2 w}{\partial z^2} +$$

$$\frac{1}{(1-2\nu)}\left(\frac{1}{\lambda_1}\frac{\partial^2 u}{\partial x \partial z} + \frac{1}{\lambda_2}\frac{\partial^2 v}{\partial z \partial y} + \frac{\partial^2 w}{\partial z^2} \right) + w\omega^2 = 0, \qquad (1.178)$$

from which the following characteristic equation is obtained

$$\left\{ 8g^2 rs \left(r^2 + g^2\right)^2 (1 - chrchs) + \left[16g^4 r^2 s^2 \left(r^2 + g^2\right)^4 \right] shrshs \right\} shr = 0, \qquad (1.179)$$

$$g = \sqrt{\tilde{\alpha}_m^2 + \tilde{\beta}_n^2}, \quad r = \sqrt{g^2 - \omega^2}, \quad s = \sqrt{g^2 - (1-2\nu)\omega^2/(2-2\nu)}.$$

The solution of that transcendental equation consists of an infinite set of eigenvalues each corresponding to a different pair (m,n). In reference [222] the authors conclude, that equation (1.177) include eigenvalues corresponding to symmetric and antisymmetric plate vibrations in relation to a mean surface (in reference [168] it has been shown, that such a distribution is always possible).

Another challenging method has been presented in reference [257], where after introducing potentials, the solution is computed for two classes of motion, and their particular solutions correspond to flat and antiflat vibrations.

1.4.3 Three Dimensional Problem

In this section, the orthotropic plate vibrations with the added masses M^i ($i = \overline{1, N}$) will be analysed. It should be emphasized, that using the three dimensional theory for an orthotropic plate in the general case, we do not know a priori the behaviour of the roots of (1.164), because an unknown frequency is included there. In addition, the coefficients are defined by complex expressions dependent on the physical and geometrical parameters, and even an analytical description for the roots does not simplify the problem. The classification of possible types of roots of the characteristic equation as well as the method of their corresponding solution should be outlined. We assume, following section 1.4.2, that physical and geometrical parameters of the plate allow for three simple roots of (1.164), and the corresponding matrix of the linear equations (1.166) is of rank equal to 2.

The following solution, to equations (1.59), is being sought (we take also $k_1 = k_2 = 0$ in (1.59))

$$u = \sin \omega t \sum_{m,n}^{\infty} \phi(z) \cos \alpha_m x \sin \beta_n y,$$

$$v = \sin \omega t \sum_{m,n}^{\infty} \psi(z) \sin \alpha_m x \cos \beta_n y,$$

$$w = \sin \omega t \sum_{m,n}^{\infty} \chi(z) \sin \alpha_m x \sin \beta_n y, \qquad (1.180)$$

where ϕ, ψ and χ are defined by (1.172).

In order to compute $A_{mn}, B_{mn}, C_{mn}, D_{mn}, E_{mn}, F_{mn}$ we substitute (1.180) to the boundary conditions on the surfaces $z = -0.5; 0.5$, which for $z = 0.5$ have the following nondimensional form

$$A_{1313}\left(\frac{\partial U}{\partial z} + \frac{1}{\lambda_1}\frac{\partial W}{\partial x}\right) = \omega^2 \sum_{i=1}^{N} M^i \left[U - \frac{1}{\lambda_1}(z_i - 0.5)\frac{\partial W}{\partial x}\right]\delta(x - x_i) \times$$

$$\delta(y - y_i), \quad (\overrightarrow{1,2}), \quad (\overleftrightarrow{x,y}),$$

$$A_{1133}\frac{1}{\lambda_1}\frac{\partial U}{\partial x} + A_{2233}\frac{1}{\lambda_2}\frac{\partial V}{\partial y} + A_{3333}\frac{\partial W}{\partial z} = \omega^2 \sum_{i=1}^{N}\left\{M^i\left[W - \frac{(z_i - 0.5)^2}{\lambda_1^2}\times\right.\right.$$

$$\frac{\partial}{\partial x}\left[\frac{\partial W}{\partial x}\delta(x-x_i)\right]\delta(y-y_i) + \frac{(z_i-0.5)}{\lambda_1}\frac{\partial}{\partial x}\left[U\delta(x-x_i)\right]\delta(y-y_i) -$$

$$\frac{(z_i-0.5)^2}{\lambda_2^2}\frac{\partial}{\partial y}\left[\frac{\partial W}{\partial y}\delta(y-y_i)\right]\delta(x-x_i) + \frac{(z_i-0.5)}{\lambda_2} \times$$

$$\frac{\partial}{\partial y}\left[V\delta(y-y_i)\right]\delta(x-x_i)\Big\} - \sum_{i=1}^{N}\left\{J_{xx}^i\frac{\partial}{\partial y}\left[\frac{\partial W}{\partial y}\delta(y-y_i)\right]\delta(x-x_i) -\right.$$

$$\left. -J_{yy}^i\frac{\partial}{\partial x}\left[\frac{\partial W}{\partial x}\delta(x-x_i)\right]\delta(y-y_i)\right\}, \qquad (1.181)$$

and for $z = -0.5$

$$\frac{\partial U}{\partial z} + \frac{1}{\lambda_1}\frac{\partial W}{\partial x} = 0, \quad (\overleftrightarrow{1,2}), \quad (\overleftrightarrow{x,y}),$$

$$A_{1133}\frac{1}{\lambda_1}\frac{\partial U}{\partial x} + A_{2233}\frac{1}{\lambda_2}\frac{\partial V}{\partial y} + A_{3333}\frac{\partial W}{\partial z} = 0. \qquad (1.182)$$

In the above equations U, V, W are the amplitudes of the displacements

$$u = U\sin\omega t, \quad v = V\sin\omega t, \quad w = W\sin\omega t$$

Substituting (1.180) to (1.181) and (1.182) and developing the right-hand sides of (1.181), consisting of the singular coefficients of the δ-type into the trigonometric series we get the following linear algebraic equations:

$$LA = Q, \qquad (1.183)$$

where

$$L = \begin{bmatrix} e_{11} & e_{12} & \dots & e_{16} \\ e_{21} & e_{22} & \dots & e_{26} \\ \dots & \dots & \dots & \dots \\ e_{61} & e_{62} & \dots & e_{66} \end{bmatrix}, \quad A = \begin{bmatrix} A_{mn} \\ B_{mn} \\ C_{mn} \\ D_{mn} \\ E_{mn} \\ F_{mn} \end{bmatrix}, \quad Q = \begin{bmatrix} \xi_1 \\ \xi_2 \\ \xi_3 \\ 0 \\ 0 \\ 0 \end{bmatrix}.$$

The L matrix coefficients and the Q rows have the following form

$$e_{1k} = p_j\varphi_k^+, \quad e_{1m} = p_j\varphi_m^+, \quad e_{2k} = r_j\varphi_k^+, \quad e_{2m} = r_j\varphi_m^+,$$

$$e_{3k} = s_j\varphi_m^+, \quad e_{3m} = s_j\varphi_k^+, \quad e_{4k} = p_j\varphi_k^-, \quad e_{4m} = p_j\varphi_m^-,$$

$$e_{5k} = r_j\varphi_k^-, \quad e_{5m} = r_j\varphi_m^-, \quad e_{6k} = s_j\varphi_m^-, \quad e_{6m} = s_j\varphi_k^-,$$

$$\varphi_k^\pm = sh\left(\pm 0.5\tau_j\right), \quad \varphi_m^\pm = ch\left(\pm 0.5\tau_j\right),$$

$$p_j = A_{1313}\left(\tau_j d_{11}^{(j)} + \tilde{\alpha}_m d_{13}^{(j)}\right), \quad r_j = A_{2323}\left(\tau_j d_{12}^{(j)} + \tilde{\beta}_n d_{13}^{(j)}\right),$$

$$s_j = -A_{1133}\tilde{\alpha}_m d_{11}^{(j)} - A_{2233}\tilde{\beta}_n d_{12}^{(j)} + A_{3333}\tau_j d_{13}^{(j)},$$

1.4 Dynamical Characteristics of Plates with Added Masses

$$\tilde{\alpha}_m = \alpha_m/\lambda_1, \quad \tilde{\beta}_n = \beta_n/\lambda_2, \quad (k = 2j-1, \; m = 2j, \; j = \overline{1,3}),$$

$$\xi_1 = \omega^2 \sum_{i=1}^{N} \left(g_{11}^i U_i + g_{12}^i \frac{\partial W_i}{\partial x} \right), \quad \xi_2 = \omega^2 \sum_{i=1}^{N} \left(g_{21}^i V_i + g_{22}^i \frac{\partial W_i}{\partial y} \right),$$

$$\xi_3 = \omega^2 \sum_{i=1}^{N} \left(g_{31}^i U_i + g_{32}^i V_i + g_{33}^i W_i + g_{34}^i \frac{\partial W_i}{\partial x} + g_{35}^i \frac{\partial W_i}{\partial y} \right). \quad (1.184)$$

The parameters g_{ij}^k are defined below

$$g_{11}^i = 4M^i \cos \alpha_m x_i \sin \beta_n y_i,$$

$$g_{12}^i = -\frac{4M^i}{2\lambda_1}(z_i - 0.5) \cos \alpha_m x_i \sin \beta_n y_i,$$

$$g_{21}^i = 4M^i \sin \alpha_m x_i \cos \beta_n y_i,$$

$$g_{22}^i = -\frac{4M^i}{2\lambda_2}(z_i - 0.5) \sin \alpha_m x_i \cos \beta_n y_i,$$

$$g_{31}^i = -\frac{4M^i}{2\lambda_1}(z_i - 0.5)\alpha_m \cos \alpha_m x_i \sin \beta_n y_i,$$

$$g_{32}^i = -\frac{4M^i}{2\lambda_2}(z_i - 0.5)\beta_n \sin \alpha_m x_i \cos \beta_n y_i,$$

$$g_{33}^i = 4M^i \sin \alpha_m x_i \sin \beta_n y_i,$$

$$g_{34}^i = \frac{4M^i}{\lambda_1^2}\alpha_m \cos \alpha_m x_i \sin \beta_n y_i + 4J_{xx}^i \alpha_m \cos \alpha_m x_i \sin \beta_n y_i,$$

$$g_{35}^i = \frac{4M^i}{\lambda_2^2}\beta_n \sin \alpha_m x_i \cos \beta_n y_i + 4J_{xx}^i \beta_n \sin \alpha_m x_i \cos \beta_n y_i. \quad (1.185)$$

$$(i = \overline{1,N}).$$

Finding the solutions of (1.183) we define $A_{mn}, B_{mn}, C_{mn}, D_{mn}, E_{mn}, F_{mn}$, which implies the following definition of the displacement components vector

$$U = \omega^2 \sum_{m,n}^{\infty} \sum_{i=1}^{N} \frac{1}{\Delta(m,n)} \left(q_{11}^i U_i + q_{12}^i V_i + q_{13}^i W_i + q_{14}^i \frac{\partial W_i}{\partial x} + q_{15}^i \frac{\partial W_i}{\partial y} \right) \times$$

$$\cos \alpha_m x \sin \beta_n y,$$

$$V = \omega^2 \sum_{m,n}^{\infty} \sum_{i=1}^{N} \frac{1}{\Delta(m,n)} \left(q_{21}^i U_i + q_{22}^i V_i + q_{23}^i W_i + q_{24}^i \frac{\partial W_i}{\partial x} + q_{25}^i \frac{\partial W_i}{\partial y} \right) \times$$

$$\sin \alpha_m x \cos \beta_n y,$$

$$W = \omega^2 \sum_{m,n}^{\infty} \sum_{i=1}^{N} \frac{1}{\Delta(m,n)} \left(q_{31}^i U_i + q_{32}^i V_i + q_{33}^i W_i + q_{34}^i \frac{\partial W_i}{\partial x} + q_{35}^i \frac{\partial W_i}{\partial y} \right) \times$$

$$\sin \alpha_m x \sin \beta_n y. \qquad (1.186)$$

The first plate's deflection derivatives are as follows

$$\frac{\partial W}{\partial x} = \omega^2 \sum_{m,n}^{\infty} \sum_{i=1}^{N} \frac{1}{\Delta(m,n)} \left(q_{31}^i U_i + q_{32}^i V_i + q_{33}^i W_i + q_{34}^i \frac{\partial W_i}{\partial x} + \right.$$

$$\left. q_{35}^i \frac{\partial W_i}{\partial y} \right) \alpha_m \cos \alpha_m x \sin \beta_n y,$$

$$\frac{\partial W}{\partial y} = \omega^2 \sum_{m,n}^{\infty} \sum_{i=1}^{N} \frac{1}{\Delta(m,n)} \left(q_{31}^i U_i + q_{32}^i V_i + q_{33}^i W_i + q_{34}^i \frac{\partial W_i}{\partial x} + \right.$$

$$\left. q_{35}^i \frac{\partial W_i}{\partial y} \right) \beta_n \sin \alpha_m x \cos \beta_n y, \qquad (1.187)$$

where

$$q_{jk}^i = p_{1k}^i d_{1j}^{(1)} ch 0.5\tau_1 + p_{2k}^i d_{1j}^{(1)} sh 0.5\tau_1 + \ldots + p_{5k}^i d_{1j}^{(3)} ch 0.5\tau_3 + p_{6k}^i d_{1j}^{(3)} sh 0.5\tau_3,$$

$$(j = 1, 2; \quad k = \overline{1, 5}),$$

$$q_{3k}^i = p_{1k}^i d_{13}^{(1)} sh 0.5\tau_1 + p_{2k}^i d_{13}^{(1)} ch 0.5\tau_1 + \ldots + p_{5k}^i d_{13}^{(3)} sh 0.5\tau_3 + p_{6k}^i d_{13}^{(3)} ch 0.5\tau_3,$$

$$p_{j1}^i = g_{11}^i \Delta_{1j} + g_{31}^i \Delta_{3j}, \quad p_{j2}^i = g_{21}^i \Delta_{2j} + g_{33}^i \Delta_{3j},$$

$$p_{j3}^i = \Delta_{3j} g_{33}^i, \quad p_{j4}^i = g_{12}^i \Delta_{1j} + g_{31}^i \Delta_{3j},$$

$$p_{j5}^i = g_{22}^i \Delta_{2j} + g_{35}^i \Delta_{3j}, \quad (j = \overline{1, 6}; \quad i = \overline{1, N}). \qquad (1.188)$$

Δ is the determinant of (1.183), whereas Δ_{ij} is the co-factor of the term located at the cross point of the i-th row and the j-th column.

After the fulfilment of the continuous conditions of the jointed masses, we get the algebraic linear equations with the unknowns being the displacement vector components and the first deflection derivatives at the points of jointed masses

$$U_i, V_i, W_i, \frac{\partial W_i}{\partial x}, \frac{\partial W_i}{\partial y}. \qquad (1.189)$$

A nontrivial solution exists if

$$\det [a_{ij}]_{5N \times 5N} = 0. \qquad (1.190)$$

The determinant elements are given in Table 1.3. The solution of (1.190) does not exist in closed form similarly to the equations (1.131) and (1.142). It is a transcendental equation, because the sought frequency is in the argument of a hyperbolic function. limiting the considerations to the first term we get an approximate equation defining the frequency of the vibrational "plate - mass" system. This equation has infinitely many solutions which are the approximate frequencies of the free vibrations. Taking into account in (1.190) two terms we get more accurate results of the frequency determination and we define a new frequency series corresponding to the second vibration mode, and so on. limiting ourselves in (1.190) to 's' terms, we can achieve the required accuracy in a frequency determination.

1.4.4 Algorithms

Here we consider a certain aspect of the problem related to the numerical solutions to the characteristic equations (1.177) and (1.179) corresponding to the free unloaded isotropic plates vibrations.

In reference [168] it has been shown, that the solutions set of (1.179) can be divided into two groups. The corresponding vibration modes are related to the symmetric and antisymmetric state of stress.

The antisymmetric vibration modes correspond to the roots of the following equations

$$-4sh\frac{r}{2}ch\frac{s}{2}prs + sh\frac{s}{2}ch\frac{r}{2}\left(r^2 + p\right)^2 = 0, \qquad (1.191)$$

$$ch\frac{r}{2} = 0, \qquad (1.192)$$

where

$$p = \tilde{\alpha}_m^2 + \tilde{\beta}_n^2, \quad r = \sqrt{p - \omega^2}, \qquad (1.193)$$

and ω is the sought frequency.

Either symmetric or flat vibration modes correspond to the second group of the solutions (1.177). They are represented by the following equations

$$-4sh\frac{s}{2}ch\frac{r}{2}prs + sh\frac{r}{2}ch\frac{s}{2}\left(r^2 + p\right)^2 = 0, \qquad (1.194)$$

$$sh\frac{r}{2} = 0. \qquad (1.195)$$

Therefore, the procedure of determining the frequencies of free vibration is simplified and instead of using equations (1.177) they are computed from (1.191)–(1.195).

Equations (1.192) and (1.195) have the explicit solutions

$$\gamma = p + (\pi + 2\pi n)^2,$$

$$\gamma = p + 4\pi^2 n^2. \qquad (1.196)$$

It should be mentioned that the procedure of finding the roots of transcendental equations (1.191) and (1.194) is difficult without any additional information. In the latter case, we can sometimes find good enough first approximation for a sought root.

Table 1.3. Terms of determinant (1.190).

$\omega^2\sum\limits_{m,n}^{\infty}\dfrac{q_{11}^1\lambda_1^1}{\Delta(m,n)}-1$	$\omega^2\sum\limits_{m,n}^{\infty}\dfrac{q_{12}^1\lambda_1^1}{\Delta(m,n)}$	$\omega^2\sum\limits_{m,n}^{\infty}\dfrac{q_{15}^1\lambda_1^1}{\Delta(m,n)}$	$\omega^2\sum\limits_{m,n}^{\infty}\dfrac{q_{11}^2\lambda_1^1}{\Delta(m,n)}$	$\omega^2\sum\limits_{m,n}^{\infty}\dfrac{q_{15}^2\lambda_1^1}{\Delta(m,n)}$	$\omega^2\sum\limits_{m,n}^{\infty}\dfrac{q_{11}^N\lambda_1^1}{\Delta(m,n)}$	$\omega^2\sum\limits_{m,n}^{\infty}\dfrac{q_{11}^N\lambda_1^1}{\Delta(m,n)}$	$\omega^2\sum\limits_{m,n}^{\infty}\dfrac{q_{11}^N\lambda_1^1}{\Delta(m,n)}$
$\omega^2\sum\limits_{m,n}^{\infty}\dfrac{q_{21}^1\lambda_1^2}{\Delta(m,n)}$	$\omega^2\sum\limits_{m,n}^{\infty}\dfrac{q_{22}^1\lambda_1^2}{\Delta(m,n)}-1$	$\omega^2\sum\limits_{m,n}^{\infty}\dfrac{q_{25}^1\lambda_1^2}{\Delta(m,n)}$	$\omega^2\sum\limits_{m,n}^{\infty}\dfrac{q_{21}^2\lambda_1^2}{\Delta(m,n)}$	$\omega^2\sum\limits_{m,n}^{\infty}\dfrac{q_{25}^2\lambda_1^2}{\Delta(m,n)}$	$\omega^2\sum\limits_{m,n}^{\infty}\dfrac{q_{21}^N\lambda_1^2}{\Delta(m,n)}$	$\omega^2\sum\limits_{m,n}^{\infty}\dfrac{q_{21}^N\lambda_1^2}{\Delta(m,n)}$	$\omega^2\sum\limits_{m,n}^{\infty}\dfrac{q_{25}^N\lambda_1^2}{\Delta(m,n)}$
$\omega^2\sum\limits_{m,n}^{\infty}\dfrac{q_{31}^1\lambda_1^5}{\Delta(m,n)}$	$\omega^2\sum\limits_{m,n}^{\infty}\dfrac{q_{32}^1\lambda_1^5}{\Delta(m,n)}$	$\omega^2\sum\limits_{m,n}^{\infty}\dfrac{q_{35}^1\lambda_1^5}{\Delta(m,n)}-1$	$\omega^2\sum\limits_{m,n}^{\infty}\dfrac{q_{31}^2\lambda_1^5}{\Delta(m,n)}$	$\omega^2\sum\limits_{m,n}^{\infty}\dfrac{q_{35}^N\lambda_1^5}{\Delta(m,n)}$	$\omega^2\sum\limits_{m,n}^{\infty}\dfrac{q_{35}^N\lambda_1^5}{\Delta(m,n)}$	$\omega^2\sum\limits_{m,n}^{\infty}\dfrac{q_{35}^N\lambda_1^5}{\Delta(m,n)}$	$\omega^2\sum\limits_{m,n}^{\infty}\dfrac{q_{35}^N\lambda_1^5}{\Delta(m,n)}$
$\omega^2\sum\limits_{m,n}^{\infty}\dfrac{q_{11}^1\lambda_2^1}{\Delta(m,n)}$	$\omega^2\sum\limits_{m,n}^{\infty}\dfrac{q_{12}^1\lambda_2^1}{\Delta(m,n)}$	$\omega^2\sum\limits_{m,n}^{\infty}\dfrac{q_{15}^1\lambda_2^1}{\Delta(m,n)}$	$\omega^2\sum\limits_{m,n}^{\infty}\dfrac{q_{11}^2\lambda_2^1}{\Delta(m,n)}-1$	$\omega^2\sum\limits_{m,n}^{\infty}\dfrac{q_{15}^2\lambda_2^1}{\Delta(m,n)}$	$\omega^2\sum\limits_{m,n}^{\infty}\dfrac{q_{11}^N\lambda_2^1}{\Delta(m,n)}$	$\omega^2\sum\limits_{m,n}^{\infty}\dfrac{q_{11}^N\lambda_2^1}{\Delta(m,n)}$	$\omega^2\sum\limits_{m,n}^{\infty}\dfrac{q_{11}^N\lambda_2^1}{\Delta(m,n)}$
$\omega^2\sum\limits_{m,n}^{\infty}\dfrac{q_{31}^1\lambda_2^5}{\Delta(m,n)}$	$\omega^2\sum\limits_{m,n}^{\infty}\dfrac{q_{32}^1\lambda_2^5}{\Delta(m,n)}$	$\omega^2\sum\limits_{m,n}^{\infty}\dfrac{q_{35}^1\lambda_2^5}{\Delta(m,n)}$	$\omega^2\sum\limits_{m,n}^{\infty}\dfrac{q_{31}^2\lambda_2^5}{\Delta(m,n)}$	$\omega^2\sum\limits_{m,n}^{\infty}\dfrac{q_{35}^2\lambda_2^5}{\Delta(m,n)}-1$	$\omega^2\sum\limits_{m,n}^{\infty}\dfrac{q_{31}^N\lambda_2^5}{\Delta(m,n)}$	$\omega^2\sum\limits_{m,n}^{\infty}\dfrac{q_{31}^N\lambda_2^5}{\Delta(m,n)}$	$\omega^2\sum\limits_{m,n}^{\infty}\dfrac{q_{31}^N\lambda_2^5}{\Delta(m,n)}$
$\omega^2\sum\limits_{m,n}^{\infty}\dfrac{q_{11}^1\lambda_N^1}{\Delta(m,n)}$	$\omega^2\sum\limits_{m,n}^{\infty}\dfrac{q_{12}^1\lambda_N^1}{\Delta(m,n)}$	$\omega^2\sum\limits_{m,n}^{\infty}\dfrac{q_{15}^1\lambda_N^1}{\Delta(m,n)}$	$\omega^2\sum\limits_{m,n}^{\infty}\dfrac{q_{11}^2\lambda_N^1}{\Delta(m,n)}$	$\omega^2\sum\limits_{m,n}^{\infty}\dfrac{q_{15}^2\lambda_N^1}{\Delta(m,n)}$	$\omega^2\sum\limits_{m,n}^{\infty}\dfrac{q_{11}^N\lambda_N^1}{\Delta(m,n)}-1$	$\omega^2\sum\limits_{m,n}^{\infty}\dfrac{q_{11}^N\lambda_N^1}{\Delta(m,n)}$	$\omega^2\sum\limits_{m,n}^{\infty}\dfrac{q_{11}^N\lambda_N^1}{\Delta(m,n)}$
$\omega^2\sum\limits_{m,n}^{\infty}\dfrac{q_{31}^1\lambda_N^1}{\Delta(m,n)}$	$\omega^2\sum\limits_{m,n}^{\infty}\dfrac{q_{32}^1\lambda_N^1}{\Delta(m,n)}$	$\omega^2\sum\limits_{m,n}^{\infty}\dfrac{q_{35}^1\lambda_N^1}{\Delta(m,n)}$	$\omega^2\sum\limits_{m,n}^{\infty}\dfrac{q_{31}^2\lambda_N^1}{\Delta(m,n)}$	$\omega^2\sum\limits_{m,n}^{\infty}\dfrac{q_{35}^2\lambda_N^1}{\Delta(m,n)}$	$\omega^2\sum\limits_{m,n}^{\infty}\dfrac{q_{31}^N\lambda_N^1}{\Delta(m,n)}$	$\omega^2\sum\limits_{m,n}^{\infty}\dfrac{q_{31}^N\lambda_N^1}{\Delta(m,n)}$	$\omega^2\sum\limits_{m,n}^{\infty}\dfrac{p_{31}^N\lambda_N^1}{\Delta(m,n)}-1$

$\lambda_i^1 = \cos\alpha_m x_i \sin\beta_n y_i, \quad \lambda_i^2 = \sin\alpha_m x_i \cos\beta_n y_i, \quad \lambda_i^3 = \sin\alpha_m x_i \sin\beta_n y_i,$

$\lambda_i^4 = \alpha_m \lambda_i^1, \quad \lambda_i^5 = \beta_n \lambda_i^2, \quad i = \overline{1,N}.$

1.4 Dynamical Characteristics of Plates with Added Masses

The solutions of equations (1.191) and (1.194) are sought in the following form

$$\gamma = \sum_{n=0}^{\infty} a_n p^n.$$

It can be proved (see reference [257]) that there exist approximate relationships for the unknown frequencies (with the accuracy related to small 'p'):

1. $$\gamma = p\left[p/6(1-\nu) + o\left(p^2\right)\right],$$

2. $$\gamma = p,$$

3. $$\gamma = p\left[2/(1-\nu) + o\left(p\right)\right],$$

4. $$\gamma = p\left[(2n-1)^2 \pi^2/p + 1\right],$$

5. $$\gamma = p\left[(2n-1)^2 \pi^2/p + k_n\right],$$

$$k_n = 1 + [(1-2\nu)/2(1-\nu)]^{\frac{1}{2}} 16ctg(s/2)/(2n-1)\pi + o(p),$$

6. $$\gamma = p\left[4n^2\pi^2/p + 1\right],$$

7. $$\gamma = 2p(1-\nu)/(1-2\nu)\left[(2n-1)^2 \pi^2/p + k_n\right],$$

$$k_n = 1 + [(1-2\nu)/2(1-\nu)]^{\frac{1}{2}} 16ctg(r/2)/(2n-1)\pi + o(p), \quad (1.197)$$

8. $$\gamma = p\left[4n^2\pi^2/p + k_n\right],$$

$$k_n = 1 - [(1-2\nu)/2(1-\nu)]^{\frac{1}{2}} 8tg(s/2)/n\pi + o(p),$$

9. $$\gamma = p\left[4n^2\pi^2/p + k_n\right] 2(1-\nu)/(1-2\nu),$$

$$k_n = 1 - [(1-2\nu)/2(1-\nu)]^{\frac{3}{2}} 8tg(r/2)/n\pi,$$

and so on. The above expressions are ordered in a sequence of increasing frequencies.

The importance of the obtained decompositions is mainly expressed during the asymptotical estimation of the frequencies delivered by different applied theories.

From a practical point of view, those relationships may be used for small p (for relatively thin plates and low vibration modes) as the approximate expressions for the free vibration isotropic plate frequencies determination. If higher accuracy is needed, then, the obtained values may serve for accuracy improvement using, for example, Newton's method. However, for large values, Newton's method may be divergent and the other numerical methods are recommended instead.

72   1 Vibration of Plates and Shells with Added Masses

The relationships analysis of (1.197) shows that equations (1.191) and (1.194) have only isolated roots. It means that for each root there exists a neighbourhood that does not include other roots. Therefore, the first step of roots determination consists in their isolation by finding intervals consisting of only one root of either (1.191) or (1.194). The numerical investigation has shown that it we take p = 0 in the relationships (1.197) then the obtained set $\{\bar{\omega}_i(0)\}_{i=1}^k$ defines the boundaries of the sought intervals with the isolated roots. To conclude, the problem of finding isolated roots has been solved. The second step includes the direct calculation of the roots.

The subroutine realizing the algorithm described in subchapter 1.3.5 is a part of a program for vibration analysis of the isotropic plates with the added masses. In subsection 1.4.2, the algorithm of finding the analytical solutions leading to the characteristic equation (1.177) has been described. A key part of that investigation lies in the analysis of the equation (1.164).

The free vibration frequency determination of the transversal-isotropic plates consists of the following steps.

1. First, the free vibration frequency of plates and shells are calculated with the geometrical parameters, using equations (1.192) and (1.195).
2. The interval $(\tilde{G}', \tilde{G})$, is devided into k subintervals $\left\{\tilde{G}_i\right\}_{i=1}^k$. $\tilde{G}$ is the relative stiffness parameter $\tilde{G} = G/E$ of an isotropic plate, and $\tilde{G}'$ is the relative stiffness parameter of the transversal isotropic plate.
3. On each step of Newton's method for a fixed $\tilde{G}_i$ the free vibration transversal isotropic plate frequencies are found. In the beginning, the frequencies for free vibration of an isotropic plate serve as the approximate values and then at each next step the $\tilde{G}$ values from the previous step are taken.
4. After computing the values for whcih $\tilde{G} = \tilde{G}'$ the procedure is finished.

It should be emphasized that in spite of the simplicity and clarity of the described procedure, its practical realisation meets many problems. Among others, they are related to the optimal choice of the optimal $(\tilde{G}', \tilde{G})$ partition, the analysis of the obtained results and the investigation of a convergence rate.

Numerical experiments have shown that the algorithm works stably if the interval $[0.01; 0.384]$ is divided into more than 20 parts. In order to ensure the required calculation accuracy ($\varepsilon = 10^{-6}$) 10–12 iterations are needed.

To summarise, it can be seen that the problem analytically formulated by Lamé in the previous century, can be solved through numerical techniques.

Finally, the presented program of numerical calculations may be successfully used with other materials whose physical and geometrical properties lead to equations similar to (1.177).

1.4.5 Frequency Spectra Comparison of Three Dimensional and Approximate Theories

The three dimensional theory is used to solve special problems in order to estimate the validity intervals of two dimensional theories approximate. This estimation may be made only when the results of two and three dimensional theories are compared.

In this chapter various theories yielding frequencies isotropic and transversal-isotropic plates in comparison to solutions obtained using the 3D theory are addressed. The comparison between two and three dimensional theories, is carried out on the basis of the following considerations. To each frequency from a spectrum corresponds a vibration mode. The mode analysis possesses additional advantage. The modes may be considered as the characteristics used for the comparison between exact and approximate theories. However, three dimensional body vibration modes (infinitely many degrees of freedom in the normal direction) and two dimensional body modes (finite degrees of freedom) differ essentially. In reference [257] it is recommended to compare drawings obtained using the two dimensional theory with the ones obtained when the three dimensional theory is used. In [258] it has been shown that for the described technique the parameters of two- and three dimensional theories are related as follows

$$2h(\bar{u} + \bar{w}) = \int_{-h}^{h} \bar{u}^* dz; \quad \frac{2h^3}{3}\bar{\gamma} = \bar{n}\int_{-h}^{h} \bar{u}^* z dz, \quad (1.198)$$

where $\bar{u} = u\bar{i} + v\bar{j}$, $\bar{w} = w\bar{n}$, $\bar{\gamma} = \gamma_x\bar{i} + \gamma_y\bar{j}$. Parameters related to the three dimensional theory are marked by an asterisk. It appears that such a averaging of the three dimensional characteristics is very suitable for a qualitative asymptotic estimation of the accuracy of the different approximate theories applied to plates analysis. For instance, if the following criterion is used

$$\lim h^{-n} \left| \frac{R^* - R}{h^m} \right| = O(1),$$

(R, R^* are the compared characteristics obtained using two and three dimensional theories, respectively and 'm' is the two dimensional theory order of accuracy in comparison to the three dimensional theory), then 'm' will be equal to 2 for all compared quantities.

Exact analysis realised during the isotropic plate vibrations is much more difficult and practically useless in the case of free vibrations of orthotropic cubicoid. If we assume that the displacement distribution along the plate's thickness can be predicted on the basis of two dimensional theory then it is possible to compare normal and tangential distribution obtained for the certain characteristic modes.

On the other hand, in references [221] argued that in some cases a 2D theory can be applied instead of 3D-one. This information is most practically needed.

Let us consider the influence of the geometric parameters λ_1, λ_2 (relative thicknesses) on a free vibration frequencies spectrum of an isotropic plate.

In the Table 1.4, some calculation results of the vibration frequencies using the classical theory, Timoshenko's type theory (two variants - with and without rotational inertia) for λ_1, λ_2 equal to 5, 10, 50, 100 are given. The frequencies are ordered in an increasing manner (m, n are the integers characterizing the halfwaves number in the x and y directions, respectively).

Because the equations (1.149) govern only the antisymmetric stress-deformation plate states, the corresponding improved theory (with the rotational inertia effects) leads to three frequency spectra. Using either Kirchhoff's or Timoshenko's theories only one spectrum, related to bending, can be found.

In Table 1.5, ten first frequency spectra, using three dimensional theory, are given. In the first row the mode numbers are given, which characterise a wave process in the x and y directions. For specific m, n the free vibration frequencies are ordered in an increasing manner.

The dimensionless vibration frequency was computed from the following expression valid for both two and three dimensional theories

$$\omega^* = 2h\sqrt{2p(1+\nu)/E}\omega.$$

Comparing the results given in the tables, the attention focuses on the corrections introduced to Kirchhoff's and improved theories (more accurate) for a spectrum related to bending with the different, for each theory errors.

It is remarkable that the approximation carried out in a neighbourhood of the frequency values obtained for all theories are applied for $\lambda_1 = 50, 100$. In the case of λ_1, the error has been smaller than 2%, whereas in the case of the Timoshenko model - 0.5%. The general Timoshenko model gives the most accurate results. In the latter case predictably the results were the same (the difference has been noticed on the fourth place after the point). The obtained results show that there exists no essential influence of the transversal displacement on the frequencies of the considered modes related to bending.

With increasing plate thickness increase (λ_1 decreases) Kirchhoff's theory is less accurate regarding the frequencies related to bending. For $\lambda_1 \geq 10$ the fundamental bending vibration frequency, found using the classical model, does not cross the threshold of 5%, whereas for $\lambda_1 = 5\div$ the corresponding thereshold is 12%.

During calculations of higher frequencies the area of validity of the classical theory is defined on the basis of comparison of the corresponding modes with the corresponding results of three dimensional theory. With an increase of m, n, an increase of an error is obtained: from 12% ($m = n = 2$) to 34% ($m = 3, n = 4$). For $\lambda_1 = 5$ an error of 10% is admitted only during the

1.4 Dynamical Characteristics of Plates with Added Masses

Table 1.4. Eigenfrequencies of isotropic plate obtained using Kirchhoff and Timoshenko theories.

m, n	Kirchhof's theory	Timoshenko's model				Generalized Timoshenko's model		
		without inertia	with inertia					
	I A	I A	I A	II A	III A	I A	II A	III A
1	*2*	*3*	*4*	*5*	*6*	*7*	*8*	*9*
			$G/E = 0{,}384$; $\lambda_1 = 5$, $\lambda_2 = 5$; $\nu = 0.3$					
1 1	0.38527	0.35347	0.34545	3.5762	3.8634	0.34095	2.9380	2.1961
1 2	0.96317	0.79441	0.76487	3.7382	4.3622	0.74789	3.0950	3.6426
2 2	1.5411	1.1643	1.1146	3.8934	4.7895	1.0847	3.2445	4.0184
1 3	1.9264	1.3831	1.3220	3.9935	5.0477	1.2840	3.3404	4.2434
2 3	2.5043	1.6801	1.6048	4.1391	5.4056	1.5556	3.4793	4.5533
1 4	3.2748	2.0317	1.9423	4.3257	5.8407	1.8797	3.6564	4.9276
3 3	3.4674	2.1134	2.0210	4.3711	5.9435	1.9554	3.6993	5.0156
2 4	3.8527	2.2702	2.1726	4.4605	6.1428	2.1012	3.7837	5.1861
3 4	4.8159	2.6312	2.5236	4.6765	6.6106	2.4390	3.9869	5.5848
			$G/E = 0.384$; $\lambda_1 = 10$, $\lambda_2 = 10$; $\nu = 0.3$					
1 1	0.096317	0.094131	0.093432	3.4925	3.5712	0.093037	2.8562	2.9282
1 2	0.24079	0.22778	0.22411	3.5346	3.7220	0.22205	2.8974	3.0672
2 2	0.38527	0.35347	0.34545	3.5763	3.8634	0.34095	2.9380	3.1961
1 3	0.48159	0.43335	0.42199	3.6037	3.9533	0.41562	2.9647	3.2777
2 3	0.62606	0.54794	0.53123	3.6446	4.0825	0.52182	3.0044	3.3935
1 4	0.81870	0.69205	0.66799	3.6983	4.2456	0.65429	3.0565	3.5391
3 3	0.86686	0.72669	0.70079	3.7117	4.2840	0.68599	3.0694	3.5741
2 4	0.96317	0.79442	0.76487	3.7382	4.3622	0.74789	3.0950	3.6426
3 4	1.2039	0.95557	0.91722	3.8036	4.5771	0.89478	3.1582	3.8058
			$G/E = 0.384$; $\lambda_1 = 50$, $\lambda_2 = 50$; $\nu = 0.3$					
1 1	0.0038527	0.0038490	0.0038478	3.4652	3.4685	0.0038471	2.8195	2.8325
1 2	0.0096317	0.0096091	0.0096013	3.4669	3.4751	0.0095970	2.8312	2.8387
2 2	0.015411	0.015353	0.015333	3.4687	3.4816	0.015322	2.8329	2.8448
1 3	0.019263	0.019173	0.015142	3.4698	3.4860	0.019125	2.8340	2.8489
2 3	0.025043	0.024890	0.024839	3.4715	3.4925	0.024809	2.8357	2.8550
1 4	0.032748	0.032489	0.032401	3.4738	3.5011	0.032352	2.8379	2.8631
3 3	0.034674	0.034384	0.034286	3.4743	3.5033	0.034231	2.8385	2.8651
2 4	0.038527	0.038170	0.038049	3.4755	3.5076	0.037981	2.8396	2.8691
3 4	0.048159	0.047602	0.047417	3.4783	3.5183	0.047311	2.8424	2.8791
			$G/E = 0.384$; $\lambda_1 = 100$, $\lambda_2 = 100$; $\nu = 0.3$					
1 1	0.00096317	0.00096295	0.00096287	3.4644	3.4652	0.00096283	2.8287	2.8295
1 2	0.0024079	0.0024065	0.0024060	3.4648	3.4668	0.0024057	2.8291	2.8310
2 2	0.0038527	0.0038491	0.0038478	3.4652	3.4685	0.0038471	2.8295	2.8325
1 3	0.0048159	0.0048102	0.0048082	3.4655	3.4696	0.0048071	2.8298	2.8336
2 3	0.0062606	0.0062511	0.0062477	3.4660	3.4712	0.0062459	2.8302	2.8351
1 4	0.0081870	0.0081650	0.0081650	3.4665	3.4734	0.0081618	2.8308	2.8371
3 3	0.0086686	0.0086503	0.0086440	3.4667	3.4740	0.0086403	2.8309	2.8377
2 4	0.0096317	0.0096092	0.0096014	3.4669	3.4751	0.0095969	2.8312	2.8387
3 4	0.012040	0.012001	0.011992	3.4677	3.4778	0.011985	2.8319	2.8412

Table 1.5. Eigenfrequencies of isotropic plate obtained using 3D theory.

m,n	Three dimensional theory									
	I A	I S	II S	II A	III A	III S	IV S	V S	IV A	V A
1	*2*	*3*	*4*	*5*	*6*	*7*	*8*	*9*	*10*	*11*
G/E = 0,384; $\lambda_1 = 5$, $\lambda_2 = 5$; $\nu = 0.3$										
1 1	0.34207	0.88858	1.4922	3.2648	3.5298	5.6010	6.3457	6.7563	9.4174	9.466
1 2	0.75110	1.4050	2.3320	3.4414	4.0037	5.4795	6.4383	7.1824	9.4184	9.5289
2 2	1.0888	1.7772	2.9070	3.6094	4.0412	5.4635	6.5297	7.5179	9.4315	9.5909
1 3	1.2881	1.9869	3.2123	3.7172	4.6376	5.4894	6.5897	7.7151	9.4460	9.6319
2 3	1.5589	2.2654	3.5870	3.8732	4.9600	5.5727	6.6791	7.9847	9.4753	9.6932
1 4	1.8908	2.5906	3.9654	4.0720	5.3430	5.7549	6.7963	8.3090	9.5269	9.7743
3 3	1.9557	2.6657	4.0423	4.1202	5.4319	5.8112	6.8253	8.3852	9.5419	9.7945
2 4	2.1000	2.8099	4.1789	4.2149	5.6024	5.9346	6.8829	8.5326	9.5741	9.8347
3 4	2.4332	3.1416	4.3718	4.4429	5.9918	6.1870	7.0248	8.8767	9.6672	9.9346
G/E = 0.384; $\lambda_1 = 10$, $\lambda_2 = 10$; $\nu = 0.3$										
1 1	0.093150*	0.44429	0.74983	3.1728	3.2465	5.7632	6.2989	6.4461	9.4222	9.4352
1 2	0.22260*	0.70248	1.1827	3.2192	3.3932	5.6652	6.3223	6.6178	9.4193	9.4509
2 2	0.34207*	0.88858	1.4922	3.2648	3.5298	5.6010	6.3457	6.7563	9.4174	9.4666
1 3	0.41714*	0.99346	1.6654	3.2949	3.6160	5.5688	6.3612	6.8384	9.4166	9.4770
2 3	0.52391*	1.1327	1.8936	3.3396	3.7393	5.5315	6.3845	6.9512	9.4162	9.4926
1 4	0.65708	1.2953	2.1569	3.3982	3.8939	5.4969	6.4153	7.0877	9.4170	9.5134
3 3	0.68893*	1.3329	2.2171	3.4126	3.9310	5.4903	6.4230	7.1199	9.4174	9.5185
2 4	0.75110*	1.4050	2.3320	3.4414	4.0037	5.4795	6.4383	7.1824	9.4184	9.5289
3 4	0.89853	1.6019	2.5925	3.5264	4.1766	5.4636	6.4842	7.3294	9.4226	9.5599
G/E = 0.384; $\lambda_1 = 50$, $\lambda_2 = 50$; $\nu = 0.3$										
1 1	0.0038474	0.088858	0.15019	3.1428	3.1459	5.8710	6.2838	6.2915	9.4247	9.4252
1 2	0.0095982	0.14050	0.23745	3.1447	3.1524	5.8618	6.2848	6.3036	9.4245	9.4258
2 2	0.015325	0.17772	0.30032	3.1466	3.1588	5.8532	6.2857	6.3151	9.4243	9.4264
1 3	0.019130	0.19869	0.33575	3.1479	3.1631	5.8477	6.2863	6.3226	9.4242	9.4269
2 3	0.024818	0.22654	0.38278	3.1457	3.1695	5.8398	6.2873	6.3334	9.4241	9.4275
1 4	0.032366	0.25906	0.43767	3.1522	3.1780	5.8297	6.2885	6.3474	9.4239	9.4283
3 3	0.034247	0.26657	0.45034	3.1529	3.1801	5.8273	6.2884	6.3508	9.4238	9.4285
2 4	0.038001	0.28099	0.47467	3.1541	3.1843	5.8226	6.2895	6.3575	9.4237	9.4290
3 4	0.047342	0.32038	0.53062	3.1579	3.1948	5.8112	6.2913	6.3737	9.4234	9.4302
G/E = 0.384; $\lambda_1 = 100$, $\lambda_2 = 100$; $\nu = 0.3$										
1 1	0.0009628	0.044429	0.07510	3.1419	3.1427	5.8758	6.2833	6.2859	9.4247	9.4249
1 2	0.0024058	0.070248	0.11874	3.1424	3.1443	5.8733	6.2836	6.2884	9.4247	9.4250
2 2	0.0038473	0.088858	0.15019	3.1428	3.1459	5.8710	6.2838	6.2915	9.2447	9.4252
1 3	0.0048075	0.099345	0.16791	3.1432	3.1470	5.8694	6.2840	6.2936	9.4246	9.4253
2 3	0.0062464	0.11327	0.19145	3.1436	3.1486	5.8671	6.2842	6.2966	9.4246	9.4255
1 4	0.0081628	0.12953	0.21892	3.1443	3.1508	5.8641	6.2845	6.3006	9.4245	9.4257
3 3	0.0086414	0.13329	0.22526	3.1444	3.1513	5.8633	6.2846	6.3016	9.4245	9.4257
2 4	0.0095982	0.14050	0.23745	3.1447	3.1524	5.8618	6.2848	6.3036	9.4245	9.4258
3 4	0.011987	0.15708	0.26546	3.1455	3.1551	5.8582	6.2852	6.3084	9.4244	9.4261

* according to the values obtained in [223]

1.4 Dynamical Characteristics of Plates with Added Masses

fundamental frequency estimation, whereas for $m = 3$, $n = 4$ (nine mode) an error is equal to 100%.

The enhanced accuracy of Timoshenko's theory allows to get practically exact values of the isotropic plate vibration frequencies.

In the considered intervals of λ_1, λ_2 the error introduced by Timoshenko's theory reached 0.5%. A little bit larger error characterises vibration frequencies related to bending (without rotation inertial effects): for $\lambda_1 = 5$ we have 3% ($m = n = 1$) and 8% ($m = 3, n = 4$). If the governing equations introduce the rotation inertial effects then the frequencies are obtained with an error of 4%.

It should be emphasized that the classical and improved Timoshenko's theories give higher frequency estimates compareds to the 3D exact results. The general Timoshenko's model gives lower estimates in comparison with exact results. Therefore, with increasing plate thickness, the difference between the frequencies obtained using classical theory and the improved Timoshenko's theory also increases. In the case of the general Timoshenko's theory, an error decreases with decreasing thickness and then slightly increases. However, the error oscillations are of the order of 0.3%.

Now we carry out an analysis of the successive frequency spectra computed by the improved two dimensional theories. The frequencies related to thickness-rotational modes (II A) and thickness-displacements modes (III A) of a free vibrating isotropic plate are studied. Here more evident numerical error is observed, in comparison with the exact three dimensional theory results as well as in relation to the "improved" two dimensional theories described in this work. The calculation error does not practically depend on the magnitude of the relative thickness parameter λ_1 and oscillates in the interval from 4% to 10% for both vibration modes. In reference [157] during the formulation of the two dimensional improved theory a displacement coefficient was equal to $\pi^2/12$. This value secures overlap of the low frequencies spectra. The corresponding modes are thickness-rotational with $w = 0$. Therefore, the displacement coefficient $\pi^2/12$ gives the best approximation for the stress deformation state, where the transversal displacements play a key role. There exist also additional criteria to choose the displacement coefficient [86].

Using "improved" theories in this work the displacement coefficient was equal to 1 (Timoshenko's model) and 2/3 (general Timoshenko's model). Therefore, the following inequalities hold for the "improved" frequencies 'ω_T' and 'ω_{0T}' and the exact one, ω

$$\omega_{0T} < \omega < \omega_T. \tag{1.199}$$

It means that the improved theories anable are to find intervals, where "real" vibration frequencies occur. A comparison between exact and approximate frequency values, corresponding to thickness-rotational and thickness-displacement modes, found by two and three dimensional theories indicates validity of the following relationship

$$\omega' = (\omega_T + \omega_{0T})/2, \tag{1.200}$$

where a frequency value close to the exact one is obtained.

The frequencies obtained from (1.200) differ from the exact values by less than 2.5%.

As it has been mentioned earlier, the essential difference between two and three dimensional theories concerns the qualitative difference of their spectra. The three dimensional theory allows for an infinite set of spectra definition, whereas the two dimensional theory predicts a finite number of spectra, which is equal to the degrees of freedom of a two dimensional model of the continuous medium. The free vibration frequencies corresponding to the modes with the numbers III S, IV S, V S, V A are not "captured" by the two dimensional theory. In order to find them, the applicational theories, characterised by less requirements with regard to the kinematic hypotheses and having more degrees of freedom (they are sometimes called higher order theories), should be formulated.

Consider the fundamental results of comparing the vibration frequency distributions using three dimensional theories (Table 1.5). For $\lambda_1 \leq 10$ and for higher vibration modes the frequency distribution is more uniform. With decreasing plate thickness (increasing λ_1 and λ_2) the distribution picture becomes fundamentally changed. For $p \to 0$ ($p = (\alpha_m/\lambda_1)^2 + (\beta_n/\lambda_2)^2$) veering of frequency spectra is observed; for instance (II A) and (III A), (IV S) and (V S), and so on. The frequencies related to bending and frequency spectra of the symmetric modes tend to approach the zero limit, and quick convergence of the asymptotic series is observed. The convergence effect can be foreseen using the modified Timoshenko's theories for frequency spectra (II A) and (III A).

The material properties influence on the free vibration frequencies will be analysed using an isotropic-transversal body, whose isotropic plane coincides with the $(x0y)$ plane.

Consider the influence of G'/E on the vibration frequency spectrum of a transversal-isotropic plate for different values of λ_1, λ_2 characterising the relative thickness.

In Table 1.6 the frequencies obtained using two dimensional classical theory, Timoshenko's type theory (with and without inertia rotational effects) are presented. The parameters used were $\lambda_1 = \lambda_2 = 5, 10, 50, 100$. For these parameters and the modes given in Table 1.7 the frequency values obtained using three dimensional theory are given (the first ten spectra arranged in increasing order for each pair of the wave numbers).

The essential influence of the transverse displacement of the frequency spectrum of the transversal-isotropic free vibrations (thick or thin) is evident. The mentioned behaviour is more evident owing to increase of a mode number and relative plate thickness. Rotation inertial effects slightly decrease the frequencies (to 0.1%) for different vibrations modes of thin ($\lambda_1 = 50, 100$), average ($\lambda_1 = 10$) and thick ($\lambda_1 = 5$) plates. This leads to the conclusion

1.4 Dynamical Characteristics of Plates with Added Masses

Table 1.6. Eigenfrequencies of transversal-isotropic plate obtained using Kirchhoff and Timoshenko theories.

m, n	Kirchhof's theory	Timoshenko's model				Generalized Timoshenko's model		
		without inertia	with inertia					
	I A	I A	I A	II A	III A	I A	II A	III A
1	2	3	4	5	6	7	8	9
$G'/E = 0{,}01$; $\lambda_1 = 5$, $\lambda_2 = 5$; $\nu = 0{,}3$								
1 1	0.38527	0.13429	0.13423	1.0496	1.6033	0.12390	0.91633	1.4181
1 2	0.96317	0.22053	0.22047	1.5119	2.4402	0.20243	1.3368	2.1710
2 2	1.5412	0.28173	0.28169	1.8629	3.0559	0.25811	1.6537	2.7231
1 3	1.9264	0.31604	0.31600	2.0639	3.4050	0.28939	1.8347	3.0358
2 3	2.5042	0.36146	0.36143	2.3333	3.8702	0.33083	2.0769	3.4523
1 4	3.2748	0.41437	0.41434	2.6502	4.4148	0.37910	2.3616	3.9396
3 3	3.4674	0.42657	0.42654	2.7236	4.5407	0.39024	2.4275	4.0523
2 4	3.8527	0.44999	0.44996	2.8649	4.7827	0.41161	2.5543	4.2688
3 4	4.8159	0.50379	0.50376	3.1909	5.3398	0.46073	2.8467	4.7672
$G'/E = 0.01$; $\lambda_1 = 10$, $\lambda_2 = 10$; $\nu = 0.3$								
1 1	0.096317	0.057482	0.057422	0.71372	0.93692	0.054084	0.60491	0.81221
1 2	0.24079	0.10249	0.10243	0.89749	1.3131	0.095029	0.77639	1.1566
2 2	0.38527	0.13429	0.13423	1.0496	1.6032	0.12390	0.91633	1.4181
2 3	0.62606	0.17534	0.17528	1.2629	1.9951	0.16121	1.1110	1.7711
1 4	0.81870	0.20238	0.20233	1.4106	2.2602	0.18582	1.2451	2.0094
3 3	0.86686	0.20860	0.20855	1.4452	2.3217	0.19148	1.2764	2.0646
2 4	0.96317	0.22053	0.22048	1.3119	2.4402	0.20234	1.3368	2.1710
3 4	1.2039	0.24786	0.24781	1.6672	2.7138	0.22724	1.4771	2.4164
$G'/E = 0.01$; $\lambda_1 = 50$, $\lambda_2 = 50$; $\nu = 0.3$								
1 1	0.0038527	0.0037205	0.0037194	0.56559	0.57858	0.0036947	0.46294	0.47557
1 2	0.0096317	0.0088639	0.0088586	0.57597	0.60732	0.0087266	0.47307	0.50337
2 2	0.015411	0.013573	0.013562	0.58616	0.63472	0.013269	0.48298	0.52968
1 3	0.019263	0.016509	0.016494	0.59286	0.65234	0.016076	0.48947	0.54650
2 3	0.025043	0.020655	0.020634	0.60276	0.67790	0.020009	0.49906	0.57078
1 4	0.032749	0.025772	0.025745	0.61573	0.71052	0.024825	0.51156	0.60162
3 3	0.034674	0.026988	0.026959	0.61892	0.718444	0.025963	0.51463	0.60908
2 4	0.038527	0.029351	0.029328	0.62526	0.73402	0.028171	0.52074	0.62374
3 4	0.048159	0.034903	0.034863	0.64075	0.77158	0.033332	0.53568	0.65893
$G'/E = 0.01$; $\lambda_1 = 100$, $\lambda_2 = 100$; $\nu = 0.3$								
1 1	0.00096317	0.0009546	0.00095451	0.56033	0.56364	0.00095282	0.45780	0.46103
1 2	0.0024079	0.0023533	0.0023548	0.56297	0.57116	0.0023448	0.46038	0.46836
2 2	0.0038527	0.0037205	0.0037194	0.56559	0.57858	0.0036947	0.46294	0.47557
1 3	0.0048159	0.0046120	0.0046104	0.56733	0.58347	0.0045727	0.46465	0.48032
2 3	0.0062606	0.0059224	0.0059199	0.56994	0.59073	0.0058587	0.46719	0.48735
1 4	0.0081870	0.0076223	0.0076183	0.57339	0.60026	0.0075192	0.47056	0.49657
3 3	0.0086686	0.0080393	0.0080348	0.57425	0.60262	0.0079252	0.47139	0.49885
2 4	0.0096317	0.0088630	0.0088546	0.57597	0.60732	0.0087267	0.47307	0.50337
3 4	0.012040	0.010874	0.010866	0.58024	0.61889	0.010673	0.47722	0.51450

80   1 Vibration of Plates and Shells with Added Masses

Table 1.7. Eigenfrequencies of transversal-isotropic plate obtained using 3D theory.

m,n	Three dimensional theory									
	I A	I S	II S	II A	III A	III S	IV S	V S	IV A	V A
1	*2*	*3*	*4*	*5*	*6*	*7*	*8*	*9*	*10*	*11*
$G'/E = 0{,}01;\ \lambda_1 = 5,\ \lambda_2 = 5;\ \nu = 0{,}3$										
1 1	0.12695	0.14328	0.88858	1.0288	1.3476	1.4922	1.5823	1.7604	1.8031	2.1314
1 2	0.21058	0.22654	0.4050	1.4935	1.7322	1.0696	2.3309	2.4179	2.4657	2.5615
2 2	0.27072	0.28656	1.7772	1.8479	2.0457	2.3383	2.6952	2.9023	3.0257	3.0941
1 3	0.30459	0.32038	1.9869	2.0505	2.2303	2.4766	2.7881	3.1468	3.2123	3.3958
2 3	0.34955	0.36529	2.2654	2.3214	2.4817	2.7279	3.0394	3.3982	3.5755	3.7908
1 4	0.40203	0.41773	2.5906	2.6397	2.7817	3.0035	3.2889	3.6231	3.9550	3.9936
3 3	0.41415	0.42984	2.6657	2.7134	2.8518	3.0685	3.3484	3.6771	4.0332	4.0428
2 4	0.43742	0.45809	2.8099	2.8552	2.9870	3.1946	3.4643	3.7830	4.1392	4.1732
3 4	0.49092	0.50657	3.1416	3.1822	3.3009	3.4899	3.7384	4.0355	4.3712	4.4429
$G'/E = 0.01;\ \lambda_1 = 10,\ \lambda_2 = 10;\ \nu = 0.3$										
1 1	0.054467	0.071639	0.44429	0.67380	0.74983	0.90580	1.1063	1.2587	1.5833	1.6937
1 2	0.056670	0.11327	0.70248	0.86608	1.1827	1.2329	1.2896	1.5553	1.6742	1.9252
2 2	0.12695	0.14328	0.88858	1.0228	1.3476	1.4922	1.5823	1.7604	1.8031	2.1314
1 3	0.14397	0.16019	0.99346	1.1152	1.4189	1.6653	1.7501	1.8156	1.9506	2.6256
2 3	0.16653	0.18265	1.1327	1.2408	1.5197	1.8936	1.9901	2.0636	2.2113	2.4134
1 4	0.19285	0.20886	1.2953	1.3908	1.6445	1.9968	2.1562	2.2391	2.3949	2.4049
3 3	0.19892	0.21492	1.3329	1.4259	1.6742	2.0214	2.2163	2.3003	2.4253	2.4517
2 4	0.21058	0.22654	1.4059	1.4935	1.7322	2.0696	2.3309	2.4179	2.4657	2.5615
3 4	0.23738	0.25328	1.5708	1.6505	1.8992	2.1150	2.5101	2.6201	2.6607	2.7321
$G'/E = 0.01;\ \lambda_1 = 50,\ \lambda_2 = 50;\ \nu = 0.3$										
1 1	0.0036950	0.014329	0.88858	0.15019	0.51430	0.52847	1.0170	1.0241	2.5344	2.5372
1 2	0.0087293	0.022654	0.14050	0.23745	0.52569	0.55968	1.0228	1.0403	2.5367	2.5438
2 2	0.013277	0.028656	0.17772	0.30032	0.53684	0.58921	1.0286	1.0563	1.5301	1.5489
1 3	0.016089	0.032038	0.19869	0.33575	0.54414	0.60808	1.0324	1.0668	1.5326	1.5561
2 3	0.020034	0.036529	0.22654	0.38278	0.55492	0.63533	1.0382	1.0824	1.5365	1.5668
1 4	0.024870	0.041773	0.25906	0.43767	0.56897	0.66991	1.0457	1.1028	1.5416	1.5810
3 3	0.026014	0.042984	0.26657	0.45034	0.57243	0.67828	1.0476	1.1079	1.5429	1.5846
2 4	0.018234	0.045309	0.28099	0.47468	0.57928	0.69471	1.0514	1.1179	1.5455	1.5916
3 4	0.033433	0.050657	0.31416	0.53062	0.59608	0.73415	1.0607	1.1426	1.5518	1.6091
$G'/E = 0.01;\ \lambda_1 = 100,\ \lambda_2 = 100;\ \nu = 0.3$										
1 1	0.00095288	0.0071639	0.044429	0.75097	0.50851	0.51213	1.0141	1.0159	2.5332	2.5339
1 2	0.0023449	0.011327	0.070248	0.11874	0.51141	0.52037	1.0156	1.0200	2.5334	2.5356
2 2	0.0036950	0.014328	0.088858	0.15019	0.51143	0.52847	1.0170	1.0241	2.5344	2.5372
1 3	0.0045732	0.016019	0.099346	0.16791	0.51622	0.53380	1.0180	1.0268	2.5348	2.5383
2 3	0.0058597	0.018264	0.11327	0.19144	0.51908	0.54170	1.0195	1.0309	2.5354	2.5400
1 4	0.0075210	0.020886	0.12953	0.21892	0.52286	0.55205	1.0214	1.0363	2.5361	2.5422
3 3	0.0079273	0.021492	0.13329	0.22526	0.52381	0.55460	1.0210	1.0376	2.5363	2.5427
2 4	0.0087293	0.022654	0.14050	0.23745	0.52569	0.55968	1.0228	1.0403	2.5367	2.5438
3 4	0.010677	0.025328	0.15708	0.26546	0.53036	0.57217	1.0252	1.0470	2.5377	2.5466

1.4 Dynamical Characteristics of Plates with Added Masses

that rotation inertia influence is smaller than in the case of isotropic plate material.

Consider now the drawbacks and the advantages of the classical and improved Timoshenko's theories.

In the considered case, the classical theory possesses a narrower application area than during the calculation of an isotropic plate frequency. For $\lambda_1 = 100$ an error less than 5% can be obtained for the calculations of the first three frequencies corresponding to the numbers $m = n = 1$; $m = 1$, $n = 2$; $m = n = 2$. For $s > 8$ (where 's' is the ordinal frequencies number in the bending spectrum) the error is greater than 10%. For $\lambda_1 = 50$ the error is smaller than 5% using classical methods, whereas for $\lambda_1 = 5, 10$ Kirchhoff's theory leads to incorrect results. For instance, for $\lambda_1 = 5$ the error in the fundamental frequency determination is equal to 200%, and for the ninth mode it exceeds 900%!

The results obtained using Timoshenko's kinematic model differ slightly from the exact results obtained using three dimensional theory. The error does not exceed 6% for the fundamental frequency ($\lambda_1 = 5$), and even decreases to 3% for increasing mode number. Also, the results obtained using the general Timoshenko's model are close to the exact ones. For $\lambda_1 = 100$ the latter theory gives practically the exact free vibration frequency values. With an increase of thickness and mode number the error increases, but for the considered changes in parameter intervals it does not exceed 6%.

Another important and interesting problem is related to the possibility of applying the modified theories to the free vibration frequencies determination of the spectrum series. If we take (1.200), then we get frequency values close to the real values, but the interval (ω_{0T}, ω_T) includes the frequencies corresponding to the different vibration modes.

The analysis of the frequency results using the three dimensional theory shows that the frequency spectra of a free vibration transversal plate are more continuous and more uniform in comparison to to those an isotropic plate. For instance, the frequency values of the first ten spectra of a three dimensional plate lie in the interval $(0, 5)$ - for a transversal-isotropic material; for an isotropic plate that interval corresponds to $(0, 10)$ and the frequencies are shifted to the left side.

In the transversal-isotropic plate frequency interval there are also frequencies of the isotropic plate vibrations. It seems to be strange at the first glance, however, according to the physical and geometrical choice of the parameters of the transversal-isotropic material, low transversal stiffness does not influence the frequencies related to the stretching-compression vibrations.

In Fig. 1.6 the frequency dependence on G'/E for $\lambda_1 = \lambda_2 = 5$ is presented. Curve 1 corresponds to the fundamental frequency found using the three dimensional theory. The curves obtained using the improved Timoshenko's theory are not practically distinguished from curve 2.

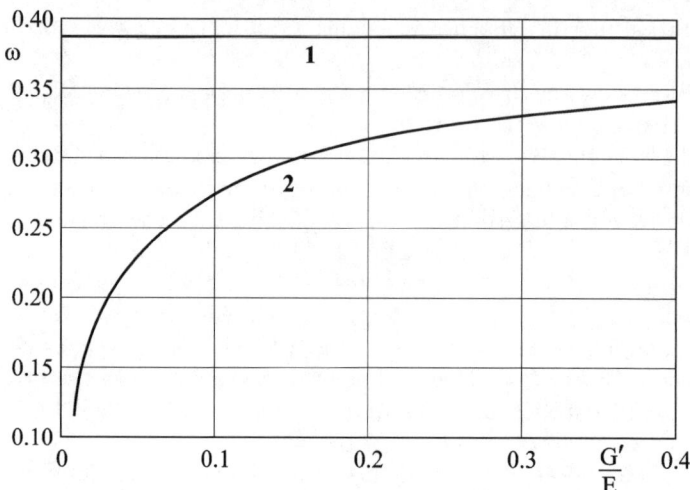

Figure 1.6. Fundamental eigenfrequency versus G'/E for transversal-isotropic plate with the relative thickness $\lambda_1 = \lambda_2 = 5$.

The presented curves clearly characterise the influence of G'/E on the fundamental frequency for free vibration. For G'/E close to 0.384 (isotropic material) the obtained results indicate a slight influence of transversal stiffness on the fundamental frequency. With a decrease of G'/E the fundamental frequency is observed to fall.

1.4.6 Influence of the Added Masses - Numerical Investigation

In the previous sections, a comparison between the results obtained using modified and three dimensional theories of the free isotropic and transversal-isotropic plates vibrations has been conducted. A similar approach is highly required for plates with the added masses. Here a comparison between two dimensional modified and three dimensional theories is also recommended.

While searching for the solution to the problems related to plates and shells, there are principally two possibilities: exact problem formulation in the frame of the three dimensional elasticity theory and definition of the possible steps of solutions or definition of the approximated calculation models for the different problems. The second approach is of more interest in technical applications.

In chapter 1.2 it has been concluded that recent fundamental results relating to free vibrations of plates and shells vibrations with added masses have been achieved using the second approach realised by the following model: a shell (Kirchhoff-Love model) being absolutely hard mass (concentrated or distributed) and vertically vibrating. Application of this model was made for the case of forced vibrations [162, 252], parametric vibrations [116, 123],

1.4 Dynamical Characteristics of Plates with Added Masses

stochastic vibrations [56], nonlinear vibrations [100, 131], as well as in dynamic stability analysis [114, 176, 177] and optimization [147, 224] of plates and shells with added masses. On the other hand, the first approach gives a better possibility of physical interpretation of the problem and seems to be a strong tool for building new models as well as, for developing calculation algorithms. It allows for the analysis of existing models and development of approximate methods applied directly by engineers. It also allows for the development of the investigated objects. However, it requires a high development of computer techniques.

If one takes into account that the numerical investigation of the dynamical characteristics with the added masses is not a simple task, then one can imagine how difficult it is to solve similar problems in the framework of three dimensional theories.

Problems related to analytical results concern the lack of optimal numerical investigations, and high consuming time of the numerical and the required symbolic computations.

Therefore, a compromise solution is to change the complex analytic formulas used the numerical calculations.

Using the assumptions of section 1.3.5 the concentrated masses with the structure we get the following transcendental equation (connections of the $N = 1$):

$$4\omega^2 M \sum_{m,n}^{\infty} w_z^0 \sin^2 \alpha_m x_1 \sin^2 \beta_n y_1 + 1 = 0, \qquad (1.201)$$

where

$$w_z^0 = \frac{\Delta_1}{\Delta} d_{13}^{(1)} sh 0.5\tau_1 + \frac{\Delta_2}{\Delta} d_{13}^{(1)} ch 0.5\tau_1 + \frac{\Delta_3}{\Delta} d_{13}^{(2)} ch 0.5\tau_2 +$$

$$\frac{\Delta_4}{\Delta} d_{13}^{(2)} sh 0.5\tau_2 + \frac{\Delta_5}{\Delta} d_{13}^{(3)} ch 0.5\tau_3 + \frac{\Delta_6}{\Delta} d_{13}^{(3)} sh 0.5\tau_3, \qquad (1.202)$$

Δ is the determinant of the linear equations (1.175), Δ_i are the minors obtained from Δ by cancellation of the l-th row and i-th column.

In the case of the isotropic plate we get the following explicit formula for w_z^0

$$w_z^0 = \left[\omega^2 s \left(shr chs \left(r^2 + p\right)^2 - 4pr s ch r chs\right)\right] / \Delta, \qquad (1.203)$$

where

$$\Delta = 8\left(r^2 + p\right)^2 prs \left(1 - chr\, chs\right) + shr\, shs \left[16p^2 r^2 s^2 + \left(r^2 + p\right)^4\right],$$

$$p = \tilde{\alpha}_m^2 + \tilde{\beta}_n^2, \quad \tilde{\alpha}_m = \alpha_m/\lambda_1, \quad \tilde{\beta}_n = \beta_n/\lambda_2,$$

$$r = \sqrt{p - \omega^2}, \quad s = \sqrt{p - (1 - 2\nu)\omega^2/(2 - 2\nu)},$$

and ω is the sought frequency.

84   1 Vibration of Plates and Shells with Added Masses

The assumptions made enable one to retain only the components of essential influence. The obtained equations are more simplified.

Finally the obtained equations are solved using the algorithm described in 1.3.5. The only one difference is that in the subroutine with the numerical calculations of the roots of equation (1.201), there is no second step included and the frequencies of the orthotropic plate free vibrations are taken as the initial values. The suitability of writing two independent programs is substantiated by lower computational time far of the free vibration frequencies of an isotropic plate with discrete added masses.

Consider the influence of the geometric and physical parameters as well as the value of the added masses on the frequency spectra of the free vibrations of an isotropic plate and compare the results obtained using classical and three dimensional theories.

In Fig. 1.7 frequency $\tilde{\omega}$ dependence

$$\tilde{\omega} = \omega_M/\omega_0, \tag{1.204}$$

against M/M_0 for lowest mode of an isotropic plate is given. Here ω_0 is the free vibration frequency of the isotropic plate (without any additional mass), ω_M is the frequency (corresponding to a plate with an added mass) with values given in Table 1.8 for the different geometric parameters λ_1, λ_2. The additional mass position has following coordinates $x_1 = 0.476$ and $y_1 = 0.476$.

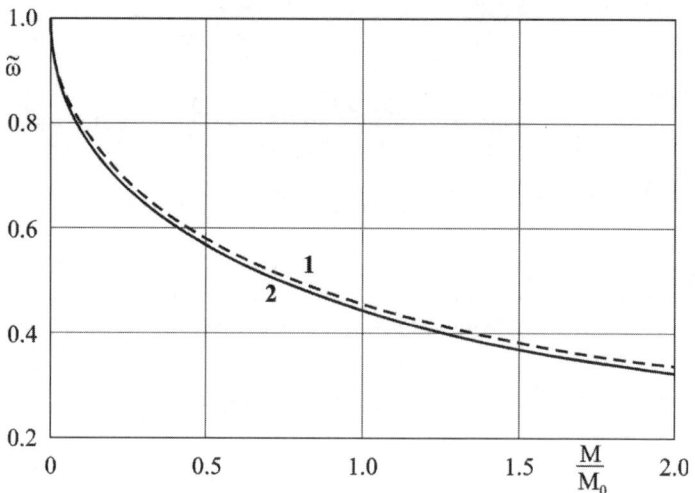

Figure 1.7. Frequency parameter $\tilde{\omega} = \omega_M/\omega_0$ versus M/M_0 for low frequency of isotropic plate for $\lambda_1 = 5$.

The results, given in Fig. 1.7, indicate a change of the parameter $\tilde{\omega}$ corresponding to the fundamental plate vibration frequency for a change in the

1.4 Dynamical Characteristics of Plates with Added Masses

Table 1.8. Fundamental frequencies ω_M of isotropic plate with added mass obtained using different theories.

M/M_0	TK	MTT		OMT	TT
		a	b		
\multicolumn{6}{c}{$G/E = 0{,}384$; $\lambda_1 = 5$, $\lambda_2 = 5$; $\nu = 0{,}3$}					
0.01	0.37785	0.34666	0.33908	0.33464	0.33602
0.1	0.32538	0.29955	0.29316	0.28906	0.29164
0.5	0.22052	0.20314	0.19897	0.19575	0.19822
1.0	0.17005	0.15709	0.15331	0.15069	0.15262
2.0	0.12631	0.11627	0.11378	0.11177	0.11318
\multicolumn{6}{c}{$G/E = 0.384$; $\lambda_1 = 10$, $\lambda_2 = 10$; $\nu = 0.3$}					
0.01	0.094463	0.092318	0.091658	0.091270	0.091404
0.1	0.081344	0.079457	0.079037	0.078690	0.078942
0.5	0.055130	0.053733	0.053605	0.053443	0.053664
1.0	0.042512	0.041389	0.041332	0.041120	0.041407
2.0	0.031577	0.030719	0.030696	0.030534	0.030766
\multicolumn{6}{c}{$G/E = 0.384$; $\lambda_1 = 50$, $\lambda_2 = 50$; $\nu = 0.3$}					
0.01	0.0037785	0.0037750	0.0037739	0.0037731	0.0037735
0.1	0.0032538	0.0032502	0.0032499	0.0032493	0.0032501
0.5	0.0022052	0.0022013	0.0022026	0.0022022	0.0022034
1.0	0.0017005	0.0016969	0.0016985	0.0016981	0.0017004
2.0	0.0012631	0.0012602	0.0012616	0.0012613	0.0012661
\multicolumn{6}{c}{$G/E = 0.384$; $\lambda_1 = 100$, $\lambda_2 = 100$; $\nu = 0.3$}					
0.01	0.00094463	0.00094441	0.00094434	0.00094429	0.00094431
0.1	0.00081344	0.00081323	0.00081320	0.00081316	0.00081320
0.5	0.00055130	0.00055107	0.00055114	0.00055111	0.00055113
1.0	0.00042512	0.00042491	0.00042499	0.00042497	0.00042501
2.0	0.00031577	0.00031560	0.00031568	0.00031566	0.00031572

TK - Kirchhof's theory
MTT - Timoshenko's model
    a) including inertia of rotation
    b) without inertia of rotation
OMT - generalized Timoshenko's model
TT - three dimensional theory

added mass M for relative thickness parameter $\lambda_1 = 5$. The results obtained using Kirchhoff's theory (curve 1), and improved Timoshenko's theory (curve 2) defined by $\tilde{\omega} = (M/M_0)$ are practically undistinguished in Fig. 1.7. From the data of Fig. 1.7, a conclusion can be drawn that both approximate and three dimensional theories give similar results. This conclusion is also true for $\lambda_1 \geq 5$.

Carrying out the analysis of the parameter related to frequency (Fig. 1.7) against the added mass the following conclusions are derived:

1. $\tilde{\omega}$ decreases with an increase of the relative thickness parameter. Therefore, the fundamental free vibration frequency of a plate with and without the added masses for a given λ_1 decreases.
2. For the case of added masses the frequencies $\tilde{\omega}$ decrease with increasing of M/M_0 for a given λ_1. With the change of M/M_0 in the interval 0.01 to 2.0 the frequencies $\tilde{\omega}$ decrease correspondingly: 1.02 times for $M/M_0 = 0.01$; 1.18 times for $M/M_0 = 0.1$; 1.74 times for $M/M_0 = 0.5$; 2.25 times for $M/M_0 = 1.0$; 3.03 times for $M/M_0 = 2.0$. From the given data one can conclude that during the change of the physical and geometrical plate parameters the added mass influence on the fundamental frequency of free vibration of an isotropic plate is essential.

Now let us focus on a composite plate made from a transversal isotropic material with each point having an isotropic plane parallel to the $x0y$ plane. In order to decrease the numbers of control parameters only the specific properties characterizing the material with a low transversal stiffness are considered.

For a plate made from a transversal isotropic material the following physical and geometrical parameters are taken: $G'/E = 0.01$; $\lambda_1 = \lambda_2 = 5; 10; 50; 100$. The added mass has the coordinates: $x_1 = y_1 = 0.476$.

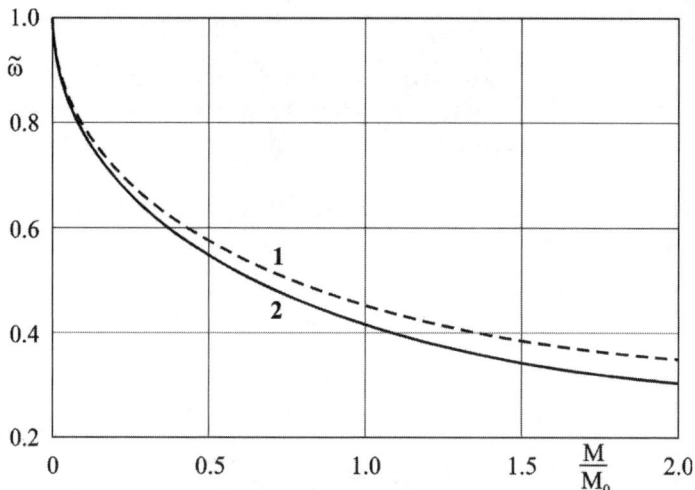

Figure 1.8. Frequency parameter $\tilde{\omega} = \omega_M/\omega_0$ versus M/M_0 for transversal-isotropic plate.

1.4 Dynamical Characteristics of Plates with Added Masses

Table 1.9. Fundamental frequencies ω_M of transversal-isotropic plate with added mass obtained using different theories.

M/M_0	TK	MTT		OMT	TT
		a	b		
\multicolumn{6}{c}{$G'/E = 0{,}01$; $\lambda_1 = 5$, $\lambda_2 = 5$; $\nu = 0{,}3$}					
0.01	0.37785	0.13168	0.13159	0.12147	0.12446
0.1	0.32538	0.11212	0.11209	0.10341	0.10615
0.5	0.22052	0.072811	0.072802	0.067081	0.068994
1.0	0.17005	0.055128	0.055124	0.050771	0.052242
2.0	0.12631	0.040445	0.040443	0.037240	0.038444
\multicolumn{6}{c}{$G'/E = 0.01$; $\lambda_1 = 10$, $\lambda_2 = 10$; $\nu = 0.3$}					
0.01	0.094463	0.056368	0.056311	0.053036	0.053399
0.1	0.081344	0.048155	0.048120	0.045288	0.045770
0.5	0.055130	0.031628	0.031619	0.029691	0.030092
1.0	0.042512	0.024049	0.024045	0.022560	0.022789
2.0	0.031577	0.017692	0.017690	0.016588	0.016810
\multicolumn{6}{c}{$G'/E = 0.01$; $\lambda_1 = 50$, $\lambda_2 = 50$; $\nu = 0.3$}					
0.01	0.0037785	0.0036489	0.0036479	0.0036235	0.0036225
0.1	0.0032538	0.0031397	0.0031391	0.0031177	0.0031183
0.5	0.0022052	0.0021209	0.0021207	0.0021050	0.0020995
1.0	0.0017005	0.0016328	0.0016327	0.0016201	0.0016184
2.0	0.0012631	0.0012114	0.0012114	0.0012018	0.0012023
\multicolumn{6}{c}{$G'/E = 0.01$; $\lambda_1 = 100$, $\lambda_2 = 100$; $\nu = 0.3$}					
0.01	0.00094463	0.00093621	0.00093613	0.00093414	0.00093415
0.1	0.00081344	0.00080603	0.00080598	0.00080452	0.00080459
0.5	0.00055130	0.00054580	0.00054578	0.00054470	0.00054478
1.0	0.00042512	0.00042069	0.00042069	0.00041982	0.00041988
2.0	0.00031577	0.00031239	0.00031239	0.00031172	0.00031176

TK - Kirchhof's theory
MTT - Timoshenko's model
    a) including inertia of rotation
    b) without inertia of rotation
OMT - generalized Timoshenko's model
TT - three dimensional theory

In Fig. 1.8 the change of the control parameter $\tilde{\omega} = \omega_M/\omega$, w_0 is the free vibration frequency of an isotropic transversal plate (without an added mass), ω_M is the frequency of the loaded plate (they are given in Table 1.9 for the different geometric parameters λ_1, λ_2) for different values of the ratio M/M_0 for the fundamental vibration frequency of a transversal isotropic

88   1 Vibration of Plates and Shells with Added Masses

Table 1.10. High frequencies of the isotropic plate obtained using Kirchhoff and Timoshenko theories (top values for $M = 0.1M_0$, bottom values for $M = 2.0M_0$). $G/E = 0.384$; $\lambda_1 = 5$, $\lambda_2 = 5$; $n = 0.3$

s	Kirchhof's theory	Timoshenko's model				Generalized Timoshenko's model		
		without inertia	with inertia					
	I A	I A	I A	II A	III A	I A	II A	III A
1	2	3	4	5	6	7	8	9
2	0.95994	0.79160	0.76236	3.7382	4.3615	0.74539	3.0950	3.6422
	0.95174	0.78294	0.75454	3.7382	4.3614	0.73729		3.6419
3	1.5409	1.1640	1.1144	3.8934	4.7895	1.0844	3.24045	4.0184
	1.3902	0.99358	0.96207	3.8934	4.7895	0.92746	3.2445	4.0184
4	1.6945	1.2104	1.1686	3.9935	5.0076	1.1299	3.3404	4.2228
	1.5416	1.1645	1.1148	3.9935	4.9975	1.0848	3.3404	4.2087
5	2.4987	1.6763	1.6015	4.1391	5.4046	1.5523	3.4793	4.5528
	2.4933	1.6727	1.5981	4.1391	5.4040	1.5488	3.4793	4.5526
6	3.1372	1.9335	1.8553	4.3257	5.8340	1.7902	3.6564	4.9256
	2.9608	1.8488	1.7699	4.3257	5.8240	1.7085	3.6564	4.9238
7	3.3085	2.0432	2.9537	4.3711	5.9026	1.8904	3.6993	4.9997
	3.2985	2.0409	1.9511	4.3711	5.8815	1.8882	3.6993	4.9898
8	3.8523	2.2701	2.1725	4.4605	6.1428	2.1010	3.7837	5.1861
	3.8522	2.2700	2.1724	4.4605	6.1427	2.1010	3.7837	5.1861
9	4.7866	2.6172	2.5105	4.6761	6.6065	2.4257	3.9869	5.5838
	4.7711	2.6112	2.5042	4.6761	6.6040	2.4196	3.9869	5.5832

plate with a concentrated mass is reported. In Fig. 1.8 the following notations are used: 1 - the $\tilde{\omega}$ the curve obtained using classical theory; 2 - the curve corresponding to Timoshenko's type theory, generalized Timoshenko's model and three dimensional theory (practically in the figure scale they are not distinguished). The calculation results given in Fig. 1.8 enable are to draw conclusions about the frequency spectra of the transversal-isotropic plate with added mass as well as some conclusions about the relative dependencies obtained using three dimensional and approximate theories for certain values of the physical and geometrical parameters and the added mass values.

The free vibration frequencies of the transversal-isotropic plate with and without added mass are lower than for similar isotropic plate. Such results are given by three dimensional theory and improved theories. An application of Kirchhoff theory leads to the determination of only one of the frequencies, which do not depend on the ratio G/E.

The effect of the additional mass is to decrease the fundamental frequency of the transversal-isotropic plate. If we take as control parameter ω_M^*/ω_0^*, then for the plates with small thickness ($\lambda_1 = 50, 100$) the dependence (transversal-isotropic material) will be practically undistinguishable from the corresponding dependence at the isotropic plate with the added mass (for both two and three dimensional theories applied). With an increase of the added mass error introduced by classical and improved theories (for $\lambda_1 < 50$) slightly increases in comparison with the corresponding case

1.4 Dynamical Characteristics of Plates with Added Masses

Table 1.11. High frequencies of the isotropic plate obtained using 3D theory (top values for $M = 0.1M_0$, bottom values for $M = 2.0M_0$). $G/E = 0.384$; $\lambda_1 = 5$, $\lambda_2 = 5$; $n = 0.3$

s	Three dimensional theory								
	1	2	3	4	5	6	7	8	9
2		0.74869	0.88858	1.0886	1.1334	1.4050	1.4857	1.5558	1.7772
		0.73891	0.88858	0.8985	1.0890	1.4050	1.4792	1.5518	1.7772
3	1.7817	1.8907	1.9869	2.0998	2.6554	2.3314	2.5906	2.6657	2.8099
	1.6751	1.8884	2.9869	2.0997	2.6554	2.3308	2.5906	2.6657	2.8099
4	2.9668	2.9847	3.2648	3.4246	3.4414	3.5820	3.6094	3.6653	3.7172
	2.7151	2.9071	3.2648	3.3554	3.4414	3.5762	3.6094	3.6103	3.7172
5	3.8732	3.9716	4.0040	4.0720	4.1202	4.1785	4.2149	4.3710	4.4017
	3.8732	3.9710	4.0040	4.0720	4.1202	4.1784	4.2149	4.3068	4.4014
6	4.7949	4.9632	5.3455	5.4470	5.4635	5.4798	5.5376	5.5731	5.6024
	4.7545	4.9623	5.3454	5.4469	5.4635	5.4798	5.5364	5.5731	5.6024
7	5.7493	5.7817	5.9345	5.3457	6.4384	6.5217	6.5297	6.5899	6.6791
	5.7487	5.7801	5.9345	5.3457	6.4384	6.4759	6.5297	6.5899	6.6791
8	6.7963	6.8253	6.8829	7.1706	7.3410	7.5180	7.9767	8.1133	8.3170
	6.7963	6.8253	6.8829	7.1635	7.2852	7.5180	7.9746	8.0875	8.3167
9	8.5323	9.3196	9.4184	9.4258	9.4315	9.4747	9.5067	9.5289	9.5741
	8.5323	9.2744	9.4184	9.4252	9.4315	9.4745	9.5025	9.5287	9.5741

without the mass. However, for the analysed variations of the ratio M/M_0 it does not achieve the following values: 5% - classical theory; 1% - Timoshenko's like theory; 0.05% - generalized Timoshenko theory. From the given examples it can be concluded that for a low stiffness transversal material the difference between results obtained using three dimensional and classical theories increases with an increase of M and it practically remains unchanged using the improved theories. Thus, the results obtained using the improved Timoshenko's type theories may be applied in all cases of variaion of the parameters λ_1, λ_2, M/M_0, as well as for plates made from different composite materials with more stiff properties.

Let us now analyze how the added masses influence higher frequencies of the isotropic plate.

In Table 1.10 the frequencies ($2 \leq s \leq 9$, where s denotes a frequency order in a corresponding spectrum) obtained using classical theory, Timoshenko's type theory (with and without rotational inertia effects) and a generalized Timoshenko's model (with a rotational inertia effect) are given. For each value of s number two corresponding frequencies are obtained: higher values are obtained for $M = 0.1M_0$ and lower for $M = 2M_0$. For the values of added mass listed in Table 1.11 higher frequencies for the first ten spectra are obtained, in the increasing sequence obtained using the three dimensional theory.

The results given in the Tables, allow for conclusions to be drawn about higher free isotropic plate vibration frequencies with concentrated mass attached at the point $x_1 = y_1 = 0.476$. It occurs that the added mass does

Table 1.12. High frequencies of the transversal-isotropic plate obtained using Kirchhoff and Timoshenko theories (top values for $M = 0.1M_0$, bottom values for $M = 2.0M_0$). $G'/E = 0.01$; $\lambda_1 = 5$, $\lambda_2 = 5$; $n = 0.3$

s	Kirchhof's theory	Timoshenko's model				Generalized Timoshenko's model		
		without inertia	with inertia					
	I A	I A	I A	II A	III A	I A	II A	III A
1	2	3	4	5	6	7	8	9
2	0.95994	0.21966	0.21961	1.5119	2.4402	0.20155	1.3368	2.1710
	0.95174	0.21608	0.21602	1.5119	2.4402	0.19826	1.3368	2.1710
3	1.5409	0.27620	0.27616	1.8629	3.0558	0.25291	1.6537	2.7231
	1.3902	0.23802	0.23795	1.8629	3.0558	0.21819	1.6537	2.7231
4	1.6945	0.28183	0.28178	2.0639	3.4050	0.25820	1.8347	3.0363
	1.5416	0.28175	0.28171	2.0639	3.4049	0.25813	1.8347	3.0363
5	2.4987	0.36072	0.36068	2.3333	3.8702	0.33015	2.0769	4.4523
	2.4933	0.36012	0.36008	2.3333	3.8702	0.32960	2.0769	4.4523
6	3.1372	0.39491	0.39487	2.6502	4.4148	0.36130	2.3616	3.9397
	2.9605	0.38336	0.38333	2.6502	4.4147	0.35080	2.3616	3.9397
7	3.3085	0.41591	0.41588	2.7236	4.5407	0.38057	2.4275	4.0529
	3.2985	0.41568	0.41565	2.7236	4.5406	0.38031	2.4275	4.0529
8	3.8523	0.44995	0.44992	2.8649	4.7827	0.41159	2.5543	4.2688
	3.8522	0.44995	0.44992	2.8649	4.7827	0.41158	2.5543	4.2688
9	4.78666	0.50143	0.50140	3.1909	5.3398	0.45857	2.8467	4.7672
	4.77117	0.50058	0.50055	3.1909	5.3398	0.45779	2.8467	4.7672

Table 1.13. High frequencies of the transversal-isotropic plate obtained using Kirchhoff and Timoshenko theories (top values for $M = 0.1M_0$, bottom values for $M = 2.0M_0$). $G'/E = 0.01$; $\lambda_1 = 5$, $\lambda_2 = 5$; $n = 0.3$

Three dimensional theory (frequencies spectrum related to bending)							
s							
2	3	4	5	6	7	8	9
0.20977	0.26613	0.27083	0.34883	0.38294	0.40357	0.43739	0.48858
0.20640	0.22827	0.27074	0.34822	0.37122	0.40334	0.43738	0.48771

not influence the higher frequency spectrum corresponding to bending (for approximate and three dimensional theories). That influence is even smaller in the case related to thickness jump. For instance, with a change of the ratio M/M_0 in the interval $0.1 \div 2.0$ the higher frequencies related to bending decrease less than 0.4%. The added mass does not influence the frequencies of the spectrum related to rotation (II A), because in that case $w = 0$. The three dimensional theory shows no influence on the other spectra (not outlined by the two dimensional theories); for example the frequencies corresponding to symmetric vibration modes of an unloaded plate.

The influence of the added mass on the higher frequencies of transversal-isotropic plate vibrations ($G'/E = 0.01$) with one added mass is shown in Table 1.12 (two-dimensional theory) and in Table 1.13 (three dimensional

1.4 Dynamical Characteristics of Plates with Added Masses

theory). Intervals of changes of the geometric parameters λ_1, λ_2 and M/M_0 are as in the previous Tables.

The analysis of the results given in the Tables leads to the conclusion that the concentrated mass leads to a frequency decrease (up to 14%) for different modes related to bending (for improved theories and the three dimensional theory). The results prove the lack of influence of the added mass on the higher frequencies of the spectra II A and III A (using Timoshenko's type theory and the generalised Timoshenko model with inertia rotational effects).

1.4.7 Fundamental Conclusions

1. The generalized Timoshenko type model and Timoshenko's model with the inertia rotational effects allow for getting practically exact results of frequency spectrum calculations related to bending in the considered geometrical ($5 \leq \lambda_1, \lambda_2 \leq 100$) and physical ($0.01 \leq G'/E \leq 0.38$) parameters of the isotropic and transversal-isotropic plates.
2. The effect of the concentrated mass is to decrease the vibration frequencies corresponding to the analysed unloaded plates using the improved and three dimensional theories as well as the Timoshenko type shells. This effect is more visible for larger values of added masses, larger geometric parameters values k_1, k_2 (for shells), and larger order modes related to bending, λ_1, λ_2 and G'/E.
3. For the high frequency spectra of the isotropic and transversal-isotropic plates the added mass influence may be omitted (except for bending modes).

2 Rational Design of Plates and Shells

2.1 Introduction

There are many works devoted to dynamics of plates and shells, however, relatively few deal with problems of varying thickness. A solution of this class of problems is perhaps the first but necessary step for addressing a wide range of more up-to-date class applications involving optimization *of plates and shells design according to required characteristics.*

Practical aspects of optimization to be addressed include realisation of construction with possibly minimal mass, but sustaining a given dynamic load; realisation of construction with an a priori given frequency or other dynamical characteristics.

Recently, an increase of interest in optimization problems has been observed, which is related to increased requirements of development of analytical as well as computational tools for dealing with this class of problems.

There are many valid optimization techniques, including variational methods. Numerical methods of solving various optimization problems have also been considered.

However, as it has been mentioned by many researchers [8, 138, 172] the application of necessary optimal conditions, in order to obtain and solve a system of closed equations, is very complicated even for simple designs, especially when strongly nonlinear equations are involved. Such systems are highly individualistic, and no universal methods for solving them exist to date.

Recently, more attention has been paid to methods which directly search for a minimum of the corresponding target function. Such techniques are particularly useful for complicated problems, such as optimization of rectangular plates and shells. In this case, we do not need to solve complicated nonlinear partial differential equations, and the search for a minimum of a target function may be realised by a wide choice of one of nonlinear programming methods.

On the other hand, nowadays, in real constructions with shell elements, new materials are applied, for example polymers or composites. This results from the observation that using, for example, polymers it is much simpler to construct a shell with variable thickness. These remarks lead to the conclusion

2.2 Orthotropic Plates and Shells with Variable Thickness and Low Stiffness

In this Chapter, different frequency spectra properties of Timoshenko-like shells will be described and the methods for free vibration solution of rectangular shells will be given. The theoretical results are applied to the analysis of plates with either constant or varying thickness and low transversal stiffness.

2.2.1 Frequency Spectra Properties of Orthotropic Shells with Variable Thickness

It is known that a shell is a system with an infinite number of degrees of freedom. It means that the number of frequencies is infinite. Stiffness and curvature coefficients as well as a thickness distribution along a plate surface have essential influence on the frequency spectra.

Below we consider linear and harmonic vibrations of a shallow, Timoshenko-like shell with varying thickness from the point of view of the spectrum distribution [234].

Suppose that the shell has a finite area Ω with a border S. For an isotropic shell with constant thickness, the system of equations using the Timoshenko kinematic model for the displacements, has been introduced by Naghdi [163]. In the case of an orthotropic shell with variable thickness, free vibrations have been described by a system of equations, which (after a slight generalisation) may be obtained on the basis of [119]:

$$K_y \frac{\partial^2 F}{\partial x^2} + K_x \frac{\partial^2 F}{\partial y^2} + \frac{2}{3}\frac{\partial}{\partial x}\left[A_{1313}h\left(\gamma_x + \frac{\partial w}{\partial x}\right)\right] +$$

$$\frac{2}{3}\frac{\partial}{\partial y}\left[A_{2323}h\left(\gamma_y + \frac{\partial w}{\partial y}\right)\right] + \omega^2 \rho h w = 0,$$

$$\frac{2}{3}\frac{\partial}{\partial x}\left[h^3\left(A_{1111}\frac{\partial \gamma_x}{\partial x} + A_{1122}\frac{\partial \gamma_y}{\partial y}\right)\right] + \frac{2}{3}\frac{\partial}{\partial y}\left[A_{1212}h^3\left(\frac{\partial \gamma_x}{\partial y} + \frac{\partial \gamma_y}{\partial x}\right)\right] -$$

$$\frac{2}{3}A_{1313}h\left(\gamma_x + \frac{\partial w}{\partial x}\right) + \frac{2}{3}\omega^2 \rho h^3 \gamma_x = 0,$$

$$\frac{2}{3}\frac{\partial}{\partial y}\left[h^3\left(A_{2222}\frac{\partial \gamma_y}{\partial y} + A_{1122}\frac{\partial \gamma_x}{\partial x}\right)\right] + \frac{2}{3}\frac{\partial}{\partial x}\left[A_{1212}h^3\left(\frac{\partial \gamma_x}{\partial y} + \frac{\partial \gamma_y}{\partial x}\right)\right] -$$

$$\frac{2}{3}A_{2323}h\left(\gamma_y + \frac{\partial w}{\partial y}\right) + \frac{2}{3}\omega^2 \rho h^3 \gamma_y = 0,$$

2.2 Orthotropic Plates and Shells with Variable Thickness and Low Stiffness

$$\frac{\partial^2}{\partial x^2}\left(a_{1111}h^{-1}\frac{\partial^2 F}{\partial x^2}\right)+\frac{\partial^2}{\partial x^2}\left(a_{1122}h^{-1}\frac{\partial^2 F}{\partial y^2}\right)+\frac{\partial^2}{\partial y^2}\left(a_{1122}h^{-1}\frac{\partial^2 F}{\partial x^2}\right)+$$
$$\frac{\partial^2}{\partial y^2}\left(a_{2222}h^{-1}\frac{\partial^2 F}{\partial y^2}\right)-\frac{\partial^2}{\partial x\partial y}\left(a_{1212}h^{-1}\frac{\partial^2 F}{\partial x\partial y}\right)+\frac{\partial^2}{\partial y^2}(k_x w)+\frac{\partial^2}{\partial x^2}(k_y w)=0.$$
(2.1)

Shearing forces, moments and membrane forces F, w, γ_x, γ_y fulfil the following conditions

$$M_{11}=\frac{2}{3}h^3\left(A_{1111}\frac{\partial\gamma_x}{\partial x}+A_{1122}\frac{\partial\gamma_y}{\partial y}\right),\quad (\overleftrightarrow{x,y})$$

$$M_{12}=\frac{2}{3}h^3 A_{1212}\left(\frac{\partial\gamma_x}{\partial y}+\frac{\partial\gamma_y}{\partial x}\right),\quad S=-\frac{\partial^2 F}{\partial x\partial y}=\frac{1}{2h}a_{1212}\varepsilon_{12},$$

$$Q_1=\frac{2}{3}A_{1313}h\left(\gamma_x+\frac{\partial w}{\partial x}\right),\quad T_1=\frac{1}{2h}(a_{2222}\varepsilon_1+a_{1122}\varepsilon_2),\quad (\overleftrightarrow{x,y}).$$

The boundary conditions on the boundary S will be formulated for a general case. Let $S=S_1+S_2+S_3$, where S_1 is the free support boundary, S_2 is the roller support and S_3 is the travelling support. Then we obtain

$$Q_n=M_n=M_\tau=0\quad\text{for}\quad S_1,\tag{2.2}$$

$$w=M_n=\gamma_\tau=0\quad\text{for}\quad S_2,\tag{2.3}$$

$$w=\gamma_n=\gamma_\tau=0\quad\text{for}\quad S_3,\tag{2.4}$$

$$F=\frac{\partial^2 F}{\partial n^2}=0\quad\text{for}\quad S.\tag{2.5}$$

We assume the following conditions relating to the stiffness coefficients and the introduced functions:

a) The following inequalities are satisfied:

$$0<A_H\le A_{ijmk}(x,y)\le A_B, 0<a_H\le a_{ijmk}(x,y)\le a_B,\tag{2.6}$$

$A_{ijmk}(x,y)$ and $a_{ijmk}(x,y)$ are boundary functions in the Ω space $(i,j,m,k=1,2,3)$;

b) For an arbitrarily $(x,y)\in\Omega$ and $\xi,\eta\in R^1$ there exists a constant $c_0>0$ such that

$$A_{1111}(x,y)\xi^2+2A_{1122}(x,y)\xi\eta+A_{2222}(x,y)\eta^2\ge c_0\left(\xi^2+\eta^2\right);\tag{2.7}$$

c) There exists a constant $c_1>0$, such that for all $(x,y)\in\Omega$ and $\xi,\eta\in R^1$ the following inequality is fulfilled

$$a_{1111}(x,y)\xi^2+(2a_{1122}(x,y)-a_{1212}(x,y))\xi\eta+a_{2222}(x,y)\eta^2\ge c_1\left(\xi^2+\eta^2\right);\tag{2.8}$$

96   2 Rational Design of Plates and Shells

d) $h(x, y)$ is a function limited on Ω, and for an arbitrarily taken $(x,y) \in \Omega$ it satisfies the inequalities

$$0 < h_H \leq h(x,y) \leq h_B . \tag{2.9}$$

It is seen, that the frequency Ω does not occur in the fourth equation of (2.1). Therefore, we can separately consider this equation with the boundary conditions (2.5). It can be expressed in the following form

$$\frac{\partial^2}{\partial x^2}\left(a_{1111}h^{-1}\frac{\partial^2 F}{\partial x^2}\right) + \frac{\partial^2}{\partial x^2}\left(a_{1122}h^{-1}\frac{\partial^2 F}{\partial y^2}\right) + \frac{\partial^2}{\partial y^2}\left(a_{1122}h^{-1}\frac{\partial^2 F}{\partial x^2}\right) +$$

$$\frac{\partial^2}{\partial y^2}\left(a_{2222}h^{-1}\frac{\partial^2 F}{\partial y^2}\right) - \frac{\partial^2}{\partial x \partial y}\left(a_{1212}h^{-1}\frac{\partial^2 F}{\partial x \partial y}\right) = -\frac{\partial^2}{\partial x^2}(k_y w) - \frac{\partial^2}{\partial y^2}(k_x w) . \tag{2.10}$$

We denote the differential operator on the left hand side of (2.10) by $G(\cdot)$. We introduce the following functional space $H_0^2(\Omega)$, which is the closure of the function set

$$V = \left\{ F \in C^\infty(\Omega) \,|\, F = \frac{\partial^2 F}{\partial n^2} = 0 \text{ on } S \right\}.$$

If we are able to prove that for an arbitrarily taken function W from the energetic space of the problem, equation (2.10) is solvable in F on the space $H_0^2(\Omega)$, then F may be deleted from the system (2.1) and therefore the dimension problem may be reduced.

THEOREM 2.1 *If the conditions (2.6) and (2.8) are fulfilled, then the G-th operator is symmetric, continuous and positively definite on $H_0^2(\Omega)$.*

Proof. Indeed, because G is a linear operator, then its continuity results directly from its restriction on $H_0^2(\Omega)$. Symmetry of G is obvious, whereas the positive definiteness results from [61] on the basis of the assumptions (2.6) and (2.8).

According to theorem 2.1, G transforms homeomorphically $H_0^2(\Omega)$ into $H^{-2}(\Omega)$. The inverse operator G^{-1} is a continuous and symmetric.

Let us introduce the functional-vectorial space

$$V_0 = \left\{ \vec{u} = (w, \gamma_x, \gamma_y) \in \left(H^4(\Omega)\right)^3 \Big| w_{(S_1+S_2)} = 0, \gamma_{\tau(S_2+S_3)} = 0, \gamma_{n(S_3)} = 0 \right\}.$$

If $\vec{u} \in V_0$, then $W \in H^1(\Omega)$. Therefore, for the finite curvatures k_x, k_y on Ω we have

$$\nabla^2_{\cdot,k} w = \frac{\partial^2}{\partial y^2}(k_x w) + \frac{\partial^2}{\partial x^2}(k_y w) \in H^{-2}(\Omega) .$$

Therefore, there exists a solution of equation (2.10), which can be written in the following form

2.2 Orthotropic Plates and Shells with Variable Thickness and Low Stiffness

$$F = G^{-1}\left(\nabla^2_{,k}w\right), \quad F \in H_0^2\left(\Omega\right).$$

The systems of equations (2.1) may be cast in the following operators form

$$Z[h]\vec{u} - \omega^2 M[h]\vec{u} = 0, \tag{2.11}$$

$$G(F) = -\nabla^2_{,k}w, \tag{2.12}$$

where $Z[h]$, $M[h]$ are linear differential operators defined by the expressions

$$Z[h] = \begin{pmatrix} L_{11} & L_{12} & L_{13} \\ L_{21} & L_{22} & L_{23} \\ L_{31} & L_{32} & L_{33} \end{pmatrix}, \tag{2.13}$$

$$M[h] = \begin{pmatrix} m_{11} & 0 & 0 \\ 0 & m_{22} & 0 \\ 0 & 0 & m_{33} \end{pmatrix}. \tag{2.14}$$

$$L_{11} = -\frac{2}{3}\left[\frac{\partial}{\partial x}\left(A_{1313}h\frac{\partial(\bullet)}{\partial x}\right) + \frac{\partial}{\partial y}\left(A_{2323}h\frac{\partial(\bullet)}{\partial y}\right)\right] + \nabla^2_k G^{-1}\left(\nabla^2_{,k}(\bullet)\right),$$

$$L_{12} = -\frac{2}{3}\frac{\partial}{\partial x}\left(A_{1313}h\bullet\right), \quad L_{13} = -\frac{2}{3}\frac{\partial}{\partial y}\left(A_{2323}h\bullet\right), \quad L_{21} = \frac{2}{3}A_{1313}h\frac{\partial(\bullet)}{\partial x},$$

$$L_{22} = -\frac{2}{3}\left[\frac{\partial}{\partial x}\left(A_{1111}h^3\frac{\partial(\bullet)}{\partial x}\right) + \frac{\partial}{\partial y}\left(A_{1212}h^3\frac{\partial(\bullet)}{\partial y}\right)\right] + \frac{2}{3}A_{1313}h(\bullet),$$

$$L_{23} = -\frac{2}{3}\left[\frac{\partial}{\partial x}\left(A_{1122}h^3\frac{\partial(\bullet)}{\partial y}\right) + \frac{\partial}{\partial y}\left(A_{1212}h^3\frac{\partial(\bullet)}{\partial x}\right)\right], \quad L_{31} = \frac{2}{3}A_{2323}h\frac{\partial(\bullet)}{\partial y},$$

$$L_{32} = -\frac{2}{3}\left[\frac{\partial}{\partial x}\left(A_{1212}h^3\frac{\partial(\bullet)}{\partial y}\right) + \frac{\partial}{\partial y}\left(A_{2211}h^3\frac{\partial(\bullet)}{\partial x}\right)\right],$$

$$L_{33} = -\frac{2}{3}\left[\frac{\partial}{\partial x}\left(A_{1212}h^3\frac{\partial(\bullet)}{\partial x}\right) + \frac{\partial}{\partial y}\left(A_{2222}h^3\frac{\partial(\bullet)}{\partial y}\right)\right] + \frac{2}{3}A_{2323}h(\bullet).$$

For m_{ii} $(i = 1, 2, 3)$ we get $m_{11} = h$, $m_{22} = m_{33} = 2h^3/3$. Let us introduce a bilinear form defining work on the virtual displacement $\vec{v} = (\tilde{w}, \tilde{\gamma}_x, \tilde{\gamma}_y)$:

$$a_h\left(\vec{u}, \vec{v}\right) = \int_\Omega \left\{\frac{2}{3}\left[A_{1111}h^3\frac{\partial\gamma_x}{\partial x}\frac{\partial\tilde{\gamma}_x}{\partial x} + A_{1122}h^3\frac{\partial\gamma_y}{\partial y}\frac{\partial\tilde{\gamma}_x}{\partial x} + A_{1122}h^3\frac{\partial\gamma_x}{\partial x}\frac{\partial\tilde{\gamma}_y}{\partial y} + \right.\right.$$

$$A_{2222}h^3\frac{\partial\gamma_y}{\partial y}\frac{\partial\tilde{\gamma}_y}{\partial y} + A_{1212}h^3\left(\frac{\partial\gamma_y}{\partial x} + \frac{\partial\gamma_x}{\partial y}\right)\left(\frac{\partial\tilde{\gamma}_y}{\partial x} + \frac{\partial\tilde{\gamma}_x}{\partial y}\right)\right] +$$

$$\frac{2}{3}\left[A_{1313}h\left(\gamma_x + \frac{\partial w}{\partial x}\right)\left(\tilde{\gamma}_x + \frac{\partial\tilde{w}}{\partial x}\right) + A_{2323}h\left(\gamma_y + \frac{\partial w}{\partial y}\right)\left(\tilde{\gamma}_y + \frac{\partial\tilde{w}}{\partial y}\right)\right] +$$

$$\nabla^2_k G^{-1}\left(\nabla^2_{,k}w\right)\tilde{w}\right\} d\Omega. \tag{2.15}$$

The following results related to $a_h(\vec{u}, \vec{v})$ hold.

THEOREM 2.2 Let the assumptions (2.6)–(2.9) be fulfilled and let k_x, k_y be finite on Ω. Then, the linear form (2.15) is continuous, symmetric, and positive in the $V_0 \times V_0$ area. It means that the following relations are true:

$$|a_h(\vec{u}, \vec{v})| \leq c_2 \|\vec{u}\|_{V_0} \|\vec{v}\|_{V_0}, \quad \forall \vec{u}, \vec{v} \in V_0, \tag{2.16}$$

$$a_h(\vec{u}, \vec{v}) = a_h(\vec{v}, \vec{u}) \quad \forall \vec{u}, \vec{v} \in V_0, \tag{2.17}$$

$$a_h(\vec{u}, \vec{u}) \geq 0 \quad \forall \vec{u} \in V_0, \quad c_2 = const > 0. \tag{2.18}$$

Proof.

A. Continuity. From (2.15) one gets:

$$|a_h(\vec{u}, \vec{v})| \leq \frac{2}{3} A_B h_B^3 \left(\left\|\frac{\partial \gamma_x}{\partial x}\right\|_{0,\Omega} \left\|\frac{\partial \tilde{\gamma}_x}{\partial x}\right\|_{0,\Omega} + \left\|\frac{\partial \gamma_y}{\partial y}\right\|_{0,\Omega} \left\|\frac{\partial \tilde{\gamma}_x}{\partial x}\right\|_{0,\Omega} + \right.$$

$$\left\|\frac{\partial \tilde{\gamma}_y}{\partial y}\right\|_{0,\Omega} \left\|\frac{\partial \gamma_x}{\partial x}\right\|_{0,\Omega} + \left\|\frac{\partial \gamma_y}{\partial y}\right\|_{0,\Omega} \left\|\frac{\partial \tilde{\gamma}_y}{\partial y}\right\|_{0,\Omega} \right) + |(G^{-1}(\nabla^2_{,k} w), \nabla^2_{,k} \tilde{w})| +$$

$$\frac{2}{3} A_B h_B \left(\left\|\gamma_x + \frac{\partial w}{\partial x}\right\|_{0,\Omega} \left\|\tilde{\gamma}_x + \frac{\partial \tilde{w}}{\partial x}\right\|_{0,\Omega} + \left\|\gamma_y + \frac{\partial w}{\partial y}\right\|_{0,\Omega} \left\|\tilde{\gamma}_y + \frac{\partial \tilde{w}}{\partial y}\right\|_{0,\Omega} \right) +$$

$$\frac{2}{3} A_B h_B^3 \left\|\frac{\partial \gamma_y}{\partial x} + \frac{\partial \gamma_x}{\partial y}\right\|_{0,\Omega} \left\|\frac{\partial \tilde{\gamma}_y}{\partial x} + \frac{\partial \tilde{\gamma}_x}{\partial y}\right\|_{0,\Omega}. \tag{2.19}$$

The last factor of (2.19) may be transformed into the following form

$$|(G^{-1}(\nabla^2_{,k} w), \nabla^2_{,k} \tilde{w})| = |-(F, \nabla^2_{,k} \tilde{w})| = |(\nabla^2_k F, \tilde{w})|.$$

Therefore

$$|(G^{-1}(\nabla^2_{,k} w), \nabla^2_{,k} \tilde{w})| \leq \|\nabla^2_k F\|_{0,\Omega} \cdot \|\tilde{w}\|_{0,\Omega}$$

$$\|\nabla^2_k F\|_{0,\Omega} \leq const. \|F\|_{0,\Omega}.$$

Because the operator G^{-1} is finite on $H^{-2}(\Omega)$, then

$$\|F\|_{2,\Omega} \leq const. \|\nabla^2_{,k} w\|_{-2,\Omega}.$$

Taking into account, that $\|\nabla^2_{,k} w\|_{-2,\Omega} \leq const \|w\|_{0,\Omega}$, and for finite k_x, k_y we get

$$|(G^{-1}(\nabla^2_{,k} w), \nabla^2_{,k} \tilde{w})| \leq const. \|w\|_{0,\Omega} \cdot \|\tilde{w}\|_{0,\Omega}. \tag{2.20}$$

Then, from (2.19) and (2.20) we get (2.16). This proves finiteness. Because the form $a_h(\vec{u}, \vec{v})$ is bilinear, then this property is equivalent to the continuity one.

2.2 Orthotropic Plates and Shells with Variable Thickness and Low Stiffness

B. Symmetry. The first seven terms of (2.15) are symmetric. Consider the last expression

$$\int_\Omega \nabla^2_k G^{-1}\left(\nabla^2_{,k} w\right) \tilde{w} d\Omega = \int_\Omega G^{-1}\left(\nabla^2_{,k} w\right) \nabla^2_{,k} \tilde{w} d\Omega = -\int_\Omega F \nabla^2_{,k} \tilde{w} d\Omega.$$

Equation (2.12) yields

$$\int_\Omega F \nabla^2_{,k} \tilde{w} d\Omega = -\int_\Omega FG\left(\tilde{F}\right) d\Omega = -\int_\Omega G(F) \tilde{F} d\Omega,$$

and in consequence one gets

$$\int_\Omega \nabla^2_k G^{-1}\left(\nabla^2_{,k} w\right) \tilde{w} d\Omega = \int_\Omega G(F) \tilde{F} d\Omega =$$

$$\int_\Omega \nabla^2_{,k} w G^{-1}\left(\nabla^2_{,k} \tilde{w}\right) d\Omega = \int_\Omega w \nabla^2_k G^{-1}\left(\nabla^2_{,k} \tilde{w}\right) d\Omega.$$

This proves the symmetry of the last term of (2.15).

C. Positiveness. Let $\overrightarrow{u} = \overrightarrow{v}$. Then $a_h(\overrightarrow{u}, \overrightarrow{u})$ presents the deformation energy

$$a_h(\overrightarrow{u}, \overrightarrow{u}) = \frac{2}{3}\int_\Omega \left\{ h^3\left[A_{1111}\left(\frac{\partial \gamma_x}{\partial x}\right)^2 + 2A_{1122}\frac{\partial \gamma_x}{\partial x}\frac{\partial \gamma_y}{\partial y} + \right.\right.$$

$$A_{2222}\left(\frac{\partial \gamma_y}{\partial y}\right)^2 \Bigg] + A_{1212} h^3\left(\frac{\partial \gamma_y}{\partial x} + \right.$$

$$\left.\frac{\partial \gamma_x}{\partial y}\right)^2 + A_{1313} h\left(\gamma_x + \frac{\partial w}{\partial x}\right)^2 + A_{2323} h\left(\gamma_y + \frac{\partial w}{\partial y}\right)^2 \Bigg\} d\Omega +$$

$$\int_\Omega \nabla^2_k G^{-1}\left(\nabla^2_{,k} w\right) w d\Omega.$$

However, the last term is equal to $\int_\Omega G(F)F d\Omega$ and is positive according to (2.8). The positiveness of the other terms results from (2.6) and (2.8). In order to analyse the vibration of the Timoshenko-like shell and taking into account a rotation energy, the following bilinear form is introduced

$$b_h(\overrightarrow{u}, \overrightarrow{v}) = \int_\Omega \left[hw\tilde{w} + \frac{2}{3}h^3 \gamma_x \tilde{\gamma}_x + \frac{2}{3}h^3 \gamma_y \tilde{\gamma}_y \right] d\Omega, \tag{2.21}$$

which is proportional to the kinetic energy of the shell.

It is easy to check, using (2.21), that the form is also symmetric, positive and continuous.

Consider now the problem related to the free vibrations of the Timoshenko-like shell from another point of view. Let us find the nontrivial functions w, γ_x, γ_y, for which the total energy on an arbitrarily taken and kinematically allowed virtual displacement is equal to zero. It means that

$$a_h(\vec{u}, \vec{v}) = \omega^2 b_h(\vec{u}, \vec{v}) \quad \text{for every } \vec{v} \in V_0. \tag{2.22}$$

Therefore, the problem of finding the free vibrations frequency spectra of a shell is reduced to a classical problem of eigenvalues [109]. We can formulate the following result, which is analogical to that given in [155] for a plate.

THEOREM 2.3 *The problem (2.22) has a discrete series of non negative eigenvalues $\{\lambda_i\}_{i=1}^{\infty}$. The eigenfunctions (modes) related to district eigenvalues are mutually orthogonal.*

Proof. The discreteness of the spectra of (2.22) results from the symmetry of the bilinear forms $a_h(\vec{u}, \vec{v})$ and $b_h(\vec{u}, \vec{v})$, and also from the existence of the invertibility of operator $M[h]$. It is proved in a way that is analogous to the theorem proof given in reference [155].

Let us prove the positiveness of the eigenvalues λ_i, $i = 1, 2, ...$ Let λ_i be a certain eigenvalue and let $\vec{u}_i$ be the corresponding eigenfunction. Then, taking $\vec{v} = \vec{u}_i$ we obtain

$$a_h(\vec{u}_i, \vec{u}_i) - \lambda_i b_h(\vec{u}_i, \vec{u}_i) = 0,$$

or

$$\lambda_i = \frac{a_h(\vec{u}_i, \vec{u}_i)}{b_h(\vec{u}_i, \vec{u}_i)}.$$

Because $a_h(\vec{u}_i, \vec{u}_i) \geq 0$ and $b_h(\vec{u}_i, \vec{u}_i) > 0$, then $\lambda_i \geq 0$ for all $i = 1, 2, ...$. Let $\vec{u}_i$ and $\vec{u}_j$ be the eigenfunctions corresponding to λ_i and λ_j, and $\lambda_i \neq \lambda_j$. Then we get

$$a_h(\vec{u}_i, \vec{u}_j) - \lambda_i b_h(\vec{u}_i, \vec{u}_j) = 0, \tag{2.23}$$

$$a_h(\vec{u}_j, \vec{u}_i) - \lambda_j b_h(\vec{u}_j, \vec{u}_i) = 0. \tag{2.24}$$

Subtracting (2.23) from (2.24) and taking into account the symmetry of the bilinear forms we get

$$(\lambda_j - \lambda_i) b_h(\vec{u}_i, \vec{u}_j) = 0.$$

Because $\lambda_i \neq \lambda_j$, then

$$b_h(\vec{u}_i, \vec{u}_j) = 0, \tag{2.25}$$

which proves orthogonality of the eigenfunctions $\vec{u}_j$ and $\vec{u}_i$.

Finally, one observes that for the eigenvalues of the problem (2.22) the following Rayleigh's quotient is true [9]:

$$\lambda_i = \min_{\substack{\vec{u} \in V_0 \\ \vec{u} \perp u_j}} \frac{a_h(\vec{u}, \vec{u})}{b_h(\vec{u}, \vec{u})}, \quad i = 1, 2, ..., \quad j = 1, 2, ..., i-1. \quad (2.26)$$

The sign $\perp$ denotes generalised orthogonality in the sense of (2.25), i.e.

$$\vec{u} \perp \vec{u}_j \Leftrightarrow b_h(\vec{u}, \vec{u}_j) = 0.$$

The ration (2.26) yields that the lowest eigenvalue λ_1 or the lowest vibration frequency $\omega_1^2 = \lambda_1$ is defined by,

$$\omega_1^2 = \min_{\vec{u} \in V_0} \frac{a_h(\vec{u}, \vec{u})}{b_h(\vec{u}, \vec{u})}, \quad \vec{u} \neq 0. \quad (2.27)$$

The values of the tension function F, corresponding to a certain vibration mode, are defined by (2.12) and have the form

$$F_i = -G^{-1}\left(\nabla^2_{,k} w_i\right),$$

where w_i is the first component of the displacement vector $\vec{u}_i$ corresponding to $\lambda_i = \omega_i^2$.

Numerical methods for computing the eigensolutions using (2.22) and (2.26) will be described in the sections 2.2.2 and 2.2.3.

2.2.2 Finite Dimensional Approximation of Free Vibrations of the Timoshenko Like Shell

General remarks and comparison of the methods. Computing the dynamics, and particularly the frequency spectra of free vibrations of plates and shells on the basis of the Kirchhoff-Love model is a rather complicated problem [84]. Further transition to more accurate models, such as Timoshenko's model, increase the difficulties. These are caused by the complex nature of the differential equations describing Timoshenko's model. This model, contrary to the Kirchhoff-Love equations, has two unknown functions, namely γ_z and γ_y.

Nowadays, the problem of the dynamics of symmetric vibrations of plates and Timoshenko-like shells posses no technical challenges [203]. However, for other types of shells (including rectangular ones) the computational problem is still open, since it corresponds to the problem of non-constant elasticity coefficients, and of varying thickness. The first aspect is addressed through discretization. It seems that the finite difference method gives some hope [112]. Indeed, this method may be applied independently of the class of governing equations of motion. However, in order to get more or less reasonable results, the required number of algebraic equations is high [41]. This particularly holds regarding the second aspect, the problem of variable thickness. Besides, for Timoshenko-like equations with four unknowns, the number of algebraic equations increases four times. This can lead to serious numerical problems.

It appears that the finite element method is more effective although it has similar drawbacks. Its advantages occur in the area of complicated shapes. For rectangular plates and shells, we need simple and economical methods from the point of view of calculations.

For all considered problems in this chapter we assume that the coefficients A_{ijmk}, a_{ijmk} and the curvatures k_x, k_y do not depend on $(x, y) \in \Omega$, and the thickness function $h(x, y)$ is symmetric in relation to the coordinates axes which originate from the centre of the surface plate.

It seems that the most suitable methods for solving these problems are the Kantorovitch method, the method of variational iterations (MVI) and the Bubnov-Galerkin method with high order approximates (MB) [155]. The first two methods allow for reduction of the original two dimensional problem to one dimensional. Therefore, it is possible to decrease the order of the system of algebraic equations that govern the dynamics. In order to solve one dimensional problems, the finite difference as well as the finite element methods may be used. The Bubnov-Galerkin method is very suitable because for Timoshenko-like equations, on rectangular areas, it is easy to construct the functions fulfilling boundary conditions of rolling and stiff supports [119]. The occurrence of variable thickness $h(x, y)$ in the equations complicates the use of this method. However as shown in section 2.2.4 an appropriate choice of the test functions in this case, leads to quick convergence.

Further, we consider two methods of analysis of plates and shells with variable thickness namely MVI and MB.

Approximation by MVI. The method of variational iterations applied to the problems related to dynamics consists of the following steps.

Suppose, that we want to determine free vibration frequencies of a shell from the equations

$$Z[h]\overrightarrow{u} - \omega^2 M[h]\overrightarrow{u} = 0, \qquad (2.28)$$

using the corresponding boundary conditions. The variables $\overrightarrow{u} \in V_0$ are kinematically allowed displacements as discussed in section 2.2.1. Let $\Omega = [0, l_1] \times [0, l_2]$. Assume, that the sought solution has the following form

$$\overrightarrow{u}(x, y) = \sum_{i=1}^{N} \left(\overrightarrow{u}_i^1(x) \otimes \overrightarrow{u}_i^2(y) \right), \quad N = 1, 2, ..., \qquad (2.29)$$

where $\left(\overrightarrow{u}_i^1(x) \otimes \overrightarrow{u}_i^2(y) \right)$ is a vector with components $w_i^1(x)w_i^2(y)$, $\gamma_{xi}^1(x)\gamma_{xi}^2(y)$, $\gamma_{yi}^1(x)\gamma_{yi}^2(y)$.

Substituting (2.29) to (2.28) we get

2.2 Orthotropic Plates and Shells with Variable Thickness and Low Stiffness

$$\sum_{i=1}^{N} Z[h] \left(\vec{u}_i^{\,1}(x) \otimes \vec{u}_i^{\,2}(y) \right) = \omega^2 \sum_{i=1}^{N} M[h] \left(\vec{u}_i^{\,1}(x) \otimes \vec{u}_i^{\,2}(y) \right). \tag{2.30}$$

Applying the Bubnov-Galerkin method we project the equations (2.30) on the functions $\{\vec{u}_j^{\,1}(x)\}_{j=1}^N$ and $\{\vec{u}_j^{\,2}(y)\}_{j=1}^N$. As a result, we get the $2N$ differential equations of the form

$$\sum_{i=1}^{N} \left(\int_0^{l_1} Z[h] \left(\vec{u}_i^{\,1}(x) \otimes \vec{u}_i^{\,2}(y) \right) \cdot \vec{u}_j^{\,1}(x) dx - \right.$$

$$\left. \omega^2 \int_0^{l_1} M[h] \left(\vec{u}_i^{\,1}(x) \otimes \vec{u}_i^{\,2}(y) \right) \cdot \vec{u}_j^{\,1}(x) dx \right) = 0, \tag{2.31}$$

$$\sum_{i=1}^{N} \left(\int_0^{l_2} Z[h] \left(\vec{u}_i^{\,1}(x) \otimes \vec{u}_i^{\,2}(y) \right) \cdot \vec{u}_j^{\,2}(y) dy - \right.$$

$$\left. \omega^2 \int_0^{l_2} M[h] \left(\vec{u}_i^{\,1}(x) \otimes \vec{u}_i^{\,2}(y) \right) \cdot \vec{u}_j^{\,2}(y) dy \right) = 0, \tag{2.32}$$

$$(j = 1, 2, ..., N).$$

In order to solve the equations (2.31) and (2.32) the iteration method will be used. As the first step, assuming certain initial approximations for $\vec{u}_i^{\,1}(x)$, $i = 1, 2, ..., N$ and substituting them to (2.31), a set of N linear ordinary differential equations in regard to the unknown functions $\vec{u}_i^{\,2}(y)$, $i = 1, 2, ..., N$ is obtained. The obtained system has a nonzero solution only for certain discrete values of ω_k^2, which can be treated as first approximations to the sought free vibration frequencies of the shell. The corresponding modes are denoted by $\{\vec{u}_{ik}^{\,2}(y)\}_{i=1}^N$. The obtained functions $\{\vec{u}_{ik}^{\,2}(y)\}_{i=1}^N$ for (k temporarily fixed) are substituted to (2.32). As a result, we get a set of N linear ordinary homogeneous equations in terms of $\{\vec{u}_i^{\,1}(x)\}_{i=1}^N$, which provides the estimates ω_{km}^2. For a fixed 'm' the corresponding functions $\{\vec{u}_{ik}^{\,2}(y)\}_{i=1}^N$ are substituted to (2.31) and then the process proceeds to the next iteration.

The iterations are managed in a two-step way. The iteration convergence process may be controlled during each step, or after the step when the values for ω_{km}^2 are obtained.

In each iteration for fixed 'k' and 'm' we get the estimates ω_{km}^2 and the corresponding vibration modes

$$\vec{u}_{km}(x,y) = \sum_{i=1}^{N} \left(\vec{u}_{im}^{\,1}(x) \otimes \vec{u}_{ik}^{\,2}(y) \right). \tag{2.33}$$

104   2 Rational Design of Plates and Shells

The operator $Z[h]$ includes the operation $G^{-1}(\nabla^2_{,k}w)$. It means that during each step of the variational iterations we should take into account the relation defined by the equation (2.39). For the approximation of that equation the same variational iteration algorithm in used.

Approximation using the Bubnov-Galerkin method. In the variational iteration method the basis functions $\vec{u}^1_i(x)$ and $\vec{u}^2_i(y)$ are not assumed a priori, and they are found by means of optimal solutions. By contrast, in the Bubnov-Galerkin method we assume that

$$\vec{u}(x,y) = \sum_{i,j=1}^{N} \vec{f}_{ij}\left(\vec{u}^1_i(x) \otimes \vec{u}^2_j(y)\right), \qquad (2.34)$$

where $\vec{u}^1_i(x)$ and $\vec{u}^2_j(y)$ are a priori given systems of independent linear functions fulfilling boundary conditions and having closure properties. Substituting (2.34) to (2.28) and projecting the obtained system of equations on $\left(\vec{u}^1_l(x) \otimes u^2_p(y)\right)$ we get the following system of homogeneous algebraic equations

$$\sum_{i,j=1}^{N} \vec{f}_{ij} \int_{\Omega} Z[h]\left(\vec{u}^1_i(x) \otimes \vec{u}^2_j(y)\right) \cdot \left(\vec{u}^1_p(x) \otimes \vec{u}^2_p(y) d\Omega\right) -$$

$$\omega^2 \sum_{i,j=1}^{N} \vec{f}_{ij} \int_{\Omega} M[h]\left(\vec{u}^1_i(x) \otimes \vec{u}^2_j(y)\right) \cdot \left(\vec{u}^1_p(x) \otimes \vec{u}^2_p(y) d\Omega\right) = 0, \quad (2.35)$$

$$(l, p = 1, 2, ..., N).$$

The solvability condition for the system (2.35) leads to the N^2 approximate values of the ω_{km} frequencies ($k, m = 1, 2, ..., N$) of the free vibrations of the plate.

convergence of the MVI and MB approximations. convergence of the MVI method applied in the free vibration analysis problems has not been investigated earlier. The MVI convergence for positively defined and symmetric operators has been proved [106]. We illustrate the MVI convergence in problems related to free vibration analysis. Let the operators $Z[h]$ and $M[h]$ in (2.28) be positively defined. According to the theorem (2.2), this condition is fulfilled. The approximation of this problem using the MVI method has been considered earlier. As a result, the problem has been reduced to computing the eigenfunctions of the set of ordinary differential equations (2.31) and (2.32).

Similarly to reference [106] we define an MVI step as a calculation on the basis of the system $\{\vec{u}^1_i(x)\}^{(p)}$ or $\{\vec{u}^2_i(y)\}^{(p)}$, with the next iterates being the systems $\{\vec{u}^2_i(y)\}^{(p+1)}$ or $\{\vec{u}^1_i(x)\}^{(p+1)}$, respectively.

2.2 Orthotropic Plates and Shells with Variable Thickness and Low Stiffness

For a given and fixed 'N' after 'p' steps we get, in analogy to (2.33), the free vibration frequencies $\omega_{km}^{(p,N)}$ and the corresponding eigenfunctions, of the form

$$\vec{u}_{km}^{(p,N)}(x,y) = \sum_{i=1}^{N} \left(\vec{u}_{im}^{1(p-1)}(x) \otimes \vec{u}_{ik}^{2(p)}(y) \right). \tag{2.36}$$

which fulfil the following equations

$$\sum_{i=1}^{N} \left\{ \int_0^{l_1} Z[h] \left(\vec{u}_{im}^{1(p-1)}(x) \otimes \vec{u}_{ik}^{2(p)}(y) \right) \cdot \vec{u}_{jm}^{1(p-1)}(x) dx - \left(\omega_{km}^{(p,N)} \right)^2 \int_0^{l_1} M[h] \left(\vec{u}_{im}^{1(p-1)}(x) \otimes \vec{u}_{ik}^{2(p)}(y) \right) \cdot \vec{u}_{jm}^{1(p-1)}(x) dx \right\} = 0, \tag{2.37}$$

$$(j = 1, 2, ..., N).$$

THEOREM 2.4 *Let $Z[h]$ and $M[h]$ be positively defined operators $\forall h \in U_\delta$ in the corresponding energetic space $H(\Omega)$ (where $D(M) \subset D(Z) \subset H(\Omega)$). Then, the frequency series $\omega_{11}^{(p,N)}$ does not increase in relation to 'p' for an arbitrarily fixed number 'N'. In addition it is bounded from below by the first free vibration frequency ω_{11}^T of the shell (the case of higher frequencies is not considered).*

Proof. Consider the system (2.37) for $k = m = 1$. Multiply it by $\vec{u}_{j1}^{2(p)}(y)$, then perform the sum with respect to 'j' and integrate it in terms of 'y' in the interval $[0, l_2]$. We can arbitrarily take [109]:

$$\int_0^{l_1} \int_0^{l_2} M[h] \vec{u}_{11}^{(p,N)} \vec{u}_{11}^{(p,N)} dxdy = 1. \tag{2.38}$$

Then we get

$$\left(\omega_{11}^{(p,N)} \right)^2 = \int_0^{l_1} \int_0^{l_2} Z[h] \vec{u}_{11}^{(p,N)} \vec{u}_{11}^{(p,N)} dxdy.$$

On the other hand, according to the minimal properties satisfied by the fundamental frequency FRC we have

$$\left(\omega_{11}^T \right)^2 \leq \left(\omega_{11}^{(p,N)} \right)^2 \leq \int_0^{l_1} \int_0^{l_2} Z[h] \vec{v}(x,y) \vec{v}(x,y) dxdy,$$

for an arbitrarily taken $\vec{v}(x,y)$ function with the shape (2.36), only if $\vec{v}(x,y)$ satisfies the norm condition (2.38). In particular, when we take $\vec{v}(x,y) = \vec{u}_{11}^{(p-1,N)}(x,y)$ then

$$\left(\omega_{11}^{(p,N)}\right)^2 \leq \int_0^{l_1}\int_0^{l_2} Z[h]\vec{u}_{11}^{(p-1,N)}\vec{u}_{11}^{(p-1,N)}dxdy = \left(\omega_{11}^{(p-1,N)}\right)^2,$$

and

$$\left(\omega_{11}^T\right)^2 \leq \left(\omega_{11}^{(p,N)}\right)^2 \leq \left(\omega_{11}^{(p-1,N)}\right)^2. \qquad (2.39)$$

This proves the theorem.

Corollary 2.1 *If each element of the base space system $H(\Omega)$ has the form $\vec{\Theta}_i(x,y) = \vec{\varphi}_i(x) \otimes \vec{\psi}_i(y)$, where $\vec{\varphi}_i \in H([0,l_1])$ and $\vec{\psi}_i \in H([0,l_2])$, and $\{\vec{\varphi}_i(x)\}$ or $\{\vec{\phi}_i(y)\}$ are taken as the initial functions, then for the arbitrary MVI iteration step the following inequality is satisfied*

$$\left(\omega_{11}^{(p,N)}\right)^2 \leq \left(\omega_{11}^{(\sigma,N)}\right)^2, \qquad (2.40)$$

where $\omega_{11}^{(\sigma,N)}$ is the fundamental frequency, obtained using the MB method for a projection on the basis system $\left\{\vec{\Theta}_i(x,y)\right\}_{i=1}^N$.

Proof. According to (2.39) it is sufficient to prove (2.40) for $p=1$. Let us take

$$\vec{u}_{11}^{(1,N)}(x,y) = \sum_{i=1}^N \left(\vec{\varphi}_i(x) \otimes \vec{u}_{i1}^2(y)\right).$$

Then, in order to define $\omega_{11}^{(1,N)}$ and $\vec{u}_{il}^2(y)$ we use the following equations

$$\sum_{i=1}^N \left\{\int_0^{l_1} Z[h]\left(\vec{\varphi}_i(x) \otimes \vec{u}_{i1}^2(y)\right) \cdot \vec{\varphi}_j(x)dx - \left(\omega_{11}^{(1,N)}\right)^2 \int_0^{l_1} M[h]\left(\vec{\varphi}_i(x) \otimes \vec{u}_{i1}^2(y)\right) \cdot \vec{\varphi}_j(x)dx\right\} = 0,$$

$$(j = 1,2,...,N).$$

From the above we get that:

$$\left(\omega_{11}^{(1,N)}\right)^2 \leq \int_0^{l_1}\int_0^{l_2} Z[h] \sum_{i=1}^N \left(c_i\vec{\varphi}_i(x) \otimes \vec{\psi}_i(y)\right) \cdot \sum_{j=1}^N \left(c_j\vec{\varphi}_j(x) \otimes \vec{\psi}_j(y)\right) dxdy,$$

for the arbitrary series $\{c_i\}_{i=1}^N$ whose combination $\sum_{i=1}^N \left(c_i\vec{\varphi}_i(x) \otimes \vec{\psi}_i(y)\right)$ satisfies (2.38). Let $c_i = c_i^\sigma$ be the coefficients defined by the MB method. Then we get

2.2 Orthotropic Plates and Shells with Variable Thickness and Low Stiffness

$$\left(\omega_{11}^{(1,N)}\right)^2 \le \left(\omega_{11}^{(\sigma,N)}\right)^2,$$

and the theorem (2.40) is proved.

Corollary 2.2 *If the MB convergence is proved for the problem (2.28), then for fixed 'p' the MVI method also converges with $N \to \tau$.*

Proof. Linking the inequalities (2.40) and (2.39) we get

$$\left(\omega_{11}^T\right)^2 \le \left(\omega_{11}^{(p,N)}\right)^2 \le \left(\omega_{11}^{(\sigma,N)}\right)^2, \qquad \forall p = 1, 2, \dots . \qquad (2.41)$$

If the convergence of the MB method is proved, then $\omega_{11}^{(\sigma,N)} \to \omega_{11}^T$ for $N \to \equiv$ and, irrespective of the iteration number, one obtains

$$\lim_{N \to infty} \omega_{11}^{(p,N)} = \omega_{11}^T .$$

Therefore, in all cases, when MB is convergent, MVI is also convergent but slower then the MB.

For an approximation based on the MB method, detailed results are considered in [155, 237]. Theorem (2.2) allows for the application of those results to our problem, where according to corollary 2.2, MVI is also convergent.

2.2.3 Algorithms for the MB and MVI Methods

As mentioned in section 2.2.2, in order to solve the problem of the free vibrations of a conical Timoshenko-like shell $\Omega = [0, l_1] \times [0, l_2]$, the MB and MVI methods will be used and the corresponding algorithms will be given. It is assumed that the stiffness coefficients are defined by the coordinates function $h = h(x, y)$.

Consider the system of equations (2.28), and introduce the following dimensionless quantities [119]

$$\bar{x} = x/l_1, \; \bar{y} = y/l_2, \; \lambda = l_1/l_2, \; \bar{h} = h/h_0, \; \lambda_1 = l_1/2h_0, \; \lambda_2 = l_2/2h_0,$$

$$\bar{w} = w/2h_0, \; \bar{\gamma}_x = \lambda_1 \gamma_x, \; \bar{\gamma}_y = \lambda_2 \gamma_y, \; \bar{A}_{ijkl} = A_{ijkl}/A_{1111},$$

$$\bar{a}_{ijkl} = a_{ijkl} \cdot A_{1111}, \; \bar{\lambda}_1 = \lambda_2^2 \bar{A}_{1313}, \; \bar{\lambda}_2 = \lambda_1^2 \bar{A}_{2323}, \; \bar{k}_x = k_x l_1^2/2h_0,$$

$$\bar{k}_y = k_y l_2^2/2h_0, \; \bar{F} = F/(8h_0^3 A_{1111}), \; \bar{\omega}^2 = l_1^2 l_2^2 \omega^2/(8h_0^3 A_{1111}).$$

The obtained nondimensional set of equations has the following form, where bars have been omitted:

$$K_y \frac{\partial^2 F}{\partial x^2} + K_x \frac{\partial^2 F}{\partial y^2} + \frac{2}{3}\bar{\lambda}_1 \frac{\partial}{\partial x}\left[h\left(\gamma_x + \frac{\partial w}{\partial x}\right)\right] +$$

$$\frac{2}{3}\bar{\lambda}_2 \frac{\partial}{\partial y}\left[h\left(\gamma_y + \frac{\partial w}{\partial y}\right)\right] + \omega^2 hw = 0, \tag{2.42}$$

$$\frac{1}{12}\frac{\partial}{\partial x}\left[\lambda^{-2}h^3 A_{1111}\frac{\partial \gamma_x}{\partial x} + h^3 A_{1122}\frac{\partial \gamma_y}{\partial y}\right] + \frac{1}{12}A_{1212}\frac{\partial}{\partial y}\left[h^3\left(\frac{\partial \gamma_x}{\partial y} + \right.\right.$$
$$\left.\left.\frac{\partial \gamma_y}{\partial x}\right)\right] - \frac{2}{3}\bar{\lambda}_1 h\left(\gamma_x + \frac{\partial w}{\partial x}\right) + \frac{1}{12}\lambda_1^{-2}\omega^2 h^3 \gamma_x = 0, \tag{2.43}$$

$$\frac{1}{12}\frac{\partial}{\partial y}\left[\lambda^2 h^3 A_{2222}\frac{\partial \gamma_y}{\partial y} + h^3 A_{1122}\frac{\partial \gamma_x}{\partial x}\right] + \frac{1}{12}A_{1212}\frac{\partial}{\partial x}\left[h^3\left(\frac{\partial \gamma_y}{\partial x} + \right.\right.$$
$$\left.\left.\frac{\partial \gamma_x}{\partial y}\right)\right] - \frac{2}{3}\bar{\lambda}_2 h\left(\gamma_y + \frac{\partial w}{\partial y}\right) + \frac{2}{3}\lambda_2^{-2}\omega^2 h^3 \gamma_y = 0, \tag{2.44}$$

$$\frac{\partial^2}{\partial x^2}(k_y w) + \frac{\partial^2}{\partial y^2}(k_x w) + \frac{\partial^2}{\partial x^2}\left[\lambda^{-4}a_{1111}\frac{\partial^2 F}{\partial x^2} + h^{-1}a_{1122}\frac{\partial^2 F}{\partial y^2}\right] +$$
$$\frac{\partial^2}{\partial y^2}\left[\lambda^4 h^{-1}a_{2222}\frac{\partial^2 F}{\partial y^2} + h^{-1}a_{1122}\frac{\partial^2 F}{\partial x^2}\right] - a_{1212}\frac{\partial^2}{\partial x \partial y}\left(h^{-1}\frac{\partial^2 F}{\partial x \partial y}\right) = 0. \tag{2.45}$$

Consider the following boundary conditions:

a) Movable support: $w = \gamma_x = \gamma_y = F = \left.\dfrac{\partial^2 F}{\partial n^2}\right|_S = 0$;

b) Rolling support: $w = F = \left.\dfrac{\partial^2 F}{\partial n^2}\right|_S = 0$, $\gamma_y = M_{11} = 0$ for $x = \text{const.}$, $\gamma_x = M_{22} = 0$ for $y = \text{const.}$

Assume that

$$w = w_1(x) \cdot w_2(y), \quad \gamma_x = \varphi_1(x) \cdot \varphi_2(y), \quad \gamma_y = \psi_1(x) \cdot \psi_2(y),$$
$$F = F_1(x) \cdot F_2(y). \tag{2.46}$$

Substituting (2.46) to (2.42)–(2.45) we get

$$k_x F_1 F_{2,yy} + k_y F_2 F_{1,xx} + \tfrac{2}{3}\bar{\lambda}_1 \varphi_2 (h\varphi_1)_{,x} + \tfrac{2}{3}\bar{\lambda}_1 w_2 (hw_{1,x})_{,x} +$$
$$+ \tfrac{2}{3}\bar{\lambda}_2 \psi_1 (h\psi_2)_{,y} + \tfrac{2}{3}\bar{\lambda}_2 w_1 (hw_{2,y})_{,y} + \omega^2 h w_1 w_2 = 0, \tag{2.47}$$

$$\tfrac{1}{12}\lambda^{-2} A_{1111} \varphi_2 (h^3 \varphi_{1,x})_{,x} + \tfrac{1}{12} A_{1122} \psi_{2,y} (h^3 \psi_1)_{,x} +$$
$$\tfrac{1}{12} A_{1212} \psi_{1,x}(h^3 \psi_2)_{,y} + \tfrac{1}{12} A_{1212} \varphi_1 (h^3 \varphi_{2,y})_{,y} -$$
$$\tfrac{2}{3}\bar{\lambda}_1 h w_{1,x} w_2 - \tfrac{2}{3}\bar{\lambda}_1 h \varphi_1 \varphi_2 + \tfrac{1}{12}\lambda_1^{-2}\omega^2 h^3 \varphi_1 \varphi_2 = 0, \tag{2.48}$$

$$\tfrac{1}{12}\lambda^2 A_{2222} \psi_1 (h^3 \psi_{2,y})_{,y} + \tfrac{1}{12} A_{1122} \varphi_{1,x}(h^3 \varphi_2)_{,y} +$$
$$\tfrac{1}{12} A_{1212} \varphi_{2,y} (h^3 \varphi_1)_{,x} + \tfrac{1}{12} A_{1212} \psi_2 (h^3 \psi_{1,x})_{,x} -$$
$$\tfrac{2}{3}\bar{\lambda}_2 h w_1 w_{2,y} - \tfrac{2}{3}\bar{\lambda}_2 h \, \psi_1 \psi_2 + \tfrac{1}{12}\lambda_2^{-2}\omega^2 h^3 \psi_1 \psi_2 = 0, \tag{2.49}$$

2.2 Orthotropic Plates and Shells with Variable Thickness and Low Stiffness

$$w_1(k_x w_2)_{,yy} + w_2(k_y w_1)_{,xx} + \lambda^{-4} a_{1111} F_2 (h^{-1} F_{1,xx})_{,xx} +$$
$$a_{1122} F_{2,yy}(h^{-1} F_1)_{,xx} + a_{1122} F_{1,xx}(h^{-1} F_2)_{,yy} +$$
$$\lambda^4 a_{2222} F_1 (h^{-1} F_{2,yy})_{,yy} - a_{1212}(h^{-1} F_{1,x} F_{2,y})_{,xy} = 0. \quad (2.50)$$

Let w_1, φ_1, ψ_1 and F_1 be known. Then, multiplying (2.47)–(2.50) correspondingly by w_1, φ_1, y_1, F_1 and integrating in the interval $[0, 1]$ with respect to x we get the following differential equations governing w_2, φ_2, ψ_2 and F_2

$$F_{2,yy} \int_0^1 k_x F_1 w_1 dx + F_2 \int_0^1 k_y F_{1,xx} w_1 dx + \frac{2}{3} \bar{\lambda}_1 \varphi_2 \int_0^1 (h\varphi_1)_{,x} w_1 dx +$$
$$\frac{2}{3} \bar{\lambda}_1 w_2 \int_0^1 (h w_{1,x})_{,x} w_1 dx + \frac{2}{3} \bar{\lambda}_2 (\psi_2 \int_0^1 h \psi_1 w_1 dx)_{,y} +$$
$$\frac{2}{3} \bar{\lambda}_2 (w_{2,y} \int_0^1 h w_1 w_1 dx)_{,y} + \omega^2 w_2 \int_0^1 h w_1 w_1 dx = 0, \quad (2.51)$$

$$\frac{1}{12} \lambda^{-2} A_{1111} \varphi_2 \int_0^1 (h^3 \varphi_{1,x})_{,x} \varphi_1 dx + \frac{1}{12} A_{1122} \psi_{2,y} \int_0^1 (h^3 \psi_1)_{,x} \varphi_1 dx +$$
$$\frac{1}{12} A_{1212} (\psi_2 \int_0^1 h^3 \psi_{1,x} \varphi_1 dx)_{,y} + \frac{1}{12} A_{1212} (\varphi_{2,y} \int_0^1 h^3 \varphi_1 \varphi_1 dx)_{,y} -$$
$$\frac{2}{3} \bar{\lambda}_1 \varphi_2 \int_0^1 h \varphi_1 \varphi_1 dx - \frac{2}{3} \bar{\lambda}_1 w_2 \int_0^1 h w_{1,x} \varphi_1 dx + \frac{\omega^2}{12} \lambda_1^{-2} \varphi_2 \int_0^1 h^3 \varphi_1 \varphi_1 dx = 0,$$
$$(2.52)$$

$$\frac{1}{12} \lambda^2 A_{2222} (\psi_{2,y} \int_0^1 h^3 \psi_1 \psi_1 dx)_{,y} + \frac{1}{12} A_{1122} (\varphi_2 \int_0^1 h^3 \varphi_{1,x} \psi_1 dx)_{,y} +$$
$$\frac{1}{12} A_{1212} \varphi_{2,y} \int_0^1 (h^3 \varphi_1)_{,x} \psi_1 dx + \frac{1}{12} A_{1212} \psi_2 \int_0^1 (h^3 \psi_{1,x})_{,x} \psi_1 dx -$$
$$\frac{2}{3} \bar{\lambda}_2 \psi_2 \int_0^1 h \psi_1 \psi_1 dx - \frac{2}{3} \bar{\lambda}_2 w_{2,y} \int_0^1 h w_1 \psi_1 dx + \frac{\omega^2}{12} \lambda_2^{-2} \psi_2 \int_0^1 h^3 \psi_1 \psi_1 dx = 0,$$
$$(2.53)$$

$$(w_2 \int_0^1 k_x w_1 F_1 dx)_{,yy} + w_2 \int_0^1 (k_y w_1)_{,xx} F_1 dx + \lambda^{-4} a_{1111} F_2 \int_0^1 (h^{-1} F_{1,xx})_{,xx} F_1 dx +$$

$$a_{1122}F_{2,yy}\int_0^1 (h^{-1}F_1)_{,xx}F_1 dx + a_{1122}(F_2\int_0^1 h^{-1}F_{1,xx}F_1 dx)_{,yy}+$$

$$\lambda^4 a_{2222}(F_{2,yy}\int_0^1 h^{-1}F_1 F_1 dx)_{,yy} - a_{1212}(F_{2,y}\int_0^1 (h^{-1}F_{1,x})_{,x}F_1 dx)_{,y} = 0.$$

(2.54)

Similarly, if w_2, φ_2, ψ_2 and F_2 are known, then we get the following equations system for w_1, φ_1, ψ_1 and F_1 determination

$$F_1\int_0^1 k_x F_{2,yy}w_2 dy + F_{1,xx}\int_0^1 k_y F_2 w_2 dy + \frac{2}{3}\bar{\lambda}_1(\varphi_1\int_0^1 h\varphi_2 w_2 dy)_{,x}+$$

$$\frac{2}{3}\bar{\lambda}_1(w_{1,x}\int_0^1 hw_2 w_2 dy)_{,x} + \frac{2}{3}\bar{\lambda}_2\psi_1\int_0^1 (h\psi_2)_{,y}w_2 dy+$$

$$\frac{2}{3}\bar{\lambda}_2 w_1\int_0^1 (hw_{2,y})_{,y}w_2 dy + \omega^2 w_1\int_0^1 hw_2 w_2 dy = 0, \qquad (2.55)$$

$$\frac{1}{12}\lambda^{-2}A_{1111}(\varphi_{1,x}\int_0^1 h^3\varphi_2\varphi_2 dy)_{,x} + \frac{1}{12}A_{1122}(\psi_1\int_0^1 h^3\psi_{2,y}\varphi_2 dy)_{,x}+$$

$$\frac{1}{12}A_{1212}\psi_{1,x}\int_0^1 (h^3\psi_2)_{,y}\varphi_2 dy + \frac{1}{12}A_{1212}\varphi_1\int_0^1 (h^3\varphi_{2,y})_{,y}\varphi_2 dy-$$

$$\frac{2}{3}\bar{\lambda}_1\varphi_1\int_0^1 h\varphi_2\varphi_2 dy - \frac{2}{3}\bar{\lambda}_1 w_{1,x}\int_0^1 hw_2\varphi_2 dy + \frac{\omega^2}{12}\lambda^{-2}\varphi_1\int_0^1 h^3\varphi_2\varphi_2 dy = 0,$$

(2.56)

$$\frac{1}{12}\lambda^2 A_{2222}\psi_1\int_0^1 (h^3\psi_{2,y})_{,y}\psi_2 dy + \frac{1}{12}A_{1122}\varphi_{1,x}\int_0^1 (h^3\varphi_2)_{,y}\psi_2 dy+$$

$$\frac{1}{12}A_{1212}(\varphi_1\int_0^1 h^3\varphi_{2,y}\psi_2 dy)_{,x} + \frac{1}{12}A_{1212}(\psi_{1,x}\int_0^1 h^3\psi_2\psi_2 dy)_{,x}-$$

$$\frac{2}{3}\bar{\lambda}_2\psi_1\int_0^1 h\psi_2\psi_2 dy - \frac{2}{3}\bar{\lambda}_2 w_1\int_0^1 hw_{2,y}\psi_2 dy + \frac{\omega^2}{12}\lambda^{-2}\psi_1\int_0^1 h^3\psi_2\psi_2 dy = 0,$$

(2.57)

2.2 Orthotropic Plates and Shells with Variable Thickness and Low Stiffness

$$w_1 \int_0^1 (k_x w_2)_{,yy} F_2 dy + (w_1 \int_0^1 k_y w_2 F_2 dy)_{,xx} + \lambda^{-4} a_{1111} (F_{1,xx} \int_0^1 h^{-1} F_2 F_2 dy)_{,xx} +$$

$$a_{1122}(F_1 \int_0^1 h^{-1} F_{2,yy} F_2 dy)_{,xx} + a_{1122} F_{1,xx} \int_0^1 (h^{-1} F_2)_{,yy} F_2 dy +$$

$$\lambda^4 a_{2222} F_1 (h^{-1} F_{2,yy})_{,yy} F_2 dy - a_{1212}(F_{1,x} \int_0^1 (h^{-1} F_{2,y})_{,y} F_2 dy)_{,x} = 0. \quad (2.58)$$

The eight equations (2.51)–(2.58) consist nonlinear system of integro-differential equations in terms of the unknown functions $w_1, \varphi_1, \psi_1, F_1, w_2, \varphi_2, \psi_2, F_2$. In order to abbreviate the equations the following notation in used:

$$A(f, u, v, k, m) = \int_0^1 f \frac{d^k u}{dx^k} \frac{d^m v}{dx^m} dx. \quad (2.59)$$

Integrating by parts (when this is possible) and taking into account the boundary conditions (2.59), we rewrite the equations (2.51)–(2.58) into the following suitable form

$$F_1 A(k_x, F_2, w_2, 2, 0) + F_{1,xx} A(k_y, F_2, w_2, 0, 0) + \tfrac{2}{3}\bar\lambda_1(\varphi_1 A(h, \varphi_2, w_2, 0, 0))_{,x} +$$

$$\tfrac{2}{3}\bar\lambda_1(w_{1,x} A(h, w_2, w_2, 0, 0))_{,x} - \tfrac{2}{3}\bar\lambda_2 \psi_1 A(h, \psi_2, w_2, 0, 1) -$$

$$\tfrac{2}{3}\bar\lambda_2 w_1 A(h, w_2, w_2, 1, 1) + \omega^2 w_1 A(h, w_2, w_2, 0, 0) = 0, \quad (2.60)$$

$$\tfrac{1}{12}\lambda^{-2} A_{1111}(\varphi_{1,x} A(h^3, \varphi_2, \varphi_2, 0, 0))_{,x} + \tfrac{1}{12} A_{1122}(\psi_1 A(h^3, \psi_2, \varphi_2, 1, 0))_{,x} -$$

$$\tfrac{1}{12} A_{1212}\psi_{1,x} A(h^3, \psi_2, \varphi_2, 0, 1) - \tfrac{1}{12} A_{1212}\varphi_1 A(h^3, \varphi_2, \varphi_2, 1, 1) -$$

$$\tfrac{2}{3}\bar\lambda_1 \varphi_1 A(h, \varphi_2, \varphi_2, 0, 0) - \tfrac{2}{3}\bar\lambda_1 w_{1,x} A(h, w_2, \varphi_2, 0, 0) +$$

$$\tfrac{1}{12}\lambda_1^{-2}\omega^2 \varphi_2 A(h^3, \varphi_2, \varphi_2, 0, 0) = 0, \quad (2.61)$$

$$\tfrac{1}{12} A_{1212}(\psi_{1,x} A(h^3, \psi_2, \psi_2, 0, 0))_{,x} - \tfrac{1}{12} A_{1122}\varphi_{1,x} A(h^3, \varphi_2, \psi_2, 0, 1) +$$

$$\tfrac{1}{12} A_{1212}(\varphi_1 A(h^3, \varphi_2, \psi_2, 1, 0))_{,x} - \tfrac{1}{12}\lambda^2 A_{2222}\psi_1 A(h^3, \psi_2, \psi_2, 1, 1) -$$

$$\tfrac{2}{3}\bar\lambda_2 \psi_1 A(h, \psi_2, \psi_2, 0, 0) - \tfrac{2}{3}\bar\lambda_2 w_1 A(h, w_2, \psi_2, 1, 0) +$$

$$\tfrac{1}{12}\lambda_2^{-2}\omega^2 \psi_1 A(h^3, \psi_2, \psi_2, 0, 0) = 0, \quad (2.62)$$

$$w_1 A(k_x, w_2, F_2, 0, 2) + (w_1 A(k_y, w_2, F_2, 0, 0))_{,xx} +$$

$$\lambda^{-4} a_{1111}(F_{1,xx} A(h^{-1}, F_2, F_2, 0, 0))_{,xx} +$$

$$a_{1122}(F_1 A(h^{-1}, F_2, F_2, 2, 0))_{,xx} + a_{1122} F_{1,xx} A(h^{-1}, F_2, F_2, 0, 2) +$$

$$\lambda^4 a_{2222} F_1 A(h^{-1}, F_2, F_2, 2, 2) - a_{1212}(F_{1,x} A(h^{-1}, F_2, F_2, 1, 1))_{,x} = 0. \quad (2.63)$$

$$F_{2,yy}A\left(k_x,F_1,w_1,0,0\right)+F_2A\left(k_y,F_1,w_1,2,0\right)-\tfrac{2}{3}\bar{\lambda}_1\varphi_2A(h,\varphi_1,w_1,0,1)-$$
$$\tfrac{2}{3}\bar{\lambda}_1w_2A(h,w_1,w_1,1,1)+\tfrac{2}{3}\bar{\lambda}_2(\psi_2A(h,\psi_1,w_1,0,0))_{,y}+$$
$$\tfrac{2}{3}\bar{\lambda}_2(w_{2,y}A(h,w_1,w_1,0,0))_{,y}+\omega^2w_2A\left(h,w_1,w_1,0,0\right)=0, \quad (2.64)$$
$$-\tfrac{1}{12}\lambda^{-2}A_{1111}\varphi_2A(h^3,\varphi_1,\varphi_1,1,1)-\tfrac{1}{12}A_{1122}\psi_{2,y}A(h^3,\psi_1,\varphi_1,0,1)+$$
$$\tfrac{1}{12}A_{1212}(\psi_2A(h^3,\psi_1,\varphi_1,1,0))_{,y}+$$
$$\tfrac{1}{12}A_{1212}(\varphi_{2,y}A(h^3,\varphi_1,\varphi_1,0,0))_{,y}-\tfrac{2}{3}\bar{\lambda}_1\varphi_2A(h,\varphi_1,\varphi_1,0,0)-$$
$$\tfrac{2}{3}\bar{\lambda}_1w_2A(h,w_1,\varphi_1,1,0)+\tfrac{1}{12}\omega^2\lambda_1^{-2}\varphi_2A(h^3,\varphi_1,\varphi_1,0,0)=0, \quad (2.65)$$
$$-\tfrac{1}{12}A_{1212}\psi_2A(h^3,\psi_1,\psi_1,1,1)+\tfrac{1}{12}A_{1122}(\varphi_2A(h^3,\varphi_1,\psi_1,1,0))_{,y}-$$
$$\tfrac{1}{12}A_{1212}\varphi_{2,y}A(h^3,\varphi_1,\psi_1,0,1)+$$
$$\tfrac{1}{12}\lambda^2A_{2222}(\psi_{2,y}A(h^3,\psi_1,\psi_1,0,0))_{,y}-\tfrac{2}{3}\bar{\lambda}_2\psi_2A(h,\psi_1,\psi_1,0,0)-$$
$$\tfrac{2}{3}\bar{\lambda}_2w_{2,y}A(h,w_1,\psi_1,0,0)+\tfrac{1}{12}\omega^2\lambda_2^{-2}\psi_2A(h^3,\psi_1,\psi_1,0,0)=0, \quad (2.66)$$
$$(w_2A(k_x,w_1,F_1,0,0))_{,yy}+w_2A(k_y,w_1,F_1,0,2)+$$
$$\lambda^{-4}a_{1111}F_2A(h^{-1},F_1,F_1,2,2)+a_{1122}F_{2,yy}A(h^{-1},F_1,F_1,0,2)+$$
$$a_{1122}(F_2A(h^{-1},F_1,F_1,2,0))_{,yy}+\lambda^4a_{2222}(F_{2,y}A(h^{-1},F_1,F_1,0,0))_{,yy}-$$
$$a_{1212}(F_{2,y}A(h^{-1},F_1,F_1,1,1))_{,y}=0. \quad (2.67)$$

The finite element method of the second order [214] is used to solve the derived at of ordinary differential equations. For this purpose, we divide the interval $[0,1]$ into N finite elements (Fig. 2.1). On each element ('e' denotes the element number) the functions w_i, φ_i, ψ_i and F_i ($i=1,2$) are expressed in the forms

$$w_i^e(x_i)=\sum_{k=1}^{3}w_{ik}^e\xi_k^e(x_i), \quad (2.68)$$

$$\varphi_i^e(x_i)=\sum_{k=1}^{3}\varphi_{ik}^e\xi_k^e(x_i), \quad (2.69)$$

$$\psi_i^e(x_i)=\sum_{k=1}^{3}\psi_{ik}^e\xi_k^e(x_i), \quad (2.70)$$

$$F_i^e(x_i)=\sum_{k=1}^{3}F_{ik}^e\xi_k^e(x_i), \quad (2.71)$$

where $x_i=x$ for $i=1$ and $x_i=y$ for $i=2$.

The quantities $\xi_k^e(x_i)$ in the above equations are the approximations of the eigenfunctions defined by the relations

2.2 Orthotropic Plates and Shells with Variable Thickness and Low Stiffness

$$\xi_1^e(x_i) = \left(1 - \frac{2x_i}{\Delta_e}\right)\left(1 - \frac{x_i}{\Delta_e}\right),$$

$$\xi_2^e(x_i) = \frac{4x_i}{\Delta_e}\left(1 - \frac{x_i}{\Delta_e}\right), \quad \xi_3^e(x_i) = -\frac{x_i}{\Delta_e}\left(1 - \frac{2x_i}{\Delta_e}\right),$$

where Δ_e is the interval length corresponding to element 'e'.

The first function corresponds to the beginning of the intervals, the second to the middle of the interval, whereas the third to the end of the interval.

In the expressions (2.68)–(2.71) governing w_{ik}^e, φ_{ik}^e, ψ_{ik}^e, F_{ik}^e the sought functions w_i, φ_i, ψ_i and F_i are evaluated at the 'k' node of the e-th interval. Substituting (2.68)–(2.71) to equations (2.60)–(2.67) and projecting on $\xi_m^e(x_i)$, the following system of algebraic in terms of w_{ik}^e, φ_{ik}^e, ψ_{ik}^e, F_{ik}^e for the finite element 'e' is obtained:

$$\tfrac{2}{3}\bar{\lambda}_2 w_{1k}^e \int_{\Delta e} A(h,w_2,w_2,1,1)\xi_k^e\xi_m^{e'}dy + \tfrac{2}{3}\bar{\lambda}_1 w_{1k}^e \int_{\Delta e} A(h,w_2,w_2,0,0)\xi_{k,y}^e\xi_{m,y}^{e'}dy+$$

$$\tfrac{2}{3}\bar{\lambda}_1 \varphi_{1k}^e \int_{\Delta e} A(h,\varphi_2,w_2,0,0)\xi_k^e\xi_{m,y}^{e'}dy + \tfrac{2}{3}\bar{\lambda}_2 \psi_{1k}^e \int_{\Delta e} A(h,\psi_2,w_2,0,1)\xi_k^e\xi_m^{e'}dy-$$

$$F_{1k}^e \int_{\Delta e} A(k_y,F_2,w_2,0,0)\xi_{k,yy}^e\xi_m^{e'}dy - F_{1k}^e \int_{\Delta e} A(k_x,F_2,w_2,2,0)\xi_k^e\xi_m^{e'}dy-$$

$$\omega^2 w_{1k}^e \int_{\Delta e} A(h,w_2,w_2,0,0)\xi_k^e\xi_m^{e'}dy = 0, \qquad (2.72)$$

$$\tfrac{2}{3}\bar{\lambda}_1 w_{1k}^e \int_{\Delta e} A(h,w_2,\varphi_2,0,0)\xi_{k,y}^e\xi_m^{e'}dy + \tfrac{1}{12}A_{1212}\varphi_{1k}^e \int_{\Delta e} A(h^3,\varphi_2,\varphi_2,1,1)\xi_k^e\xi_m^{e'}dy+$$

$$\tfrac{2}{3}\bar{\lambda}_1 \varphi_{1k}^e \int_{\Delta e} A(h,\varphi_2,\varphi_2,0,0)\xi_k^e\xi_m^{e'}dy + \tfrac{1}{12}\lambda^{-2}A_{1111}\varphi_{1k}^e \times$$

$$\int_{\Delta e} A(h^3,\varphi_2,\varphi_2,0,0)\xi_{k,y}^e\xi_{m,y}^{e'}dy + \tfrac{1}{12}A_{1122}\psi_{1k}^e \int_{\Delta e} A(h^3,\psi_2,\varphi_2,1,0)\xi_k^e\xi_m^{e'}dy+$$

$$\tfrac{1}{12}A_{1212}\psi_{1k}^e \int_{\Delta e} A(h^3,\psi_2,\varphi_2,0,1)\xi_{k,y}^e\xi_m^{e'}dy-$$

$$\tfrac{1}{12}\lambda_1^{-2}\omega^2\varphi_{1k}^e \int_{\Delta e} A(h^3,\varphi_2,\varphi_2,0,0)\xi_k^e\xi_m^{e'}dy = 0, \qquad (2.73)$$

$$\tfrac{2}{3}\bar{\lambda}_2 w_{1k}^e \int_{\Delta e} A(h,w_2,\psi_2,1,0)\xi_k^e\xi_m^{e'}dy + \tfrac{1}{12}A_{1122}\varphi_{1k}^e \int_{\Delta e} A(h^3,\varphi_2,\psi_2,0,1)\xi_{k,y}^e\xi_m^{e'}dy+$$

$$\frac{1}{12}A_{1212}\varphi_{1k}^e \int_{\Delta e} A(h^3,\varphi_2,\psi_2,1,0)\xi_k^e\xi_{m,y}^{e'}dy + \frac{1}{12}\lambda^2 A_{2222}\psi_{1k}^e \times$$

$$\int_{\Delta e} A(h^3,\psi_2,\psi_2,1,1)\xi_k^e\xi_m^{e'}dy + \frac{2}{3}\bar\lambda_2\psi_{1k}^e \int_{\Delta e} A(h,\psi_2,\psi_2,0,0)\xi_k^e\xi_m^{e'}dy +$$

$$\frac{1}{12}A_{1212}\psi_{1k}^e \int_{\Delta e} A(h^3,\psi_2,\psi_2,0,0)\xi_{k,y}^e\xi_{m,y}^{e'}dy -$$

$$\frac{1}{12}\lambda_2^{-2}\omega^2\psi_{1k}^e \int_{\Delta e} A(h^3,\psi_2,\psi_2,0,0)\xi_k^e\xi_m^{e'}dy = 0\,, \tag{2.74}$$

$$w_{1k}^e \int_{\Delta e} A(k_x,w_2,F_2,0,2)\xi_k^e\xi_m^{e'}dy + w_{1k}^e \int_{\Delta e} A(k_y,w_2,F_2,0,0)\xi_k^e\xi_{m,yy}^{e'}dy +$$

$$\lambda^{-4}a_{1111}F_{1k}^e \int_{\Delta e} A(h^{-1},F_2,F_2,0,0)\xi_{k,yy}^e\xi_{m,yy}^{e'}dy + a_{1122}F_{1k}^e \times$$

$$\int_{\Delta e} A(h^{-1},F_2,F_2,2,0)\xi_k^e\xi_{m,yy}^{e'}dy + a_{1122}F_{1k}^e \int_{\Delta e} A(h^{-1},F_2,F_2,0,2)\xi_{k,yy}^e\xi_m^{e'}dy +$$

$$\lambda^4 a_{2222}F_{1k}^e \int_{\Delta e} A(h^{-1},F_2,F_2,2,2)\xi_k^e\xi_m^{e'}dy -$$

$$a_{1212}F_{1k}^e \int_{\Delta e} A(h^{-1},F_2,F_2,1,1)\xi_{k,y}^e\xi_{m,y}^{e'}dy = 0\,, \tag{2.75}$$

$$\frac{2}{3}\bar\lambda_1 w_{2k}^e \int_{\Delta e} A(h,w_1,w_1,1,1)\xi_k^e\xi_m^{e'}dx + \frac{2}{3}\bar\lambda_2 w_{2k}^e \int_{\Delta e} A(h,w_1,w_1,0,0)\xi_{k,x}^e\xi_{m,x}^{e'}dx +$$

$$\frac{2}{3}\bar\lambda_1\varphi_{2k}^e \int_{\Delta e} A(h,\varphi_1,w_1,0,1)\xi_k^e\xi_m^{e'}dx + \frac{2}{3}\bar\lambda_2\psi_{2k}^e \int_{\Delta e} A(h,\psi_1,w_1,0,0)\xi_k^e\xi_{m,x}^{e'}dx -$$

$$F_{2k}^e \int_{\Delta e} A(k_x,F_1,w_1,0,0)\xi_{k,xx}^e\xi_m^{e'}dx - F_{2k}^e \int_{\Delta e} A(k_y,F_1,w_1,2,0)\xi_k^e\xi_m^{e'}dx -$$

$$\omega^2 w_{2k}^e \int_{\Delta e} A(h,w_1,w_1,0,0)\xi_k^e\xi_m^{e'}dx = 0\,, \tag{2.76}$$

$$\frac{2}{3}\bar\lambda_1 w_{2k}^e \int_{\Delta e} A(h,w_1,\varphi_1,1,0)\xi_k^e\xi_m^{e'}dx + \frac{1}{12}\lambda^{-2}A_{1111}\varphi_{2k}^e \int_{\Delta e} A(h^3,\varphi_1,\varphi_1,1,1)\xi_k^e\xi_m^{e'}dx +$$

$$\frac{1}{12}A_{1212}\varphi_{2k}^e \int_{\Delta e} A(h^3,\varphi_1,\varphi_1,0,0)\xi_{k,x}^e\xi_{m,x}^{e'}dx + \frac{2}{3}\bar\lambda_1\varphi_{2k}^e \times$$

2.2 Orthotropic Plates and Shells with Variable Thickness and Low Stiffness

$$\int_{\Delta e} A(h, \varphi_1, \varphi_1, 0, 0) \xi_k^e \xi_m^{e'} dx + \tfrac{1}{12} A_{1212} \psi_{2k}^e \int_{\Delta e} A(h^3, \psi_1, \varphi_1, 1, 0) \xi_k^e \xi_{m,x}^{e'} dx +$$

$$\tfrac{1}{12} A_{1122} \psi_{2k}^e \int_{\Delta e} A(h^3, \psi_1, \varphi_1, 0, 1) \xi_{k,x}^e \xi_m^{e'} dx -$$

$$\tfrac{1}{12} \lambda_1^{-2} \omega^2 \varphi_{2k}^e \int_{\Delta e} A(h^3, \varphi_1, \varphi_1, 0, 0) \xi_k^e \xi_m^{e'} dx = 0, \qquad (2.77)$$

$$\tfrac{2}{3} \bar{\lambda}_2 w_{2k}^e \int_{\Delta e} A(h, w_1, \psi_1, 0, 0) \xi_{k,x}^e \xi_m^{e'} dx + \tfrac{1}{12} A_{1122} \varphi_{2k}^e \times$$

$$\int_{\Delta e} A(h^3, \varphi_1, \psi_1, 1, 0) \xi_k^e \xi_{m,x}^{e'} dx + \tfrac{1}{12} A_{1212} \varphi_{2k}^e \int_{\Delta e} A(h^3, \varphi_1, \psi_1, 0, 1) \xi_{k,x}^e \xi_m^{e'} dx +$$

$$\tfrac{1}{12} \lambda^2 A_{2222} \psi_{2k}^e \int_{\Delta e} A(h^3, \psi_1, \psi_1, 0, 0) \xi_{k,x}^e \xi_m^{e'} dx + \tfrac{1}{12} A_{1212} \psi_{2k}^e \times$$

$$\int_{\Delta e} A(h^3, \psi_1, \psi_1, 1, 1) \xi_k^e \xi_m^{e'} dx + \tfrac{2}{3} \bar{\lambda}_2 \psi_{2k}^e \int_{\Delta e} A(h, \psi_1, \psi_1, 0, 0) \xi_k^e \xi_m^{e'} dx -$$

$$\tfrac{1}{12} \lambda_2^{-2} \omega^2 \psi_{2k}^e \int_{\Delta e} A(h^3, \psi_1, \psi_1, 0, 0) \xi_k^e \xi_m^{e'} dx = 0, \qquad (2.78)$$

$$w_{2k}^e \int_{\Delta e} A(k_x, w_1, F_1, 0, 0) \xi_k^e \xi_{m,xx}^{e'} dx + w_{2k}^e \int_{\Delta e} A(k_y, w_1, F_1, 0, 2) \xi_k^e \xi_m^{e'} dx +$$

$$\lambda^{-4} a_{1111} F_{2k}^e \int_{\Delta e} A(h^{-1}, F_1, F_1, 2, 2) \xi_k^e \xi_m^{e'} dx + a_{1122} F_{2k}^e \times$$

$$\int_{\Delta e} A(h^{-1}, F_1, F_1, 0, 2) \xi_{k,xx}^e \xi_m^{e'} dx + a_{1122} F_{2k}^e \int_{\Delta e} A(h^{-1}, F_1, F_1, 0, 0) \xi_{k,xx}^e \xi_{m,xx}^{e'} dx -$$

$$a_{1212} F_{2k}^e \int_{\Delta e} A(h^{-1}, F_1, F_1, 1, 1) \xi_{k,x}^e \xi_{m,x}^{e'} dx +$$

$$\lambda^4 a_{2222} F_{2k}^e \int_{\Delta e} A(h^{-1}, F_1, F_1, 0, 0) \xi_{k,xx}^e \xi_{m,xx}^{e'} dx = 0. \qquad (2.79)$$

In (2.72)–(2.79) a summation is denoted by repeated indices. In addition, $k, m = 1, 2, 3, ...$, $e = e_1, e_2, ..., e_N$, are the numbers of those finite elements, which touch the 'm' node of the 'e' finite element. Let us introduce a vector of the unknown local variables on the 'e' finite element

$$\vec{u}_i^e = \{w_{i1}^e, w_{i2}^e, w_{i3}^e, \varphi_{i1}^e, \varphi_{i2}^e, \varphi_{i3}^e, \psi_{i1}^e, \psi_{i2}^e, \psi_{i3}^e, F_{i1}^e, F_{i2}^e, F_{i3}^e\}.$$

Then, equations (2.72)–(2.79) can be expressed in the form

$$C_i^e \vec{u}_i^e - \omega^2 D_i^e \vec{u}_i^e = 0, \quad (i = 1, 2). \tag{2.80}$$

For $i = 1$ equation (2.80) models the system of equations (2.72)–(2.75), whereas for $i = 2$ it models the system of equations (2.76)–(2.79). The matrices and may be presented as follows:

$$C_i^e = \begin{pmatrix} Cww_i^e & Cw\varphi_i^e & Cw\psi_i^e & CwF_i^e \\ C\varphi w_i^e & C\varphi\varphi_i^e & C\varphi\psi_i^e & 0 \\ C\psi w_i^e & C\psi\varphi_i^e & C\psi\psi_i^e & 0 \\ CFw_i^e & 0 & 0 & CFF_i^e \end{pmatrix}, \tag{2.81}$$

$$D_i^e = \begin{pmatrix} Dww_i^e & 0 & 0 & 0 \\ 0 & D\varphi\varphi_i^e & 0 & 0 \\ 0 & 0 & D\psi\psi_i^e & 0 \\ 0 & 0 & 0 & DFF_i^e \end{pmatrix}. \tag{2.82}$$

The submatrices of C_i^e and D_i^e are defined below (where, $j = 1$ if $i = 2$, and $j = 2$ if $i = 1$):

$$Cww_{ikm}^e = ww1i \int_{\Delta e} A(h, w_j, w_j, 1, 1) \xi_{k,x_j}^e \xi_m^{e'} dx_j +$$

$$ww2i \int_{\Delta e} A(h, w_j, w_j, 0, 0) \xi_k^e \xi_{m,x_j}^{e'} dx_j,$$

$$Cw\varphi_{ikm}^e = w\varphi 1i \int_{\Delta e} A(h, \varphi_j, w_j, 0, 0) \xi_k^e \xi_{m,x_j}^{e'} dx_j +$$

$$w\varphi 2i \int_{\Delta e} A(h, \varphi_j, w_j, 0, 1) \xi_k^e \xi_m^{e'} dx_j,$$

$$Cw\psi_{ikm}^e = w\psi 1i \int_{\Delta e} A(h, \psi_j, w_j, 0, 1) \xi_k^e \xi_m^{e'} dx_j +$$

$$w\psi 2i \int_{\Delta e} A(h, \psi_j, w_j, 0, 0) \xi_k^e \xi_{m,x_j}^{e'} dx_j,$$

$$CwF_{ikm}^e = \int_{\Delta e} A(k_i, F_j, w_j, 2, 0) \xi_k^e \xi_m^{e'} dx_j +$$

$$\int_{\Delta e} A(k_j, F_j, w_j, 0, 0) \xi_{k,x_j}^e \xi_m^{e'} dx_j,$$

2.2 Orthotropic Plates and Shells with Variable Thickness and Low Stiffness

$$C\varphi w_{ikm}^e = \varphi w1i \int_{\Delta e} A(h, w_j, \varphi_j, 0, 0) \xi_{k,x_j}^e \xi_m^{e'} dx_j +$$

$$\varphi w2i \int_{\Delta e} A(h, w_j, \varphi_j, 1, 0) \xi_k^e \xi_m^{e'} dx_j,$$

$$C\varphi\varphi_{ikm}^e = \varphi\varphi1i \int_{\Delta e} A(h, \varphi_j, \varphi_j, 0, 0) \xi_k^e \xi_m^{e'} dx_j +$$

$$\varphi\varphi2i \int_{\Delta e} A(h^3, \varphi_j, \varphi_j, 1, 1) \xi_k^e \xi_m^{e'} dx_j +$$

$$\varphi\varphi3i \int_{\Delta e} A(h^3, \varphi_j, \varphi_j, 0, 0) \xi_{k,x_j}^e \xi_{m,x_j}^{e'} dx_j,$$

$$C\varphi\psi_{ikm}^e = \varphi\psi1i \int_{\Delta e} A(h^3, \psi_j, \varphi_j, 1, 0) \xi_k^e \xi_{m,x_j}^{e'} dx_j +$$

$$\varphi\psi2i \int_{\Delta e} A(h^3, \psi_j, \varphi_j, 0, 1) \xi_{k,x_j}^e \xi_m^{e'} dx_j, \qquad (2.83)$$

$$C\psi w_{ikm}^e = \psi w1i \int_{\Delta e} A(h, w_j, w_j, 1, 0) \xi_k^e \xi_m^{e'} dx_j +$$

$$\psi w2i \int_{\Delta e} A(h, w_j, w_j, 0, 0) \xi_{k,x_j}^e \xi_m^{e'} dx_j,$$

$$C\psi\varphi_{ikm}^e = \psi\varphi1i \int_{\Delta e} A(h^3, \varphi_j, \psi_j, 0, 1) \xi_{k,x_j}^e \xi_m^{e'} dx_j +$$

$$\psi\varphi2i \int_{\Delta e} A(h^3, \varphi_j, \psi_j, 1, 0) \xi_k^e \xi_{m,x_j}^{e'} dx_j,$$

$$C\psi\psi_{ikm}^e = \psi\psi1i \int_{\Delta e} A(h, \psi_j, \psi_j, 0, 0) \xi_k^e \xi_m^{e'} dx_j +$$

$$\psi\psi2i \int_{\Delta e} A(h^3, \psi_j, \psi_j, 1, 1) \xi_k^e \xi_m^{e'} dx_j + \psi\psi3i \int_{\Delta e} A(h^3, \psi_j, \psi_j, 0, 0) \xi_{k,x_j}^e \xi_{m,x_j}^{e'} dx_j,$$

$$CFw_{ikm}^e = \int_{\Delta e} A(k_i, w_j, F_j, 0, 2) \xi_k^e \xi_m^{e'} dx_j + \int_{\Delta e} A(k_j, w_j, F_j, 0, 0) \xi_k^e \xi_{m,x_jx_j}^{e'} dx_j,$$

$$CFF_{ikm}^e = FF1i \int_{\Delta e} A(h^{-1}, F_j, F_j, 2, 2) \xi_k^e \xi_m^{e'} dx_j +$$

$$FF2i \int_{\Delta e} A\left(h^{-1}, F_j, F_j, 0, 2\right) \xi^e_{k,x_jx_j} \xi^{e'}_m dx_j +$$

$$FF3i \int_{\Delta e} A\left(h^{-1}, F_j, F_j, 2, 0\right) \xi^e_k \xi^{e'}_{m,x_jx_j} dx_j +$$

$$FF4i \int_{\Delta e} A\left(h^{-1}, F_j, F_j, 1, 1\right) \xi^e_{k,x_j} \xi^{e'}_{m,x_j} dx_j +$$

$$FF5i \int_{\Delta e} A\left(h^{-1}, F_j, F_j, 0, 0\right) \xi^e_{k,x_jx_j} \xi^{e'}_{m,x_jx_j} dx_j,$$

$$Dww^e_{ikm} = \int_{\Delta e} A(h, w_j, w_j, 0, 0) \xi^e_k \xi^{e'}_m dx_j,$$

$$D\varphi\varphi^e_{ikm} = \frac{1}{12}\bar{\lambda}_1^{-2} \int_{\Delta e} A\left(h^3, \varphi_j, \varphi_j, 0, 0\right) \xi^e_k \xi^{e'}_m dx_j,$$

$$D\psi\psi^e_{ikm} = \frac{1}{12}\bar{\lambda}_2^{-2} \int_{\Delta e} A\left(h^3, \psi_j, \psi_j, 0, 0\right) \xi^e_k \xi^{e'}_m dx_j.$$

In the above expressions the various coefficients are defined according to:

$$ww1i = \frac{2}{3}\begin{cases} \bar{\lambda}_2, & i=1 \\ \bar{\lambda}_1, & i=2 \end{cases}, \quad ww2i = \frac{2}{3}\begin{cases} \bar{\lambda}_1, & i=1 \\ \bar{\lambda}_2, & i=2 \end{cases},$$

$$\varphi w1i = w\varphi1i, \quad \varphi w2i = w\varphi2i,$$

$$w\varphi1i = \frac{2}{3}\begin{cases} \bar{\lambda}_1, & i=1 \\ 0, & i=2 \end{cases}, \quad w\varphi2i = \frac{2}{3}\begin{cases} 0, & i=1 \\ \bar{\lambda}_2, & i=2 \end{cases},$$

$$\varphi\varphi1i = \frac{2}{3}\begin{cases} \bar{\lambda}_1, & i=1 \\ \bar{\lambda}_2, & i=2 \end{cases}, \quad \varphi\varphi2i = \frac{1}{12}\begin{cases} A_{1212}, & i=1 \\ \lambda^{-2}A_{1111}, & i=2 \end{cases},$$

$$\varphi\varphi3i = \frac{1}{12}\begin{cases} \lambda^{-2}A_{1111}, & i=1 \\ A_{1212}, & i=2 \end{cases}, \quad \varphi\psi1i = \frac{1}{12}\begin{cases} A_{1122}, & i=1 \\ A_{1212}, & i=2 \end{cases},$$

$$\varphi\psi2i = \frac{1}{12}\begin{cases} A_{1212}, & i=1 \\ A_{1122}, & i=2 \end{cases}, \quad \psi w1i = \frac{2}{3}\begin{cases} \bar{\lambda}_2, & i=1 \\ 0, & i=2 \end{cases},$$

$$\psi w2i = \frac{2}{3}\begin{cases} 0, & i=1 \\ \bar{\lambda}_2, & i=2 \end{cases}, \quad \psi\varphi1i = \frac{1}{12}\begin{cases} A_{1122}, & i=1 \\ A_{1212}, & i=2 \end{cases},$$

$$\psi\varphi2i = \frac{1}{12}\begin{cases} A_{1212}, & i=1 \\ A_{1122}, & i=2 \end{cases}, \quad \psi\psi1i = \frac{2}{3}\begin{cases} \bar{\lambda}_2, & i=1 \\ \bar{\lambda}_1, & i=2 \end{cases},$$

2.2 Orthotropic Plates and Shells with Variable Thickness and Low Stiffness

$$\psi\psi 2i = \frac{1}{12} \begin{cases} A_{1212}, & i=1 \\ \lambda^2 A_{2222}, & i=2 \end{cases}, \quad \psi\psi 3i = \frac{1}{12} \begin{cases} \lambda^2 A_{2222}, & i=1 \\ A_{1212}, & i=2 \end{cases},$$

$$FF1i = \begin{cases} \lambda^4 a_{2222}, & i=1 \\ \lambda^{-4} a_{1111}, & i=2 \end{cases}, \quad FF2i = \begin{cases} a_{1122}, & i=1 \\ a_{1122}, & i=2 \end{cases},$$

$$FF3i = \begin{cases} a_{1122}, & i=1 \\ a_{1122}, & i=2 \end{cases}, \quad FF4i = \begin{cases} -a_{1212}, & i=1 \\ -a_{1212}, & i=2 \end{cases},$$

$$FF5i = \begin{cases} \lambda^{-4} a_{1111}, & i=1 \\ \lambda^4 a_{2222}, & i=2 \end{cases}.$$

Thus, in order to get the C_i^e and D_i^e matrices for different $i = 1, 2$ it is sufficient to change the coefficients according to the expressions (2.83).

During the construction of the stiffness and mass matrices (C_i^e and D_i^e) we need to calculate the integrals in the rectangular spaces $\Omega^e = [0,1] \times \Delta e$:

$$\int_{\Delta e} A(f, u, v, p, q)\, \xi_k^e \xi_m^{e'} dy = \int_{\Delta e} \int_0^1 f(x, y)\, \frac{\partial^p u(x)}{\partial x^p} \frac{\partial^q v(x)}{\partial x^q} \xi_k^e(y) \xi_m^{e'}(y)\, dy\, dx,$$

and the corresponding change of the variables in $f(x, y)$. For this purpose, a two dimensional formula (similar to Simpson's rule) may be used.

According to the algorithm described above, the elements of the local matrices [170] of stiffness C_i^e and mass D_i^e are calculated.

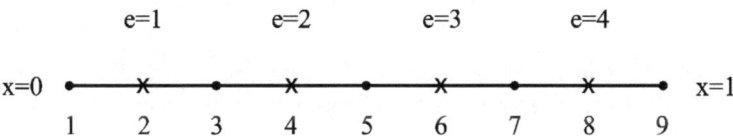

Figure 2.1. Local and global numeration of the [0, 1] partition into 4 elements.

In Fig. 2.1 the distribution for four elements on the interval [0, 1], where 'e' is the element number, is shown. For each element local and global numbers are defined above and below, respectively. As shown (in the one dimensional case) the global numbers are related to the local numbers and to the finite element number by the expression:

$$i_g = i_1 + 2(e - 1). \tag{2.84}$$

The above formula allows for the arrangement of the obtained elements of the local matrices C_i^e and D_i^e on the corresponding places in the global matrices C_i and D_i according to their calculation sequence. The algorithm is composed of one iteration variational step matched with FEM, and it yields matrices C_i and D_i for $i = 1$ or $i = 2$. The boundary conditions are realised by cancelling the columns and the rows from the C_i and D_i matrices corresponding to the variables w, γ_x, γ_y, F defined on the boundaries. Boundary conditions of the type $M_{11} = M_{22} = 0$, $\frac{\partial^2 F}{\partial n^2} = 0$ are introduced automatically, because the algorithm is realised on the basis of the stationary energy deformation. It means that the algorithm is variational in nature.

The eigenvalues of the generalised problems are obtained from the equations

$$C_i \vec{u}_i - \omega^2 D_i \vec{u}_i = 0, \quad i = 1 \text{ or } i = 2, \tag{2.85}$$

where $\vec{u}_i = \left(w_i^1, w_i^2, ..., w_i^N, \varphi_i^1, \varphi_i^2, ..., \varphi_i^N, \psi_i^1, ..., \psi_i^N, F_i^1, ..., F_i^N \right)$ are the sought values for $w_i, \varphi_i, \psi_i, F_i$ on the global nodes $j = 1, 2, ..., N$.

In order to solve (2.85) the Schwartz iterational method has been used [108]. This method is particularly effective for low frequency vibrations.

Suppose that we wish to determine the fundamental (lowest) free vibration frequency. Suppose that $\vec{u}_i^{(n)}$ has been computed we then solve for the next iterate using the equation

$$C_i \vec{u}_i^{(n+1)} = D_i \vec{u}_i^{(n)}. \tag{2.86}$$

We substitute the obtained solution again to the right hand side of (2.86), and find $\vec{u}_i^{(n+2)}$, and so on. In each step the approximate value of the fundamental frequency ω_n^2 may be obtained from the Rayleigh quotient:

$$\omega_n^2 = \frac{\vec{u}_i^{(n+1)} C_i \vec{u}_i^{(n+1)}}{\vec{u}_i^{(n+1)} D_i \vec{u}_i^{(n+1)}}. \tag{2.87}$$

The process is controlled because of the difference between ω_n^2 and ω_{n+1}^2. The algorithm has been described in detail in [109], where its convergence has been proved.

The convergence of the finite elements method results from the properties of the bilinear forms $a_h(\vec{u}, \vec{v})$ and $b_h(\vec{u}, \vec{v})$ proved on the basis of the corollary 2.2 and the respective theorems for second order finite elements [214].

convergence of the MB is illustrated in Fig. 2.3, where the solid line corresponds to rolling support, and dashed line - to stiff support.

It is seen from the Fig. 2.2 and 2.3 that the convergence is practically achieved for $N = 12$ for the MVI and for $M = 18$ for the MB. However, the error in the frequency estimation for the MVI method compared to the MB method (Fig. 2.2) does not tend towards zero (it is 1.9% for the stiff support, and 2.3% for the rolling support). The reason is that in the MVI only one factor in the function distribution (2.29) has been taken into account.

In order to to provide a comparison between the free vibration frequencies of a plate with a rolling support calculated using the MVI and MB methods a

2.2 Orthotropic Plates and Shells with Variable Thickness and Low Stiffness

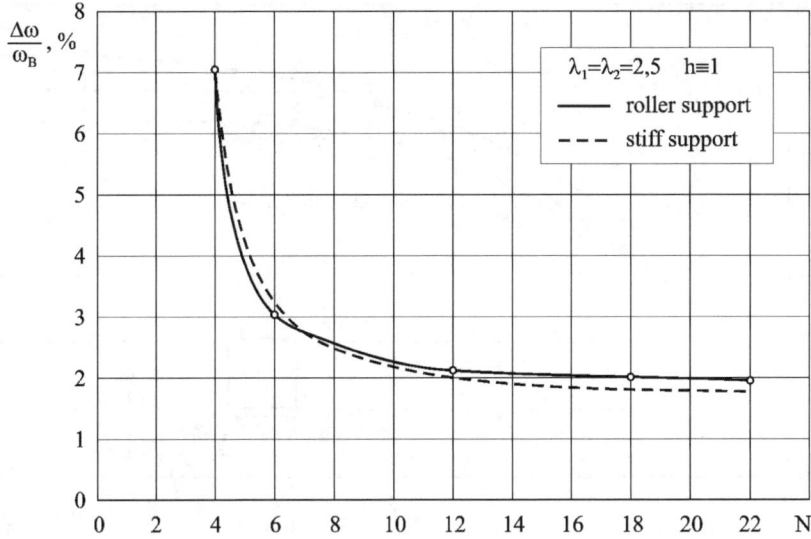

Figure 2.2. Estimation error of fundamental frequency vibration using MVI method in relation to the frequency ω_B defined by the MB method.

computation has been carried out for the nondimensional parameter $(\bar\lambda/0.25)$ in the range $[0.25; 2500]$, for $\nu = 0.25$, and $\lambda_1 = \lambda_2$ in the range $[1 - 100]$. Representative results are depicted in Fig. 2.4.

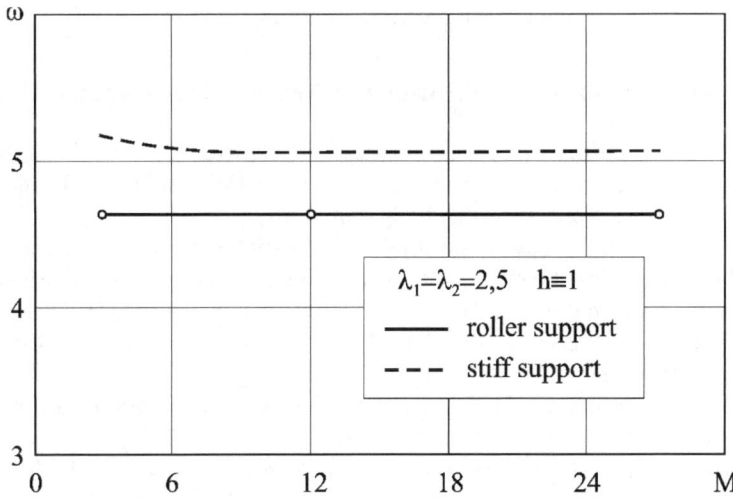

Figure 2.3. Illustration of convergence of the MB method for stiffly and free supported plate.

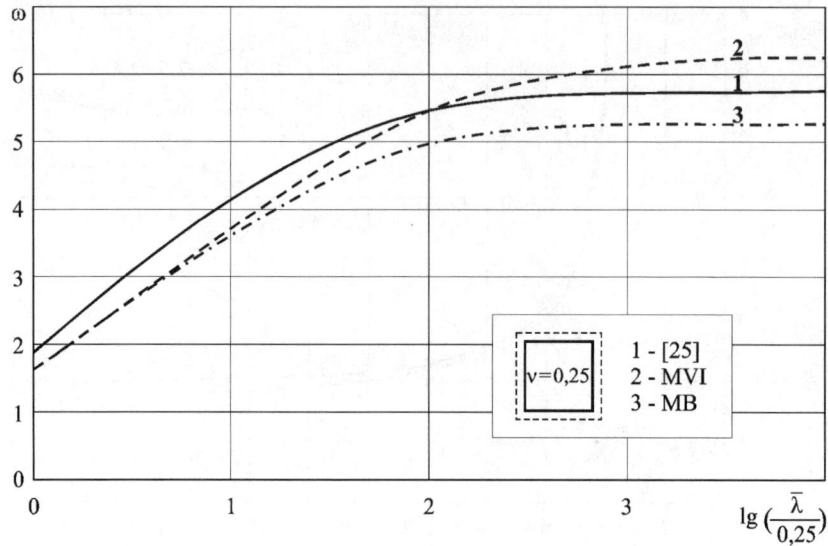

Figure 2.4. Comparison of different methods of roller support for $\bar{\lambda} \in (0.25; 2500)$.

Analysis of the obtained dependencies leads to the following conclusions. The frequencies obtained using the first approximation of the MVI for $\bar{\lambda}_1 = \bar{\lambda}_2 = 25$ differ on amount not greater than 5%. However, for $\bar{\lambda}_1 = \bar{\lambda}_2 \leq 25$ they are very close to each other and for $\bar{\lambda}_1 = \bar{\lambda}_2 = 0.25$ they fully overlap. It follows, that for plates with low transversal stiffness MVI leads to accurate results, with lower computational time in comparison to the MB.

2.2.4 Analysis of Free Vibrations of Transversal-Isotropic Plates

Test of algorithms. In order to carry out algorithm tests relating to the free vibration frequencies obtained using the MVI and the MB methods, first plates with constant thickness (equal to 1) for $\bar{\lambda}_1 = \bar{\lambda}_2 = 2.5$, and for $G/E \in [0.4; 0.01]$ have been analysed. The considered parameters correspond to properties of composite materials used nowadays [256]. For instance, for plastic-glass material $G/E = 1/7$ [250], for borplastic $G/E = 1/25$, for graphitoplastic $G/E = 1/40$ [250]. There also exists a material for which $G/E = 1/100$ [228].

During the analysis of the algorithmic convergence, fundamental tones of a vibrating plate with constant thickness for $n = 0.25$, $\lambda_1 = \lambda_2 = 2.5$ for isotropic material ($G/E = 0.4$) have been found ($\bar{\lambda}_1 = \bar{\lambda}_2 = 2.34375$).

In Fig. 2.2 the error dependence on the fundamental tone calculation, using the MVI in the first approximation versus the finite elements number N along the x and y axes, has been shown; ω_B denotes the frequency of

2.2 Orthotropic Plates and Shells with Variable Thickness and Low Stiffness 123

the fundamental tone obtained using the MB for $M = 27$ (9 terms for each function).

Comparing the two results obtained with those given in reference [119] (the dashed line in Fig. 2.4) obtained using the MB method with higher approximations and without inertial effects related to rotation, the following conclusions are drawn. The inertial effect related to rotation decreases the frequency by 12% for $\bar{\lambda}_1 = \bar{\lambda}_2 = 0.25$ and by 9% for $\bar{\lambda}_1 = \bar{\lambda}_2 = 2500$. This means that this effect should be taken into account in the analysis during transversal-isotropic plates.

Similar results have been obtained for plates with varying thickness according to the exponential law

$$h(x,y) = z_1 + z_2 e^{z_3(x^2+y^2)}. \tag{2.88}$$

In Fig. 2.5 the curves of such dependencies with the rolling plate support for $\bar{\lambda}_1 = \bar{\lambda}_2$, $z_1 = 0.4$, $z_2 = 0.02$, $z_3 = 3$ obtained using the MB (dashed line) and the MVI (solid line) for $\lambda_1 = \lambda_2 = 2.5$ and $\bar{\lambda}_1 = \bar{\lambda}_2 \in [2.5; 0.0625]$ have been shown.

Investigation of the influence of boundary conditions for plates possessing low transversal stiffness. Decreasing the transversal stiffness should cause a decrease of the area of boundary effects. Therefore, the influence of the boundary conditions on the free vibration frequencies should also decrease. We analyse this behaviour for the example of the fundamental tone of plates with $\bar{\lambda}_1 = \bar{\lambda}_2 \in [0.0625; 0.25]$, for different thickness values of a plate, and for $\lambda_1 = \lambda_2 = 2.5$. In Fig. 2.6 the dependencies ω versus $\bar{\lambda}_1 = \bar{\lambda}_2$ for different values of h have been shown. The curves 1 are related to plates with $h = 1$, curves 2 to plates with $h = 0.25$ (the solid lines correspond to stiffly supported plate, whereas the dashed lines to rolling supported plate). The results obtained using the MVI correspond to $N = 12$.

As it can be seen from Fig. 2.6, decreasing the transversal stiffness for both supports, the fundamental frequencies approach each other independently of the thickness. For $\bar{\lambda} = 0.25$ they differ by 10%, for $h = 0.25$, and by 1% for $h = 1$. Therefore, one can conclude that for thick plates with low transversal stiffness the support conditions only slightly influence the free vibration frequencies obtained using Timoshenko's model.

Investigation of the effect of the plate's thickness on free vibration frequencies for different transversal stiffness. Contrary to the previous example, in Fig. 2.6 we provide, an example of approaching tune frequencies of the fundamental tones for different thickness and low transversal stiffness. We analyse this effect in more detail. In Fig. 2.7 the dependencies of 'ω' versus 'h' for different $\bar{\lambda}_1 = \bar{\lambda}_2$ and for $\lambda_1 = \lambda_2 = 2.5$, $n = 0.25$ have been shown. This case corresponds to a plate with rolling support.

Similar dependencies are shown in Fig. 2.8 for a stiffly supported plate.

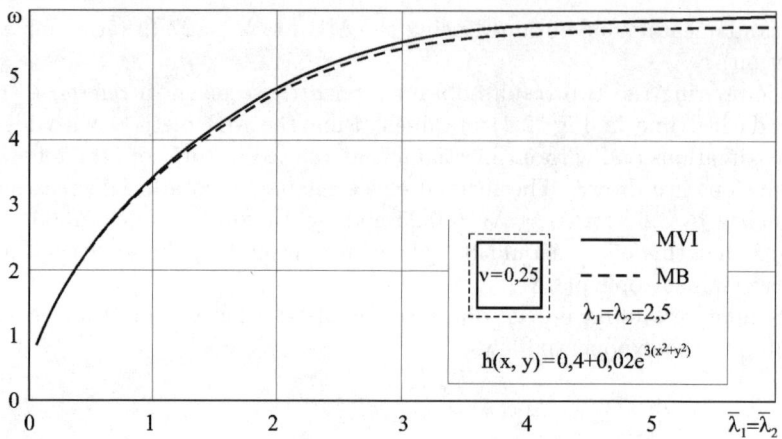

Figure 2.5. Frequency of vibrations of rolling supported plate and exponentially variable thickness h using MVI and MB methods.

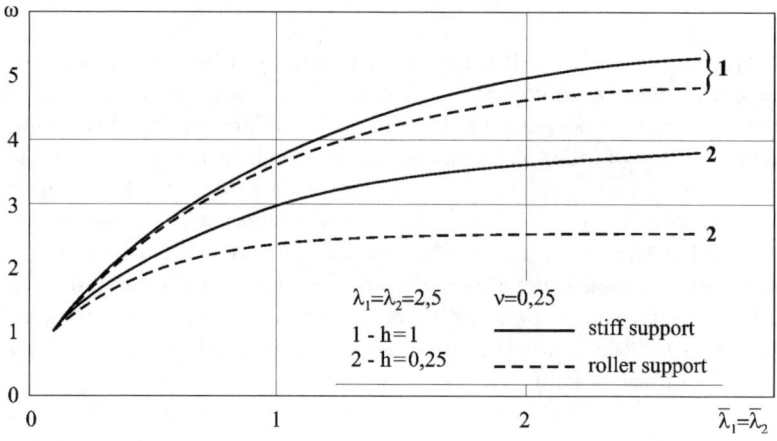

Figure 2.6. Frequency of vibrations of a plate with different thickness h defined by the MVI method for $N = 12$.

Analysis of the obtained diagrams leads to the conclusion that for plates with low transversal stiffness $\bar{\lambda}_1 = \bar{\lambda}_2 \leq 0.25$ the fundamental tone frequency weakly depends on the plate's thickness. For instance, for $\bar{\lambda}_1 = \bar{\lambda}_2 = 0.25$ the frequencies for $h = 2$ and $h = 0.5$ for plates with a rolling support differ by 4%, whereas for plates with a stiff support they differ by 2%.

In problems related to weight optimization and for a priori given frequencies the so called ε-optimization can be used. In that case because of the required accuracy results the constraints are given with ε-precision. Because of the 'slight' dependence of 'ω' versus 'h' in the given interval $|\Delta \omega| \leq \varepsilon$

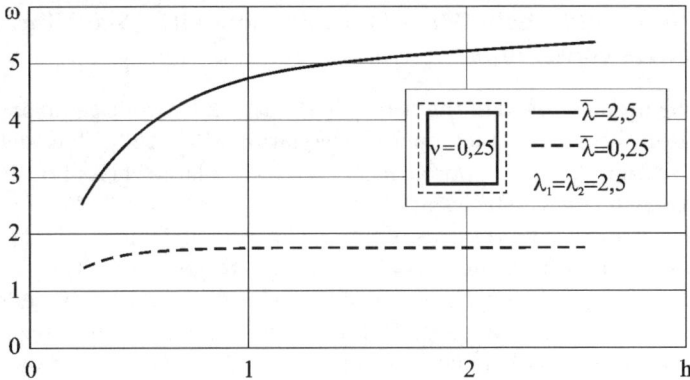

Figure 2.7. Frequency of a roller supported plate versus its thickness h.

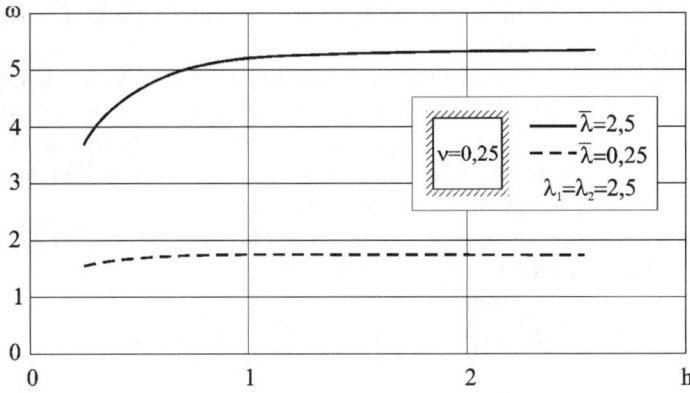

Figure 2.8. Frequency of a stiffly supported plate versus its thickness h.

there exist solutions with plates with constant stiffness and smaller volume. They are referred to as 'quasi-ε-optimal', which are of considerable practical importance.

2.3 Rational Design of Plates with Finite Transversal Stiffness

In this chapter general problems related to optimization of plates weight as a consideration will be considered. Then an engineering approach will be illustrated, further called the rational design. Solving rational design problems the methods described earlier will be used.

2.3.1 Plates and Shells Mass Optimization with Free Vibration Frequencies constraints

The problems mentioned in the title only slightly differ from a similar problem formulation for an abstract operator equation [134, 236]. The details are given in section 2.6. This approach characterizes a large class of optimization problems dealing with statics and dynamics.

In this chapter the problems of minimalization of Timoshenko-type shells using fundamental frequencies constraints are considered.

Assume, that a shell with changeable thickness occupies a bounded space Ω with a piecewise smooth edge S. Free vibrations of this shell are governed by equations (2.1) with boundary conditions (2.2) and (2.5). Consider the following problem. Among all possible material distributions along the shell, described by the thickness function $h(x, y)$ from a given set of functions U_∂ (defined further) we want to find the function $h^*(x, y)$ minimises the shell's mass and leads to a fundamental frequency equal to that of a shell with constant thickness equal to 1.

This problem may be formulated also in a slightly different manner [37]. For given mass of the shell m^* we want to find the distribution $h_\partial^*(x, y)$ that leads to maximum frequency of the fundamental tone. This can be proved in a similar way to the one presented in reference [236]. For Timoshenko-like shells, these two approaches are equivalent.

Therefore, the problem is reduced to that of minimalization of a functional subject to the constraints

$$h^*(x, y) :\to \min_{h \in U_\partial} \int_\Omega \rho d\Omega, \quad \omega = \omega_0, \tag{2.89}$$

where 'ω' denotes the fundamental frequency tone found from (2.1)–(2.5), and ω_0 the free vibration frequency of the plate with constant thickness $h(x, y) \equiv 1$.

Using the penalty method we avoid the explicit constraint $\omega = \omega_0$. As a result, problem (2.89) is reduced to the following one

$$h_\partial^*(x, y) :\to \min_{h \in U_\partial} \left[\int_\Omega \rho h d\Omega + \frac{1}{\varepsilon} (\omega - \omega_0)^2 \right], \tag{2.90}$$

where 'ω' again is defined by (2.1)–(2.5). Using the approach presented in reference [83] we can show that for a defined choice of the set U_∂ the solution $h_\partial^*(x, y)$ of equation (2.90) tends to $h^*(x, y)$ for $\varepsilon \to 0$.

However, in the majority of cases that occur in practice we do not need to have the exact relation $\omega = \omega_0$, and therefore without the condition of $\varepsilon \to 0$ we still have a practically valid relation. Similar problems are called ε-optimization problems or rational design problems and they are considered further in this section.

2.3 Rational Design of Plates with Finite Transversal Stiffness

Define a set of admissible functions $h(x,y)$, called the set of admissible control

$$U_\partial = \left\{ h(x,y) | h \in H^1(\Omega), \ \|h\|_{1,\Omega} \leq const, \ 0 < h_H \leq h(x,y) \leq h_B \right\}. \tag{2.91}$$

As admissible functions we take, functions with piecewise continuous first derivatives and bounded from above and below.

As it has been mentioned in reference [134], a proper choice of admissible control is defined by the $h(x,y)$ and its derivatives. It should be mentioned that for wider sets, the rational design defined by (2.90) may either not exist or may not be achieved [138]. A choice of the admissible control of (2.91) guarantees a solution to the problem (2.90), which defines the following theorem.

THEOREM 2.5 *If the assumptions (2.6)–(2.8) are fulfilled and k_x and k_y are bounded on Ω, then the rational design (2.90) defined on U_∂ according to (2.91) has at least one solution for an arbitrarily taken $\varepsilon > 0$.*

Proof. Let $\{h_n\}_{n=1}^\infty \in U_\partial$ be the series which minimises the functional

$$J(h) = \int_\Omega h d\Omega + \frac{1}{\varepsilon}(\omega - \omega_0)^2, \tag{2.92}$$

or

$$\{h_n\}_{n=1}^\infty \in U_\partial, \quad J(h_n) \xrightarrow[n \to \infty]{} \inf_{h \in U_\partial} J(h).$$

A solution to the problem (2.1) with respect to $\vec{u}$ as well as with respect to ω, $a(\vec{u}, \vec{v})$, $b(\vec{u}, \vec{v})$ depends on $h(x,y)$. We denote this dependence by $a_h(\vec{u}, \vec{v})$, $b_h(\vec{u}, \vec{v})$, $\vec{u}_h$, ω_h, and for $h = h_n(x,y)$ in the form $a_n(\vec{u}, \vec{v})$, $b_n(\vec{u}, \vec{v})$, $\vec{u}_n$, ω_n. Then, according to the Rayleigh formula (2.26), we get the following estimation for $h = h_n(x,y)$:

$$\omega_n^2 = \frac{a_n(\vec{u}_n, \vec{u}_n)}{b_n(\vec{u}_n, \vec{u}_n)} = \inf_{\vec{u} \in V_0} \frac{a_n(\vec{u}, \vec{u})}{b_n(\vec{u}, \vec{u})}. \tag{2.93}$$

Without any restrictions for generality of the considerations we can assume that $b_n(\vec{u}_n, \vec{u}_n) = 1$ [155]. Then, repeating the procedure in a set of points we get

$$\omega_n^2 = a_n(\vec{u}, \vec{u}). \tag{2.94}$$

Besides, from (2.93) it results that because of the first free vibration frequency optimization for in arbitrary $\vec{v} \in V_0$ the following inequality holds

$$\omega_n^2 \leq \frac{a_n(\vec{v}, \vec{v})}{b_n(\vec{v}, \vec{v})}. \tag{2.95}$$

However, from (2.15) and (2.21) we get

$$a_n\left(\vec{v},\vec{v}\right) \leq a_{h_B}\left(\vec{v},\vec{v}\right), \quad b_n\left(\vec{v},\vec{v}\right) \geq b_{h_H}\left(\vec{v},\vec{v}\right),$$

because $h_H \leq h(x,y) \leq h_B$. Then, from (2.95) we obtain

$$\omega_n^2 \leq \frac{a_{h_B}\left(\vec{v},\vec{v}\right)}{b_{h_H}\left(\vec{v},\vec{v}\right)}.$$

Because $\vec{v} \in V_0$ is already fixed, then

$$\omega_n^2 \leq const. \tag{2.96}$$

From (2.94) and (2.96) it results, that also $a_n(\vec{u}_n, \vec{u}_n) \leq const.$ and we get

$$\|\vec{u}_n\|_{V_0} \leq const. \tag{2.97}$$

Therefore, from (2.96), (2.97) and the U_∂ set definition, it results that from the series $\{\vec{u}_n\}$, $\{h_n\}$, $\{\omega_n^2\}$ a series may be detached $\{\vec{u}_\mu\}$, $\{h_\mu\}$, $\{\omega_\mu^2\}$ satisfying the following properties [110]:

$$\omega_\mu^2 \to \omega_p^2 \quad \text{is weak in R,} \tag{2.98}$$

$$\vec{u}_\mu \to \vec{u}_p \quad \text{is weak in } V_0, \tag{2.99}$$

$$h_\mu \to h_p \quad \text{is weak in } H^1(\Omega), \tag{2.100}$$

$$h_\mu^\alpha \to h_p^\alpha \quad \text{is strong in } L^2(\Omega) \text{ for } \forall \alpha \geq 1. \tag{2.101}$$

It is seen that from the conditions (2.98)–(2.101) it results that also $a_\mu(\vec{u}_\mu, \vec{v}) \to a_{h_p}(\vec{u}_p, \vec{v})$ for any $\vec{v} \in V_0$. Consider an arbitrary term from equations (2.1), for instance $\frac{\partial}{\partial x}\left(h_\mu^3 \frac{\partial \gamma_{x\mu}}{\partial x}\right)$. For the arbitrary $\vec{v} \in V_0$ we have

$$\int_\Omega h_\mu^3 \frac{\partial \gamma_{x\mu}}{\partial x} \frac{\partial \tilde{\gamma}_x}{\partial x} d\Omega = \int_\Omega h_\mu^3 \frac{\partial \gamma_{xp}}{\partial x} \frac{\partial \tilde{\gamma}_x}{\partial x} d\Omega + \int_\Omega h_\mu^3 \frac{\partial (\gamma_{x\mu} - \gamma_{xp})}{\partial x} \frac{\partial \tilde{\gamma}_x}{\partial x} d\Omega. \tag{2.102}$$

Because $h_\mu^3 \to h_p^3$ is strong in $L^2(\Omega)$ and $\gamma_{x\mu} \to \gamma_{xp}$ is weak in $H^1(\Omega)$, then the first term approaches a constant, whereas the second tends to zero. In a similar way we realise a limit transition in the last terms of the system (2.1). For an arbitrary $\vec{v} \in V_0$ we get

$$a_\mu(\vec{u}_\mu, \vec{v}) \to a_p(\vec{u}_p, \vec{v}),$$

$$b_\mu(\vec{u}_\mu, \vec{v}) \to b_p(\vec{u}_p, \vec{v}).$$

Then, taking into account (2.98), we get

$$a_p(\vec{u}_p, \vec{v}) = \omega_p^2 b_p(\vec{u}_p, \vec{v}).$$

It shows that $\vec{u}_p$ and ω_p^2 are solutions to (2.1).

Besides, $h_p(x,y) \in U_\partial$. Therefore, the minimising series has a limit, for which all conditions are fulfilled. It means that $h_p(x,y)$ is the solution of the problem (2.90), $h_\varepsilon^*(x,y) = h_p(x,y)$.

The obtained result is valid not only from the point of view of different designs, but also from the point of view of approximation to $h(x,y)$. If $h(x,y) \in U_\partial$ then during a construction of the finite approximation $h(x,y)$ we require the approximation function to belong to the interior of U_∂, otherwise, the process may not be convergent. As shown in references [135, 138] it is of no use to look looking for piecewise continuous thickness for the Kirchhoff-Love model.

The obtained results are used in subchapter 2.3.2 for the numerical examples.

2.3.2 Finite Dimensional Approximations of the Rational Design of Plates and Shells

Finite dimensional approximation. The problem dealing with the finite dimensional approximation is particularly important for optimal design of plates and shells. It is connected with the evidence of lack of the analytical solutions in the majority of design tasks [44]. The only possible solutions are approximate ones obtained using numerical methods.

In further considerations we omit the problem (2.89) and we focus in the rational design defined by (2.90) for a given ε. A solution to the problem (2.89), as has been mentioned earlier, may be obtained for $\varepsilon \to 0$. For the purpose of engineering calculation it is sufficient to take ε as a small number. This problem will be discussed further. Now we consider the questions concerned with the occurrence of approximated solutions on the finite dimensional subspaces U_∂ and their convergence. Those questions have been initially addressed in subchapter 2.3.1.

It will be shown that the frequency of the fundamental tone of the Timoshenko-like shell is a weakly continuous functional for all $h \in U_\partial$ (U_∂ is defined by (2.91)). Taking into account that ω^2 is defined by the Rayleigh formula (2.26), this result will be formulated as the following theorem.

THEOREM 2.6 *The functional*

$$\varphi(h) = \omega^2 = \inf_{\overrightarrow{u} \in V_0} \frac{a_h(\overrightarrow{u}, \overrightarrow{u})}{b_h(\overrightarrow{u}, \overrightarrow{u})}$$

is a weakly continuous transformation from U_∂ into R^1.

Proof. We need to prove that an arbitrary series of functions $\{h_n\}_{n=1}^\infty$ integrated together with their first order derivatives ($\{h_n\}_{n=1}^\infty \in U_\partial$), and being slowly convergent to $h_p \in U_\partial$ with respect to the norm $H^1(\Omega)$, the following relation holds

$$\lim_{n \to \infty} \varphi(h_n) = \lim_{n \to \infty} \omega_n^2 = \varphi(h_p) = \omega_p^2,$$

where ω_p^2 is the fundamental tone frequency of a shell with thickness $h_p(x,y)$. □

It has been shown during the proof of theorem (2.5) that this is true for the minimising series $\{h_\mu\}_{\mu=1}^\infty$. The proof does not change for an arbitrary slowly convergent series.

Therefore, we can approach the arbitrary function $h_p(x,y) \subset U_\partial$ using the function $h_n(x,y) \in U_\partial$. According to (3.89) the series ω_n^2 converges to ω_p^2.

Thus, the problem is reduced to the construction of finite dimensional subspaces of functions from U_∂ that have the required convergence properties. Otherwise, as it has been shown in subchapter 2.3.1.1, a situation may occur for which it will be extremely difficult to achieve the minimum of the functional

$$J(h) = \int_\Omega \rho h d\Omega + \frac{1}{\varepsilon}\left(\omega^2 - \omega_0^2\right)^2. \qquad (2.103)$$

Let us formally choose a series of the finite dimensional subspaces $\{H_n\}_{n=1}^\infty$ in such a way that for an arbitrary n we have $H_n \subset H^1(\Omega)$ and

$$\lim_{n \to \infty} \inf_{g \in H_n \cap U_\partial} \|g - h_1\|_{1,\Omega} = 0 \quad \forall h_1 \in U_\partial. \qquad (2.104)$$

In other words, an arbitrary function $h_1 \in U_\partial$ should approach the norm of $H^1(\Omega)$ through the functions $g_n \in H_n \cap U_\partial$ with an arbitrarily given accuracy for $n \to \infty$. Concrete approximation methods having the property (2.104) will be given below.

Now we reduce the problem of the rational design of a plate in regard to its mass with a given constraint for its fundamental frequency ω to the following finite dimensional programming problem: ind $h_n^* \in H_n \cap U_\partial$ and the corresponding $\omega_n, \vec{u}_n \in V_0$ such that

$$J(h_n^*) = \inf_{h \in H_n \cap U_\partial} J(h), \quad n = 1, 2, ..., \qquad (2.105)$$

where $J(h)$ is defined by (2.103).

THEOREM 2.7 *If the series of closed finite dimensional subspaces $\{H_n\}_{n=1}^\infty \in H^1(\Omega)$ fulfils the condition (2.104), then for an arbitrary 'n' the problem of the finite dimensional nonlinear programming (2.105) has the solution h_n^* and*

$$\lim_{n \to \infty} J(h_n^*) = J(h_p), \qquad (2.106)$$

where h_p is the solution to the problem (2.90). Besides, from the functional series $\{h_n^*\}_{n=1}^\infty$ minimising the functional $J(h)$ on $H_n \cap U_\partial$ we may choose the series $\{h_\mu\}_{\mu=1}^\infty$ that $h_\mu \to h_p$ weakly on $H^1(\Omega)$ and strongly on $L^q(\Omega)$, $q \geq 1$.

2.3 Rational Design of Plates with Finite Transversal Stiffness

Proof. For an arbitrarily fixed n the set $H_n \cap U_\partial$ is bounded, closed and compact [110]. The functional $J(h)$ is weakly discontinuous in U_∂ because the first term of (2.103) is continuous, and $\omega = \varphi(h)$ is a weakly discontinuous functional according to theorem (2.103). From the second generalised Weierstrass theorem it follows that for an arbitrary 'n' the functional $J(h)$ is bounded on the set $H_n \cap U_\partial$ and achieves the lower limit $J(h_n^*)$ for a certain h_n^*. Thus, the first part of the theorem has been proved.

We now prove that (2.106) is true. Because $h_p \in U_\partial$, then, from (2.104) it follows [135], that there exists a series $\{g_n\}_{n=1}^\infty$ minimising $\|g - h_p\|_{1,\Omega}$ in $H_n \cap U_\partial$ for every n, which converges to h_p because of the norm of $H^1(\Omega)$. However, the series $\{h_n\}_{n=1}^\infty$ obtained during the solution of problem (2.105) consists of minimising functions $J(h)$ for every n. Therefore, $J(h_n^*) \leq H(g_n)$. Besides, for every n the minimum (2.105) is obtained in a more narrow set $H_n \cap U_\partial$ in comparison to the problem (2.90). Thus, it cannot be less than $J(h_p)$, i.e. it is the minimum of (2.90). Then the inequality $J(h_p) \leq J(h_n^*)$ is true for every n, and we have also

$$J(h_p) \leq J(h_n^*) \leq J(g_n).$$

However, $g_n \to h_p$ is a strong transformation in $H^1(\Omega)$, but $J(h)$ is weakly discontinuous in U_∂. Therefore $J(g_n) \to J(h_p)$, which implies that $J(h^*) \to J(h_p)$. It means that the series of the minimal values of the finite dimensional problems converges to a minimal value of the rational design problem (2.90).

From the solution series $\{h_n^*\}_{n=1}^\infty$ of the finite dimensional problems one can always choose the series $\{h_\mu^*\}_{\mu=1}^\infty$, which converges to h_p, because $h_n^* \in U_\partial \subset H^1(\Omega)$. This guarantees the possibility of getting solutions of rational design problems (2.90) applying subsequent solutions to the finite dimensional problems of nonlinear programming (2.105).

Very often, from the point of view of engineering practice and design and second order constraints, the set U_∂ the thickness distribution functions is chosen as finite dimensional. It consists of functions depending on finite numbers of parameters or of piecewise smooth functions [212]. At a first glance, the cases considered here already finite dimensional, and therefore, questions related to convergence and approximation should not be of concern. However, this is not entirely true. We take into account the function $h(x, y)$ in one of the earlier discussed forms and substitute it to system (2.1). Applying one of the numerical methods for the determination of ω and $\vec{u}$, one deals with approximation of $h(x,y)$. What sort of conditions should the assumed approximation scheme fulfil? The approximation should fulfil the condition (2.104).

The methods which satisfy these conditions are the finite element method, the mesh method with $O(h^\alpha)$ approximation, $\alpha \geq 1$ and other computational methods. In the case of the Kirchhoff-Love model, we need to apply finite elements satisfying not only $h(x, y)$ continuity but also continuity of its first

2 Rational Design of Plates and Shells

partial derivatives, a requirement that introduces numerical difficulties. Similar problems appear while using the finite difference method [112].

Engineering approach - choice of ε. In order to solve the rational design problem (2.90) we need a choice of the ε interval changes characterising the penalty method when the constraint $\omega = \omega_0$ is violated. From the engineering point of view, we do not need to achieve absolutely exact results, because not precisely stated products of plates and shells already include errors.

We illustrate how ε may be chosen considering a shell with minimal mass rational design and using the constraints due to the fundamental frequency.

Assume, that we want to find a construction with minimal weight m_0 and accuracy of d. It means, that the first term of (2.90) has to be minimised with the accuracy $\Delta_1 = \left| \delta \left(\int_\Omega \rho h d\Omega \right) \right| \leq m_0 \Delta$. Because the weight is unknown, the error $\Delta_1 = \Delta \cdot \int_\Omega \rho h_H d\Omega = \Delta \cdot h_H S(\Omega)\rho$ (where ρ is constant and $S(\Omega)$ denotes the shell's surfaces). Thus, the accuracy of the $J(h)$ minimisation should be less than $\Delta_3 = \min(\Delta_1, \Delta_2)$ where Δ_2 is the absolute accuracy of the second term minimalization defined by

$$\Delta_2 = \frac{1}{\varepsilon} \left(\omega^2 - \omega_0^2 \right)^2.$$

Assume that we need $\omega^2 = \omega_0^2 \pm \Delta_4$. Then $\Delta_2 = D_4^2/\varepsilon$ and

$$\Delta_3 = \min \left(\Delta \cdot h_H \rho S(\Omega), \Delta_4^2/\varepsilon \right). \tag{2.107}$$

Giving the weight minimisation accuracy Δ we define the accuracy Δ_2 of the functional minimalization. However, in practice it seems irrational to use significantly different Δ_1 and Δ_2 [83]. Usually, they are assumed to be equal.

$$\Delta \cdot h_H \rho S(\Omega) = \Delta_4^2/\varepsilon. \tag{2.108}$$

If the weight optimal calculation accuracy and frequency (Δ_1) are given, then from (2.108) one gets

$$\varepsilon = \Delta_4^2/(\Delta \cdot h_H \}S(\Omega)), \tag{2.109}$$

and $\Delta_3 = D_4^2/\varepsilon = \Delta \cdot h_H \rho S(\Omega)$.

Thus, assuming real required values of Δ and Δ_1 it is sufficient to solve the problem (2.90) for a given ε defined by (2.109) and not necessarily equal to zero, which requires the classical penalty method.

In practice, while seeking good initial approximations, we can use larger values ε in the initial steps, then those defined by (2.109). convergence changes with the change of ε [83]. The process begins with large ε values which decrease gradually at the accuracy of the computation increases.

2.3 Rational Design of Plates with Finite Transversal Stiffness

Engineering approach - choice of U_∂. A condition for the admissible control U_∂ is formulated in chapter 2.3.1. It defines the limiting set of thickness distribution functions which we need in order to solve the problems (2.89) and (2.90). Of course, all derived results still hold for narrower sets of $h(x,y)$ functions.

The possibilities of getting the solutions to the problem (2.89) in the whole set U_∂ are strictly limited by existing methods which deal with frequency spectrum determination of plates and shells with arbitrary functions $h(x,y)$. The investigations described in chapter 2.2.4 have shown that even such universal methods as the Bubnov-Galerkin (MB) do not have sufficient convergence speed, and can not be applied to optimization problems while finding the solutions. The method of variational iterations does not lead to sufficiently accurate results for plates or shells with arbitrary $h(x,y)$ functions.

Besides, from an engineering point of view, there is no point in seeking $h(x,y)$ in a very wide class of functions.

The analysis of the optimal solutions obtained for beams [172, 186] and circled plates [87, 172] on the basis of the Kirchhoff-Love model has shown that very often thickness distributions may be approximated by the function

$$h(x,y) = h_H + z_1^2 + z_2^2 e^{z_3(x^2+y^2)}. \quad (2.110)$$

Exceptional cases deal with plates with rectangular surfaces [60] where some singularities occur. This seems to be caused by the Kirchhoff-Love model. In Timoshenko's model such problems do not occur and the approximation (2.110) is well-founded.

The function (2.110) has one more important property. For given z_1, z_2, z_3 it gives the possibility of approximating both convex and concave functions. Introducing the constant h_H into (2.110) and using z_1 and z_2 enables one to satisfy the constraint $h(x,y) \geq h_H$ (implicitly) for arbitrary z_1, z_2, z_3. The constraint $h(x,y) \leq h_B$ is satisfied explicitly by applying the constraints in $|z_1|$ and $|z_2|$, and it is controlled at edge points by setting $z_3 > 0$ and at the plate's centre by $z_3 < 0$.

It is easy to check that for arbitrary z_1, z_2, z_3 the function (2.110) belongs to U_∂ defined by (2.91), and that the problem of rational design has a solution in a certain cubicoid

$$U'_\partial = [-a_1, a_1] \times [-a_2, a_2] \times [-a_3, a_3] .$$

For purposes of approximation it is sufficient to use the finite elements method of the first and higher orders. In the algorithm described in chapter 2.2.3, the second order finite element method was used. This approximation fulfils the conditions (2.110). It means that for solutions to the problems (2.90) defined on the subset U'_∂ of U_∂, the MVI method described in chapter 2.2.3 may be used.

The results presented here will be used in section 2.3.3.

2.3.3 Timoshenko-Like Rectangular Plates with Stiff and Rolling Supports

Below the analysis of the rational design of stiffly supported transversal-isotropic plates for different $\bar\lambda_1 = \bar\lambda_2$ will be carried out. Rolling supported plates are further considered at the end of this subchapter. In the considered cases the adaptive method of stochastic search for a minimum has been used [195].

Consider an isotropic-transversal plate with cubic surface ($\lambda = 1$), varying thickness according to the law (2.110), and stiffly supported on its edges. The boundary conditions have the form

$$w = \gamma_x = \gamma_y|_S = 0.$$

Free vibrations, in the frame of the Timoshenko kinematic model, are governed by the following dimensionless equations

$$\frac{2}{3}\bar\lambda_1 \frac{\partial}{\partial x}\left[h\left(\gamma_x + \frac{\partial w}{\partial x}\right)\right] + \frac{2}{3}\bar\lambda_2 \frac{\partial}{\partial y}\left[h\left(\gamma_y + \frac{\partial w}{\partial y}\right)\right] + \omega^2 h w = 0,$$

$$\frac{1}{12}\frac{\partial}{\partial x}\left[h^3\left(A_{1111}\frac{\partial \gamma_x}{\partial x} + A_{1122}\frac{\partial \gamma_y}{\partial y}\right)\right] + \frac{1}{12}A_{1212}\frac{\partial}{\partial y}\left[h^3\left(\frac{\partial \gamma_x}{\partial y} + \frac{\partial \gamma_y}{\partial x}\right)\right] -$$

$$\frac{2}{3}\bar\lambda_1\left(\gamma_x + \frac{\partial w}{\partial x}\right) + \frac{1}{12}\bar\lambda_1^{-2}\omega^2 h^3 \gamma_x = 0,$$

$$\frac{1}{12}\frac{\partial}{\partial y}\left[h^3\left(A_{2222}\frac{\partial \gamma_y}{\partial y} + A_{1122}\frac{\partial \gamma_x}{\partial x}\right)\right] + \frac{1}{12}A_{1212}\frac{\partial}{\partial x}\left[h^3\left(\frac{\partial \gamma_x}{\partial y} + \frac{\partial \gamma_y}{\partial x}\right)\right] -$$

$$\frac{2}{3}\bar\lambda_2\left(\gamma_y + \frac{\partial w}{\partial y}\right) + \frac{1}{12}\bar\lambda_2^{-2}\omega^2 h^3 \gamma_y = 0. \quad (2.111)$$

The physical-geometrical plate parameters are the following $\nu = 0.25$, $\lambda_1 = \lambda_2 = 2.5 - 25$, $G/E = 0.01 - 0.4$.

In all cases an inertia rotational effect is included (the last terms in the second and third equations (3.111)).

The material density is assumed to be constant. Thus, the first integral of (2.90) describes the full of the plate weight and can be replaced by $\int_\Omega h d\Omega$.

Let us denote the fundamental frequency of the plate with the thickness 1 by ω_0, and by ω_h again the fundamental frequency for the same parameters but for an arbitrary thickness function from the family defined by (2.110).

Consider the problem of rational plate design of minimal volume on the given set

$$U'_\partial = [-a_1, a_1] \times [-a_2, a_2] \times [-a_3, a_3].$$

The aim is to find the vector $\vec{z}^* = (z_1^*, z_2^*, z_3^*)$ which satisfies the requirement,

2.3 Rational Design of Plates with Finite Transversal Stiffness

$$J(\vec{z}^*) = \inf_{\vec{z} \in U'_{\partial}} \left[\int_{\Omega} h d\Omega + \frac{1}{\varepsilon} \left(\omega_h^2 - \omega_0^2 \right)^2 \right], \quad (2.112)$$

for $h(x, y)$ is defined by (2.110).

For the physical-geometrical parameters the choice of a_i ($i = 1, 2, 3$) has been realised in order to satisfy $h_{\max} \leq 2$.

The problem of rational design has been solved for different $\lambda_1 = \lambda_2$, $\bar{\lambda}_1 = \bar{\lambda}_2$ and h_H. As the lower limit of the plate's thickness the values $h_H = 0.25; 0.4; 0.8$ have been taken.

In Table 2.1 the optimization results for $\lambda_1 = \lambda_2 = 2.5$ for different h_H and $G_{13}/E = G_{23}/E$ are given ($\varepsilon = 0.001$). The approximation of the problem (2.111) was realised using the MVI and the algorithms described in Chapter 2.2.3. A condition of the MVI algorithm end related to the fundamental frequency has the form

$$\frac{|\omega_h^{n+1} - \omega_h^n|}{\omega_h^{n+1}} \leq \varepsilon_B, \quad \text{where } \varepsilon_B = 10^{-4}.$$

In all cases the number of finite elements was restricted to $N = 12$. In the Table 2.1 K_u denotes the number of steps needed by the adaptive stochastic search method to minimise the target function (2.112) with a prescribed accuracy $\Delta_3 = 10^{-3}$.

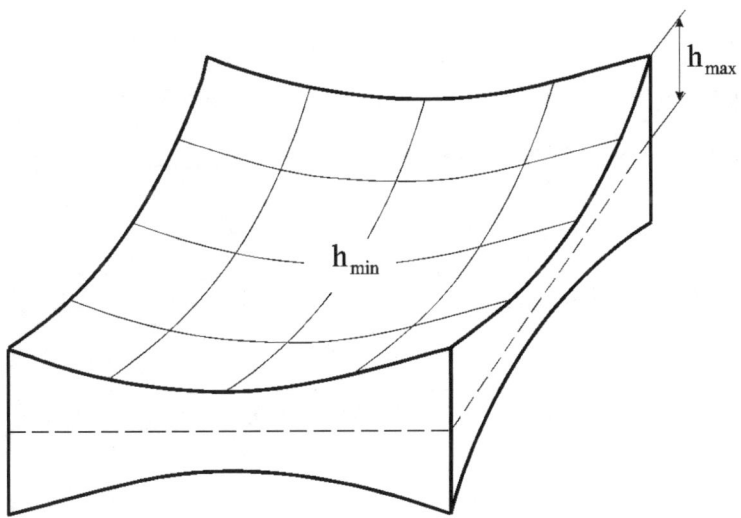

Figure 2.9. Rational Design shape of a stiffly supported plate obtained for $h_H = 0.4$.

136   2 Rational Design of Plates and Shells

Similar results are given in Table 2.2 for $\lambda_1 = \lambda_2 = 5$, in Table 2.3 for $\lambda_1 = \lambda_2 = 10$, and in Table 2.4 for $\lambda_1 = \lambda_2 = 25$.

In Fig. 2.9 the shape satisfying the rational design for a plate made of isotropic material is shown ($h_H = 0.40$).

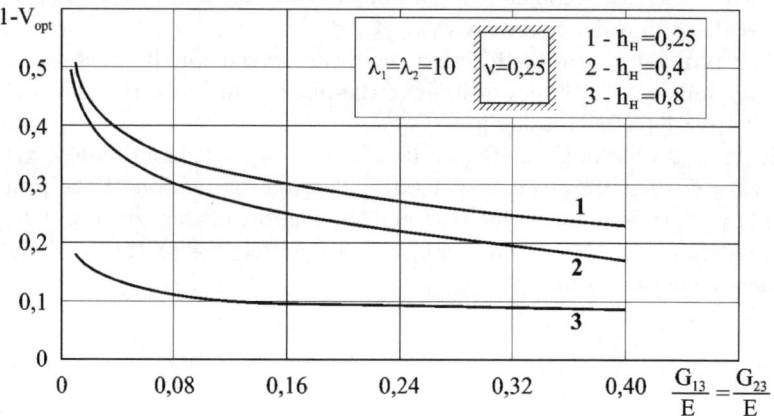

Figure 2.10. Dependence of $(1 - V_{opt})$ versus the modulus $G_{13}/E = G_{23}/E$ for a stiffly supported plate; $\lambda_1 = \lambda_2 = 10$.

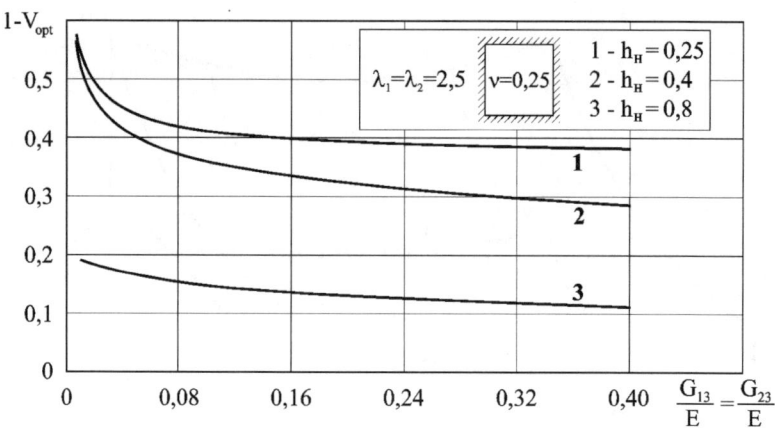

Figure 2.11. Dependence of $(1 - V_{opt})$ versus the modulus $G_{13}/E = G_{23}/E$ for a stiffly supported plate; $\lambda_1 = \lambda_2 = 2.5$.

---

[1] for isotropic material (see Tables 2.1–2.5)

2.3 Rational Design of Plates with Finite Transversal Stiffness 137

Table 2.1. Optimization results for a stiffly supported plate; $\lambda_1 = \lambda_2 = 2.5$.

h_H	$\frac{G_{13}}{E} = \frac{G_{23}}{E}$	ω_0	ω_{opt}	$\frac{\Delta\omega}{\omega_0}100\%$	V_{opt}	z_1	z_2	z_3	K_u
	0.40[1]	5.21	5.20	0.19	0.618	0.475	0.246	2.10	138
0.25	0.04	1.73	1.73	0.00	0.585	0.565	0.082	2.08	62
	0.02	1.23	1.23	0.000	0.486	0.448	0.079	2.16	29
	0.01	0.869	0.867	0.23	0.424	0.413	0.037	1.99	25
	0.40[1]	5.21	5.15	0.15	0.715	0.322	0.224	2.03	86
0.40	0.04	1.73	1.73	0.00	0.615	0.299	0.109	1.99	40
	0.02	1.23	1.22	0.81	0.508	0.002	0.066	1.90	25
	0.01	0.869	0.867	0.23	0.434	0.001	0.048	1.87	21
	0.40[1]	5.21	5.25	0.77	0.894	0.050	0.169	1.88	28
0.80	0.04	1.73	1.73	0.00	0.801	0.002	0.021	1.85	25
	0.02	1.23	1.23	0.00	0.801	0.035	0.004	1.83	21
	0.01	0.869	0.869	0.00	0.801	0.035	0.004	1.83	5

Table 2.2. Optimization results for a stiffly supported plate; $\lambda_1 = \lambda_2 = 5$.

h_H	$\frac{G_{13}}{E} = \frac{G_{23}}{E}$	ω_0	ω_{opt}	$\frac{\Delta\omega}{\omega_0}100\%$	V_{opt}	z_1	z_2	z_3
	0.40[1]	9.54	9.55	0.10	0.733	0.482	0.330	2.08
0.25	0.04	3.43	3.39	1.17	0.606	0.567	0.125	2.03
	0.02	2.44	2.42	0.82	0.400	0.314	0.149	2.09
	0.01	1.73	1.73	0.00	0.325	0.183	0.132	2.12
	0.40[1]	9.54	9.48	0.63	0.790	0.480	0.249	2.19
0.40	0.04	3.43	3.42	0.29	0.562	0.303	0.171	2.12
	0.02	2.44	2.45	0.41	0.498	0.192	0.159	2.14
	0.01	1.73	1.73	0.00	0.437	0.016	0.124	2.12
	0.40[1]	9.54	9.52	0.21	0.913	0.153	0.219	1.83
0.80	0.04	3.43	3.41	0.58	0.803	0.032	0.029	1.88
	0.02	2.44	2.43	0.41	0.801	0.014	0.016	1.86
	0.01	1.73	1.73	0.00	0.800	0.11	0.013	1.85

Table 2.3. Optimization results for a stiffly supported plate; $\lambda_1 = \lambda_2 = 10$.

h_H	$\frac{G_{13}}{E} = \frac{G_{23}}{E}$	ω_0	ω_{opt}	$\frac{\Delta\omega}{\omega_0}100\%$	V_{opt}	z_1	z_2	z_3
	0.40[1]	15.21	15.22	0.06	0.772	0.513	0.391	2.11
0.25	0.04	6.61	6.56	0.76	0.597	0.402	0.270	2.18
	0.02	4.79	4.78	0.21	0.568	0.430	0.236	2.12
	0.01	3.43	3.45	0.58	0.486	0.364	0.202	2.18
	0.40[1]	15.21	15.22	.06	0.827	0.470	0.292	2.13
0.40	0.04	6.61	6.64	0.45	0.645	0.207	0.293	2.10
	0.02	4.79	4.76	0.63	0.580	0.149	0.231	2.12
	0.01	3.43	3.42	0.29	0.500	0.134	0.182	2.15
	0.40[1]	15.21	15.23	0.12	0.922	0.012	0.248	1.90
0.80	0.04	6.61	6.59	0.45	0.865	0.028	0.181	1.88
	0.02	4.79	4.77	0.42	0.842	0.055	0.149	1.75
	0.01	3.43	3.42	0.29	0.818	0.029	0.094	1.86

Table 2.4. Optimization results for a stiffly supported plate; $\lambda_1 = \lambda_2 = 25$.

h_H	$\dfrac{G_{13}}{E} = \dfrac{G_{23}}{E}$	ω_0	ω_{opt}	$\dfrac{\Delta\omega}{\omega_0} 100\%$	V_{opt}	z_1	z_2	z_3
	0.40^1	22.97	23.04	0.30	0.368	0.002	0.245	1.89
0.25	0.04	13.54	13.57	0.22	0.683	0.131	0.471	1.83
	0.02	10.65	10.56	0.48	0.651	0.167	0.441	1.86
	0.01	8.05	8.06	0.13	0.598	0.168	0.393	1.95
	0.40^1	22.97	22.80	0.74	0.526	0.080	0.239	196
0.40	0.04	13.54	13.52	0.15	0.753	0.194	0.392	1.95
	0.02	10.65	10.66	0.10	0.727	0.136	0.389	1.94
	0.01	8.05	8.00	0.62	0.657	0.087	0.347	1.96
	0.40^1	22.97	22.87	0.44	0.873	0.015	0.193	188
0.80	0.04	13.54	13.52	0.15	0.918	0.002	0.245	1.89
	0.02	10.65	10.66	0.10	0.909	0.006	0.235	1.90
	0.01	8.05	8.03	0.25	0.886	0.005	0.206	1.88

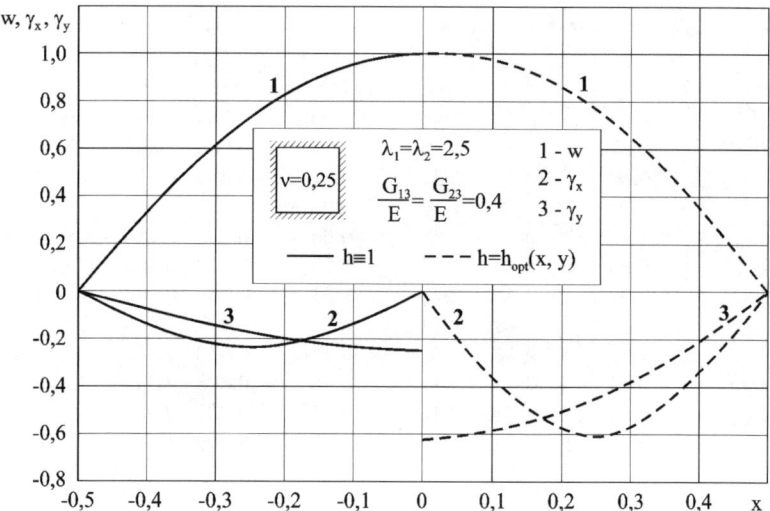

Figure 2.12. Vibrational modes of a stiffly supported plate with constant thickness (solid line) and optimal one (dashed line).

For plates with different ratio $G_{13}/E = G_{23}/E$ the shape of the plate will be similar, and only $h_{\min}$ and $h_{\max}$ will be different. In Table 2.5 the values of $h_{\min}$ and $h_{\max}$ for different h_H and $G_{13}/E = G_{23}/E$ and for $\lambda_1 = \lambda_2 = 10$ are given. For other values of $\lambda_1 = \lambda_2$ the results are given in Tables 2.1–2.4.

For the purpose of further analysis we construct the dependency of $(1 - V_{opt})$ versus $G_{13}/E = G_{23}/E$ for different h_H values. These dependencies for $\lambda_1 = \lambda_2 = 10$ are given in Fig. 2.10. Curve 1 corresponds to $h_H = 0.25$; curve 2 to $h_H = 0.4$ and curve 3 to $h_H = 0.8$. Similar dependencies, with the

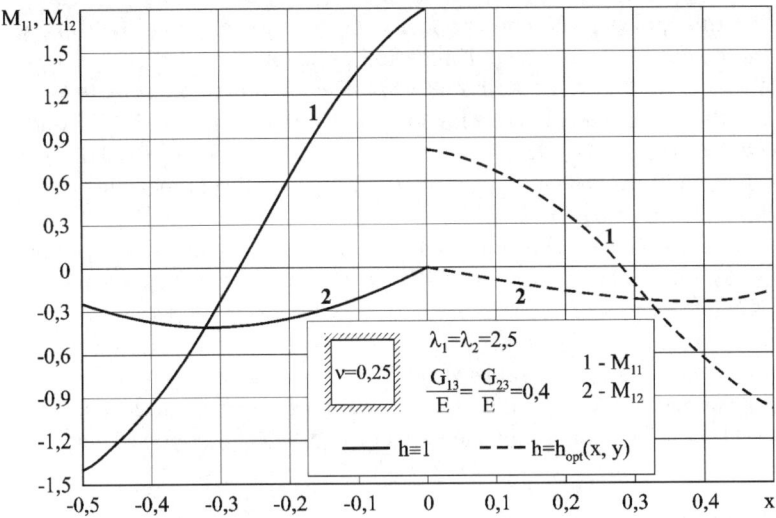

Figure 2.13. Moments of forces for a stiffly supported plate with constant thickness (solid line) and optimal one (dashed line).

Table 2.5. Values of $h_{\min}$ and $h_{\max}$ for an optimal stiffly supported plate; $\lambda_1 = \lambda_2 = 10$

h_H	$\dfrac{G_{13}}{E} = \dfrac{G_{23}}{E}$	$h_{\min}$	$h_{\max}$	$(1 - V_{opt}) \cdot 100$
	0.40[1]	0.67	1.93	23%
0.25	0.04	0.49	1.20	40%
	0.02	0.49	0.96	43%
	0.01	0.42	0.82	51%
	0.40[1]	0.71	1.45	17%
0.40	0.04	0.53	1.22	35%
	0.02	0.48	0.92	42%
	0.01	0.45	0.75	50%
	0.40[1]	0.86	1.18	8%
0.80	0.04	0.83	0.99	13%
	0.02	0.82	0.90	15%
	0.01	0.81	0.85	18%

same numbering scheme for the presented curves are depicted in Fig. 2.11 for $\lambda_1 = \lambda_2 = 2.5$.

The analysis of the obtained results shows that the most suitable result is obtained in relation to the volume (weight). It is achieved in all considered cases for low values of the transversal stiffness $G_{13}/E = G_{23}/E$.

On one hand, for small values of $\bar{\lambda}_1 = \bar{\lambda}_2$ the fundamental frequency slightly depends on the plate thickness (as shown in section 2.2.4); on the other hand, anisotropic constructions (such as transversal-isotropic plates) are more suitable from the optimization point of view.

As seen in Fig. 2.10 and 2.11 that the boundeness from below of h_H essentially influences the results for $h_H > 0.4$. For $h_H < 0.4$ and small $G_{13}/E = G_{23}/E$ the results nearly overlap. This leads to the conclusion that for small values of $G_{13}/E = G_{23}/E$ the minimum is located in the neighbourhood of $h_H = 0.25$.

In order to check the obtained rational design results, we performed a series of numerical tests using the MB method with higher approximations (27 terms). The results of the two methods are compared in Table 2.6 for a plate with $n = 0.25$; $\lambda_1 = \lambda_2 = 2.5$; $h_H = 0.40$. For the sake of convenience, the shape parameters z_1, z_2, z_3 are also given. The largest difference of 1.73% has been obtained for $G/E = 0.04$ with economy of volume of 28% for isotropic material, and 57% for a material with $G_{13}/E = G_{23}/E = 0.01$.

Table 2.6. Frequency ω_{opt} of stiffly supported rationally designed plates for $\lambda_1 = \lambda_2 = 2.5$ and $h_H = 0.40$.

$\frac{G_{13}}{E} = \frac{G_{23}}{E}$	z_1	z_2	z_3	ω_0	ω_{opt}, by MB	$\Delta\omega$, [%]
0.4	0.322	0.224	2.03	5.21	5.13	1.54
0.04	0.299	0.109	1.99	1.73	1.70	1.73
0.02	0.002	0.066	1.90	1.23	1.21	1.63
0.01	0.001	0.048	1.87	0.869	0.862	0.8

In Figs. 2.12 the vibration modes are shown for a plate with constant thickness are equal to one (solid line) and for a rationally designed plate (dashed line) made from isotropic material for $\lambda_1 = \lambda_2 = 2.5$; $h_H = 0.4$. The results are reported for $y = 0$ and $-0, 5 \leq x \leq 0.5$.

Analysing the results given in Fig. 2.12 the following conclusions are drawn. The difference in deflection between the considered plates is small even at large rotational angles. The rotational angles are two times bigger for the plate with varying thickness than for the plate with constant thickness. Taking into account results are reported

$$M_{11} = \frac{1}{12}h^3 \left(A_{1111}\frac{\partial \gamma_x}{\partial x} + A_{1122}\frac{\partial \gamma_y}{\partial y} \right),$$

$$M_{22} = \frac{1}{12}h^3 \left(A_{2222}\frac{\partial \gamma_y}{\partial y} + A_{1122}\frac{\partial \gamma_x}{\partial x} \right),$$

$$M_{12} = \frac{1}{12}h^3 A_{1212} \left(\frac{\partial \gamma_x}{\partial y} + \frac{\partial \gamma_y}{\partial x} \right),$$

---

[1] for isotropic material (see Tables 2.1–2.5)

2.3 Rational Design of Plates with Finite Transversal Stiffness

it is clear that the optimal rationally designed plate is more stiff because the bending moments on its edges increase. Both the mass of its central part and its inertia decrease. This allows for a decrease of its weight leaving at the same time its fundamental frequency [118] unchanged. The consequences os such a designed from a practical engineering point of view are clear. The design leads to economies of weight and material, while at the same time preserves the required dynamical characteristics of the system.

The plot representing bending are shown in Fig. 2.13. The solid lines correspond to plate with constant thickness, whereas those dashed lines the optimally designed plates.

The application of the optimal rational design to Timoshenko-like plates for the narrow class of functions (2.110) allows for an material economy of about 58% for plates with low transversal stiffness and with the fundamental frequencies overlapping with an error of 2%. The best economical effect is achieved for materials with low transversal stiffness.

The constraints applied to $h(x,y)$ when $h_H < 0.4$ do not lead to an essential decrease in weight for transversal-isotropic plates. However, they play a crucial role for $h_H > 0.4$.

The results, given in Tables 2.1–2.4, cover a wide area of the physical-geometrical parameters of rectangular plates and may be utilised for the design of light plates with a priori specified vibration frequencies.

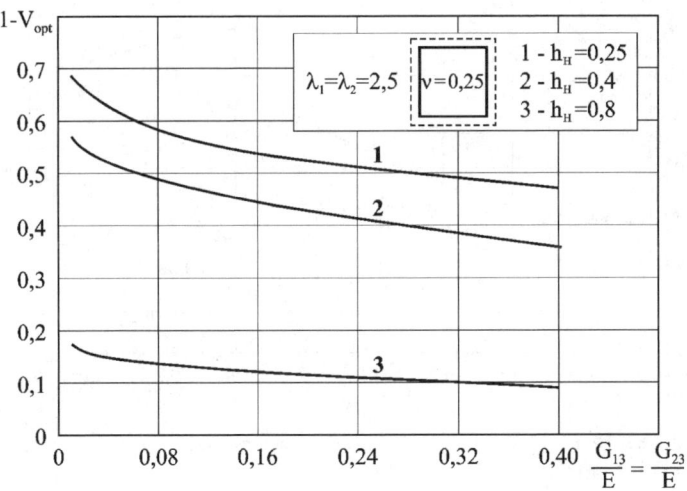

Figure 2.14. Dependence of $(1 - V_{opt})$ versus modulus $G_{13}/E = G_{23}/E$ for a plate with roller supports; $\lambda_1 = \lambda_2 = 2.5$.

In case of rolling support, the following boundary conditions are applied

$$w|_S = 0, \quad \gamma_x = M_{22}|_{y=\pm 0,5} = 0, \quad \gamma_y = M_{11}|_{x=\pm 0,5} = 0. \qquad (2.113)$$

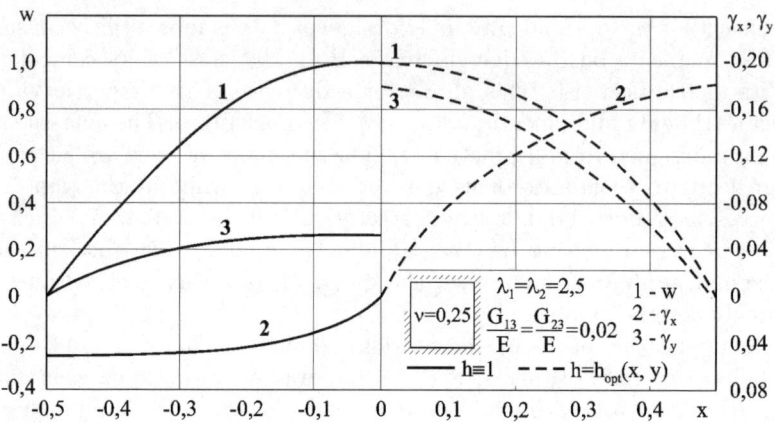

Figure 2.15. Vibrational modes for a plate with roller supports with constant (solid line) and optimal thickness (dashed line).

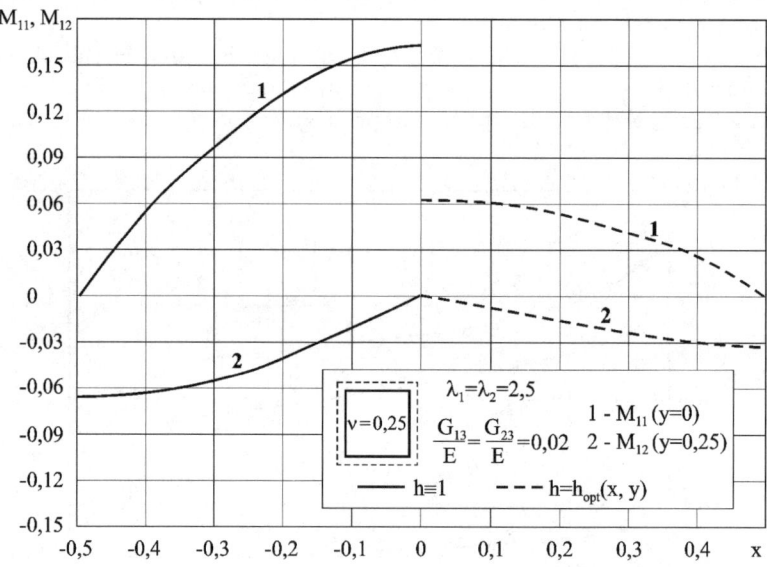

Figure 2.16. Bending moments for a plate with roller supports with constant (solid line) and optimal thickness (dashed line).

The same physical-geometrical parameters and class of functions $h(x,y)$ defined by (2.110) will be considered again. Let the target function $J(\vec{z})$ be defined by (2.104) with $\varepsilon = 0.001$.

In Table 2.7 the computational results obtained for plates with $\lambda_1 = \lambda_2 = 2.5$ and different h_H and $G_{13}/E = G_{23}/E$ are presented. Again, at each step

2.3 Rational Design of Plates with Finite Transversal Stiffness

of the frequency calculation the MVI and Schwartz methods were used, and, the so called after step optimization has been applied.

In Fig. 2.14 the material economy $(1 - V_{opt})$ versus the shear modulus $G_{13}/E = G_{23}/E$ is depicted for different h_H (curve 1 - corresponds to $h_H = 0.25$; 2 - $h_H = 0.4$; 3 - $h_H = 0.8$).

Comparing the graphical dependencies shown in Fig. 2.11 and 2.14 it is seen that the efficiency in economy is higher than in case of stiff support. Decreasing the ration $G_{13}/E = G_{23}/E$ the results. This is explained by the boundness of the boundary effects, as already mentioned in Chapter 2.2.4.

In order to verify the obtained results all rational designs have been checked once more using the MB method with higher order approximations (27 terms). The results are given in Table 2.8 for $h_H = 0.4$. The largest deviation related to the frequency was 0.63% for the case of isotropic material.

We conclude that the terms γ_x and γ_y have a more essential influence on the obtained results than 'ω' in the case of the rational designed plate. Considering, terms w, γ_x and γ_y simultaneously does not essentially improve the results - the frequency remains equal to 4.74 whereas the exact frequency value is 5.30.

In Figure 2.15 the vibration modes of a plate with constant thickness (solid lines) and a plate with rationally designed shape (dashed lines) are depicted for the $y = 0$ cross-section and $\lambda_1 = \lambda_2 = 2.5$, $G_{13}/E = G_{23}/E = 0.02$. Similarly to Figure 2.12, curves 1 are related to the deflections w, whereas curves 2 and 3 are related to the angles of rotation γ_x and γ_y. The curves corresponding to the bending moments are reported in Figure 2.16.

Table 2.7. Optimization results for a plate with roller supports ($\lambda_1 = \lambda_2 = 2.5$).

h_H	$\dfrac{G_{13}}{E} = \dfrac{G_{23}}{E}$	ω_0	ω_{opt}	$\dfrac{\Delta\omega}{\omega_0}100\%$	V_{opt}	z_1	z_2	z_3
	0.40^1	4.77	4.73	0.84	0.529	0.094	0.201	2.94
0.25	0.04	1.71	1.73	1.17	0.378	0.154	0.128	2.91
	0.02	1.22	1.22	0.00	0.338	0.127	0.094	2.96
	0.01	0.864	0.854	1.16	0.313	0.252	0.042	2.99
	0.40^1	4.77	4.73	0.84	0.641	0.082	0.186	2.95
0.40	0.04	1.71	1.71	0.00	0.478	0.127	0.094	2.96
	0.02	1.22	1.22	0.00	0.456	0.200	0.073	2.95
	0.01	0.864	0.858	0.69	0.429	0.139	0.033	3.08
	0.40^1	4.77	4.76	0.21	0.906	0.191	0.113	2.80
0.80	0.04	1.71	1.71	0.00	0.854	0.215	0.030	3.10
	0.02	1.22	1.22	0.00	0.831	0.175	0.003	3.21
	0.01	0.864	0.763	0.12	0.831	0.175	0.003	3.21

The mass distributions along the plates with rationally designed shapes have similar behavior character to that presented in Fig. 2.9. However, the

Table 2.8. Frequency ω_{opt} of rationally designed plates with roller supports for $h_H = 0.40$.

$\dfrac{G_{13}}{E} = \dfrac{G_{23}}{E}$	z_1	z_2	z_3	ω_0	ω_{opt}, by MB	$\Delta\omega$, [%]
0.4	0.082	0.186	2.95	4.77	4.74	0.63
0.04	0.127	0.094	2.96	1.71	1.707	0.18
0.02	0.200	0.073	2.95	1.22	1.215	0.41
0.01	0.139	0.033	3.08	0.864	0.862	0.23

ratio $h_{\max}/h_{\min}$ is lower in this case than in the case of the plate with stiff supports.

2.4 Optimization of Plates and Shells Surfaces with Constraints

In this chapter, the problems of synthesis and optimization of shells and plates' surfaces with constraints on the spectral lines will be considered.

2.4.1 Shells Optimization - Formulation of the Problem

In the problems dealing with shells optimization, considered in the previous chapters, the shell's surface Ω (a projection of the averaged surface into $X0Y$ plane) has been given. Assume now, that the thickness $h(x,y)$ and the plate's surface Ω are varying, so they can be treated as the control functions [19, 20].

A similar problem appears in the mechanics of constructions and electronic devices, where constraints on the free vibration frequencies are imposed. These problems play an important role in engineering and have been considered considerably in the literature [29, 31, 30, 37, 55, 139].

The problem dealing with vibrations of shells with varying thickness can be reduced to the following, equivalent eigenvalue problem

$$Z[h]\vec{u} - \omega^2 M[h]\vec{u} = 0; \quad \vec{u} = \vec{u}(x,y), \quad (x,y) \in \Omega, \qquad (2.114)$$

where $Z[h]$ and $M[h]$ are differential operators related to the elastic and mass distributions along the shell. For Timoshenko-like models they have been defined in Chapter 2.2.1. For the other models $Z[h]$ and $M[h]$ are built in a similar way using the corresponding governing differential equations.

Denote by $\Theta_k = \omega_k^2$ the frequencies squared and by $\vec{u}_k$ the corresponding modes.

Assume, that we are with to design a plate or shell with a prescribed set of N leading frequencies $\{\alpha_k\}_{k=1}^N$, by choosing, in an appropriate manner, the thickness function $h(x,y) = h^*(x,y)$ and the domain $\Omega = \Omega*$. Such, a problem will be termed the shell's synthesis. It is not known in this case, if shell exists for any given $h(x,y) \in U_{\partial 1}$ and $\Omega \in U_{\partial 2}$, $\Theta_k = \alpha_k$ for all

2.4 Optimization of Plates and Shells Surfaces with Constraints

$k = 1, 2, ..., N$. Therefore, as solutions to the above problem, we designs that in some sense are close located to the consider exact ones (i.e., those that are exact solutions of the optimization problem).

An alternative approach will be used for the optimization. Usually, we begin with a rather wide class of known functions as first iterates and then improve the design incrementally for higher iterates. However, if we are going to decrease the weight of a plate (or shell) without a change of its first (N) free vibration frequencies $\{\Theta_k^0\}_{k=1}^{\infty}$, then we need to solve the synthesis problem for $\alpha_k = \Theta_k$, $k = 1, 2, ..., N$ satisfying at the same time the corresponding constraints $U_{\partial 1}, U_{\partial 2}$.

We say that two constructions are close regarding their spectral lines if the function

$$R(N, \vec{\varepsilon}) = \sum_{k=1}^{N} \frac{1}{\varepsilon_k} |\Theta_k - \alpha_k|^2, \qquad (2.115)$$

is close to zero. Thus, the optimization (synthesis) problem can be formulated in the following manner. Determine $h^*(x, y) \in U_{\partial 1}$ and the area $\Omega^* \in U_{\partial 2}$ which minimise the functional:

$$J(h, \Omega) = \sum_{k=1}^{N} \frac{1}{\varepsilon_k} |\Theta_k - \alpha_k|^2 + I(h, \Omega), \qquad (2.116)$$

where: $\{\alpha_k\}_{k=1}^{N}$ is a given set of the positive numbers (required spectral lines); $\{\Theta_k\}_{k=1}^{N}$ is the frequency spectrum defined by (2.114), which governs the free vibration of the shell with thickness h and surface Ω; ε_k are small numbers characterizing the penalty for the disturbance of the constraint $\Theta_k = \alpha_k$ (see Chapter 2.3.2); $I(h, \Omega)$ is a positive functional for all $h \in U_{\partial 1}$, $\Omega \in U_{\partial 2}$, that includes the additional design constraints related to the shell (weight, surface area or surface shape, etc.). We assume, that $I(h, \Omega)$ depends continuously on the two constraints, which is not always realised in practice.

The sets $U_{\partial 1}$ and $U_{\partial 2}$ are defined by the design on the basis of requirements such as strength of materials and so on. However, they should satisfy the mathematical requirements, which guarantee solvability of the formulated synthesis problem (2.116).

The problems mentioned above do not possess exact solutions [120]. First, on the given sets $U_{\partial 1}$ and $U_{\partial 2}$ the possibility of designing with the given $\Theta_k = \alpha_k$, $k = 1, 2, ..., N$ may not exist. Therefore, by a solution we will mean an arbitrary design minimising the functional (2.116). Second, slight changes of $\{\alpha_k\}_{k=1}^{N}$ occurring during the application of the numerical methods or during the experimental measurements may lead to large changes of h^* and Ω^* [230]. This introduces instabilities in the h^* and Ω^* solutions and numerical techniques cannot be conveniently applied.

In order to avoid the mentioned drawbacks, i.e. in order to get a well-defined problem of synthesis (2.116) we parameterise the function $h(x, y)$ and the space Ω. Suppose, that $h(x, y)$ is the linear combination of the functions

$$h_\beta(x,y) = \sum_{i=1}^{m} \beta_i \psi_i(x,y), \tag{2.117}$$

and that the Ω space is obtained as a result of regular continuous deformation of a certain initial space Ω_0. The character of deformation is defined by a given F function continuously dependent on the parameters $\{\beta_i\}_{i=m+1}^{m+p}$

$$\Omega_\beta = [1 + F(x,y,\beta_{m+1},...,\beta_{m+p})] \cdot \Omega_0, \tag{2.118}$$

where $F(x,y,0,...,0) \equiv 0$ for all $(x,y) \in \Omega_0$. By a regular continuous deformation of the initial surface Ω_0 we mean a continuous dependence between new $(x',y') \in \Omega$ and the old $(x,y) \in \Omega_0$ coordinates describing the surface of the shell. In addition, by regularity we mean that a simply connected are a is transformed to a simply connected one, and a multiply connected area to a multiply connected one with the same number of gaps (holes). This means that distinct points in Ω_0 are transformed to district points in Ω. Additional mechanical stresses do not occur, because Ω is a relaxed surface and its shape's influence on the free vibration frequencies is investigated.

We assume that $\vec{\beta} \in U_\partial$, where U_∂ is the bounded and closed set in the Euclidean space R^{m+p} and $\vec{\beta} = (\beta_1,...,\beta_m,\beta_{m+1},...,\beta_{m+p})$.

Substituting (2.117) into (2.118) we transform the synthesis formulation (2.116).

Consider the vector $\{\alpha_k\}_{k=1}^{N}$ and the design constraints

$$\Phi_0\left(\vec{\beta}\right) = I(h_\beta, \Omega_\beta) \geq 0 \qquad \forall \vec{\beta} \in U_\partial. \tag{2.119}$$

We seek the vector $\vec{\beta}^* \in U_\partial \subset R^{m+p}$ which minimises the function

$$\Phi\left(\vec{\beta}\right) = \sum_{k=1}^{N} \frac{1}{\varepsilon_k} |\Theta_{k\beta} - \alpha_k|^2 + \Phi_0\left(\vec{\beta}\right), \tag{2.120}$$

where $\Theta_{k\beta}$ are the second powers of the free vibration frequencies of the shell with thickness h_β and surface Ω_β defined by (2.114).

Thanks to this formulation, the synthesis (or optimization) problem is well defined, expressed by the following theorem.

THEOREM 2.8 *If $F(x,y,\vec{\beta})$ secures a continuous and regular deformation of the initial space Ω_0, and $\Phi_0(\vec{\beta})$ is continuous on U_∂, which is closed and bounded in R^{m+p}, then for the arbitrary vector $\{\alpha_k\}_{k=1}^{N} \subset K$ (K is a closed bounded set in R^N) the function (2.120) achieves its lower limit for the pair of elements $\vec{\Theta}^* = \{\Theta_{k\beta^*}\}_{k=1}^{N}, \vec{\beta}^*$. In addition, $\vec{\beta}^* \in U_\partial$ is a stable solution to the synthesis problem and $\vec{\Theta}^*$ is the vector consisting of N first eigenvalues of the problem (2.114) for $h = h_{\beta^*}$ and $\Omega = \Omega_{\beta^*}$.*

2.4 Optimization of Plates and Shells Surfaces with Constraints

Proof. It is enough to prove that $\Phi(\vec{\beta})$ is continuous in U_∂. By assumption the function $F(x, y, \vec{\beta})$ secures a continuous deformation from Ω_0 to Ω. Therefore, the transformation $\vec{\beta} \to \Omega$ is continuous. Because the transformation $F(x, y, \vec{\beta})$ is smooth and analytic, using the Rayleigh quotients for the frequencies squared, and remembering that h_β depends continuously on $\vec{\beta}$, one can show [121], that the transformation $\vec{\beta} \to \vec{\Theta}_\beta$ will also be continuous for all $k = 1, ..., N$. Besides, $\Phi_0(\vec{\beta})$ is by assumption continuous on U_∂. Therefore, $\Phi(\vec{\beta})$ is continuous on U_∂ and on the basis of the Weierstrass theorem it achieves its lower limit for a certain $\vec{\beta}^* \subset U_\partial$. Because $\Theta_{k\beta}$ is continuous in relation to $\vec{\beta}$, the corresponding $\Theta_{k\beta^*}$ can be obtained from (2.114) for $h = h_{\beta^*}$, $\Omega = \Omega_{\beta^*}$.

Stability of the obtained solution $\vec{\beta}^*$ results from reference [95]. Both U_∂ and K are bounded and finite dimensional.

Note, that without the additional convexity conditions imposed the theorem (2.114) does not guarantee a unique solution $\vec{\beta}^*$ for the synthesis problem (2.120). Theorem (2.114) remains true if instead of (2.115) we take

$$R(N, \vec{\varepsilon}) = H(\Theta_1, ..., \Theta_N, \vec{\varepsilon}), \qquad (2.121)$$

where H is an arbitrary function continuous in relation to its arguments $\Theta_1, ..., \Theta_N$.

To conclude, the transformation to a finite dimensional approximation of $h(x, y)$ and Ω transforms the shell synthesis problem to a well-defined one from a mathematical point of view. These results will be used in the next chapters in order to solve different problems involving surface optimization of shell and plates using various criteria related to the spectral lines requirements.

2.4.2 Algorithm for Optimal Surface Search

The results outlined in the previous chapter pave the way for numerically solving optimization problems related to shapes of plates and shells.

In this section we describe an algorithm of optimal surface search, or more precisely a surface control algorithm in order to find optimal solutions.

As mentioned in section 2.4.1, the initial area Ω_0 is deformed during the search process in a such way that the smoothness and analyticity properties are preserved. However, this introduces challenges from a numerical point of view. Indeed, in [31] a search for the optimal surface is reduced to finding an initial estimate of the surface function and next its optimal limit. In order to avoid nonuniquess, a special coordinate system, linked with the arc length, is introduced.

From the point of view of numerical implementation of the optimization problem, parallel application of the search method for optimal surface and the method of finding solutions to (2.114) is recommended. For instance, for shells with varying thickness with a curvilinear boundary it is impossible to find an approximate function using the MB method. The application of the mesh method is difficult in case of complicated shapes. An exception is FEM, which is well adapted to problems with complicated shapes. However, here also some problems occur. First, the algorithm should be built in such a way, that for each step we do not need to build new finite elements, but only apply the initial topological information. This certainly leads to computational effectiveness. Second, during surface deformation we need to track if the finite elements do not create lines, or surfaces, do not overlap with each other, do not have gaps and so on. This task is rather complicated.

The algorithm given below addresses the aforementioned concerns, is well adapted to complicated shapes, satisfies the so called "star condition", and can be used for both simply and multiply connected surfaces. We call Ω a star surface, if it possesses at least one point where an arbitrary radius emanating from this point intersects the surface edge only once. In case of doubly connected surfaces the internal and external edges are intersected by the mentioned radius only once.

Consider an initial surface Ω_0, as presented in Figure 2.17. Choose a point, at a star surface of Ω_0, and define it as the origin of the coordinate system. Then divide the surface Ω_0 into sectors by a given number of radiuses $\Lambda_1, \Lambda_2, ..., \Lambda_n$.

Each of the sectors will be divided into one row of finite elements. Here triangle and quadrangle elements will be used. This division is called the initial division, and it is used for the numerical calculations together with the classical methods applied to regular finite element calculations. The topology defined in this way remains valid during the surface deformation of Ω_0. The role of the parameter $\vec{\beta}$ in the function $F(x, y, \vec{\beta})$ is to denote the distance between the intersection point of the radius with the surface boundary and the origin. The radii r_{10}, r_{11} denote the distances to the points I_0, I_1; the radii r_{20}, r_{21} denote distances to the point II_0, II_1, and so on. Suppose now that a distance changes but remains on the radius; then the surfaces are deformed in relation to the chosen radius. A similarly algorithm has been applied for the optimization of the holes shape in a plate with a flat stress state [236].

If the following design constraints,

$$r_{i0} \geq c_{0i}, \quad r_{i1} \leq c_{1i}, \quad |r_{i1} - r_{i0}| \geq c_{10i}, \tag{2.122}$$

characterising the sought surface, do not exist, then the only constraints being controlled during the calculations are the following:

$$r_{i0}, r_{i1} > 0; \quad r_{i1} > r_{i0} \quad \text{for all } i. \tag{2.123}$$

The conditions (2.123) secure the regularity of the edges of Ω for the arbitrary radii r_{i0}, r_{i1}. Let us analyse a regularity inside the Ω surface. Because

2.4 Optimization of Plates and Shells Surfaces with Constraints   149

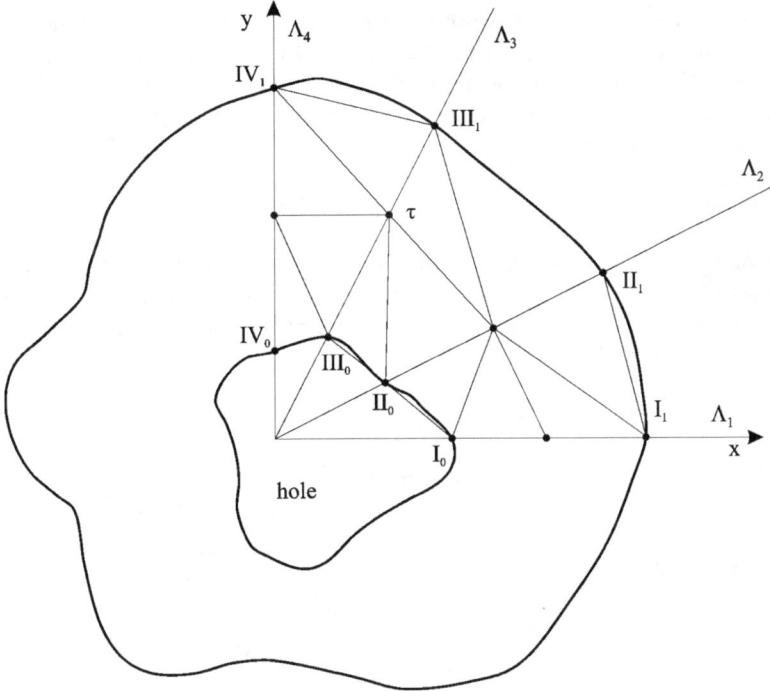

Figure 2.17. Surface Ω_0 divided for sectors by radii Λ_n and finite elements.

the radii do not move during the deformation process and between the radii, there is only one row of triangles, then no overlap between sectors occurs. If we assume that, according to Fig. 2.17, the node τ is transformed along the radius Λ_3, then only the triangles linked to it will be deformed and an overlap or gaps between them will not occur. However, if that node crosses the node III_1 or III_0, then the corresponding triangles will overturn the left boundary and regularity will be violated. In order to avoid this, we proceed in the following manner. New coordinates of the triangles' corners will be defined in a way that does not violate the initial topology. For instance, for a node numbered τ (see Fig. 2.17) we define

$$r'_\tau = r_\tau \frac{(r'_{31} - r'_{30})}{(r_{31} - r_{30})}, \qquad (2.124)$$

where r_{31}, r_{30} are the distances to the points III_0, III_1 in the initial area Ω_0; r'_{31}, r'_{30} are the distances in the deformation state; r_τ and r'_τ are the distances from points 0 to τ in the initial and deformable surfaces, correspondingly. If point τ is situated at the centre of the interval III_0, III_1, then in the deformable state it will also be situated in the centre between III_0 and III_1. The points III_0, τ, III_1 overlap only if $r'_{31} = r'_{30}$. However, the condition

(2.123) will not permit this. Another negative effect may occur, if along two neighbouring radii strong stretching occurs. In this case the triangles are stretched (Fig. 2.18) and the accuracy of the calculation of free vibration frequencies rapidly decreases [226]. In order to avoid this we need to satisfy the following inequality

$$|(r_{i1} - r_{i0}) - (r_{j1} - r_{j0})| \leq c_{ij}. \tag{2.125}$$

If during the process of optimal surface search the constraint condition (2.125) is close to be violated this is an indication that we need to increase the numbers of partition points, the values of c_{ij} and prolong the search procedure.

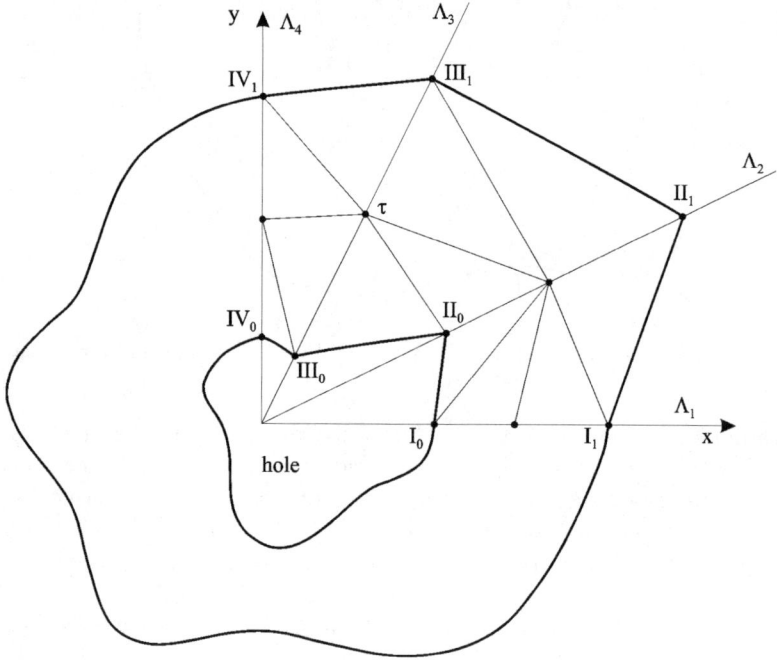

Figure 2.18. Stretching triangle effects during an optimization process of the surface Ω_0.

Note, that all constraints (2.122), (2.123) and (2.125) can be taken into account using only one constraint of the form

$$f(r_{i0}, r_{i1}) \geq 0, \qquad i = 1, 2, ..., N. \tag{2.126}$$

The described algorithm can be applied without any changes to all problems in the context of the theory of shells. It can be applied to a surface

2.4 Optimization of Plates and Shells Surfaces with Constraints   151

with arbitrary shape not only for finding a solution to the optimization problem but also in the formulation of an integrated algorithm for shells (with complicated shapes).

The algorithm may be applied in order to design plates and shells with different constraints imposed to their free vibration frequencies.

2.4.3 Design of Plates and Shells with Constant Thickness and Rolling Supports

Free vibrations of elastic plates and shells have been analysed by many researchers [112, 181]. Recently, more attention has been paid to the application of the finite elements method (FEM). As it shown in Chapter 2.4.2, the FEM is very suitable for optimization of shells surfaces. Therefore, the FEM can be considered as the basic method for optimization of shells and plates.

However, for higher order differential equations, the order of the obtained algebraic equations as a result of finite dimensional approximations is very high. This leads to numerical problems and norm economical calculations. In particular, such problems occur in the case of the Kirchhoff-Love kinematic model. In case of the Timoshenko model the order of the obtained equation is lower, but the number of unknowns is higher. Therefore, the obtained set of algebraic equations is large in both cases, which increases the computational time of the optimization problems.

As an example we consider the optimization problem of spherical shells and plates with rolling supports. In that case we can initially simplify the problem of determining the free vibration frequencies by using membrane analogy and then FEM. On the basis of those two methods the established algorithm enables the computation (using a relatively small number of the finite elements) of the frequencies with high accuracy. Regarding the modes determination, the membrane analogy does not improve the effectiveness of the method because the use of reccuring relations leads to the same result obtained directly by the FEM. Let us briefly describe the method of membrane analogy [41].

Consider a shallow spherical shell with the curvature k under a constant load leading to the following main inertial stresses

$$T_{11} = T_{22} = P. \tag{2.127}$$

Then, the bending vibration investigation of this shell is reduced to the consideration of the following equation

$$\Delta^4 F - 2\chi \Delta^3 F - 2\theta \Delta^2 F = 0, \tag{2.128}$$

where

$$2\chi = -\bar{P}, \quad 2\theta = \bar{\omega}^2 - \bar{k}^2 \frac{12\left(1-\nu^2\right)}{\bar{h}^2}, \quad \bar{w} = w/h_0,$$

$$\bar{P} = Pb^2/D, \quad \bar{\omega}^2 = \rho\omega^2 b^4/D, \quad \bar{k} = kb^2/h_0,$$
$$\bar{F} = F/D, \quad \bar{x} = x/b, \quad \bar{y} = y/b, \quad \bar{h} = h/h_0. \quad (2.129)$$

The quantity D denotes the cylindrical stiffness; h_0 - thickness; n - Poisson coefficient; ρ - material density; b - maximal dimension in the projection plane.

Consider the case rolling support boundary conditions. In this case the application of the membrane analogy leads to the essential simplification of the problem. Using other boundary conditions one may also decrease the order of the differential equation but an additional iteration procedure should be added [61, 83]. Indeed, using the rolling support, the boundary conditions are presented in the following way [41]

$$F = \Delta F = \Delta^2 F = \Delta^3 F\big|_S = 0. \quad (2.130)$$

Denoting $\Delta^2 F = u$, from (2.128) and (2.130) we get

$$\Delta^2 u - 2\chi \Delta u - 2\theta u = 0, \quad (2.131)$$

with the boundary conditions

$$u = \Delta u\big|_S = 0. \quad (2.132)$$

In the case of a plate, $\bar{k} = 0$ and we get an equation similar to (2.131) for the deflection w with the corresponding boundary conditions

$$\Delta^2 w - 2\chi \Delta w - 2\theta w = 0,$$
$$w = \Delta w\big|_S = 0.$$

Thus, for free vibration analysis of plates and shells we need to analyse only the equation (2.131) with the boundary conditions (2.132). In case of the plate replace the variable 'u' with 'w'.

Equation (2.131) can be expressed in the form

$$(\Delta + l_1)(\Delta + l_2) u = 0,$$

where:

$$l_1 = \sqrt{2\theta + \chi^2} - \chi, \quad l_2 = -\sqrt{2\theta + \chi^2} - \chi. \quad (2.133)$$

Denoting

$$\Delta u + l_2 u = v, \quad (2.134)$$

we get

$$\Delta v = l_1 v = 0. \quad (2.135)$$

According to (2.132) the boundary conditions for 'u' and 'v' can be defined independently as,

$$v\big|_S = 0, \quad u\big|_S = 0. \quad (2.136)$$

2.4 Optimization of Plates and Shells Surfaces with Constraints 153

Then, in order to determine the free vibration frequencies from equations (2.128) and (2.130) it is sufficient to find free vibration spectral lines defined by (2.135), and include the boundary conditions (2.136). The corresponding frequencies $\bar{\omega}_n^2$ are defined by (2.133) and (2.129), which yield

$$\bar{\omega}_n^2 = \left[(l_{1n} + \chi)^2 - \chi^2 \right] + \bar{k}^2 \frac{12(1 - \nu^2)}{\bar{h}^2}. \tag{2.137}$$

In order to solve the problem of membrane vibrations (2.135) we apply the second order FEM method [214]. For this purpose the surface Ω_0 is partitioned to a set of not intersecting triangles with the arbitrary shapes (see Chapter 2.4.2). On each triangle the function 'v' is approximated by

$$v(x, y) = \sum_{e=1}^{M} v^e(x, y) = \sum_{e=1}^{M} \sum_{i=1}^{6} v_i^e \xi_i^e(x, y), \tag{2.138}$$

where 'e' denotes the finite element number, 'i' the node local number, 'v_i^e' the value of the unknown function 'v' at node 'i' on element 'e', and M the total number of finite elements used.

The functions $\xi_i^e(x, y)$ are the shape functions of a finite element [226], defined by:

$$\xi_1^e(x, y) = L_1(2L_1 - 1), \quad \xi_2^e(x, y) = 4L_1 L_2,$$
$$\xi_3^e(x, y) = L_2(2L_2 - 1), \quad \xi_4^e(x, y) = 4L_2 L_3,$$
$$\xi_5^e(x, y) = L_3(2L_3 - 1), \quad \xi_6^e(x, y) = 4L_1 L_3. \tag{2.139}$$

The variables L_1, L_2, L_3 are special L-coordinates defined by the following equations

$$L_1(x_1^e - x_3^e) + L_2(x_2^e - x_3^e) = x - x_3^e,$$
$$L_1(y_1^e - y_3^e) + L_2(y_2^e - y_3^e) = y - y_3^e,$$
$$L_1 + L_2 + L_3 = 1,$$

where x_i^e, y_i^e are the corners of the triangle finite element 'e' in the rectangular coordinate system ($i = 1, 2, 3$). Substituting (2.138) to (2.135), and applying the MB method on each triangle we get

$$\sum_{i=1}^{6} v_i^e \int_{\Delta_e} \nabla \xi_i^e \nabla \xi_j^{e'} dx dy - l_1 \sum_{i=1}^{6} v_i^e \int_{\Delta_e} \xi_i^e \xi_j^{e'} dx dy = 0,$$

where $\xi_j^{e'}(x, y)$ are the test functions.

In matrix form we express the derived system of equations in compact form:

$$C^e \bar{v}^e - l_1 D^e \bar{v}^e = 0, \tag{2.140}$$

where $\bar{v}^e$ is a vector with six components, which are the unknown values v_i^e defined at the nodes of an element. The elements of the matrices C^e and D^e with the dimensions 6×6 are defined as follows:

$$C_{ij}^e = \int_{\Delta e} \nabla \xi_i^e \nabla \xi_j^{e'} dxdy, \quad D_{ij}^e = \int_{\Delta e} \xi_i^e \xi_j^{e'} dxdy,$$

where Δe is a triangle area. The matrices C^e and D^e can be determined analytically, and are considered as a priori given in the numerical algorithms.

In order to get the global matrices C (stiffness) and D (mass) we need to place the local matrices C^e and D^e on the necessary places of the corresponding global matrices using the information about the triangles, as outlined in section 2.2.3 for the one dimensional case. Next, the problem is reduced to a standard one

$$C\bar{v} - l_1 D\bar{v} = 0, \qquad (2.141)$$

where C and D are the global stiffness and mass matrices, $\bar{v}$ the vector of the unknown nodal coordinates of the function being sought. The equation (2.141) can be solved using one of the standard methods, instead of using the Schwartz method.

For an arbitrary surface Ω obtained from Ω_0 as a result of deformation, it is sufficient to calculate the coordinates x_i^e and y_i^e according to (2.124) and according to the algorithm described in section 2.4.2.

Program testing. In order to check the accuracy of the FEM algorithm the problem of the torsion of a shaft with a circular cross section and with a material relief is considered (Fig. 2.19). The problem is defined by the equation

$$\Delta \varphi = -2, \quad \varphi|_S = 0.$$

In Table 2.9 the values of the stress function φ given. They are obtained using FEM at points a, a^1, c, c^1, c^2 and so on. Column I corresponds to the solution using $N = 52$ finite elements, column II to the exact solution [9], and III to the solutions given in reference [32].

Table 2.9. Stress function values in relation to Figure 2.19.

	I	II	III	I	II	III	I	II	III
		a			a^1			c	
φ	0	0	0	0.09	0.09	0.10	0.374	0.37	0.37
		c^1			c^2			e	
	0.341	0.34	0.34	0.225	0.23	0.23	0.477	0.48	0.46
		e^1			e^2			e^3	
	0.435	0.44	0.42	0.307	0.31	0.30	0.089	0.09	0.09

2.4 Optimization of Plates and Shells Surfaces with Constraints

Already for $N > 50$, the obtained solution differs from the exact by only 2%. In order to achieve this accuracy we need to solve an equation with 1600 unknowns using the finite difference method [32].

Examples of shells design with rolling supports. According to the membrane analogy instead of ω_n^2 we take the corresponding l_{1n} (see (2.133)).

Example A. Design a plate (shell) with constant thickness, minimal surface S and a priori given fundamental frequency equal to the free vibration frequency of a square plate (shell) with side equal to 1.

According to (2.137), the results for spherical shells when the compressive force P is applied can be obtained using the results obtained for a plate on the basis of simple calculation. Thus, in what follows we consider only a plate without the compressive load ($\chi = 0$, $\bar{k} = 0$).

For a square plate with unit side we have $\omega_1^2 = 4\pi^4$. Therefore, for the optimization we need to assume that $\alpha_1 = l_{11} = 2\pi^2$. As constraint functional $I(h, \Omega) = \int_\Omega dxdy$ we take the plate's surface.

Thus, the considered problem is equivalent (see Chapter 2.4.1) to that of finding the minimum of the functional

$$J(h, \Omega) = \tfrac{1}{\varepsilon_1} \left(l_{11} - 2\pi^2\right)^2 + \int_\Omega dxdy. \qquad (2.142)$$

As the zero iteration for the surface we take the square plate Ω_0. According to the algorithm used for searching for the optimal surfaces Ω, we divide the plate into sectors and the sectors into finite elements. In Fig. 2.20 the partition of the plate with number of radiuses $N = 7$ and number of triangles $M = 30$ is shown.

The surface Ω_0 is simply connected. Therefore, it is sufficient to take the constraints only for r_{1i} and for $|r_{1i} - r_{1j}|$. We take them in the form

$$r_{1i} > 0.25; \quad |r_{1i} - r_{1j}| \le 0.25; \quad i, j = 1, 2, ..., 7. \qquad (2.143)$$

The search for a minimum of a function with seven variables r_{1i} ($i = 1, ..., 7$) obtained from (2.142) is realised using the method of adaptive stochastic search with $\varepsilon_1 = 0.001$. As the optimization result after 230 steps, the surface presented in Fig. 2.21 was found.

The first free vibration frequency of the optimal surface is $l_{11} = 19.86$ for $\alpha_1 = 2\pi^2 = 19.74$, which leads to the error of $\sim 0,6\%$. The area of the obtained surface is $S_{opt} = 0.88$. The economy coefficient is equal to 12%. For three radiuses and six triangles the obtained surface is $S_{opt} = 0.928$ and $l_{11} = 20.12$.

To conclude, regarding the surface and the free vibration frequency, the circular surface is the optimal one. This result was confirmed also by others [191].

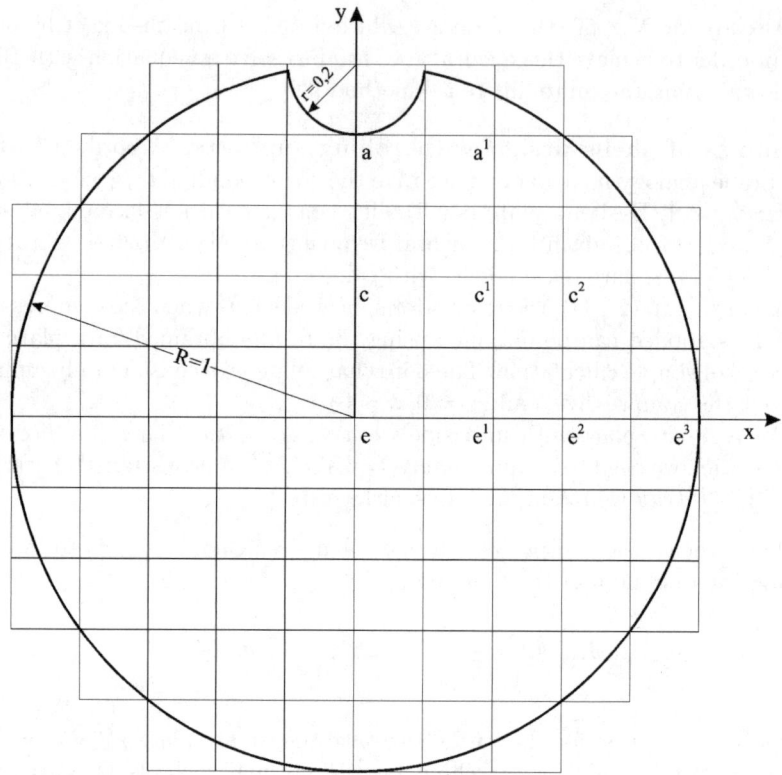

Figure 2.19. Torsioned shaft divided into finite elements.

Example B. Design a plate (shell) with constant thickness and a given hole situated in its centre, having a minimal surface, and fundamental frequency equal to that of a square plate (shell) with the same hole.

As an example the plate (shell) with the rectangular hole in its centre with dimension 0.1×0.2 (in the nondimensional quantities) has been optimized. The initial surface and its partition for the sectors and the finite elements are shown in Fig. 2.22. Similarly to the previous example A, the number of radiuses $N = 7$, and the number of triangles $M = 36$ were used. For such a plate, the exact value of l_{11} is not known. First using FEM the initial value $l_{11} = 65.52$, and then $\alpha_1 = l_{11} = 65.52$ has been assumed. The target function has the form of (2.142), where instead of $2\pi^2$ we have $\alpha_1 = 65.52$.

The constraints for r_{1i} have been taken in the form (2.143). Although in this case we have a doubly connected surface, the hole shape does not change, and therefore r_{0i} does not change during the optimization process.

Using the above mentioned constraints in all cases, the obtained surfaces have cusps on their boundaries. The attempts to avoid them using more stiff constraints did not yield good results. In what follows, the coefficients

2.4 Optimization of Plates and Shells Surfaces with Constraints

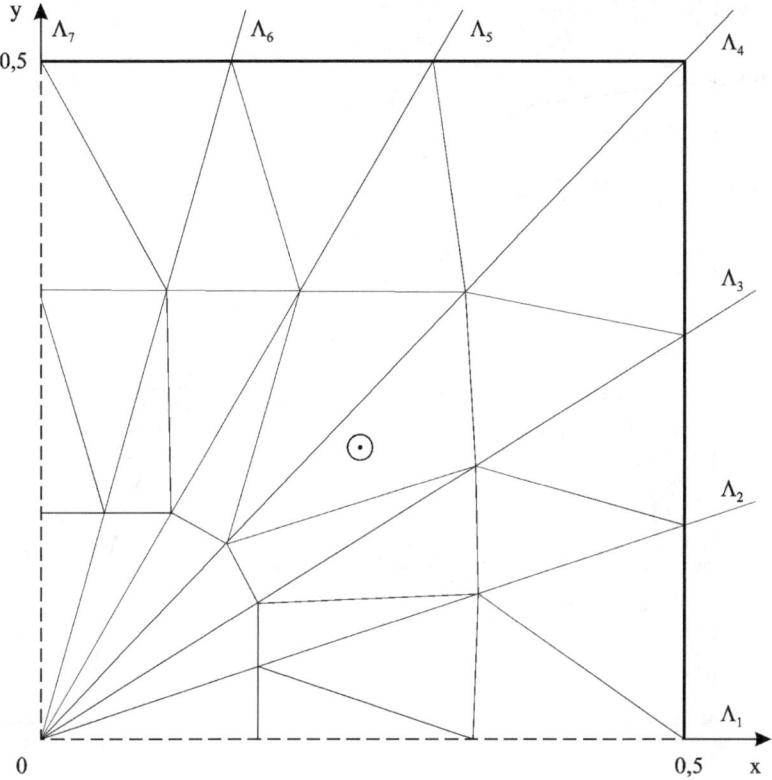

Figure 2.20. Fourth part of the squared plate from example A divided into sectors and finite elements.

of a certain smooth function describing the boundary shape in cylindrical coordinates (instead of the radius length) have been used as the variational parameters. The function is defined by a fourth order polynomial where some relations between its coefficients have been established earlier (the derivatives for $\varphi = 0$ and $\varphi = \pi/2$ have been equal to zero and $r_{1i} \geq 0.25$).

The function is given by

$$r(\varphi) = 0,25 + z_1^2 + z_2^2 \frac{16}{\pi^3} \varphi^2 (3\pi/4 - \varphi) + (z_3^2 - z_2^2) \frac{16}{\pi^4} (\pi^2/2 - \varphi^2), \quad (2.144)$$

where z_1, z_2, z_3 are the optimization parameters and φ the angle measured from $0X$ in the anti clock-wise direction.

Therefore, the use of the expression (2.144) leads to a decrease of the optimized parameters from 7 to 3 and guarantees the solution with a smooth boundary.

Using an adaptive stochastic search for $\varepsilon_1 = 0.001$, the optimized surface shown in Fig. 2.23a was obtained. The variables z_1, z_2, z_3 for the function (2.144) have the values $z_1 = 0.050504$; $z_2 = 0.189645$; $z_3 = 0.582342$.

158   2 Rational Design of Plates and Shells

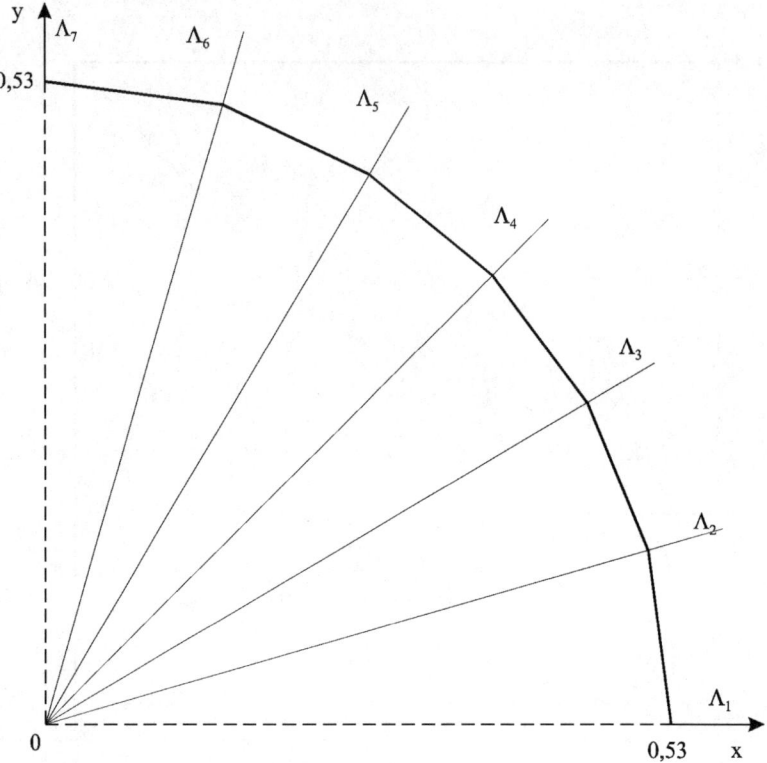

Figure 2.21. Optimization results (after 230 steps) of the surface from Figure 2.20.

The first vibration frequency $l_{11} = 65.52$ of the optimal surface overlaps with a1 with the accuracy of $\sim 0.5\%$. The surface projection is $S_{opt} = 0.708$, whereas the projection of the initial surface is $S_0 = 0.92$. The economy coefficient is equal to 23%.

Example C. Design a plate (shell) with constant thickness, with a hole situated at its centre, with a given boundary configuration, having a minimal projection surface and with the fundamental frequency equal to the frequency of a plate (shell) with a rectangular hole at its centre.

The target function has the form given in example B. The external boundary is defined, whereas the internal one is governed by the function (2.144), where instead of the lower limit equal to 0.25 we take 0.05. Instead of the constraints (2.143) we take

$$r_{0i} \leq 0.4, \qquad (2.145)$$

which guarantees lack of intersection of the hole with the external boundary.

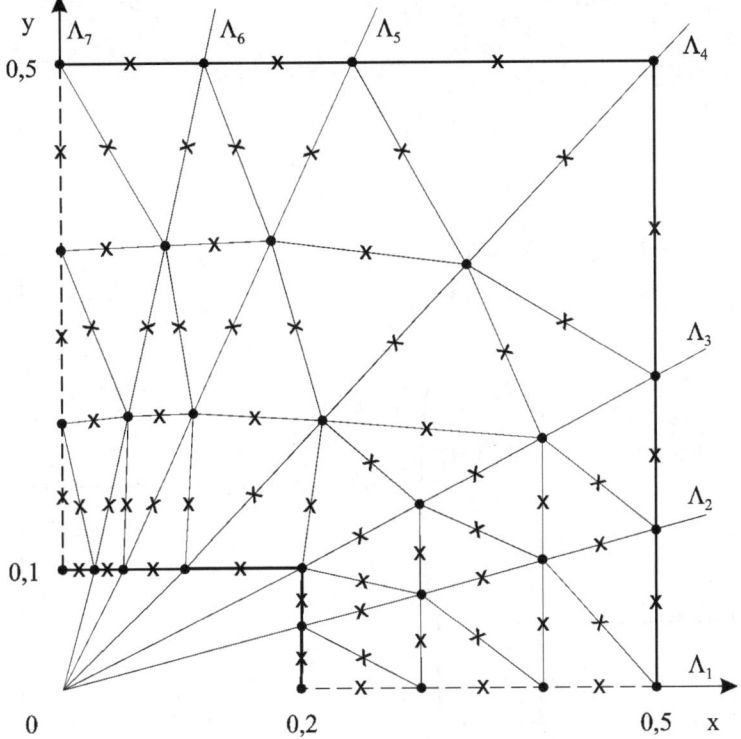

Figure 2.22. A quarter of the squared plate with the rectangular hole from example B.

As a result of the optimization process the two solutions shown in Fig. 2.23b and Fig. 2.23c, are shown. For the first one we have $l_{11} = 66.6$; $S_{opt} = 0.228$; $z_1 = 0.340251$; $z_2 = 0.214267$; $z_3 = 0.020896$.

The economical effectiveness is smaller than 1% for both cases. Therefore, all three designs shown in Fig. 2.22, Fig. 2.23b and Fig. 2.23c are equivalent because of the criterion (2.142).

2.5 Vibroisolation of a Construction with Shells Elements

In this section problems of optimal vibrationisolation of mechanical systems composed of shells elements are considered.

The problem of optimal vibroisolation requires finding the most stable (in some sense) design in relation to the external time-varying load, and is very important in practical applications involving mechanical constructions. From

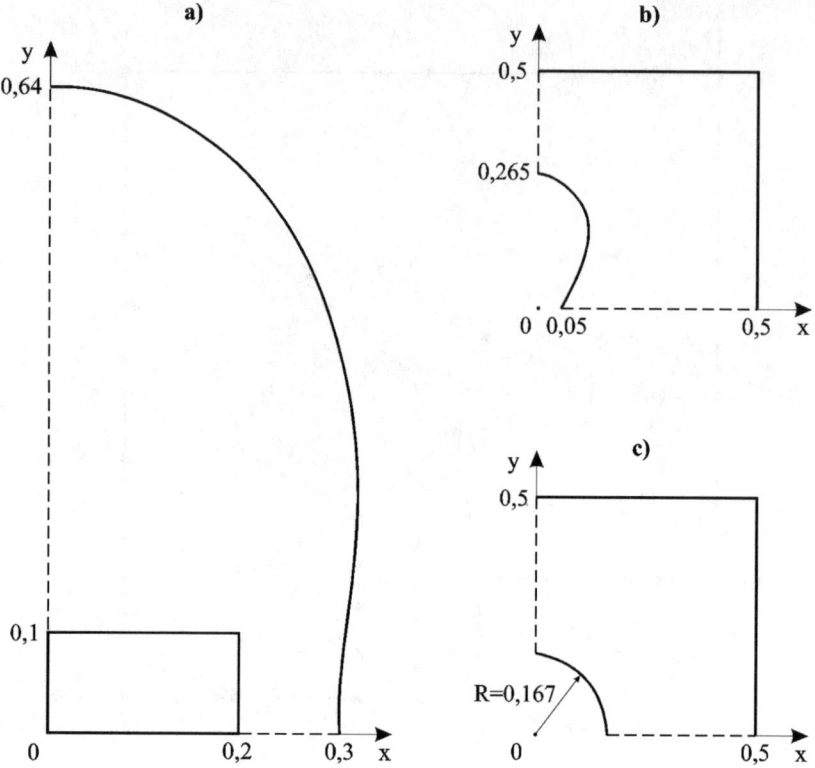

Figure 2.23. Surface optimization results: a) plate with a hole from example B; b), c) two designs of a plate with a hole and with the required configuration of the external edge from example C.

the point of view of the linear approach, these problems are closely related to those of spectrum optimization, considered earlier. However, the nonlinear vibroisolation problem has to be treated in a different way.

2.5.1 Nonlinear Forced Vibrations of the Timoshenko Like Shells

A general theory of finding solutions to the boundary value problems of non-linear shells governed by the Kirchhoff-Love equation has been formulated by I.I. Vorovič [248, 249]. N.F. Morozov has included an inertial effect caused by rotation [161].

Consider the problem of vibroisolation on the basis of the improved Timoshenko like model. We will analyse non-linear shell vibrations governed by the Timoshenko model, and formulate conditions leading to a solution of the problem [19].

2.5 Vibroisolation of a Construction with Shells Elements

Suppose that a shallow shell has the projection Ω with boundary S. The linear equations governing the vibrations on the basis of the kinematic Timoshenko model are given by (2.1). Taking into account geometrical nonlinearities, we need to modify certain terms in the first and last equations. As a result, we get the following differential equations governing the forced nonlinear vibrations of Timoshenko-like shells [119] in the following nondimensional hybrid form:

$$k_y \frac{\partial^2 F}{\partial x^2} + k_x \frac{\partial^2 F}{\partial y^2} - L(w,F) + \frac{2}{3}\bar{\lambda}_1 \frac{\partial}{\partial x}\left[h\left(\gamma_x + \frac{\partial w}{\partial x}\right)\right] +$$

$$\frac{2}{3}\bar{\lambda}_2 \frac{\partial}{\partial y}\left[h\left(\gamma_y + \frac{\partial w}{\partial y}\right)\right] - h\frac{\partial^2 w}{\partial t^2} = -q(x,y,t),$$

$$\frac{1}{12}\frac{\partial}{\partial x}\left[h^3\left(\lambda^{-2} A_{1111}\frac{\partial \gamma_x}{\partial x} + A_{1122}\frac{\partial \gamma_y}{\partial y}\right)\right] + \frac{1}{12}A_{1212}\times$$

$$\frac{\partial}{\partial y}\left[h^3\left(\frac{\partial \gamma_x}{\partial y} + \frac{\partial \gamma_y}{\partial x}\right)\right] - \frac{2}{3}\bar{\lambda}_1 h\left(\gamma_x + \frac{\partial w}{\partial x}\right) - \frac{1}{12}\lambda_1^{-2} h^3 \frac{\partial^2 \gamma_x}{\partial t^2} = 0,$$

$$\frac{1}{12}\frac{\partial}{\partial y}\left[h^3\left(\lambda^2 A_{2222}\frac{\partial \gamma_y}{\partial y} + A_{1122}\frac{\partial \gamma_x}{\partial x}\right)\right] + \frac{1}{12}A_{1212}\times$$

$$\frac{\partial}{\partial x}\left[h^3\left(\frac{\partial \gamma_x}{\partial y} + \frac{\partial \gamma_y}{\partial x}\right)\right] - \frac{2}{3}\bar{\lambda}_2 h\left(\gamma_y + \frac{\partial w}{\partial y}\right) - \frac{1}{12}\lambda_2^{-2} h^3 \frac{\partial^2 \gamma_y}{\partial t^2} = 0,$$

$$\frac{\partial^2}{\partial x^2}(k_y w) + \frac{\partial^2}{\partial y^2}(k_x w) + \lambda^{-4} a_{1111}\frac{\partial^2}{\partial x^2}\left(h^{-1}\frac{\partial^2 F}{\partial x^2}\right) +$$

$$\lambda^4 a_{2222}\frac{\partial^2}{\partial y^2}\left(h^{-1}\frac{\partial^2 F}{\partial y^2}\right) + a_{1122}\left[\frac{\partial^2}{\partial x^2}\left(h^{-1}\frac{\partial^2 F}{\partial y^2}\right) + \frac{\partial^2}{\partial y^2}\left(h^{-1}\frac{\partial^2 F}{\partial x^2}\right)\right] -$$

$$a_{1122}\frac{\partial^2}{\partial x \partial y}\left(h^{-1}\frac{\partial^2 F}{\partial x \partial y}\right) + \frac{1}{2}L(w,w) = 0. \quad (2.146)$$

Above $L(w, F)$ denotes the differential operator widely used in plate and shell theory [244], which has the form

$$L(w,F) = \frac{\partial^2 w}{\partial x^2}\frac{\partial^2 F}{\partial x^2} - 2\frac{\partial^2 w}{\partial x^2}\frac{\partial^2 F}{\partial y^2} + \frac{\partial^2 w}{\partial y^2}\frac{\partial^2 F}{\partial y^2}.$$

The deformations, forces, and bending moments have the following expressions [119]:

$$\varepsilon_{11} = \frac{\partial u}{\partial x} + \frac{1}{2}\left(\frac{\partial w}{\partial x}\right)^2 - k_x w, \quad \chi_{11} = \frac{\partial \gamma_x}{\partial x}, \quad \overrightarrow{(1,2)}\ \overleftrightarrow{(x,y)},$$

$$\varepsilon_{12} = \frac{\partial u}{\partial y} + \frac{\partial v}{\partial x} + \frac{\partial w}{\partial x}\frac{\partial w}{\partial y}, \quad \chi_{12} = \frac{\partial \gamma_x}{\partial y} + \frac{\partial \gamma_y}{\partial x},$$

$$\varepsilon_{13} = \gamma_x + \frac{\partial w}{\partial x}, \quad \varepsilon_{23} = \gamma_y + \frac{\partial w}{\partial y},$$

$$M_{11} = \frac{1}{12}h^3 \left(\lambda^{-2} A_{1111} \frac{\partial \gamma_x}{\partial x} + A_{1122} \frac{\partial \gamma_y}{\partial y} \right), \quad T_{11} = \frac{\partial^2 F}{\partial y^2}, \quad \overleftrightarrow{(1,2)} \; \overleftrightarrow{(x,y)},$$

$$M_{12} = \frac{1}{12}h^3 A_{1212} \left(\frac{\partial \gamma_x}{\partial y} + \frac{\partial \gamma_y}{\partial x} \right), \quad T_{12} = -\frac{\partial^2 F}{\partial x \partial y},$$

$$T_{12} = 2h, \quad A_{1212}\varepsilon_{12}, \quad T_{11} = 2h\left(A_{1111}\varepsilon_{11} + A_{1122}\varepsilon_{12} \right),$$

$$T_2 = 2h\left(A_{2222}\varepsilon_{22} + A_{1122}\varepsilon_{12} \right),$$

$$Q_1 = \frac{2}{3}\bar{\lambda}_1 h \left(\gamma_x + \frac{\partial w}{\partial x} \right), \quad Q_2 = \frac{2}{3}\bar{\lambda}_2 h \left(\gamma_y + \frac{\partial w}{\partial y} \right). \tag{2.147}$$

The boundary conditions are taken in the form (2.2)–(2.5). The initial conditions are

$$\overrightarrow{u}(x,y,0) = \overrightarrow{u}^0(x,y), \quad \left. \frac{\partial \overrightarrow{u}}{\partial t} \right|_{t=0} = \overrightarrow{u}^1(x,y). \tag{2.148}$$

The assumptions (2.6)–(2.9) are valid in this case.

One important property of the operator $L(w, F)$ is defined by the following theorem.

THEOREM 2.9 *If $w \in H_0^1(\Omega)$, then for an arbitrary function $F \in H_0^2(\Omega)$, with $L(w, F) \in H^{-1}(\Omega)$, the following relation is satisfied*

$$\int_\Omega L(w,F) w d\Omega = \int_\Omega L(w,w) F d\Omega, \tag{2.149}$$

which implies $L(w,w) \cdot F \in L^1(\Omega)$.

Proof. If w and F are smooth enough functions, then (2.149) is satisfied. This can be proved using the integration by parts [161]. However, in our case the functions w space is narrow, because $w \in H_0^1(\Omega)$ (and not in $H_0^2(\Omega)$, as it holds for the Kirchhoff-Love model). The complexity of the problem is the following. If $\varphi(x,y) \in C^\infty(\Omega)$ and for φ the relation (2.149) is satisfied, than in the limit $\varphi \to w$ in $H_0^1(\Omega)$ the relation (2.149) can be false. In a general case, we have $L(\varphi, F) \not\to L(w, F)$, if $\varphi \to w$ in $H_0^1(\Omega)$. We apply the property $L(w, F) \in H^{-1}(\Omega)$. In order to prove (2.149), special properties of the function's space [133] are applied.

According to the of conditions theorem 2.9, for a given $F \in H_0^2(\Omega)$ and arbitrary $w \in H^{-1}(\Omega)$ the relation $L(w, F) \in H^{-1}(\Omega)$ is true; then the following equation holds

$$\int_\Omega L(w,F) w d\Omega = \lim_{k \to \infty} \int_\Omega L(w,F) w_k d\Omega,$$

2.5 Vibroisolation of a Construction with Shells Elements

where w_k are functions in $H_0^1(\Omega) \cap H^2(\Omega)$ satisfying,

$$\|w - w_k\|_{1,\Omega} \leq \frac{1}{k}, \quad k = 1, 2, \ldots .$$

But

$$\int_\Omega L(w, F) w_k d\Omega = \int_\Omega L(w, w_k) F d\Omega,$$

and therefore

$$\int_\Omega L(w, F) w d\Omega = \int_\Omega L(w, w) F d\Omega,$$

which complete the proof. □

Now the following question will be discussed. In what sense and in what functional spaces the problem defined by (2.146)–(2.148) can be solved? Let us introduce the Hilbert space $L^2(\Omega)$ with the scalar product (u, v) defined by the relation

$$(u, v) = \int_\Omega u v d\Omega .$$

The space V_0 is the same as in section 2.2.1.

THEOREM 2.10 *Let the shell curvatures k_x, k_y be he limited functions in Ω together with its second derivatives, and let the conditions (2.6)–(2.9) be satisfied. Then, for arbitrary $q(x, y, t) \in L^2\left(0, R; L^2(\Omega)\right)$, $\vec{u}^0(x, y) \in V_0$, $\vec{u}^1(x, y) \in \left(L^2(\Omega)\right)^3$ there exists a weak solution to the problem (2.146)–(2.148) and the following relations are true*

$$\vec{u}(x, y, t) \in L^\infty\left(0, T; V_0\right), \qquad (2.150)$$

$$F(x, y, t) \in L^\infty\left(0, T; H_0^2(\Omega)\right) . \qquad (2.151)$$

Proof.

A. *Construction of the approximate solution.* Take the following basis functions

$$\left\{\vec{\psi}_i\right\}_{i=1}^\infty \in V_0,$$

defined, for instance, by the functions in $C^\infty(\Omega)$. The approximate solution is being sought in the form

$$\vec{u}_m(x, y, t) = \sum_{i=1}^m \vec{g}_{im}(t) \otimes \vec{\psi}_i(x, y),$$

where $\vec{g}_{im}(t) = (\xi_{im}(t), \eta_{im}(t), \theta_{im}(t))$ and $\otimes$ is defined in section 2.2.2. Introduce the operator $G^{-1}(\cdot)$, which exists according to the assumptions

(2.6)–(2.9). The operator G transforms $H_0^2(\Omega)$ to $H^{-2}(\Omega)$ and is a self-conjugated. The operator G^{-1}: $H^{-2}(\Omega) \tau H_0^2(\Omega)$ is also self-conjugate. The fourth equation of (2.146) can be expressed in the following equivalent form

$$F = -G^{-1}\left(\nabla_{,k}^2 w + \tfrac{1}{2}L(w,w)\right).$$

If

$$\phi(w) = \nabla_{,k}^2 w + \tfrac{1}{2}L(w,w),$$

then

$$F = -G^{-1}(\phi(w)). \qquad (2.152)$$

Taking into account (2.152) and applying the Bubnov-Galerkin method (MB) to the remaining three equations, we get the following set of ordinary differential equations in regard to the solution $\vec{g}_{im}(t)$:

$$\left(hw_m'', \psi_j^{(1)}\right) - \left(\nabla_k^2 G^{-1}(\phi(w_m)), \psi_j^{(1)}\right) + \left(L\left(w_m, G^{-1}(\phi(w_m))\right), \psi_j^{(1)}\right) -$$

$$\tfrac{2}{3}\bar{\lambda}_1 \left(\frac{\partial}{\partial x}\left[h\left(\gamma_{xm} + \frac{\partial w_m}{\partial x}\right)\right], \psi_j^{(1)}\right) -$$

$$\tfrac{2}{3}\bar{\lambda}_2 \left(\frac{\partial}{\partial y}\left[h\left(\gamma_{ym} + \frac{\partial w_m}{\partial y}\right)\right], \psi_j^{(1)}\right) = \left(q, \psi_j^{(1)}\right),$$

$$\tfrac{1}{12}\lambda_1^{-2}\left(h^3 \gamma_{xm}'', \psi_j^{(2)}\right) -$$

$$\tfrac{1}{12}\left(\frac{\partial}{\partial x}\left[h^3\left(\lambda^{-2} A_{1111} \frac{\partial \gamma_{xm}}{\partial x} + A_{1122} \frac{\partial \gamma_{ym}}{\partial y}\right)\right], \psi_j^{(2)}\right) -$$

$$\tfrac{1}{12} A_{1212}\left(\frac{\partial}{\partial y}\left[h^3\left(\frac{\partial \gamma_{xm}}{\partial y} + \frac{\partial \gamma_{ym}}{\partial x}\right)\right], \psi_j^{(2)}\right) +$$

$$\tfrac{2}{3}\bar{\lambda}_1 \left(\left[h\left(\gamma_{xm} + \frac{\partial w_m}{\partial x}\right)\right], \psi_j^{(2)}\right) = 0,$$

$$\tfrac{1}{12}\lambda_2^{-2}\left(h^3 \gamma_{ym}'', \psi_j^{(3)}\right) - \tfrac{1}{12}\left(\frac{\partial}{\partial y}\left[h^3\left(A_{2222} \frac{\partial \gamma_{ym}}{\partial y} + A_{1122} \frac{\partial \gamma_{xm}}{\partial x}\right)\right], \psi_j^{(3)}\right) -$$

$$\tfrac{1}{12} A_{1212}\left(\frac{\partial}{\partial y}\left[h^3\left(\frac{\partial \gamma_{xm}}{\partial y} + \frac{\partial \gamma_{ym}}{\partial x}\right)\right], \psi_j^{(3)}\right)$$

$$+\tfrac{2}{3}\bar{\lambda}_2 \left(\left[h\left(\gamma_{ym} + \frac{\partial w_m}{\partial y}\right)\right], \psi_j^{(3)}\right) = 0, \qquad (2.153)$$

where $j = 1, 2, ..., m$ and the marks of w_m, γ_{xm}, γ_{ym} denote the derivatives with respect to time t. ∇_k^2 denotes a differential operator of the form

$$\nabla_k^2(\bullet) = k_y \frac{\partial^2(\bullet)}{\partial x^2} + k_x \frac{\partial^2(\bullet)}{\partial y^2},$$

2.5 Vibroisolation of a Construction with Shells Elements

and $\psi_j^{(1)}, \psi_j^{(2)}, \psi_j^{(3)}$ are the components of the basis function $\vec{\psi}_j$.
The approximation of the initial conditions (2.148) is defined by the relations

$$\vec{u}_m(x,y,0) = \vec{u}^{\,0}_m(x,y) = \sum_{i=1}^{m} \left(\vec{u}^{\,0} \otimes \vec{\psi}_i\right) \otimes \vec{\psi}_i,$$

$$\vec{u}'_m(x,y,0) = \vec{u}^{\,1}(x,y) = \sum_{i=1}^{m} \left(\vec{u}^{\,1} \otimes \vec{\psi}_i\right) \otimes \vec{\psi}_i.$$

If the basis is full then

$$\vec{u}^{\,0}_m \to \vec{u}^{\,0} \text{ in } V_0, \quad \vec{u}'_m \to \vec{u}^{\,1} \text{ in } \left(L^2(\Omega)\right)^3.$$

Note, that for an arbitrary t, if $w_m \in H^1_0(\Omega)$, then $\phi(w_m) \in H^{-2}(\Omega)$ and the approximate value F_m is defined by relations (2.152) of the form

$$F_m(x,y,t) = -G^{-1}\left(\phi(w_m)\right). \tag{2.154}$$

In a general case F_m does not need to belong to a linear shell with the chosen basis.
Taking into account the assumptions (2.6)–(2.9) and the remark (2.1) about the positively defined operator G^{-1}, we can conclude [133] that a solution $\vec{g}_m(t)$ of the ordinary differential equations (2.153) exists. This implies that also $\vec{u}_m(x,y,t)$, $F_m(x,y,t)$ exist.

B. *A priori estimation.* Let us multiply the first equation of (2.153) by $\xi'_{jm}(t)$, the second by $\eta'_{jm}(t)$, the third by $\theta'_{jm}(t)$ and compose the sum with respect to j, from 1 to m. Besides, let us sum the left and right hand sides of the equations. As a result, we get

$$\frac{1}{2}\frac{d}{dt}\{(hw'_m, w'_m) + \frac{1}{12}\lambda_1^{-2}\left(h^3\gamma'_{xm}, \gamma'_{xm}\right) + \frac{1}{12}\lambda_2^{-2}\left(h^3\gamma'_{ym}, \gamma'_{ym}\right) +$$

$$\frac{1}{12}A_{1111}\lambda^{-2}\left(h^3\frac{\partial\gamma_{xm}}{\partial x}, \frac{\partial\gamma_{xm}}{\partial x}\right) + \frac{1}{6}A_{1122}\left(h^3\frac{\partial\gamma_{xm}}{\partial x}, \frac{\partial\gamma_{ym}}{\partial y}\right) +$$

$$\frac{1}{12}\lambda^2 A_{2222}\left(h^3\frac{\partial\gamma_{ym}}{\partial y}, \frac{\partial\gamma_{ym}}{\partial y}\right) + (Q_{1m},\varepsilon_{13m}) + (Q_{2m},\varepsilon_{23m}) +$$

$$(M_{12m},\chi_{12m})\} + \left(\nabla^2_k G^{-1}\left(\phi(w_m)\right), w'_m\right) +$$

$$\left(L\left(w_m, G^{-1}\left(\phi(w_m)\right)\right), w'_m\right) = (q, w'_m), \tag{2.155}$$

where $Q_{1m}, Q_{2m}, M_{12m}, \varepsilon_{13m}, \varepsilon_{23m}, \chi_{12m}$ are defined by (2.147) for $w = w_m$, $\gamma_x = \gamma_{xm}$ and $\gamma_y = \gamma_{ym}$. Consider the last two terms of the left hand side of (2.155). From (2.154) one obtains

$$B_1 = \left(\nabla^2_k G^{-1}\left(\phi(w_m)\right), w'_m\right) = -\left(\nabla^2_k F_m, w'_m\right),$$

$$B_2 = \left(L\left(w_m, G^{-1}\left(\phi(w_m)\right)\right), w'_m\right) = -\left(L\left(w_m, F_m\right), w'_m\right).$$

Taking into account (2.146) and the equation

$$\left(\nabla_{,k}^2 F_m, w'_m\right) = \left(F_m, \nabla_{,k}^2 w'_m\right),$$

where $\nabla_{,k}^2 w'_m = \frac{\partial^2}{\partial x^2}(k_y w'_m) + \frac{\partial^2}{\partial y^2}(k_x w'_m)$. It holds that,

$$B_1 + B_2 = -\left(\frac{d}{dt}\left[\nabla_{,k}^2 w_m + \frac{1}{2}L(w_m, w_m)\right], F_m\right) = \frac{1}{2}\frac{d}{dt}\left(G(F_m), F_m\right). \tag{2.156}$$

Substituting $B_1 + B_2$ from (2.156) into (2.155) we obtain

$$(hw'_m, w'_m) + \frac{1}{12}\lambda_1^{-2}\left(h^3\gamma'_{xm}, \gamma'_{xm}\right) + \frac{1}{12}\lambda_2^{-2}\left(h^3\gamma'_{ym}, \gamma'_{ym}\right) +$$

$$\frac{1}{12}A_{1111}\lambda^{-2}\left(h^3\frac{\partial \gamma_{xm}}{\partial x}, \frac{\partial \gamma_{xm}}{\partial x}\right) + \frac{1}{6}A_{1122}\left(h^3\frac{\partial \gamma_{xm}}{\partial x}, \frac{\partial \gamma_{ym}}{\partial y}\right) +$$

$$\frac{1}{12}\lambda^2 A_{2222}\left(h^3\frac{\partial \gamma_{ym}}{\partial y}, \frac{\partial \gamma_{ym}}{\partial y}\right) + (M_{12m}, \chi_{12m}) + (Q_{1m}, \varepsilon_{13m}) +$$

$$(Q_{2m}, \varepsilon_{23m}) + (G(F_m), F_m) = \left(hw_m^1, w_m^1\right) + \frac{1}{12}\lambda_1^{-2}\left(h^3\gamma_{xm}^1, \gamma_{xm}^1\right) +$$

$$\frac{1}{12}\lambda_2^{-2}\left(h^3\gamma_{ym}^1, \gamma_{ym}^1\right) + \frac{1}{12}A_{1111}\lambda^{-2}\left(h^3\frac{\partial \gamma_{xm}^0}{\partial x}, \frac{\partial \gamma_{xm}^0}{\partial x}\right) +$$

$$\frac{1}{6}A_{1122}\left(h^3\frac{\partial \gamma_{xm}^0}{\partial x}, \frac{\partial \gamma_{ym}^0}{\partial y}\right) + \frac{1}{12}\lambda^2 A_{2222}\left(h^3\frac{\partial \gamma_{ym}^0}{\partial y}, \frac{\partial \gamma_{ym}^0}{\partial y}\right) +$$

$$(M_{12m}^0, \chi_{12m}^0) + (Q_{1m}^0, \varepsilon_{13m}^0) + (Q_{2m}^0, \varepsilon_{23m}^0) +$$

$$(G(F_m^0), F_m^0) + \int_0^t (q(\tau), w'_m(\tau))\, d\tau, \tag{2.157}$$

where $w_m, \gamma_{xm}^0, \gamma_{ym}^0, w_m^1, \gamma_{xm}^1, \gamma_{ym}^1$ are the components of the initial values of the vectors $\vec{u}_m^0$ and $\vec{u}_m^1$, respectively. $M_{12m}^0, Q_{1m}^0, Q_{2m}^0, \chi_{12m}^0, \varepsilon_{13m}^0, \varepsilon_{23m}^0$ are obtained from (2.147) for $\vec{u} = \vec{u}_m^0$. F_m^0 is obtained from (2.154) for $w = w_m^0$.

By assumption we have that $h_H \leq h(x,y) \leq h_B$. Besides, all components of the left hand side of (2.157) are bounded from below (see Theorem 2.1). On the right hand side, all components, except for $(G(F_m^0), F_m^0)$, are bounded from above, because $\vec{u}_m^0$ and $\vec{u}_m^1$ are given.

Consider $(G(F_m), F_m)$. According to (2.154) we have

$$F_m^0 = -G^{-1}\left(\phi\left(w_m^0\right)\right).$$

However, $\phi(w_m^0)$ belongs to a bounded set in $H^{-2}(\Omega)$, because $L(w_m^0, w_m^0) \in H^{-2}(\Omega)$, $w_m^0 \in H^2(\Omega)$ and k_x, k_y are bounded; therefore $\nabla_{,k}^2 w_m^0 \in$

2.5 Vibroisolation of a Construction with Shells Elements

$H^{-2}(\Omega)$. It means that F_m^0 belongs also to a bounded set in $H_0^2(\Omega)$, because G is an isomorphism from $H_0^2(\Omega)$ to $H^{-2}(\Omega)$. Thus, we have

$$(G(F_m^0), F_m^0) \leq const. \tag{2.158}$$

Finally, from (2.157) we get

$$c_1 \left(\|w_m'\|_{0,\Omega}^2 + \|\gamma_{xm}'\|_{0,\Omega}^2 + \|\gamma_{ym}'\|_{0,\Omega}^2 \right) + c_2 \left(\left\|\frac{\partial \gamma_{xm}}{\partial x}\right\|_{0,\Omega}^2 + \left\|\frac{\partial \gamma_{ym}}{\partial y}\right\|_{0,\Omega}^2 \right) +$$

$$c_3 \left\|\frac{\partial \gamma_{xm}}{\partial y} + \frac{\partial \gamma_{ym}}{\partial x}\right\|_{0,\Omega}^2 + c_4 \left(\left\|\gamma_{xm} + \frac{\partial w_m}{\partial x}\right\|_{0,\Omega}^2 + \left\|\gamma_{ym} + \frac{\partial w_m}{\partial y}\right\|_{0,\Omega}^2 \right) +$$

$$c_5 \left(\left\|\frac{\partial^2 F_m}{\partial x^2}\right\|_{0,\Omega}^2 + \left\|\frac{\partial^2 F_m}{\partial y^2}\right\|_{0,\Omega}^2 \right) \leq const + \int_0^t (q(\tau), w_m'(\tau)) \, d\tau.$$

From the above inequality it results [133] that t_m can always be extended to a given finite T, therefore $t \in [0, T]$ and we have

$$\vec{u}_m \quad \text{bounded in} \quad (L^\infty(0, T; V_0))^3, \tag{2.159}$$

$$\vec{u}_m' \quad \text{bounded in} \quad (L^\infty(0, T; L^2(\Omega)))^3, \tag{2.160}$$

$$F_m \quad \text{bounded in} \quad \lambda^\infty(0, T; H_0^2(\Omega)), \tag{2.161}$$

for all m.

C. *Transition to the limits.* From equations (2.159)–(2.161) and because of the weak compactness of the bounded set in the corresponding spaces, it results that from the series $\{\vec{u}_m\}_{m=1}^\infty$ and $\{F_m\}_{m=1}^\infty$ convergent subsets $\{\vec{u}_\mu\}_{\mu=1}^\infty$ and $\{F_{\mu^*}\}_{\mu^*=1}^\infty$ can be chosen, which satisfy the conditions

$$\vec{u}_\mu \to \vec{u} \quad \text{is weak in} \quad (L^\infty(0, T; V_0))^3, \tag{2.162}$$

$$\vec{u}_\mu' \to \vec{u}' \quad \text{is weak in} \quad (L^\infty(0, T; L^2(\Omega)))^3, \tag{2.163}$$

$$\vec{u}_\mu \to \vec{u} \quad \text{is strong in} \quad (L^2(0, T; L^2(\Omega)))^3, \tag{2.164}$$

$$F_{\mu^*} \to F \quad \text{is weak in} \quad L^\infty(0, T; H_0^2(\Omega)), \tag{2.165}$$

$$F_{\mu^*} \to F \quad \text{is strong in} \quad L^2(0, T; L^2(\Omega)). \tag{2.166}$$

Let us introduce the functions $a_i(t)$, $b_i(t)$, $c_i(t)$, $(i = 1, ..., i_0)$ [133], and $a_i(T) = b_i(T) = c_i(T) = 0$. After substituting $a(x, y, t)$, $b(x, y, t)$, $c(x, y, t)$ to the respective functions we get

$$a(x, y, t) = \sum_{i=1}^{i_0} a_i(t) \psi_i^{(1)}(x, y), \quad b(x, y, t) = \sum_{i=1}^{i_0} b_i(t) \psi_i^{(2)}(x, y),$$

$$c(x,y,t) = \sum_{i=1}^{i_0} c_i(t)\psi_i^{(3)}(x,y). \qquad (2.167)$$

A function satisfying the following equation is called a weak solution of the problem (2.146),

$$\int_0^T \{(hw_m'', a) - (\nabla_k^2 G^{-1}(\phi(w_m)), a) + (L(w_m, G^{-1}(\phi(w_m))), a) +$$

$$\frac{2}{3}\bar{\lambda}_1 \left(h^3 \left[\gamma_{xm} + \frac{\partial w_m}{\partial x}\right], \frac{\partial a}{\partial x}\right) + \frac{2}{3}\bar{\lambda}_2 \left(h^3 \left[\gamma_{ym} + \frac{\partial w_m}{\partial y}\right], \frac{\partial a}{\partial y}\right)\} dt =$$

$$\int_0^T (q, a) dt,$$

$$\int_0^T \left\{\frac{1}{12}\lambda_1^{-2}(h^3 \gamma_{xm}'', b) + \frac{1}{12}\left(h^3 \left[\lambda^{-2} A_{1111} \frac{\partial \gamma_{xm}}{\partial x} + A_{1122} \frac{\partial \gamma_{ym}}{\partial y}\right], \frac{\partial b}{\partial x}\right) +$$

$$\frac{1}{12} A_{1212} \left(h^3 \left[\frac{\partial \gamma_{xm}}{\partial y} + \frac{\partial \gamma_{ym}}{\partial x}\right], \frac{\partial b}{\partial y}\right) + \frac{2}{3}\bar{\lambda}_1 \left(h \left[\gamma_{xm} + \frac{\partial w_m}{\partial x}\right], b\right)\right\} dt = 0,$$

$$\int_0^T \left\{\frac{1}{12}\lambda_2^{-2}(h^3 \gamma_{ym}'', c) + \frac{1}{12}\left(h^3 \left[\lambda^2 A_{2222} \frac{\partial \gamma_{ym}}{\partial y} + A_{1122} \frac{\partial \gamma_{xm}}{\partial x}\right], \frac{\partial c}{\partial y}\right) +$$

$$\frac{1}{12}\left(h^3 \left[\frac{\partial \gamma_{xm}}{\partial y} + \frac{\partial \gamma_{ym}}{\partial x}\right], \frac{\partial c}{\partial x}\right) + \frac{2}{3}\bar{\lambda}_2 \left(h \left[\gamma_{ym} + \frac{\partial w_m}{\partial y}\right], c\right)\right\} dt = 0,$$

$$(2.168)$$

for arbitrary functions a, b, c of the form (2.167). The approximate solutions constructed at the beginning of the proof satisfy the equations (2.168). Therefore, (2.168) is obtained also for the subseries $\{\overrightarrow{u}_m u\}_{\mu=1}^\infty$. Let $m = \mu \to \infty$. According to (2.162)–(2.163), all linear terms occurring in equations (2.168) will approach the values defined in the limit function $\overrightarrow{u}$.

Consider the following nonlinear term

$$\int_0^T (L(w_m, G^{-1}(\phi(w_m))), a) \, dt =$$

$$\int_0^T (L(w_m, F_m), a) \, dt = \int_0^T (L(a, F_m), w_m) \, dt.$$

2.5 Vibroisolation of a Construction with Shells Elements

The transformation $L(a, F_m) \to L(a, F)$ is weak in $L^\infty\left(0, T; L^2(\Omega)\right)$. On the other hand, because $w_m \to w$ is strong in $L^\infty\left(0, T; L^2(\Omega)\right)$, then from (2.164) it results that,

$$\int_0^T (L(w_m, F_m), a)\, dt \to \int_0^T (L(w, F), a)\, dt$$

and equations (2.168) are satisfied by the limit functions $\vec{u} = (w, \gamma_x, \gamma_y)$ and F for all a, b, c in the form (2.167). The fundamental difficulty occurs during the transition to the limit in the fourth equation of (2.146). For each m we have

$$F_m = -G^{-1}\left(\nabla_{,k}^2 w_m + \tfrac{1}{2} L(w_m, w_m)\right).$$

However, if $w_m \to w$ is weak in $H_0^1(\Omega)$, then $L(w_m, w_m)$ is not necessarily convergent in $H^{-2}(\Omega)$. To prove the theorem, were consider the Theorem 2.1. Multiplying the first equation of (2.153) by $\xi_{jm}(t)$, we sum the terms in respect to j and then integrate. Taking into account (2.154) we get

$$\int_0^T \left\{ (hw_m'', w_m) + (\nabla_k^2 F_m, w_m) + 2(G(F_m), F_m) - \right.$$

$$\left. \left(\frac{\partial Q_{1m}}{\partial x}, w_m\right) - \left(\frac{\partial Q_{2m}}{\partial y}, w_m\right) \right\} dt = \int_0^T (q, w_m)\, dt. \qquad (2.169)$$

In this equation, the transition to the limit is possible, because the operator $G(F_m)$ satisfies the (M) property (see [133], page 184). Besides, $\nabla_k^2 F_m \to \nabla_k^2 F$ weakly in $L^2\left(0, T; L^2(\Omega)\right)$, and strongly in $L^2\left(0, T; L^2(\omega)\right)$. Therefore

$$\int_0^T (\nabla_k^2 F_m, w_m)\, dt \to \int_0^T (\nabla_k^2 F, w)\, dt.$$

Comparing (2.169) with the first equation of (2.168) for $w_m = w$, $F_m = F$ and $a = w$ we get

$$(\nabla_k^2 F, w) + (L(w, F), w) = -(\nabla_k^2 F, w) - 2(G(F), F), \qquad (2.170)$$

which is true almost everywhere in the interval $[0, T]$. However, as this results from the first equation of (2.168), $L(w, F) \in H^{-1}(\Omega)$. According to corollary (2.146) for the limit values of w and F we obtain

$$(L(w, F), w) = (L(w, w), F). \qquad (2.171)$$

On the other hand, from (2.154) it results, that in the limiting case we get

$$\nabla^2_{,k} w + G(F) = f, \qquad (2.172)$$

where f is the weak limit value in $H^{-2}(\Omega)$. Multiplying (2.172) by F and taking into account (2.171) and (2.170) we get

$$(f, F) = -\frac{1}{2}(L(w,w), F). \qquad (2.173)$$

Taking into account the fact that the operator $G(\bullet)$ possesses the (M) property, we conclude that for the arbitrary function $d \in L^2\left(0, T; H^{-2}(\Omega)\right)$ the relation (2.173) is satisfied. Therefore

$$\int_0^T \left\{ (\nabla^2_{,k} w, d) + (G(F), d) + \tfrac{1}{2}(L(w,w), d) \right\} dt = 0$$

is true for an arbitrary function. This means that the fourth equation of (2.146) is satisfied in the weak sense for the limiting values of w and F. Thus, if the conditions (2.6)–(2.9) are satisfied, then the problem of defining the dynamical properties of the slope of the Timoshenko-like shell with bounded curvatures has a solution for arbitrary loads $q \in L^2\left(0, T; L^2(\Omega)\right)$ and arbitrary initial conditions $\vec{u}^0 \in V_0$ and $\vec{u}^1 \in \left(L^2(\Omega)\right)^3$.

The solvability conditions of the dynamical problem are necessary for the proper definition of the problem related to the vibroisolation shells for both periodic and harmonic loads.

2.5.2 Optimal Vibroisolation-Formulation of the Problem and Vibroisolation Harmonic Excitation

Let the dynamical behaviour of a shell be governed by differential equations of the form:

$$A[h]\vec{u} + B[h]\vec{u}'' = q(x, y, t), \qquad (2.174)$$

where $A[h]$ is a nonlinear differential operator, related to the shell's deformation energy, $B[h]$ an operator related to the mass distribution along the shell (it also includes the inertial forces), $\vec{q}(x, y, t)$ the external load, and $\vec{u}$ the vector of variables. Specifically, for a Timoshenko-like shell we have $\vec{u} = (w, \gamma_x, \gamma_y, F)$.

Similarly to previous investigations, we assume that the shell's thickness $h(x, y)$ and its surface projection Ω are not given a priori. They have to be chosen from the sets $U_{\partial 1}$ and $U_{\partial 2}$ from the point of view of minimalizing the dynamical action $\vec{q}(x, y, t)$ on the shell, and economizing on the material. The influence of $\vec{q}(x, y, t)$ on the shell can be characterised by different criteria: maximal deflection at a given set of points, on maximal stress, work

2.5 Vibroisolation of a Construction with Shells Elements

of the external forces, weight, and so on [152, 192]. We concentrate mainly on the following vibroisolation problem [111].

We seek to find $h^*(x,y) \in U_{\partial 1}$ and $\Omega^* \in U_{\partial 2}$ in order to minimise the shell's weight, and in addition, we want the work of the elastic forces in the period $[0,T]$ to overlap with the corresponding work for a shell with configuration Ω_0 and thickness $h(x,y) \equiv 1$. In other words, we are going to find $h^*(x,y) \in U_{\partial 1}$ and $\Omega^* \in U_{\partial 2}$ that minimise the functional

$$I_1(h,\Omega) = \int_\Omega h\, d\Omega + \frac{1}{\varepsilon} \left| \int_0^T \int_\Omega \vec{q}\vec{u}\, d\Omega dt - P_0 \right|^2, \qquad (2.175)$$

where $\vec{u}$ is the solution to the equation (2.174) for the shell with the surface projection Ω, and

$$P_0 = \int_0^T \int_{\Omega_0} \vec{q}\vec{u}_0 d\Omega dt$$

is the work of the external forces on the $\vec{u}_0$ displacements of the initial shell configuration.

The problem (2.175) is formulated in a way similar to the previous considerations. The constraint $P = P_0$ is taken into account and the penalty multiplier $1/\varepsilon$ is applied.

For the linear case, numerous conclusions related to vibroisolation can be obtained using the analytical considerations including the superposition property. Developing $\vec{q}(x,y,t)$ in Fourier series we get (for periodic excitation):

$$\vec{q}(x,y,t) = \sum_{n=1}^\infty \vec{q}_n(x,y) \sin \omega_n t.$$

Further analysis will be carried out for one harmonic excitation $\vec{q}_n \sin \omega_n t$ in a stationary state. The variables $\vec{u}_n$ can be presented in the form $\vec{u}_n = \vec{u}_n(x,y) \sin \omega_n t$, where $\vec{u}_n(x,y)$ satisfies the equation

$$A[h]\vec{u}_n - \omega_n^2 B[h]\vec{u}_n = \vec{q}_n(x,y). \qquad (2.176)$$

Consider a shell with the projection surface Ω and thickness h. If a solution to the free vibration problem

$$A[h]\vec{\psi}_i - \lambda_i^2 B[h]\vec{\psi}_i = 0$$

is known, then the solution $\vec{u}_n$ and excitation $\vec{q}_n$ can be presented in the following model series

$$\vec{u}_n(x,y) = \sum_{i=1}^\infty a_{in} \vec{\psi}_i(x,y),$$

$$\vec{q}_n(x,y) = \sum_{i=1}^{\infty} f_{in}\vec{\psi}_i(x,y), \qquad (2.177)$$

where f_{in} are defined by the equations

$$\sum_{i=1}^{\infty} f_{in}\left(\vec{\psi}_i, \vec{\psi}_j\right) = \left(\vec{q}_n, \vec{\psi}_j\right), \qquad j = 1, 2, \ldots .$$

In the relations above $(\cdot, \cdot)$ denotes the scalar product in the corresponding spaces.

According to the superposition rule, we can consider the influence of the shell on only one term $\vec{q}_{nk} = f_{nk}\vec{\psi}_k(x,y)$. Let us seek a solution to equations (2.176) for that case. Substituting (2.177) and (2.176) we get

$$\sum_{i=1}^{\infty} a_{ink}\left(A[h]\vec{\psi}_i(x,y) - \omega_n^2 B[h]\vec{\psi}_i(x,y)\right) = f_{nk}\vec{\psi}_k(x,y). \qquad (2.178)$$

Multiplying (2.178) by $\vec{\psi}_j(x,y)$ and integrating Ω we get

$$\sum_{i=1}^{\infty} a_{ink}(\lambda_i^2 - \omega_n^2)\left(B[h]\vec{\psi}_i, \vec{\psi}_j\right) = f_{nk}\left(\vec{\psi}_k, \vec{\psi}_j\right).$$

If the eigenfunctions $\vec{\psi}_i$ are normalised, $(B[h]\vec{\psi}_i, \vec{\psi}_j) = \delta_{ij}$, we get

$$a_{ink} = \frac{f_{nk}\left(\vec{\psi}_k, \vec{\psi}_i\right)}{(\lambda_i^2 - \omega_n^2)},$$

or equivalently,

$$\vec{u}_{nk}(x,y) = \sum_{i=1}^{\infty} \frac{f_{nk}\left(\vec{\psi}_k, \vec{\psi}_i\right)}{(\lambda_i^2 - \omega_n^2)}\vec{\psi}_i(x,y).$$

The external force work is given by

$$P_{nk} = \frac{\pi f_{nk}^2}{\omega_n}\sum_{i=1}^{\infty} \frac{\left(\vec{\psi}_k, \vec{\psi}_i\right)^2}{(\lambda_i^2 - \omega_n^2)}.$$

The quantities $\vec{\psi}_i, \vec{\psi}_k, f_{nk}, \lambda_i$ depend on both h and Ω. However, in practice, the vibration modes $\vec{\psi}_i$ change only slightly. Therefore, we can take f_{nk} and $\vec{\psi}_i$ as independent of h, and then the problem related to the work performed by the external forces is considered in a simple way. The factor $(\vec{\psi}_k, \vec{\psi}_k)^2/(\lambda_k^2 - \omega_n^2)$ has to be minimal. This means that λ_k^2 should be maximally shifted from ω_n^2, the excitation frequencies. In this simple framework,

2.5 Vibroisolation of a Construction with Shells Elements

the vibroisolation problem is reduced to that of spectrum optimization and the results obtained earlier are valid for that case. However, in a general case, even for harmonic excitation with frequency ω_n^2 the full work is defined by

$$P_n = \sum_{k=1}^{\infty} P_{nk} = \frac{\pi}{\omega_n} \sum_{k=1}^{\infty} \frac{\left(\vec{\psi}_k, \vec{\psi}_i\right)^2}{(\lambda_i^2 - \omega_n^2)}. \tag{2.179}$$

According to (2.179) it is not possible to put constraints on the shell's spectral lines λ_i, because they depend very strongly on f_{nk}^2. Taking into account the fact that λ_i change essentially with the change of the surface projection and that $P = \sum_{n=1}^{\infty} P_n$, it is clear, that a construction with a maximal shift of the excitation frequencies and resonance frequencies sometimes can not be realised. The construction can work on one of the resonance frequencies if its corresponding component f_{nk} does not occur in the excitation. The problem becomes more complicated when nonlinearity is included. Except for the already mentioned problems, the response frequencies ω_n depend on the vibration amplitudes, on the thickness distribution of the plate and on the shape of the projected surface.

The problem of optimal vibroisolation can be rather treated not in the framework of the constraints' formulations with respect to free vibration frequencies, but in a wider sense as a direct minimalization of the external forces work.

However, there appear numerical problems in the solution. Indeed in the numerical realization of the algorithms situations may occur in which the free vibration frequencies of the shell are close to one of the excitation frequencies. In that case, the solutions to (2.174) do not exist, and the algorithms do not converse. Formulating the conditions applied to $U_{\partial 1}$ and $U_{\partial 2}$ in order to avoid such situations is a rather complicated issue. Hence, we consider the problems of optimal vibroisolations as linear ones and consider harmonic excitation at a given frequency. Then, the problem can be reduced to that of finding solutions to equation (2.176), where the operators $A[h]$ and $B[h]$ should be replaced by $Z[h]$ and $M[h]$ introduced in Chapter 2.2.1.

For a proper formulation of the vibroisolation optimization we apply some of the general results related to the problem (2.176).

THEOREM 2.11 *If the conditions of theorem 2.2 are satisfied and the following inequality is satisfied*

$$\omega_n^2 < \frac{\min\left\{\frac{d^2 h_H^3 c}{12 h_B^3}, \frac{d^2}{4\pi}\left(\frac{1}{\sqrt{A_H h_H}} + \frac{d}{\sqrt{h_H^3 c}}\right)^{-1}\right\}}{\max\left\{h_B, \lambda_1^{-2} h_B^3/12, \lambda_2^{-2} h_B^3/12\right\}}, \tag{2.180}$$

where h_H, h_B, c, A_H are defined earlier and d is the given number, then the problem (2.180) is solved for an arbitrary function $\vec{q}_n(x,y) \in (L^2(\Omega))^3$

for the arbitrary $h \in U_\partial$ defined by (2.91) for an arbitrary space Ω with $diam(\Omega) \leq d$ with smooth enough boundary.

Proof. The system of equations is solvable if [158]

$$a(\vec{u}, \vec{u}) - \omega_n^2 b(\vec{u}, \vec{u}) \geq const. \|\vec{u}\|_{V_0}^2, \qquad (2.181)$$

where $a(\cdot, \cdot)$ and $b(\cdot, \cdot)$ are the bilinear operators introduced in section 2.2.1. We have that $a(\vec{u}, \vec{u}) \geq \alpha \|\vec{u}\|_{V_0}^2$, where

$$\alpha = \min\left\{ \frac{d^2 h_H^3 c}{12 h_B^3}, \frac{d^2}{4\pi} \left(\frac{1}{\sqrt{A_H h_H}} + \frac{d}{\sqrt{h_H^3 c}} \right)^{-1} \right\}. \qquad (2.182)$$

In the expression above the constraint c is defined for U_∂ by (2.91) and d denotes the diameter of a minimal surface covering all the surfaces from the $U_{\partial 2}$. On the other hand $b(\vec{u}, \vec{u}) \leq \beta \|\vec{u}\|_{V_0}^2$, where

$$\beta = \max\left\{ h_B, \lambda_1^{-2} h_B^3/12, \lambda_2^{-2} h_B^3/12 \right\}. \qquad (2.183)$$

In order to satisfy the inequality (2.181) it is sufficient to take

$$\alpha - \omega_n^2 \beta > 0 \quad \text{or} \quad \omega_n^2 < \alpha/\beta.$$

The condition (2.180) guarantees the above inequality and therefore also a solution to the problem (2.176). $\square$

Thus, if we define $U_{\partial 1} = U_\partial$ and $U_{\partial 2}$ and the excitation frequency ω_n satisfy (2.180), then for all $h \in U_{\partial 1}$ and $\Omega \in U_{\partial 2}$ we can realise a search for the optimal design in the sense defined by (2.175) and the question of solvability of (2.176) in not important. However, the problem of the existence of a minimum during the optimal vibroisolation defined by (2.175) is still an open question. The result given below allows for the formulation of a sufficient condition leading to the solution of this problem. Assume, in a similar way to that in section 2.4.1, that the function $h(x,y)$ and the surface Ω are parameterised. Let the vector $\vec{\beta} \in U_\partial'$ satisfy the constraints of $h(x,y) \in U$ from (2.91) and additionally $diam(\Omega) \leq d$. Then, the following theorem is true.

THEOREM 2.12 *If all conditions of the theorem (2.11) are valid, and ω_n satisfies the constraint (2.180) and $\vec{q}_n \in (L^2(\Omega))^3$, then the problem of optimal vibroisolation (2.175) possesses at least one solution in $U_\partial' \in R^{m+p}$.*

Proof. The proof is similar to that of theorem (2.8). It is enough to show that in this case the function $I_1(h, \Omega)$ is continuous because of $\vec{\beta}$. This results from the boundless of the solution

$$\|\vec{u}\|_{V_0} \leq const. \|\vec{q}_n\|_{0,\Omega_0} \qquad \text{for all } \vec{\beta} \in U_\partial'.$$

2.5.3 Algorithm of the Optimal Vibroisolation of Timoshenko-Like Shells

The results obtained in the previous sections allow for the application of the algorithms for spectral line optimization to the problems of optimal vibroisolation of Timoshenko-like shells, as well.

In case of the optimization of shell, in terms of the parameters $h(x,y)$ and Ω, the most powerful method is the FEM. In case of optimization in terms of the thickness $h(x,y)$ the MVI is recommended. However, if the load $\vec{q}_n(x,y)$ differs strongly from being constant, then we need a high number of terms in (2.29).

2.6 Generalisations

The previously considered weight optimization methods applied to shells can be generalised to a wide class of other elastic systems using different objective functions [134].

Let a state of an elastic system be governed by the equation

$$A[h]u = f, \qquad (2.184)$$

where $u \in V$ is the Hilbert space; $f \in V^*$ the space conjugate to V and h the continuous control parameter. We also need $h \in U_\partial$ to be an admissible control parameter (h denotes thickness or any other scalar of vector function). Assume, that A is a nonlinear operator with the following properties:

a) A is bounded; $V \to V^*$ for $\forall h \in U_\partial$;
b) A is monotonous and semicontinuous;
c) A is coercive, i.e., $(A[h]u, u) \geq \cup u \cup_V \gamma_h(\cup u \cup_V)$, where $\gamma_h(t) \to \infty$ for $t \to \infty$, for $\forall h \in U_\partial$;
d) if $\{h_n\}_{n=1}^\infty$ is weakly convergent for $h_0 \in U_\partial$ in a certain space U ($U_\partial \in U$), then $u_n = u(h_n)$ is also weakly convergent for $u(h_0)$ in V.

Note, that

$$Z(h) = C[h]u \qquad (2.185)$$

is true for the certain bounded operator $C[h]$ operating from V to the Hilbert space H. In the particular case, when $H \equiv R^1$, the operator $C[h]$ is a functional defined in V.

Thus, to an arbitrary thickness distribution h, there correspond certain mechanical characteristics of the elastic system $Z(h)$, such as forces, moments, deflection, critical load, spectral lines and so on.

DEFINITION 2.1 *Operator $C[h]$ is called fully continuous and uniform in relation to the control parameter if for an arbitrary series $\{h_n\}_{n=1}^\infty \in U_\partial$ it weakly converges to h_0 in U and for the arbitrary series $\{u_n\}_{n=1}^\infty \in V$ the following condition is true*

$$C[h_n]u_n \xrightarrow[n\to\infty]{} C[h_0]u_0, \qquad (2.186)$$

which is strong in H.

For each value of h assign the function value [134]:

$$J(h) = \|C[h]u - z_\partial\|_H^2 + (Nh, h)_U, \qquad (2.187)$$

where N is a linear, bounded and positively defined operator in U.

The optimization problem is defined as

$$h^* \in U_\partial :\to \inf_{h\in U_\partial} J(h) \qquad (2.188)$$

with the attached condition $A[h]u = f$.

THEOREM 2.13 *Let the conditions a) — d) be satisfied, let U_∂ be a weakly closed and compact subset of U, let $f \in V^*$, and let $C[h]$ be uniformly continuous in relation to the optimal control of the operator in U_∂. Then, for the $J(h)$ functional defined by (2.187) there exists at least one optimal solution (2.188).*

Proof. Let $\{h_n\}_{n=1}^\infty$ be the series minimising $J(h)$, i.e., $J(hn) \to \inf J(h)$. Denote $u(h_n) = u_n$.

From the coercive property of operator A it follows that $\cup u_n \cup_V \leq const$. Thus, one can take the subseries $\{u_n, h_n\}$ with the following properties

$$u_n \to u_0 \quad \text{is weak in} \quad V,$$
$$h_n \to h_0 \quad \text{is weak in} \quad U. \qquad (2.189)$$

According to the weak closure of U_∂, we have $h_0 \in U_\partial$. Besides, from property d) of operator A it results that u_0 is the weak solution of equation (2.184), i.e. $u_0 = u(h_0)$. Then, from the uniform continuity of the operator $C[h]$ in regard to the control it follows that,

$$C[h_n]u_n \to C[h_0]u_0 \quad \text{in} \quad H. \qquad (2.190)$$

Taking into account (2.189) and (2.190), and the fact that the operator N is positive definite, we get that $J(h)$ is weakly semi-continuous from below in U_∂ and therefore

$$J(h_0) \leq \lim_{n\to\infty} J(h_n) = \inf_{h\in U_\partial} J(h).$$

It means that h_0 is the optimal control.

Note that in the previous remarks, where the solvability of the optimization problems has been proved the properties a) – d) have also been proved, as well as the uniform continuity of the observation operators in relation to control parameter. However, this has been carried out in an implicit way. The provided theorem enables one to estimate the optimal solution of the optimization problem of the elastic systems considered.

3 Order Reduction by Proper Orthogonal Decomposition (POD) Analysis

3.1 The POD Method

3.1.1 Introduction

The method of proper orthogonal mode decomposition (POD) or Karhunen-Loeve decomposition (KLD) is a means of extracting spatial information from a set of time-series data available from a set of sensing locations over a domain. The POD can be used to obtain low-dimensional models or discrete or distributed dynamical systems by computing an orthogonal set of eigenfunctions through a finite-dimensional eigenvalue problem that is obtained by post processing of time-series measurements at different spatial locations. Interestingly enough, these eigenfunctions form an orthogonal basis (irrespective of the linear or nonlinear nature of the measured signals) which is optimal in the sense that fewer POD modes are needed to capture a given amount of energy of the measured signal than any other linear set of modes, including vibration modes [219]. Moreover, the eigenvalue corresponding to a given eigenfunction quantifies the amount of energy of the measured signal that is captured by the specific POD mode. Hence, the POD method not only provides a linear orthogonal basis of modes, but also a quantitative measure of the relative importance of these modes with regard to the energy of the signal captured by the POD analysis. This feature of the method makes it a valuable tool in the analysis, system identification and order reduction of the dynamics of engineering systems. As pointed by Kerschen [102] the POD analysis resembles the Singular Value Decomposition Method, with the later method providing additional information related to amplitude modulations of the identified waveforms.

An additional advantage of the method is that the POD modes corresponding to a certain set of system parameters can, in most cases, be used to reconstruct the response of the dynamical system when certain of its parameters are varied. Aubry et al. [14] studied preserved symmetries in POD. As shown in the recent Thesis by Kerschen [102], POD modes obtained from chaotic time series of a dynamical system can be used as orthogonal basis to accurately reconstruct not only regular (periodic or quasi-periodic) responses of the same system, but also its bifurcations as a system parameter is varied. This statement underlines a basic limitation of the POD calculation, that is, that the derived modes are signal dependent; hence, the computed POD orthogonal basis changes if the excitation or the initial conditions of the dynamical system change (this is not the case of vibration modes which are properties of the system itself). This limitation, however, is somehow relaxed in problems where no direct external excitation exists and where the initial conditions are prescribed; examples, are rotor dynamic systems where the varying parameter is the frequency of rotation, or steady fluid flows. In any case, a set of POD modes derived from a specific excitation or a specific set of system parameters, can be used (at least in principle) for order-reducing the dynamics of the same system for different excitation or with varied system parameters, although then the POD basis in that case is not optimal in the sense described previously. An additional limitation of the method is the lack of direct physical interpretation of the POD modes, although recent works have addressed this issue and have provided certain results to address this issue [67, 98, 66, 132, 103, 102].

The method was primarily introduced as a means of extracting dominant energy signals from stochastic random processes [99, 136]. In acoustical and random signal decomposition the POD method has been widely used to ascertain the modes and energy of the signals under consideration [1, 251]. Due to its order reduction capacity, the POD method is very important in applications where compression of data and storage of data is involved; as an example, in [92] the method is used to produce reduced order models of fluid flows. Additional indicative works in this area includes the ones by [218] on the dynamical structure of a two-dimensional axisymmetric jet, and by Aubry [12], Aubry et al. [13], Rodriguez and Sirovich [199], Rajaee et al. [194], Sirovich [217] and Moehlis et al. [159] on different contexts of fluid flow problems. A key advantage of the method lies in the fact that it can be applied not only to Hamiltonian systems, but also to dissipative ones. In [227] POD was applied to a turbulent thermal convective system and low-order models were created to study the dynamics of thermal behavior. In [85] the modes of a reaction-diffusion chemical process were captured by means of the POD method and the dynamics of the process was analyzed. Park and Cho [182] apply the method in the context of control of distributed parameter systems.

Few works exist to date on the application of the POD method to engineering dynamics. Bayly and Virgin [35] applied POD to a reduce a stability analysis problem. Additional notable works in structural dynamics are those by Mari and Glangeaud [151] and Cusumano et al. [58]. In the later work the dimensionality of the dynamics of an impacting beam is studied by means of traditional time-delay techniques and POD; the energy transfer between POD modes was also studied experimentally in that work. In [36] the method was used to perform system identification of an aeroelastic problem. Azeez [24] and Azeez and Vakakis [25] applied POD to model and reduce the order of vibro-impacting systems; Georgiou and Schwartz [77] and Georgiou et al.

[81] examined theoretically and experimentally the interaction between the slow and fast dynamics of dynamical systems by means of POD; additional applications of the POD method to decompose the dynamics of extended flexible systems with high modal densities were given by [140, 141, 143]. Some of these results will be presented in this work. In [141] POD is used to reduce the order of two nonlinearly coupled flexible rods and to study transient motion confinement phenomena in this system.

In the next Section we provide the theoretical basis for the POD method and discuss the features of the method are applicable to problems of system identification, order reduction and response reconstruction in engineering dynamics. In later Sections we provide various applications of the methods to linear and nonlinear problems involving continuous structures.

3.1.2 Theoretical Basis

The POD is in essence a method of representing a stochastic system in terms of a minimum number of degrees of freedom. The systems considered in this work are deterministic in the sense that, for a given set of inputs the outputs are always the same without any uncertainty. As outputs we consider either numerical or experimental time series measured at various sensing locations distributed along the structures considered. The definitions followed in this Section follow closely those of [182].

Define a random or deterministic field $u(x,t)$ on some domain Ω. This field can be vector-valued, but in this work we will consider or scalar fields. First, the field is decomposed into average - $U(x) = \langle u(x,t) \rangle$, and fluctuating - $v(x,t)$ parts. This is represented as:

$$u(x,t) = U(x) + v(x,t) . \tag{3.1}$$

In any practical situation (both experimental and numerical) the fields $u(x,t)$ and $v(x,t)$ are available at a finite number of points in space and time. Hence, at a give time instant t_n the system displays a *snapshot* $v(x,t_n) \equiv v_n(x)$; the collection or ensemble of snapshots is represented by $\{v_n(x)\}$. At this stage it is assumed that the ensemble of snapshots $\{v_n(x)\}$ has a continuous dependence on $x \in \Omega$.

The problem now is to obtain the most typical or characteristic structure $\phi(x)$ among the ensemble of snapshots $\{v_n(x)\}$. This is performed by minimizing the objective function $\bar{\lambda}$ defined as follows:

$$\text{Minimize} \left\{ \bar{\lambda} = \sum_{n=1}^{N} (\phi(x) - v_n(x))^2 \right\} \quad \forall x \in \Omega, \tag{3.2}$$

where it is assumed that there are N snapshots of the field on the domain Ω; moreover, the above minimization is carried out pointwise at each $x \in \Omega$.

As shown below problem (3.2) can be transformed into the form of an eigenvalue problem, which provides multiple solutions for $\phi(x)$. In order to render the computation unique it is necessary to impose the following normalization condition,

$$\int_\Omega \phi^2(x)dx = 1. \tag{3.3}$$

To simplify the following exposition, the following notations are employed for the inner product in functional space and average in finite-dimensional space:

$$(f,g) = \int_\Omega f(x)g(x)\,d\Omega, \quad \langle v_n(x) \rangle = \frac{1}{N}\sum_{n=1}^{N} v_n(x). \tag{3.4}$$

By (3.2) one requires that the difference between $\phi(x)$ and the ensemble of snapshots be minimized in least squares sense. In other words, the ensemble average of the inner products between $v_n(x)$ and $\phi(x)$ needs to be maximized. By using the above notations and expanding (3.2) while utilizing (3.3) the problem reduces to the maximization of $\langle (\phi(x), v_n(x)) \rangle$. In addition, it is required $\langle (\phi(x), v_n(x)) \rangle$ to be positive (to move closer to a true minimum), and, hence there is the need to maximize the average $\langle (\phi(x), v_n(x))^2 \rangle$. It follows that (3.2) can be reposed in the form,

$$\text{Maximize } \left\{ \lambda = \frac{\langle (\phi, v_n) \rangle}{(\phi, \phi)} \right\} \text{ with respect to } \phi(x). \tag{3.5}$$

The numerator of the right-hand-side of this equation is now expanded to yield,

$$\left\langle (\phi, v_n)^2 \right\rangle = \left\langle \int_\Omega \phi(x)v_n(x)dx \int_\Omega \phi(y)v_n(y)dy \right\rangle =$$

$$\int_\Omega \left\{ \int_\Omega \langle v_n(x)v_n(y) \rangle \phi(x)dx \right\} \phi(y)dy. \tag{3.6}$$

Introducing the two-point correlation function K defined as,

$$K(x,y) = \langle v_n(x)v_n(y) \rangle = \frac{1}{N}\sum_{n=1}^{N} v_n(x)v_n(y) \tag{3.7}$$

and the linear operator $R[\cdot]$,

$$R[\tau(x)] = \int_\Omega K(x,y)\tau(x)dy, \tag{3.8}$$

one can represent expression (3.6) compactly as,

$$\left\langle (\phi, v_n)^2 \right\rangle = \int_\Omega R[\phi(x)]\,\phi(x)\,dx \equiv (R[\phi], \phi). \tag{3.9}$$

Inspection of relations (3.5) and (3.9) reveals that the optimization problem is reduced to the following integral eigenvalue problem:

$$R[\phi] = \lambda \phi \quad \Rightarrow \quad \int_\Omega K(x,y)\phi(y)dy = \lambda \phi(x). \tag{3.10}$$

Solving the eigenvalue problem (3.10) one computes the POD modes and their corresponding eigenvalues λ. Note, however, that the formulation adopted so far assumes that x is a continuous variable in Ω. Given that in practical experimental or computational measurements only a finite number of sensing locations and measured data points is possible, there is the need to apply numerical schemes in order to convert the eigenvalue problem (3.10) defined on an infinite dimensional functional space to a reduced finite-dimensional eigenvalue problem defined on a finite vector space. There are two ways to perform this dimensionality reduction, namely, the *direct method* and the *method of snapshots*, both of which are discussed in detail in the Thesis by Azeez [23]. The choice of the method depends on the problem and on the volume of data points available. If the snapshots $v_n(x)$ are vector-valued and the structural response is slowly varying, the method of snapshots is preferable. However, if the snapshots are scalar quantities and the measured time series are oscillatory, the direct method is preferred. The direct method attempts to directly solve the eigenvalue problem (3.10) using standard numerical schemes. The method of snapshots is discussed in detail in what follows.

The method of snapshots is based on the technique used by Sirovich and Kirby [218], whereby, the sought POD mode is expressed as linear combination of snapshots:

$$\phi(x) = \sum_k \alpha_k v_n(x). \tag{3.11}$$

Substituting (3.11) into (3.10) yields,

$$\sum_n \int_\Omega \frac{1}{N} v_n(x) v_n^T(y) \sum_k \alpha_k v_k(y) dy = \lambda \sum_n \alpha_k v_k(y) \tag{3.12}$$

or, in matrix form,

$$\sum_{k=1}^N C_{nk}\alpha_k = \lambda a_n, \quad n = 1, 2, ..., N \quad \Rightarrow \quad [C]_{(n\times n)}\{\alpha\}_{(n\times 1)} = \lambda\{\alpha\}_{(n\times 1)}, \tag{3.13}$$

where: $C_{nk} = \frac{1}{N}\int_\Omega v_n(x)v_n(y)dy$. The square matrix $[C]$ is of dimensions $(n \times n)$, symmetric and positive definite, whereas the $(n \times 1)$ vector $\{\alpha\}$ contains the N unknown coefficients α_k. Hence, the reduced order eigenvalue problem (3.13) (defined on an N-dimensional vector space, where N

is the number of snapshots) yields N real-valued eigenvectors and eigenvalues. Discrete approximations to the POD modes are then obtained by means of (3.11). It should be noted that the eigenvalues λ give the strength of the participation of the corresponding eigenfunctions. Although an eigenvalue cannot be interpreted as a measure of mechanical energy contained in the corresponding POD mode, it is related to the portion of the energy of the measured time series captured by the POD mode under consideration.

Continuous POD modes can be obtained from their discrete analogs (3.11) through interpolation. In practical situations the snapshots $v_n(x)$ are available at discrete measurement points x_i, $i = 1, \ldots, M$. They can be approximated by cubic splines to create snapshots continuous in $x \in \Omega$, to be used in the computation of the entries of the matrix $[C]$ and the continuous POD mode shapes $\phi_n(x)$. Hence, one obtains N eigenfunctions $\{\phi_n(x)\}$ and eigenvalues λ_n. Suppose that the eigenvalues are ordered such that $\lambda_1 > \lambda_2 > \ldots > \lambda_N$; then, the corresponding eigenfunctions represent the most typical structure of the field in the domain Ω. The p leading POD modes are obtained by the criterion,

$$\frac{\sum_{i=1}^{p} \lambda_i}{\sum_{i=1}^{N} \lambda_i} \geq 0.999, \tag{3.14}$$

for the smallest integer p. This value is useful to estimate the dimensionality of the system under consideration. Note that in most practical applications it holds that $p \ll N$, so only a small subset of the modes of the discretized eigenvalue problem (3.13) needs to be computed; this feature makes feasible the application of the POD to practical engineering applications involving spatially extended systems or systems with high modal densities.

The POD modes can also be used as finite-dimensional basis for reducing the order of the measured dynamics. Since the kernel $K(x, y)$ in the integral eigenvalue problem (3.10) is symmetric, and taking into account the normalization condition (3.3), the eigenfunctions satisfy a first-order orthonormality condition given by,

$$\int_{\Omega} \phi_i(x) \phi_j(x) dx = \delta_{ij}, \tag{3.15}$$

where: δ_{ij} is Kronecker's symbol. This condition can be used to partially decouple the time and space dependence in the governing partial differential equations of motion, thus reducing the order of the dynamical system from infinite degrees-of-freedom (DOF) to p–DOF (where p is the number of leading POD modes according to (3.14)). Moreover, this order reduction procedure can be applied to linear as well as nonlinear measured dynamics.

In the next Sections applications are provided of the POD method to linear and nonlinear system identification and order reduction problems involving continuous oscillators. The POD modes will be computed using the method of snapshots. Among the dynamic phenomena consider are vibro-impact oscillations, bifurcations and chaotic motions, and vibrations of extended continuous with high modal densities. These examples reveal the wide applicability of the POD to challenging problems in the area of mechanics of continuous systems, where other methods of analysis are not easily applicable.

3.2 Application of POD to Vibro-impact Continuous Oscillators

In this first application, it will be shown how the POD analysis can be used to study nonlinear dynamical effects in continuous systems due to vibro-impacts. In particular, it will be shown that the POD mode shapes and the portion of energy captured by each of the leading (dominant) POD modes are valuable tools for studying the strength of nonlinear effects in these systems. Indeed, by studying POD mode shape variations and energy redistributions among leading POD modes as system parameters change can find application to diagnosis of developing defects in structural systems. Hence, the POD method can be an effective non-parametric system identification tool that can be used for diagnosis and monitoring of the performance of vibrating structural assemblies (see also [68]). In addition, the relative simplicity of the POD calculations enables the applicability of the method to a wide class of mechanical systems.

3.2.1 Vibro-impacting Beam

The first continuous system considered is the impacting beam depicted in Figure 3.1. The impacts are assumed to occur at $x = e = 0.5m$, where the length of the beam is equal to $L = 0.766m$. The values of e and L will be held fixed for all numerical simulations that follow. The mass per unit length of the beam is $m = 0.753 kg/m$, and the product of the modulus of elasticity with the moment of inertia of the beam cross section is $EI = 57.7 Nm^2$. This leads to the following leading natural frequencies, $\omega_1 = 52.45$, $\omega_2 = 328.73$, $\omega_3 = 920.45 rad/sec$ and the corresponding modal damping ratios $d_1 = 0.475$, $d_2 = 0.750$ and $d_3 = 0.375 Ns^2/m^2$ (identified through experimental modal analysis of a cantilever of similar geometry and material properties); The clearance between the beam and the rigid boundary causing the impacts is equal to c, and the system is forced by the harmonic force $F \cos \Omega t$ and is applied at the position $x = f$.

The governing partial differential equation is given by,

$$EI\, u_{xxxx}(x,t) + m u_{tt}(x,t) + d u_t(x,t) = \delta(x-f) F \cos \Omega t + \delta(x-e) P(u(e,t)), \tag{3.16}$$

where: $u(x,t)$ is the transverse displacement of the beam, $\delta(\cdot)$ is the delta function, and the term $P(u(e,t))$ represents the concentrated nonlinear force

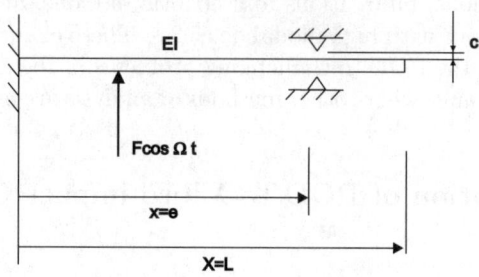

Figure 3.1. The vibro-impacting beam.

applied to the beam at $x = e$ due to the vibro-impacts; in addition, the short-hand notation for partial differentiation was utilized.

There are different ways to represent the vibro-impact term. The *exact way* (which is the most difficult one from a computational point of view) is to incorporate the vibro-impact effects during the numerical integration of the equation of motion, by continuously monitoring the displacement of the beam at the point of vibro-impact $x = e$; once the rigid boundary is reached at time $t = t^*$ (computed by the condition $|u(e, t^*)| = c$) the interaction between the beam and the rigid boundary is enforced according to the chosen impact law, and the numerical simulation is newly initiated for $t \geq t^*$ with initial conditions of identical to those encountered immediately after the impact. The computation is continued until the next impact is encountered. The main difficulty in the exact way of modelling vibro-impact effects is that the impact times depend on the solution itself, so very accurate numerical estimations of these times are required to obtain accurate numerical simulations. For works on accurately locating switching points (in this case, the exact times of impact), see [91, 57]. In addition, no explicit expression for the nonlinear term $P(u(e, t))$ in the equation of motion can be derived.

The *approximate way* of incorporating the vibro-impact terms is to replace the rigid boundary by a stiff spring-damper element; the spring simulates the local compliance of the boundary and the damper the inelastic impact effects. This scheme is easier to implement computationally, and it enables the direct expression of the vibro-impact term in the equation of motion:

$$P(u(e,t)) = \begin{cases} K(|u(e,t)| - c) \operatorname{sgn}(u(e,t)), & |u(e,t)| \geq c \\ 0, & |u(e,t)| < c \end{cases}. \quad (3.17)$$

Due to its relative simplicity, in the following simulations the approximate method for modelling the vibro-impact effects is adopted. A more detailed

3.2 Application of POD to Vibro-impact Continuous Oscillators

discussion of the convergence of the adopted numerical scheme is given in (Emaci et al., 1997a).

Table 3.1. Set I: $f = 0.1m$, $F = 500N$.

Case No	1	2	3	4	5	6	7	8	9	10	11	12	13
$c(mm)$	0	1	2	3	4	5	6	7	8	9	10	20	100

Two basic sets of numerical simulations were performed in which the forcing frequency Ω, forcing location f, and the clearance c are varied over a range of values. For each numerical simulation f is fixed and Ω and c are varied; the parameters for the two sets of simulations (labeled Sets I and II) are presented in Tables 3.1 and 3.2. Twenty different frequencies Ω were selected for the numerical simulations, and these are indexed in Table 3.3. For each case of each set the time series for the snapshots are obtained by integrating the governing partial differential equation of motion; an initial time window of $20sec$ is discarded to eliminate the transient effects from the response, and afterwards data is acquired and stored for post processing and POD analysis.

Table 3.2. Set II: $f = 0.4m$, $F = 500N$.

Case No	1	2	3	4	5	6	7	8	9	10	11	12	13
$c(mm)$	0	1	2	3	4	5	6	7	8	12	15	20	30

Table 3.3. Forcing frequency index convention for Sets I and II.

Frequency Index	1	2	3	4	5	6	7	8	9	10
$\Omega\ (rad/s)$	30	35	40	45	50	52.4	55	70	80	90
Frequency Index	11	12	13	14	15	16	17	18	19	20
$\Omega\ (rad/s)$	100	110	120	130	140	150	160	170	180	190

In Figure 3.2 the mode shapes of the leading POD modes are depicted for various cases of Set I [similar results were obtained for Set II (Azeez and Vakakis, 2001)]. It is to be noted that case 1 of Set I represents a clamped-simply supported linear beam (zero clearance), whereas case 12 of the same Set represents a different linear continuous system, namely a cantilever beam (relatively large clearance for vibro-impacts to occur). The cases in-between between these limiting linear systems represent vibro-impacting beams with strongly nonlinear dynamical effects; it is of interest to study the nonlinear transition between the two limiting linear cases using POD analysis.

For Set I the harmonic excitation is located away from the position of the clearance, and the POD mode shapes are sensitive to the clearance parameter c. Indeed, even for small clearance changes, the corresponding POD mode shapes show large variations even when the forcing frequency is held constant. Note that there is a propensity for many vibration modes of the beam to be excited due to the vibro-impacts.

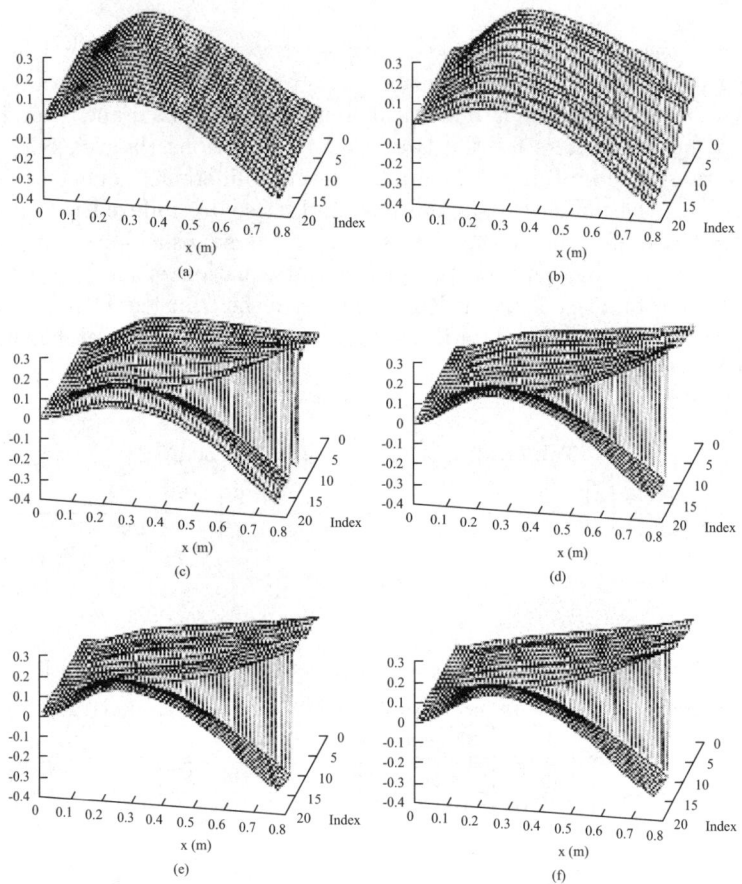

Figure 3.2. Mode shapes of the first POD mode for selected cases of Set I: (a) case 1, (b) case 3, (c) case 5, (d) case 7, (e) case 9, (f) case 11; the indices refer to those of Table 3.3.

When the forcing frequency is low the vibro-impacts are intermittent (relatively small number of impacts), and the POD mode shapes are nearly invariant even when the clearance is changed. As the forcing frequency increases the impacts become more frequent and the deformations of the POD mode

3.2 Application of POD to Vibro-impact Continuous Oscillators 187

shapes are more evident (cf. frequency indices 5 to 15 in Figure 3.2d). In this case the deformations of the POD mode shapes are caused by the strongly nonlinear effects due to vibro-impacts, which cause profound energy redistribution among the leading POD modes, with energy being 'transferred' from leading to higher order POD modes. Similar conclusions can be drawn from the plots of Figure 3.3, where the variations of the mode shapes for the second POD mode of the vibro-impacting beam are depicted for varying forcing frequencies and for different cases of Set I.

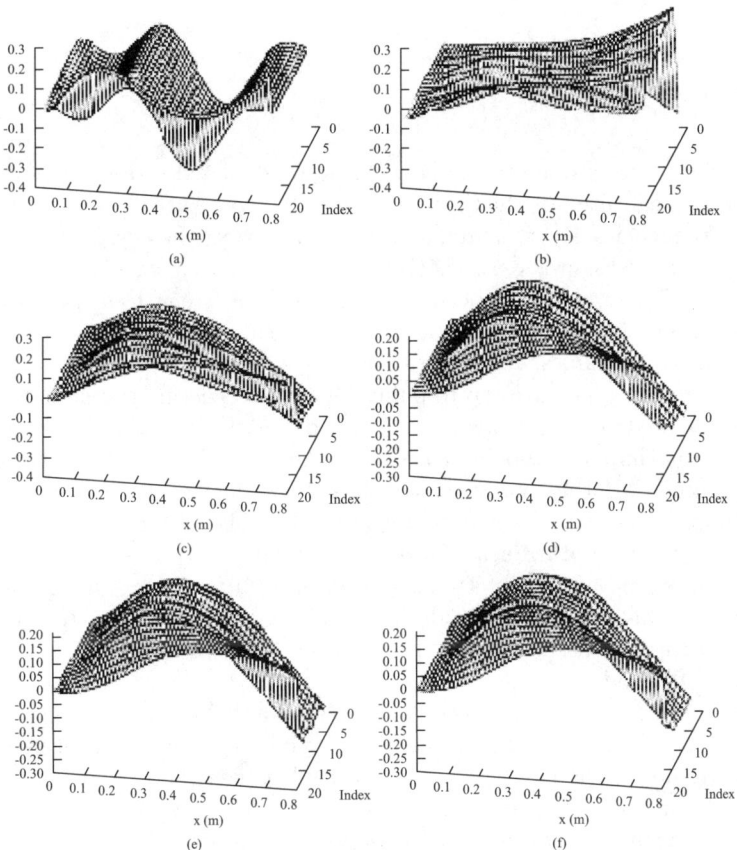

Figure 3.3. Mode shapes of the second POD mode for selected cases of Set I: (a) case 1, (b) case 3, (c) case 5, (d) case 7, (e) case 9, (f) case 11; the indices refer to those of Table 3.3.

Of interest is to study the distribution of the energy of the measured time series among the leading POD modes for Sets I and II, since this provides significant insight into the nonlinear effects due to vibro-impacts. Typical

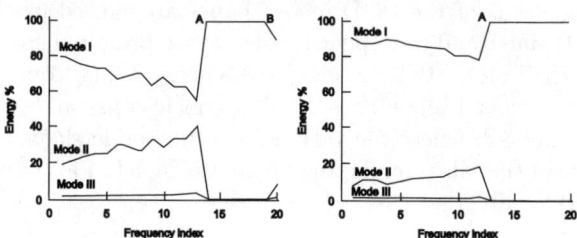

Figure 3.4. Typical energy distributions among POD modes: (a) Set I, case 6, (b) Set I, case 11.

energy distribution diagrams for two cases are depicted in Figure 3.4, where the energies captured by the three leading POD modes are plotted for varying forcing frequencies. By combining all cases one derives the energy distribution surface plots of Figures 3.5 (Set I) and 3.6 (Set II), for varying forcing and clearances. These plots can be used to deduce the strength of the nonlinear effects in the system, and to obtain an approximate dimensionality of the vibro-impact dynamics. Referring to Figure 3.4a it can be seen that as the forcing frequency increases up to point A profound energy 'transfer' from the first to the second POD mode takes place, indicating strong nonlinear effects due to vibro-impacts and increase of the dimensionality of the dynamics. In the range AB the motion of the beam is sufficiently small so that no vibro-impacts occur; as a result the response of the beam is linear and the first POD mode capture nearly all the energy of the time series. Increasing the frequency beyond point B the second natural frequency of the beam is approached leading to amplitude increase and vibro-impacts; the induced nonlinearities are manifested by the 'transfer' of energy from the first to the second POD mode.

By studying the surface plots of Figures 3.5 and 3.6 it is possible to study the nonlinear effects in the system for varying clearance. Indeed, referring to Figure 3.5, as the clearance index varies between the two limiting linear systems (corresponding to clearance indices 0 and 12 of Set I), there is energy redistribution from the first POD mode to the second, indicating strong nonlinear effects and increase in the dimensionality of the dynamics. It is of interest to note that there is a point in the plot where 'maximum' nonlinear effects occur, represented by point C in the surface plot; at that point the energy captured by the first (second) POD mode is minimized (maximized) indicating the furthest departure of the dynamics from linear behavior. Similar conclusions can be made by considering the surface plots of Figure 3.5. Hence, the POD method can be used to study the transition of the dynamics

3.2 Application of POD to Vibro-impact Continuous Oscillators   189

of a system from linear to nonlinear regimes, and to find values of system or excitation parameters where the nonlinear effects are maximized.

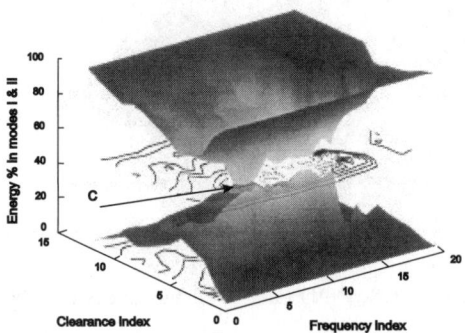

Figure 3.5. POD energy surface plots for Set I.

A key advantage of obtaining the dominant POD modes is that they can be used to construct low-order dynamical models for the system under consideration. For a large-scale structure its response can be simulated using finite-element models using traditional shape functions. Once a reasonable amount of data is obtained, the POD modes can be extracted from them and low-dimensional models can be constructed which would result in tremendous savings of computational time. These low-order models can then be used for dynamical studies of structural modifications or for control system design. This procedure will be referred to as *reconstruction*, and will be first applied to the case of the vibro-impacting beam. To this end, the deformation of the beam in equation (3.16) in the series form,

$$u(x,t) = \sum_{i=1}^{p} \phi_i(x)\, a_i(t), \qquad (3.18)$$

where only the p-dominant POD modes are considered, since they capture 99.9% of the energy of the analyzed time series (cf. relation (3.14) for the definition of p).

Where the POD modes $\phi_i(x)$ are used as admissible functions (that should satisfy at least the essential boundary conditions [69]), and $a_i(t)$ are the corresponding amplitudes. The POD modes are an ideal choice for this operation, since they satisfy both natural and essential boundary conditions, as well as orthonormality conditions.

Substituting the expansion (3.18) into (3.16) one obtains,

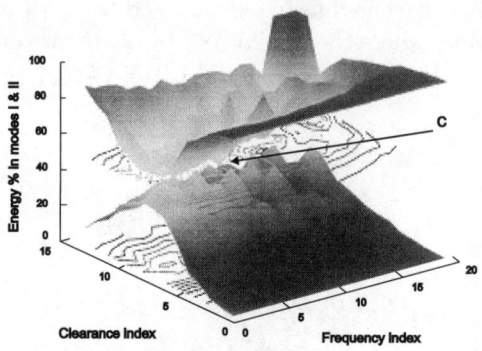

Figure 3.6. POD energy surface plots for Set II.

$$EI \sum_{i=1}^{p} \phi_i''''(x)a_i(t) + m \sum_{i=1}^{p} \phi_i(x)\ddot{a}_i(t) + d \sum_{i=1}^{p} \phi_i(x)\dot{a}_i(t) =$$

$$P\left(\sum_{i=1}^{p} \phi_i(e)a_i(t)\right)\delta(x-e) + F\cos\Omega t \delta(x-f), \quad (3.19)$$

where primes denote differentiation with respect to x, and dots differentiation with respect to t. Multiplying (3.19) by $\phi_j(x)$, integrating by parts over the domain $x \in \Omega$ (to reduce the order of the highest derivative), and employing the orthogonality condition (3.15), one obtains the following set of discretized ordinary differential equations:

$$mc_j\ddot{a}_j + dc_j\dot{a}_j + EI \sum_{i=1}^{p} \xi_{ij}a_i = \phi_j(f)F\cos\Omega t + P\left(\sum_{i=1}^{p} \phi_i(e)a_i\right)\phi_j(e), \quad (3.20)$$

where:

$$\xi_{ij} = \phi_j(L)\phi_i'''(L) - \phi_j'(L)\phi_i''(L) + \phi_j'(0)\phi_i''(0) + \int_0^L \phi_i''(x)\phi_j''(x)dx .$$

It is noted that the boundary conditions of the problem are satisfied automatically by the POD modes (by construction) and need not be explicitly satisfied.

Equation (3.20) is the $2p$–dimensional discretized model of the continuous system, and is integrated numerically by a 4th-order Runge-Kutta algorithm. Details on the routines used can be found in [25]. Since the computed POD modes are computed at discrete points in the domain $0 \leq x \leq L$, interpolations were performed (using B-spline interpolation functions of 6th order) to

3.2 Application of POD to Vibro-impact Continuous Oscillators

obtain continuous POD modes valid over the entire domain of the computation.

When vibro-impacts are involved it is not always possible to obtain pointwise time domain convergence between the simulated and reconstructed results due to sensitive dependence on initial conditions. In such a situation the responses are transformed and compared in the frequency domain. Such a comparison is shown in Figure 3.7 for a system with $f = 0.1m$, $c = 3mm$, $\Omega = 30 rad/s$, and using only the leading 3 POD for the reconstruction. It is to be noted that for continuous systems undergoing vibro-impacts very small time steps have to be used for the numerical integration in order to get repeatability of the simulations.

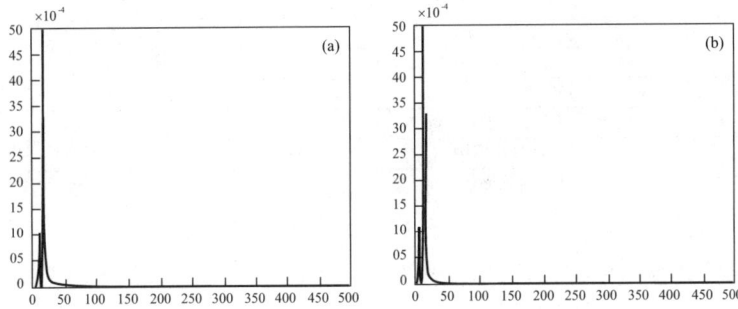

Figure 3.7. Comparisons of power spectra of the beam responses: (a) Reconstructed, (b) simulated response.

3.2.2 Overhung Rotor

The schematic of the rotor with the definition of the various parameters is shown in Figure 2.8. The global dynamics of this system was studied in [24, 25], were a variety of regular (periodic and quasi-periodic) and irregular (chaotic) orbits were investigated, together with the bifurcations of the dynamics as the frequency of rotation changes. The system parameters are listed in Table 3.4 and are held constant thought the simulations of this Section. EI and m are the flexural rigidity and mass per unit length of the rotor (shaft), M_D and I_D the mass and moment of inertia of the cross section of the attached flywheel, $d_x = d_y$ and L the viscous damping coefficients per unit length and the length of the rotor, c the clearance at the point of impact $x = a$, and m_e a mass unbalance of the flywheel at a radial location equal to e.

The governing partial differential equations of motion are given by:

$$EI\, u_{1xxxx}(x,t) + m u_{1tt}(x,t) + d_x u_{1t}(x,t) + M_D \delta(x-L) u_{1tt}(x,t)+$$

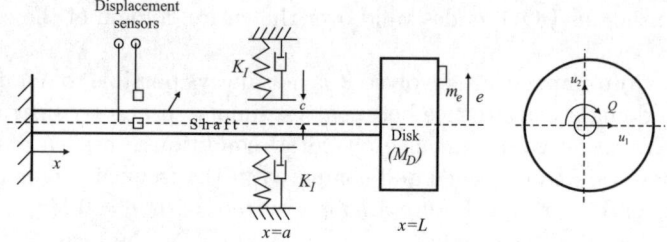

Figure 3.8. Schematic of the overhung rotor.

Table 3.4. Parameter values for the overhung rotor undergoing vibro-impacts.

Parameter	Value	Parameter	Value
a	$0 \cdot 5064 m$	c	$1 \cdot 5mm$
d_x	$1 \cdot 0 Ns/m^2$	d_y	$1 \cdot 0 Ns/m^2$
EI	$128 \cdot 97518 Nm^2$	e	$0 \cdot 05m$
I_D	$5 \cdot 4032 \times 10^{-4} kg\, m^2$	K_I	$10^8 N/m$
L	$0 \cdot 635m$	M_D	$0 \cdot 8375 kg$
m	$1 \cdot 14628 kg/m$	m_e	$0 \cdot 03 kg$

$$2\Omega I_D u_{2xt}(x,t)\delta'(x-L) = \delta(x-L)Q_1(t) + \delta(x-a)P_1(u_1(a,t)),$$

$$EI\, u_{2xxxx}(x,t) + mu_{2tt}(x,t) + d_y u_{2t}(x,t) + M_D\delta(x-L)u_{2tt}(x,t) +$$
$$2\Omega I_D u_{1xt}(x,t)\delta'(x-L) = \delta(x-L)Q_2(t) + \delta(x-a)P_2(u_2(a,t)), \quad (3.21)$$

where: $u_1(x,t)$ and $u_2(x,t)$ are the rotor displacements in the vertical and horizontal directions, respectively; P_1 and P_2 represent the vertical and horizontal forces, respectively, due to the vibro-impacts; and Q_1, Q_2 are the external forces acting on the rotor due to the mass imbalance of the flywheel. These forces are defined by the following relations:

$$Q_1(t) = m_e \Omega^2 e \cos \Omega t, \quad Q_2(t) = m_e \Omega^2 e \sin \Omega t - M_D g,$$

$$P_1(u_1(a,t)) = -\sigma \left(K_i \frac{(r-c)u_1(a,t)}{r} \right),$$

$$P_2(u_2(a,t)) = -\sigma \left(K_i \frac{(r-c)u_2(a,t)}{r} \right),$$

$$\sigma = \begin{cases} 1, & r = \sqrt{u_1^2 + u_2^2} \geq c \\ 0, & r < c \end{cases}. \quad (3.22)$$

The 'exact' numerical simulations of the system were numerically obtained by discretizing the governing partial differential equations of motion

3.2 Application of POD to Vibro-impact Continuous Oscillators   193

Table 3.5. Parameters for rotor simulations.

Case No	1	2	3	4	5	6	7
$c\ (mm)$	0.001	0.2	0.4	0.6	0.8	0.9	1.0
$L\ (m)$	0.6064	0.6064	0.6064	0.6064	0.6064	0.6064	0.6064
Case No	8	9	10	11	12	13	14
$c\ (mm)$	1.2	1.4	1.6	1.8	2.4	1.0	1.0
$L\ (m)$	0.6064	0.6064	0.6064	0.6064	0.6064	0.4064	0.5064

Table 3.6. Speed index notation for the numerical simulations.

Frequency Index	1	2	3	4	5	6	7	8	9	10
$\Omega\ (rpm)$	400	500	600	700	800	900	1000	1100	1200	1300
Frequency Index	11	12	13	14	15	16	17	18	19	20
$\Omega\ (rpm)$	1400	1500	1600	1800	1800	1900	2000	2100	2200	2300
Frequency Index	21	22	23	24	25					
$\Omega\ (rpm)$	2400	2600	2900	3200	4000					

using the leading six vibration mode shapes of the corresponding (linear) cantilever beam with a mass at its end and no vibro-impacts. These 'exact' simulation results will be compared to reconstructed results obtained from low-order models resulting from discretization of the governing partial differential equations of motion using POD modes as trial functions. The procedure follows closely that used in the previous Section.

The POD modes of the overhung rotor and their energies were computed for the 14 cases of varying clearance listed in Table 3.5. For each of these cases the speed of rotation is varied from 400 to 4000 r.p.m. as shown in Table 3.6. The different clearance and speed values are referred to by the case numbers (Table 3.5) and speed indices (Table 3.6).

There are two displacements at each point of the rotor (a vertical and a horizontal one). These data are processed independently to obtain the horizontal and vertical POD modes of the system. The leading horizontal POD mode shapes for few cases of Table 3.5 are depicted in Figure 3.9. At *low speeds* the rotor just rides on the lower surface of the bearing (with clearance) because of the weight of the flywheel. There are no violent impacts in this case, as the rotor inertia is not sufficient for the rotor to impact violently on the bearing. Hence, despite of the 'weak' impacts the rotor amplitudes are not large and the POD mode shapes resemble the first vibration modes of the corresponding cantilever beam systems. Clearly, at low speeds the rotor dynamics is approximately linear, and the leading POD mode dominates the response; distribution of POD energy to higher order modes is limited.

As *speeds increase* the harmonic force acting on the rotor also increases due to imbalance effects and vigorous vibro-impacts occur. This leads to the 'bending' of the POD mode shapes, and the leading POD resembles the second vibration mode of the corresponding cantilever beam. At the same time,

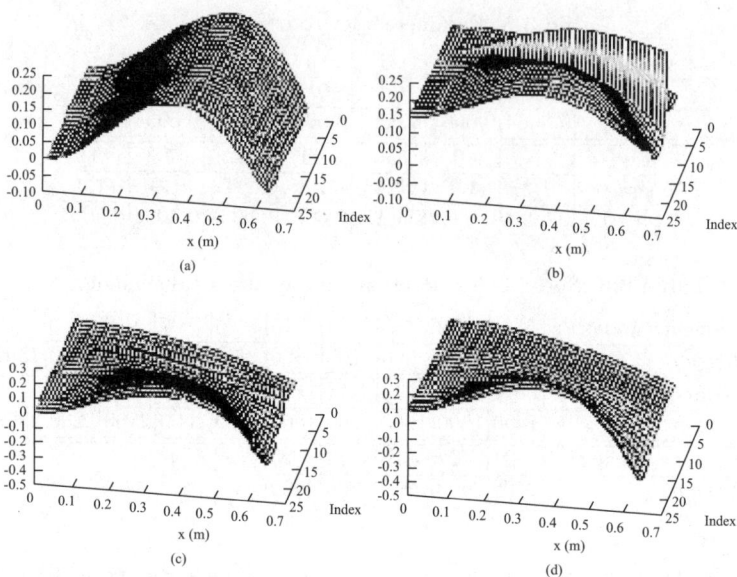

Figure 3.9. Leading horizontal POD modes of the overhung rotor: (a) Case 1, (b) case 3, (c) case 7, (d) case 10.

the second POD mode (not depicted in the Figures) changes and appears to be similar to either the first or third vibration mode of the cantilever beam. At higher speeds the first vibration mode does not dominate the response and may appear as a higher-order, less dominant POD mode. For larger clearances violent impacts do not occur until very high speeds are reached, a feature that is manifested as POD mode shapes that are less sensitive to parameter variations.

The energies of the POD modes corresponding to all cases listed in Table 3.5 are depicted in Figures 3.10 and 3.11. The time traces of a subset of the simulations (marked as numbers in Figures 3.10 and 3.11) are presented in Figure 3.12 (with the corresponding number on the right top corner of each plot). Note that there are many factors influencing the dynamics of the rotor, including, gyroscopic interactions, multi-modal response, shaft-flywheel interactions and strong nonlinearities due to violent vibro-impacts. Indeed, as shown in [25] there are sudden 'chaotic explosions', where for a small change of the speed of rotation there is a sudden transition from regular periodic (or quasi-periodic) motions to irregular, chaotic ones; moreover, no transitional sequences from regular to chaos appear to occur in the form of period doublings or invariant tori breakdowns, and instead the chaotic motions appear suddenly with no precursors. The question then is, *how to interpret such sudden transitions to chaos in terms of the POD energy distributions*; or the inverse question, namely, *what can be said about the rotor responses*

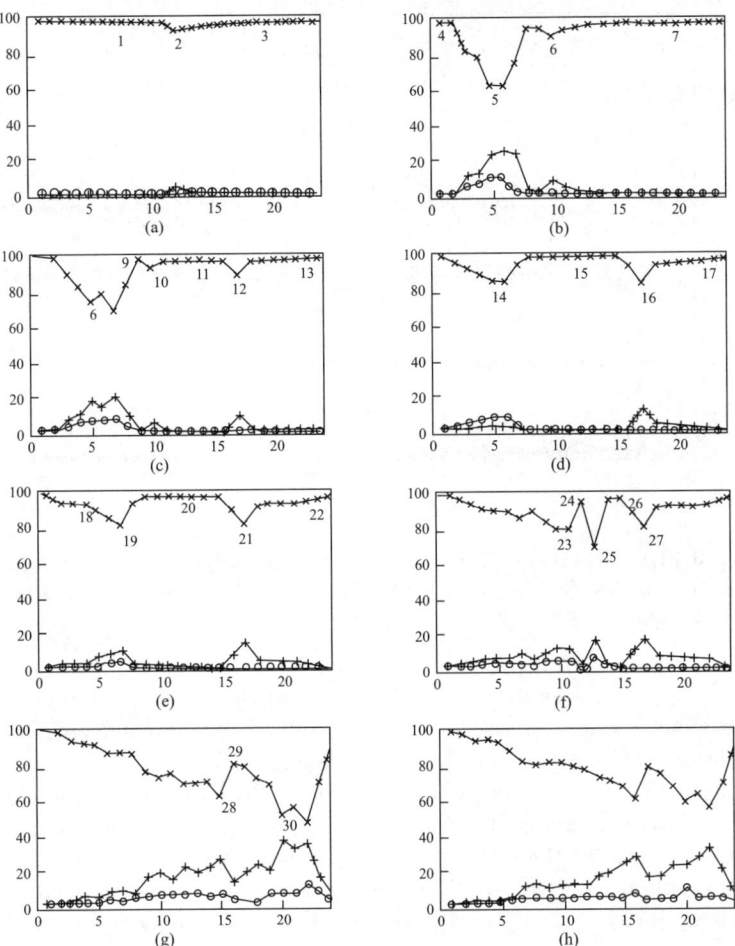

Figure 3.10. POD energy distributions for the three POD modes of the rotor: (a) Case 1, (b) case 2, (c) case 3, (d) case 4, (e) case 5, (f) case 6, (g) case 7, (h) case 8; –x– POD mode 1, –+– POD mode 2, –o– POD mode 3.

based on the corresponding POD energy plots. These questions are important if the POD method is to be used as a method for diagnosis and monitoring of defects in mechanical systems.

For almost zero clearance the system is nearly linear, and most of the energy is captured by a single POD mode at all speeds of rotation considered. As the clearance increases the distribution of energy among higher-order POD modes also increases. However, there seems to be a critical value of clearance, $c \approx 1mm$ when the energy transfer distribution is at a maximum. This seems to be the value of clearance at which the nonlinear effects are most profound;

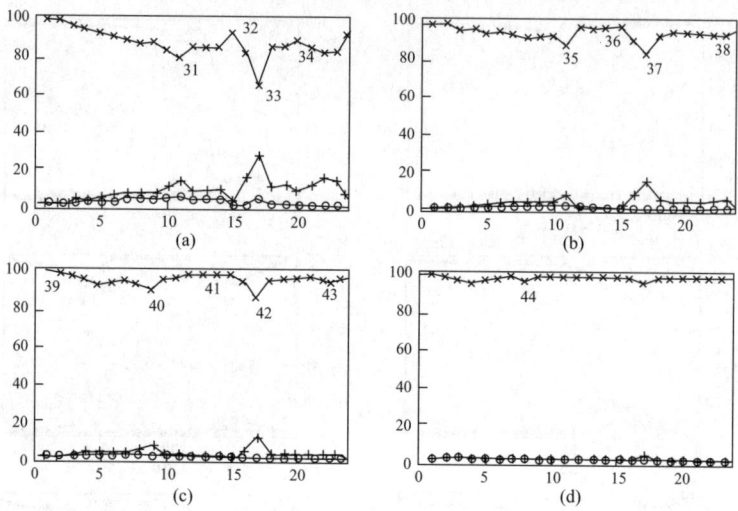

Figure 3.11. POD energy distributions for the three POD modes of the rotor: (a) Case 9, (b) case 10, (c) case 11, (d) case 12; –x– POD mode 1, –+– POD mode 2, –o– POD mode 3.

similarly to what was observed for the case of the vibro-impacting beam in the previous Section, increasing or decreasing the clearance from its critical value leads to two *linear* limiting rotor systems. Moreover, the participation of higher-order POD modes is also enhanced close to this critical clearance value. It is concluded that strong nonlinear effects in the neighborhood of the critical value of the clearance, are associated with enhanced energy distribution among multiple POD modes. Away from the critical value of the clearance, the system approaches either a linear cantilever rotor (for $c \gg 0$), or a linear clamped - simply supported rotor (for $c \to 0$), and the leading POD mode captures nearly all the energy of the measured time series and the participation of higher-order modes is minimal. Therefore, as in the previous Section *there is a direct relation between nonlinear dynamic effects and enhanced distribution of energy among multiple POD modes.*

From the orbits depicted in Figure 3.12 it is observed that whenever the rotor response is periodic (for example, point 7) there is only one dominant POD mode. However, if the motion of the rotor is chaotic (points 5, 8, 33) there is participation of higher-order POD modes in the rotor response and enhanced distribution of energy among multiple POD modes exists; in cases of chaotic motions at least three POD modes are needed to capture the energy of the time series. For other types of rotor response, such as subharmonic (points 6, 34) or quasi-periodic orbits (point 36), there is moderate energy distribution between POD modes. It can be seen from the energy distribution diagrams that there are well-defined regions where the energy distribution is high or low. Moreover, in other regions (for example, between points 23 and 27 in Figure 3.10f) the energy distribution among POD modes varies erratically; this ties well with the observation of chaotic 'explosions', and with sudden appearances and disappearances of periodic, quasi-periodic and chaotic motions [23].

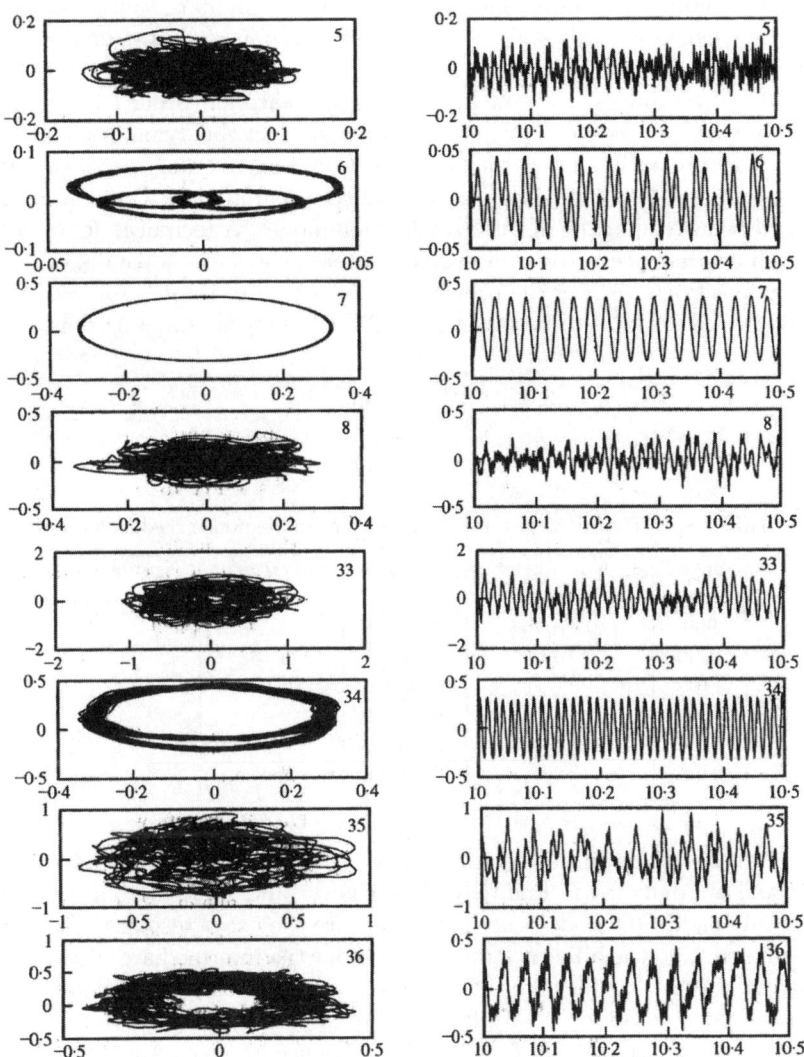

Figure 3.12. Orbits and horizontal displacements (in mm) for certain of the marked points in Figures 3.10 and 3.11, at $x = 0.302$.

From the reported results it is concluded that the dimensionality of the rotor dynamics increases as the rotor response makes the transition from regular to chaotic: whereas the dimensionality of the regular dynamics is one, in chaotic regions the dimensionality increases to (at least) three. This result, which is directly inferred from the energy distribution plots, is of significant importance in studies of order reduction and creation of the low-order models of the dynamics of continuous systems. In fact, the aforementioned results will be used in the following analysis where response reconstruction of the dynamics is considered.

The POD modes will now be utilized to create low-order models of the system that can accurately capture and reconstruct the dynamics. Following the procedure outlined in the previous Section, the governing partial differential equations (3.21) are discretized using the computed POD modes as trial functions instead of the linearized vibration modes. A technical detail of this procedure that needs to be discussed concerns the shear force discontinuity at the position $x = L$ of the rotor. This discontinuity is not captured by the POD modes and has to be manually enforced. This is represented as,

$$\phi_i'''(x = L) = 0, \tag{3.23}$$

where: ϕ_i are either horizontal or vertical POD modes.

Table 3.7. The rotor parameters used for the response reconstructions.

Case	Speed (r.p.m.)	Non-linear spring	a (m)	c (mm)	Number of $K - L$ modes used
1	900	$10^6(a/c)^3$	$0 \cdot 4064$	$0 \cdot 5$	1
2	1200	$10^6(a/c)^3$	$0 \cdot 4064$	$0 \cdot 5$	1
3	1000	↑	$0 \cdot 6064$	$0 \cdot 5$	2
4	800	|	$0 \cdot 6064$	$0 \cdot 5$	4
5	1100	$10^8(a-c)(a>c)$	$0 \cdot 6064$	$0 \cdot 5$	4
6	2300	↓	$0 \cdot 6064$	$0 \cdot 5$	4

Six cases of response reconstructions were performed as listed in Table 3.7, and are depicted in Figure 3.13. In cases 1 and 2 the motion is periodic and only one POD mode is needed to capture the exact response. Cases 3 to 6 correspond to the bilinear stiffness model for the intermediate stiffness with varying stiffness coefficients. For case 3 'weak' vibro-impacts occur and the rotor response is composed of a fast oscillation with a slowly-varying envelope; in this case eventhough three POD modes are needed to capture 99.9% of the energy, only the first two were used in the reconstruction. The reconstructed response is close to the simulated one. In case 4 the rotor response is chaotic and four POD modes are used for the reconstruction.

Cases 5 and 6 are nearly periodic and one POD mode nearly captures the entire energy; nevertheless, four POD modes were used to reconstruct

3.2 Application of POD to Vibro-impact Continuous Oscillators

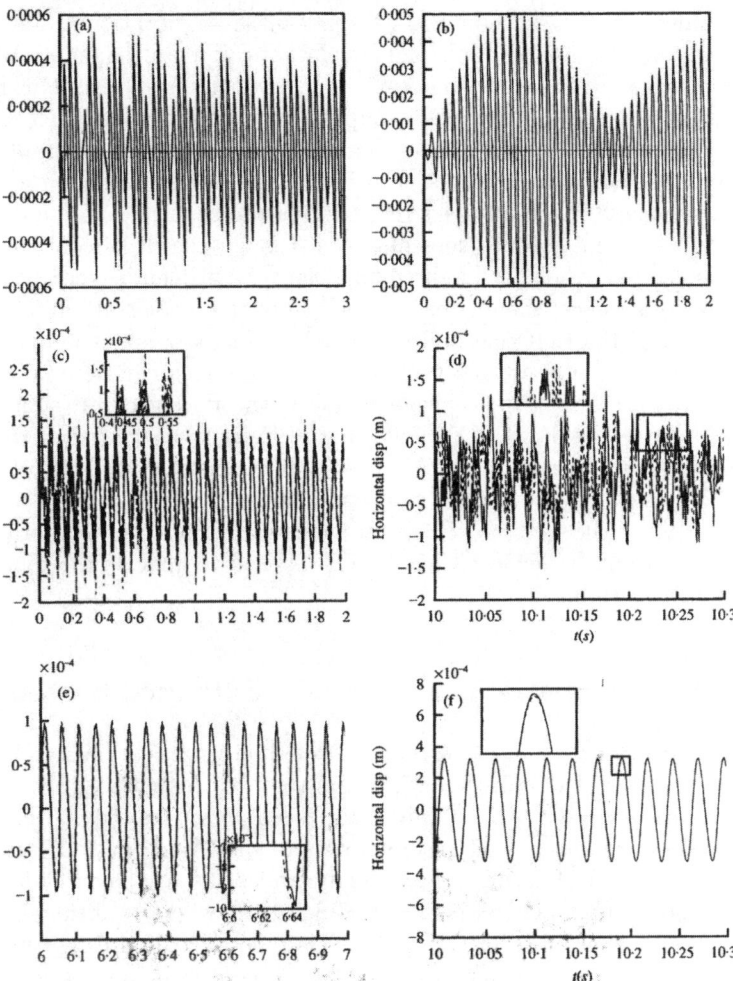

Figure 3.13. Some reconstructions for cases 1-6 of Table 3.7: (a) Case 1, (b) case 2, (c) case 3, (d) case 4, (e) case 5, (f) case 6; - - - - - reconstructed, ——— simulated responses.

the responses for these cases, in order to demonstrate that the higher POD modes are not spurious and can still be used as trial functions to discretize the response. This is especially important if the POD modes computed for a set of system parameters is used as orthogonal basis to reconstruct the response of the system with a different set of parameters; in such cases it is found that including higher-order POD modes (even though capturing small energy) increases the convergence of the numerical reconstruction.

It is noted that even though six linear cantilever vibration modes were used to obtain the 'exact' numerical simulation results, a maximum of four POD modes is need to derive an accurate reconstructed response. This might not appear to be a major saving of computational time. But consider a practical situation where a finite element model is used to simulate complicated rotor systems, for an example, a crankshaft typical in automotive applications. In that case, the rigid disks representing the cams and nose must also be discretized using several elements; even though the disks contribute to the dynamics as a whole, the individual discretized element's contribution is negligible, and, hence, the degrees of freedom of the models are unnecessarily increased. If the POD method is applied to such systems it immediately provides a dimensional estimate of the dynamics, which can be used to significantly reduce the size of the problem and yet produce accurate simulation results. Moreover, as shown in the following Sections, when spatially extended structures are considered, the POD-based reduced-order models are of much smaller dimensionality (and yet efficient) compared to the original dimensionality size of the system. These statements hint on the computational efficiency that can be achieved when POD is used to analyze the dynamics of continuous systems.

3.3 Coherent Spatial Structures in Extended Systems

To demonstrate the potential of the POD method to create efficient low dimensional models of vibrating continuous structures, a second application of the method is given with the dynamic analysis of linear truss dynamics. The main focus of the analysis of this Section is to show that the POD method is capable to effectively order reduce the dynamics of spatially extended continuous systems that possess high modal densities, where traditional modal analysis methods are difficult to apply. This type of structures is typical in certain civil engineering or aerospace applications (examples are long construction trusses; lightweight, large-radius circular communication or radar antennas; and the space station, which, in essence, is an extended periodic truss with attached modules).

There is extensive literature on periodic systems, and specifically, lightweight flexible truss structures: On computing their natural frequencies and mode shapes [154, 216]; on mode localization and passive motion confinement due to structural disorder [73, 49]; and on system identification and control strategies for eliminating unwanted disturbances caused by external excitations [246, 46, 93].

The dynamic analysis of extended multi-bay lightweight trusses poses interesting technical challenges. This is due to the fact that such structures generally possess high modal densities, with clusters of densely packed modes occurring even at relatively low frequencies. In addition, in multi-coupled trusses wavemode conversions occur [49, 63] that produce closely spaced modes corresponding to different types of overall truss vibration (e.g., predominantly bending, near shear etc.). As a result, system identification (modal analysis) of flexible extended trusses is a challenging task, and accurate low-order models of such systems are difficult to obtain. Such low-order models are essential for controlling disturbance propagation in such systems or for performing substructure synthesis and structural modification studies.

Even the task of numerically simulating the transient or steady state dynamics of extended trusses introduces serious technical difficulties due to the presence of exponential dichotomy in the corresponding transfer matrices due to the co-existence of very small and very large eigenvalues [48, 63]. This leads to numerical instabilities of numerical schemes based on direct multiplications of transfer matrices of the system. The use of finite element based schemes for computing the transient dynamics has other limitations related to the large number of elements required, and to the incomplete description of the dynamics [48]; hence, at most, finite element - based methods can provide the response at early times [166]. Similar limitations were encountered by Von Flotow [246] in his attempt to compute the transient response of a truss by transforming the dynamics into wavemode coordinates and numerically convoluting individual impulsive responses of truss components.

To overcome the aforementioned technical difficulties and to be able to simulate the transient truss dynamics with numerically stable computational algorithms, Ma and Vakakis [141] employed the direct global matrix (DGM) approach developed by Schmidt and Jensen [204]. This method is based on a single numerical inversion of a global square transfer matric of large dimension, instead of performing error-prone lower dimensional transfer matrix multiplications. For application of the same approach to a problem of axisymmetric wave propagation in layered media, the reader is referred to the Thesis by Cetinkaya [43]. Since the numerical scheme for accurately computing the truss dynamics is an interesting topic in itself in the general area of dynamics of continuous systems, an exposition of this topic will be first provided. In later Sections the results of application of POD to the truss dynamics, and the construction of low-order models will be discussed.

3.3.1 Computation of the Transient Dynamics

The truss structure considered is shown in Figure 3.14, consisting of periodic sets (bays) coupled by clamped joints. The analysis will be carried under the assumption of linearity, which is a good assumption in this type of spatially extended systems. Each periodic set consists of coupled beams undergoing combined axial and bending vibrations. Because the beam vibrations are assumed to be linear, the governing partial differential equations for bending and longitudinal of each beam are uncoupled, and the only coupling and mode conversion between bending and longitudinal vibrations takes place through the boundary conditions at the joints. The joints connecting the beams of each periodic set transmit axial and transverse forces, as well as bending moments. Moreover, to simplify the analysis only in-plane vibrations of the truss are considered, thus rendering the dynamical problem two-dimensional. The dynamics of the truss of infinite spatial extent was analyzed by constructing a linearly exact transfer matrix by [48, 63].

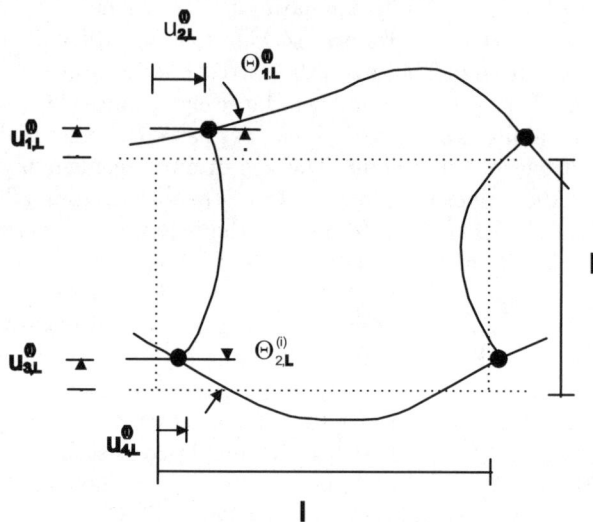

Figure 3.14. Truss configuration and adopted notation for the displacements and rotations.

Considering an 18-bay truss, assuming steady-state harmonic vibration, no structural disorder, and six displacement/rotation coordinates at the boundary of each bay (four axial/transverse displacements and two rotations), the following matrix expression defines the (12×12) transfer matrix of each periodic set:

$$\left\{\begin{array}{c} u_L^{(i+1)} \\ f_L^{(i+1)} \end{array}\right\} = \begin{bmatrix} T_{11} & T_{12} \\ T_{21} & T_{22} \end{bmatrix} \left\{\begin{array}{c} u_L^{(i)} \\ f_L^{(i)} \end{array}\right\}, \quad i = 1, ..., 18, \qquad (3.24)$$

where the elements T_{ij} represent (6×6) submatrices that depend of the system parameters and the frequency of vibration; their exact expressions were derived with computer algebra and are reported in [64]. The (6×1) vector of frequency-dependent generalized displacements at the left internal boundary of the i-th periodic set is defined as (see Figure 3.14),

$$u_L^{(i)} = \begin{bmatrix} u_{1,L}^{(i)} & u_{2,L}^{(i)} & \theta_{1,L}^{(i)} & u_{3,L}^{(i)} & u_{4,L}^{(i)} & \theta_{2,L}^{(i)} \end{bmatrix}^T \qquad (3.25)$$

and the corresponding (6×1) vector of frequency-dependent generalized forces is defined as,

3.3 Coherent Spatial Structures in Extended Systems

$$f_L^{(i)} = \begin{bmatrix} F_{1,L}^{(i)} & F_{2,L}^{(i)} & M_{1,L}^{(i)} & F_{3,L}^{(i)} & F_{4,L}^{(i)} & M_{2,L}^{(i)} \end{bmatrix}^T. \tag{3.26}$$

For a fixed frequency, expressions (3.23)–(3.26) defines internal forces and moments are transmitted through each bay at steady-state, and their relation to the displacements and rotations of the joints at the bay boundaries. A straightforward approach to computing the dynamics of the truss would be to compute an overall matrix relating the generalized displacements and forces at the left and right external boundaries of the 18-bay truss,

$$\begin{Bmatrix} u_L^{(19)} \\ f_L^{(19)} \end{Bmatrix} = \begin{bmatrix} T_{11} & T_{12} \\ T_{21} & T_{22} \end{bmatrix}^{18} \begin{Bmatrix} u_L^{(1)} \\ f_L^{(1)} \end{Bmatrix} \tag{3.27}$$

and then impose the specific boundary conditions of the problem. Unfortunately, raising the transfer matrix to a power leads to numerical instabilities [48, 43]. These are caused by exponential dichotomy, e.g., the co-existence of eigenvalues with very large and very small magnitudes, corresponding to left- and right-going attenuating (near-field) non-propagating or complex wave-modes [154]. Note that due to the six coupling coordinates between each adjacent bays there exist six independent co-existing wavemodes in the truss under consideration [64]. Hence, it is not possible to achieve numerically stable computations over wide frequency ranges in order to obtain the frequency response functions (FRFs) of the truss. It follows, that using (3.27) prevents the accurate computation of the transient dynamics as well, as this involves the numerical inversion of the corresponding FRFs.

To overcome this limitation an alternative approach is adopted, based on the concept of DGM developed by Schmidt and Jensen [204]. To demonstrate the method one needs to impose specific boundary conditions. To this end, it is assumed that the truss is free (unsupported) in space; that a prescribed (external) generalized excitation vector $f_L^{(1)}$ is applied on the left external boundary of the truss; and that no external generalized forces exist on the right external boundary of the truss $f_L^{(19)} = 0$ (e.g., both left and right external boundaries of the truss are unsupported). At this point a square DGM of large dimension is constructed that relates all unknown generalized displacements and forces at all joints to the external generalized force,

$$Ds = r \Rightarrow s = D^{-1}r, \tag{3.28}$$

where:

$$s = \begin{bmatrix} u_L^{(1)T} & u_L^{(2)T} & f_L^{(2)T} & u_L^{(3)T} & f_L^{(3)T} & \ldots & u_L^{(18)T} & f_L^{(18)T} & u_L^{(19)T} \end{bmatrix}^T,$$

$$s = \begin{bmatrix} f_L^{(1)T} & f_L^{(1)T} & 0\ 0\ 0 \ldots 0\ 0\ 0 \end{bmatrix}^{\leq T}$$

and

204   3 Order Reduction by Proper Orthogonal Decomposition (POD) Analysis

$$D = \begin{bmatrix} -T_{12}^{-1}T_{11} & T_{12}^{-1} & [0] & [0] & [0] & \cdots & [0] & [0] & [0] \\ -T_{22}^{-1}T_{21} & [0] & T_{22}^{-1} & [0] & [0] & \cdots & [0] & [0] & [0] \\ [0] & -T_{11} & -T_{12} & [I] & [0] & \cdots & [0] & [0] & [0] \\ [0] & -T_{21} & -T_{22} & [0] & [I] & \cdots & [0] & [0] & [0] \\ \cdot & \cdot & \cdot & \cdot & \cdot & & \cdot & \cdot & \cdot \\ \cdot & \cdot & \cdot & \cdot & \cdot & & \cdot & \cdot & \cdot \\ \cdot & \cdot & \cdot & \cdot & \cdot & & \cdot & \cdot & \cdot \\ [0] & [0] & [0] & [0] & [0] & \cdots & -T_{11} & -T_{12} & [I] \\ [0] & [0] & [0] & [0] & [0] & \cdots & -T_{21} & -T_{22} & [0] \end{bmatrix}$$

In the above expressions, $[0]$ and $[I]$ denote the (6×6) zero and unit matrices, respectively. In essence, the need to perform the matrix multiplication (3.27) is eliminated by expanding the size of the frequency response matrix, and by performing a single numerical inversion of a square matrix of large dimension. Fortunately, for fixed frequency this inversion can be performed free of numerical instabilities, with the result that the FRFs of the truss dynamics can be computed for varying frequency in a numerically stable operation. Once the FRFs are evaluated, the corresponding transient dynamics are computed by numerical inverse fast Fourier transform (FFT) operations.

The elastic and geometric parameters of the 18-bay truss were assigned the following values, $EA = 2.216 \times 10^6 Pa\,m^2$, $EI = 5.587 Pa\,m^4$, $m = 0.0855 kg/m$, $l = 0.903 m$, where EI is the rigidity, A the cross section, m the mass per unit length, and l the length of a longeron beam. In addition, the non dimensional frequency $\beta^2 = \omega(EI/ml^4)^{-1/2} = 0.101\omega$ is defined, where ω is the frequency in rad/s.

The computation of the transient response at a selected set of points of the truss was carried out in two steps. First, the receptance FRFs (generalized displacements over applied force) were numerically computed over a specified frequency range by performing the numerical inversion (3.28) at a discrete set of frequencies. The numerical inversion was performed using the IMSL routine DLINCG, based on lower-upper factorization. The inversion was facilitated by the fact that D is a zone matrix with the majority of terms far from a diagonal zone being zero. This was the most error-prone stage of the computation, and in order to assume numerically accurate results the following actions were undertaken: (a) A sufficiently small frequency step was selected, assuring that no truss resonances were missed due to coarse frequency sampling; (b) A small amount of artificial damping was added by transforming the modulus of elasticity to a weakly complex quantity, $E \to E(1 + 10^{-6}j)$, $j = (-1)^{1/2}$, thus ensuring moderate values of the resonance FRFs near resonance points. In the second step of the computation, the receptance FRFs were converted to inertance FRFs (accelerations over applied force), which were then multipled by the FFT of the applied external force. The resulting numerical functions were inverse-Fourier transformed to render the transient acceleration time series at the prescribed set of points of the truss. The reason for computing transient accelerations instead of displacements (or velocities)

3.3 Coherent Spatial Structures in Extended Systems

lies in the numerical difficulties associated with FFT inversions of functions with poles at $\omega = 0$, as is the case with receptance (and mobility) FRFs of the freely supported truss. The FFT and inverse-FFT operations were numerically carried out using MATLAB.

In Figure (3.15) representative numerical receptance FRFs of the truss are depicted, for vertical external forcing acting on the upper joint of the left boundary. In Figure 3.16 the same FRFs are shown but for horizontal external forcing at the same joint. Note that resonances in this system occur in dense clusters which correspond to propagation zones (PZs) of the various families of wavemodes of the corresponding truss of infinite spatial extent, e.g., with an infinite number of bays [154]. In previous works [63, 64] it was shown that the infinite truss possesses six distinct families of wavemodes, with each existing independently of the others, and possessing PZs that overlap in the frequency domain. Depending on the position and direction of the external forcing, truss resonances may correspond to predominantly shear, longitudinal or bending motions, depending on the specific family of wavemode PZ in which the specific cluster of resonances lies. In addition, mixed-mode resonances may occur, with no obvious truss type of motion, whenever two or more resonance clusters corresponding to different families of wave modes mix. In spite of this complicated resonance structure, one notes that the numerical method based on the numerical inversion (3.28) gives numerically stable results.

In Figure 3.17 some representative acceleration time series at various points of the truss are depicted, corresponding to the following trapezoidal vertical impulsive excitation acting vertically on the upper joint of the left boundary:

$$F(t) = \begin{cases} 10t, & 0 \leq t < 0.1s \\ 1.0, & 0.1s \leq t < 0.5s \\ 6 - 10t, & 0.5s \leq t \leq 0.6s \\ 0, & t > 0.6s \end{cases} . \tag{3.29}$$

Note the order of magnitude difference between the accelerations in the vertical and horizontal directions. Also note that the horizontal accelerations nearly vanish after some initial transients, whereas the vertical accelerations are slower in their decay. In Figure 3.18 the acceleration time series at the same positions but for a horizontal trapezoidal force (3.29) are shown. Similar results can be derived for displacement and velocity time series at the joints of the truss. The transient responses at beam locations inside a bay can be also evaluated by constructing appropriate transfer functions and inverting them numerically.

The preceding numerical results indicate the presence of dominant vibration modes in the transient dynamics of the impulsively loaded truss. Depending on the specific direction of the applied force, these dominant vibrations are predominantly horizontal or vertical motions of the joints of the truss. As shown in the next Section, processing the time series by POD reveals dominant coherent spatial structures of the dynamics. The interesting finding

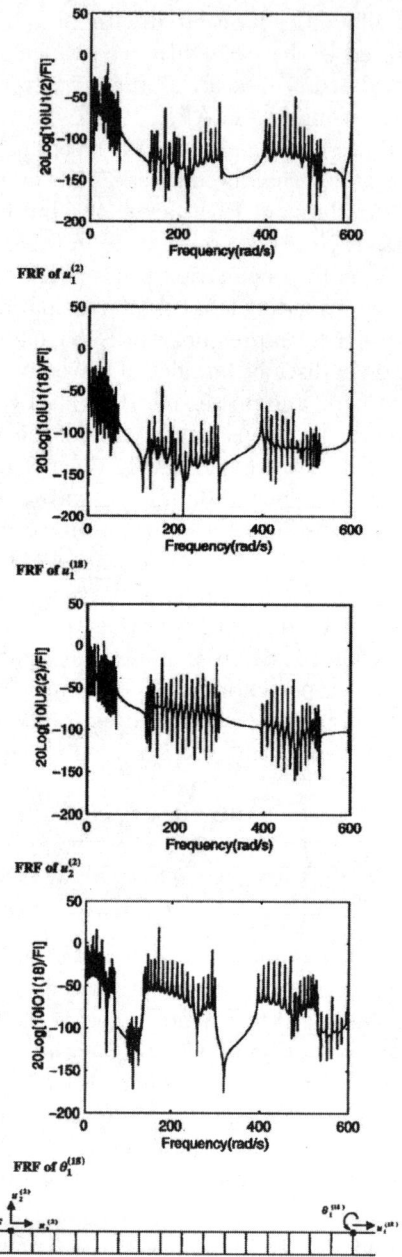

Figure 3.15. Numerical receptance FRFs of the truss for a vertical external force acting on the upper joint of the left boundary.

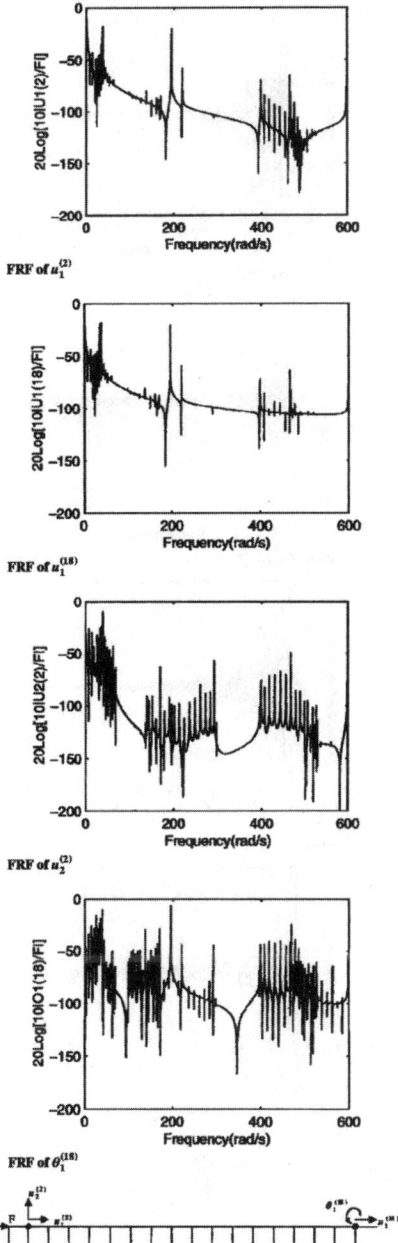

Figure 3.16. Numerical receptance FRFs of the truss for a horizontal externally force acting on the upper joint of the left boundary.

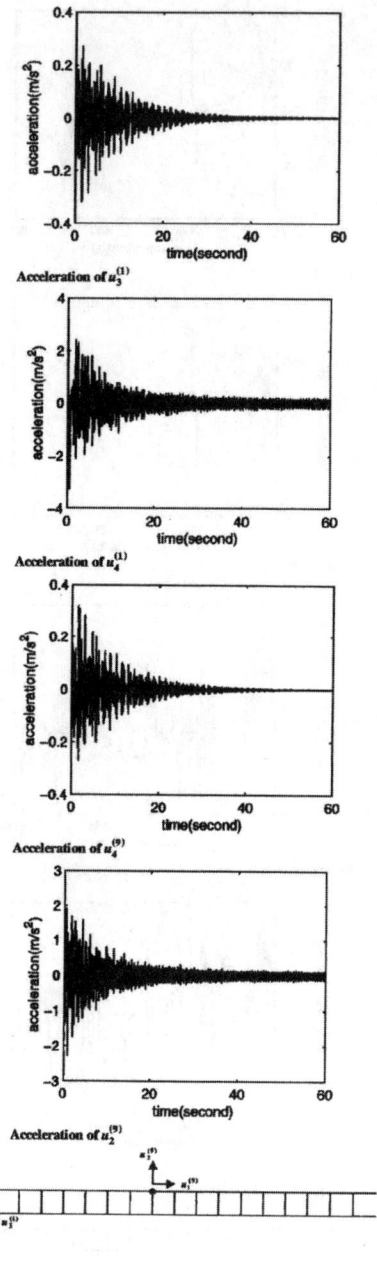

Figure 3.17. Numerical acceleration time series of the truss for a vertical external force acting on the upper joint of the left boundary.

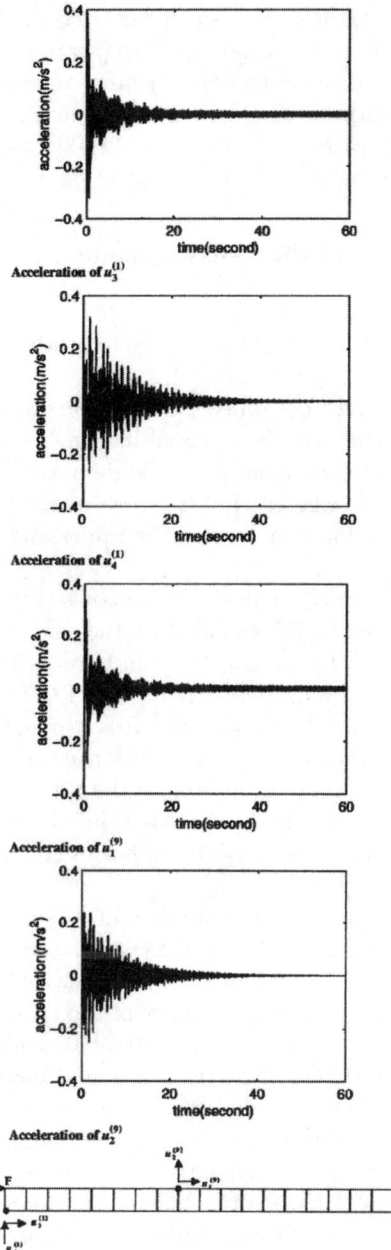

Figure 3.18. Numerical acceleration time series of the truss for a horizontal trapezoidal impulsive force acting on the upper joint of the left boundary.

is that, although the 18-bay truss is a structure described by a large number of coupled partial differential equations (two per beam), only a few dominant POD modes are needed to capture nearly all of the energy of longitudinal or transverse truss vibrations. This indicates that a POD-based reduced-order model can accurately capture the motions of the truss; this will be confirmed by the numerical computations that follow.

3.3.2 POD Analysis of the Truss Dynamics

POD of the truss dynamics was performed for three distinct loading conditions (LCs) labeled I, II and III, in order to study the dominant coherent structures of the truss under different forcing conditions. In all cases impulsive trapezoidal forces described by equation (3.29) were utilized. LC I corresponds to a single vertical force acting on the upper joint of the left truss boundary. LC II corresponds to a single horizontal force acting on the same joint of the truss, whereas LC III corresponds to two identical and simultaneous horizontal forces acting on the upper and lower joints of the left truss boundary.

The results of the POD analysis of the truss transient dynamics for the three LCs are tabulated in Tables 3.8–3.10; the corresponding first few dominant POD modes for each case are shown in Figures 3.19–3.21. The following remarks are made regarding these results. For LC I the truss executes predominantly dending vibrations, and the first three POD modes capture a significant portion of the energy of the truss time series, with the first POD mode capturing a little more than 50% of the total energy. The POD mode shapes involve predominantly vertical motions of the joints; the relatively large displacements of the joints of the right boundary are due to the lack of a vertical longeron there.

When horizontal forces are considered (LCs II and III), there is a qualitative change of the POD results. In these cases the truss vibrates predominantly in the longitudinal direction, and for each case the leading POD modes capture an essential amount of the energy of the truss and dominate over all other higher order modes. This is in contrast to what was observed in LC I, where a stronger partition of energy was noted between the leading set of POD modes. Hence, it appears that for the case of horizontal forcing most of the energy of the truss is channeled to a single dominant coherent structure (POD mode shape), that dominates the truss response. This feature is even more noticeable for LC III when two horizontal forces are symmetrically applied on the left boundary of the truss; in this case the dominant POD mode captures 96.5% of the total energy.

For LC II the forcing is applied non symmetrically, and the dominant POD mode corresponds to shear-like deformation of the truss, as one would expect from physical intuition. For LC III where symmetric forcing is applied, the deformation of the truss resembles one of a pressure wave propagating

3.3 Coherent Spatial Structures in Extended Systems

longitudinally along the truss. An interesting observation is that for LC III the fourth POD mode shape resembles a rigid-body deformation of the truss.

Table 3.8. POD Modes for LC I.

POD Mode No.	Energy captured by mode (%)	Cumulative Total Energy captured (%)
1	52.14	52.14
2	21.77	73.91
3	7.98	81.89
4	3.41	85.30
5	2.66	87.97
6	2.18	90.15
7	1.84	91.99
8	1.71	93.70
9	1.10	94.80
10	1.08	95.88

It is emphasized that the POD modes discussed herein are mathematical modes, describing the spatial coherent structures developing in the truss as energy gets partitioned among the periodic sets of the system for each specific forcing condition. As such, these modes have no resemblance to the classical vibration modes of the truss that one obtains by solving the relevant eigenvalue problem of the dynamics. However, in similarity to vibration mode shapes, the POD mode shapes form a complete orthogonal basis (if all of then are retained), and, hence, they can be used to project the dynamics into low-dimensional spaces, thus creating low-order models of the truss dynamics. By measuring the energy captured by each mode one can estimate the number of POD modes required for the creation of accurate reduced-order models, and, hence, the dimensionality of the truss dynamics. The fact that only a few leading POD modes capture nearly all of the energy of the time series, indicates that the dynamics can be described by reduced-order models with few degrees-of-freedom, a result that perhaps is unexpected, given that the 18-bay truss under consideration possesses clusters of densely packed resonances, and thus high modal density. This conjecture will be tested and confirmed in the next Section.

Modal analysis of extended lightweight continuous systems similar to the one discussed here is a challenging task due to the high modal densities involved. This feature prevents the application of traditional modal analysis methods due to the presence of numerous interacting vibration modes. By not dealing with physical vibration modes, the POD analysis encounters no such limitations and can be effectively used to perform system identification of this type of systems. Moreover, as mentioned previously for linear systems the computed POD modes are optimal in the sense that they capture more energy of the time series per mode than any other basis of orthogonal modes

Table 3.9. POD Modes for LC II.

POD Mode No.	Energy captured by mode (%)	Cumulative Total Energy captured (%)
1	90.77	90.77
2	2.11	92.88
3	1.79	94.67
4	0.91	95.59
5	0.87	96.46
6	0.87	97.34
7	0.69	98.03
8	0.55	98.59
9	0.39	98.99
10	0.28	99.27

(including the vibration modes); this optimality feature explains the low-order of the reduced models derived by projecting the truss dynamics into POD bases.

Table 3.10. POD Modes for LC III.

POD Mode No.	Energy captured by mode (%)	Cumulative Total Energy captured (%)
1	96.53	96.53
2	1.46	97.99
3	1.21	99.21
4	0.40	99.61
5	0.27	99.89
6	0.06	99.96

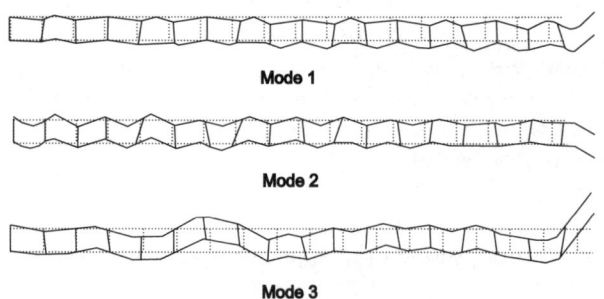

Figure 3.19. Dominant POD modes for LC I.

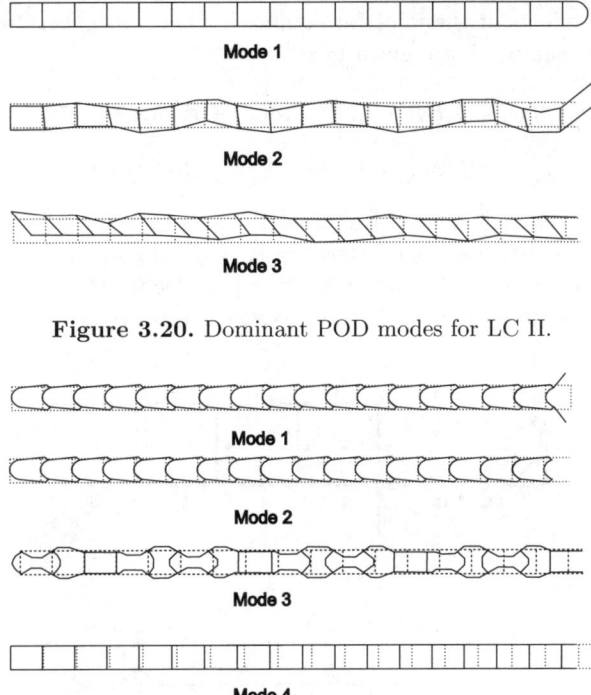

Figure 3.20. Dominant POD modes for LC II.

Figure 3.21. Dominant POD modes for LC III.

3.3.3 POD Based Reduced-Order Models

To demonstrate the capacity of using the POD modes for effectively reducing the order of the dynamics of a continuous system, a four-bay truss with free boundary conditions is now considered, as depicted in Figure 3.22. As in the truss considered in the previous Section, each bay is composed of four elastic members that are connected to adjacent members by means of clamped joints. The joints sustain forces in the $x-$ and $y-$directions, as well as bending moments. The truss is assumed to be linear, so that the bending and axial vibrations of each member are governed by the classical uncoupled second- and fourth-order partial differential equations, respectively, of linear vibration theory. Coupling between axial and bending motions takes place only through the boundary conditions at the joints, and only in-plane vibrations of the truss are considered.

Referring to the schematic of Figure 3.22, a local coordinate system $x \in [0, L_x]$ is introduced to parametrize the underformed configuration of the horizontal member 1 along the $x-$deirection. In addition, the scalar fields $u_1(x,t)$ and $v_1(x,t)$ are used to denote the axial and transverse displacements, respectively, at point x of this member. Structural damping of the member and damping effects at the joints are ignored. Then, the governing equations of motion for member 1 are given by:

$$m\,u_{1,tt} - EA\,u_{1,xx} = A_1^1 \delta(x) + A_2^1 \delta(x - L_x),$$

$$m\,v_{1,tt} + EI\,v_{1,xxxx} = T_1^1 \delta(x) + T_2^1 \delta(x - L_x) + R_1^1 \delta'(x) + R_2^1 \delta'(x - L_x). \quad (3.30)$$

The quantities A_1^j, A_2^j denote the axial forces acting on the left and right boundary of the j-th member, respectively, with similar definitions holding for the transverse forces T_i^j and the bending moments R_i^j; $\delta(\cdot)$ and $\delta'(\cdot)$ are Dirac's function and its generalized derivative, respectively.

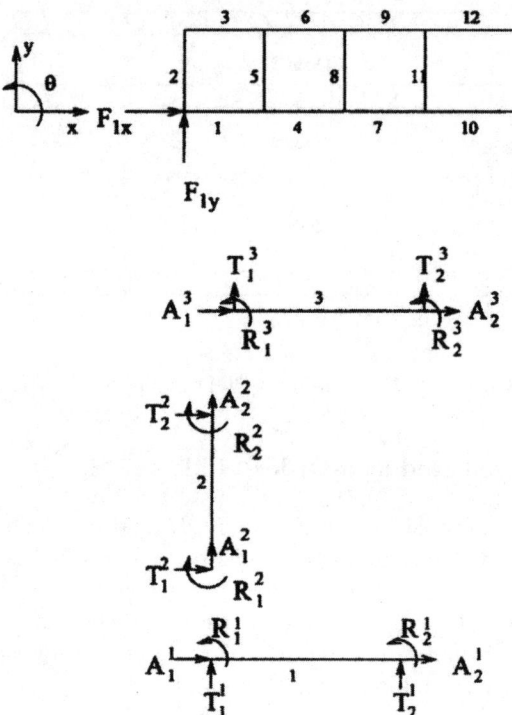

Figure 3.22. The four-bay truss, with notation for internal forces and moments at the joints.

Similarly, the equations of motion that govern the axial and bending vibrations of member 2 with underformed configuration in the y-direction are given by,

$$m\,v_{2,tt} - EA\,v_{2,yy} = A_1^2 \delta(y) + A_2^2 \delta(y - L_y),$$

$$m\,u_{2,tt} + EI\,u_{2,yyyy} = T_1^2 \delta(y) + T_2^2 \delta(y - L_y) - R_1^2 \delta'(y) - R_2^2 \delta'(y - L_y), \quad (3.31)$$

where the local coordinate system $y \in [0, L_y]$ is used for that member. Similar equations can be written for all twelve members of the truss, each time defining a local coordinate x (for members with undeformed horizontal orientations) or y (for members with undeformed vertical configurations) to parametrize the undeformed configuration of the specific member considered. Hence, one obtains a total of 24 partial differential equations of motion governing the dynamics, that are coupled through the boundary conditions (e.g., through internal forces and moments at the joints).

Suppose that the POD modes of the four-bay truss have already been computed. By construction these modes form an orthogonal basis. In addition, as the number of sensing (measurement) points tends to infinity and covers pointwise the entire truss, the POD basis becomes identical to that form by the vibration modes of the truss [102]. A discretization of the 24 governing partial differential equations of motion is now performed by projecting the dynamics into a finite-dimensional basis parametrized by the leading p POD modes. To this end, the transverse and axial displacements of the k-th elastic member are expressed as (assuming that this member has horizontal undeformed orientation):

$$u_k(x,t) = \sum_{i=1}^{p} \phi_{ki}(x) a_i(t), \quad v_k(x,t) = \sum_{i=1}^{p} \psi_{ki}(x) a_i(t), \quad k = 1, ..., 12, \quad (3.32)$$

where: $a_i(t)$ is the modal amplitude of the i-th POD mode, and $\phi_{ki}(x)$ and $\psi_{ki}(x)$ are the horizontal and vertical displacement components of the i-th POD mode for structural member k. The quantities $a_i(t)$ represent global POD modal amplitudes, that for fixed i (mode) are common for all structural members; this definition facilitates significantly the discretization since only p such modal amplitudes need to be defined, instead of defining a separate modal amplitude for each of the twelve structural members.

Substituting the preceding relations into the governing equations of motion of the k-th structural member (similar in form to expressions (3.30) and (3.31)), multiplying each of the resulting equations by either $\phi_{ki}(x)$ or $\psi_{ki}(x)$, integrating from 0 to L_x with respect to x, and making use of the orthogonality properties of the POD modes, one derives a set of discretized ordinary differential equations for structural member k. Following this procedure one derives the following discretized set for the structural member 1:

$$\ddot{a}_j \int_0^{L_x} \phi_{1j}^2(x) dx - \frac{EA}{m} \sum_{i=1}^{p} a_i(t) \int_0^{L_x} \phi_{1i}''(x) \phi_{1j}(x) \, dx = \frac{1}{m} \left[A_1^1 \phi_{1j}(0) + A_2^1 \phi_{1j}(L_x) \right],$$

$$\ddot{a}_j \int_0^{L_x} \psi_{1j}^2(x) dx + \frac{EI}{m} \sum_{i=1}^{p} a_i(t) \int_0^{L_x} \psi_{1i}''''(x) \psi_{1j}(x) \, dx =$$

$$\frac{1}{m}\left[T_1^1\psi_{1j}(0) + T_2^1\psi_{1j}(L_x) + R_1^1\psi'_{1j}(0) + R_2^1\psi'_{1j}(L_x)\right], \quad j=1,...,p. \quad (3.33)$$

Similarly, for structural member 2,

$$\ddot{a}_j\int_0^{L_x}\psi_{2j}^2(y)dy - \frac{EA}{m}\sum_{i=1}^p a_i(t)\int_0^{L_y}\psi''_{2i}(y)\psi_{2j}(y)\,dy = \frac{1}{m}\left[A_1^2\psi_{2j}(0) + A_2^2\psi_{2j}(L_y)\right],$$

$$\ddot{a}_j\int_0^{L_y}\phi_{2j}^2(y)dy + \frac{EI}{m}\sum_{i=1}^p a_i(t)\int_0^{L_y}\phi''''_{2i}(y)\phi_{2j}(y)\,dy =$$

$$\frac{1}{m}\left[T_1^2\phi_{2j}(0) + T_2^2\phi_{2j}(L_y) - R_1^2\phi'_{2j}(0) + R_2^2\phi'_{2j}(L_y)\right], \quad j=1,...,p. \quad (3.34)$$

Applying discretization to all twelve structural members of the truss, one obtains a total of twenty four ordinary differential equations governing the transverse and axial motions of the members. Because the only external forces are F_{1x} and F_{1y} (cf. Figure 3.22), by applying force equilibrium in the x- and y-directions at the joint that coupled the elastic members 1 and 2, one obtains,

$$A_1^1 + T_1^2 = F_{1x}, \quad A_1^2 + T_1^1 = F_{1y}. \quad (3.35)$$

All other internal forces and moments acting at the other joints of the truss sum to zero, since no external forces or moments act on these joints. Making use of these results, one proceeds to appropriate summations of the ordinary differential equations (3.33), (3.34),..., in order to eliminate all internal forces and moments from the discretized equations. As a result, one obtains the following set of p coupled ordinary differential equations that govern the p POD modal amplitudes:

$$\ddot{a}_j(t)\left[\sum_{k=1}^{12}\left(\int_0^L\phi_{kj}^2(z)\,dz + \int_0^L\psi_{kj}^2(z)\,dz\right)\right] -$$

$$\frac{EA}{m}\sum_{i=1}^p a_i(t)\sum_{k=0}^3\left[\int_0^L\phi''_{(3k+1)i}(z)\,\phi_{(3k+1)j}(z)\,dz + \right.$$

$$\left.\int_0^L\psi''_{(3k+2)i}(z)\,\psi_{(3k+2)j}(z)\,dz + \int_0^L\phi''_{(3k+3)i}(z)\,\phi_{(3k+3)j}(z)\,dz\right] +$$

$$\frac{EI}{m}\sum_{i=1}^p a_i(t)\sum_{k=0}^3\left[\int_0^L\psi''''_{(3k+1)i}(z)\,\psi_{(3k+1)j}(z)\,dz + \right.$$

$$\left. \int_0^L \phi''''_{(3k+2)i}(z)\,\phi_{(3k+2)j}(z)\,dz + \int_0^L \psi''''_{(3k+3)i}(z)\,\psi_{(3k+3)j}(z)\,dz \right] =$$

$$\frac{F_{1y}}{m}\psi_{1j}(0) + \frac{F_{1x}}{m}\phi_{1j}(0), \quad j=1,...,p, \qquad (3.36)$$

where depending on the structural member in the summation, $L = L_x$ or L_y, and $z = x$ or y. The system above represents the POD-based reduced-order model of the truss. Note that since the POD base is not identical to the basis formed by the vibration modes of the truss, the reduced-order modes (3.36) is coupled. Moreover, since the POD modes are excitation-dependent, the derived reduced-order model is only optimal (in the sense that the basis used captures more energy per mode) for the specific excitation that is considered for the POD computation. The same orthogonal POD basis can be used to discretize the truss dynamics for a different type of forcing, however, in that case the basis will not be optimal (in an energy-capture sense).

To simplify the analysis, one introduces the following notation,

$$c_j = \sum_{k=1}^{12}\left(\int_0^L \phi^2_{kj}(z)\,dz + \int_0^L \psi^2_{kj}(z)\,dz\right),$$

$$e_{ij} = \sum_{k=0}^{3}\left[\int_0^L \phi''_{(3k+1)i}(z)\,\phi_{(3k+1)j}(z)\,dz + \right.$$

$$\left. \int_0^L \psi''_{(3k+2)i}(z)\,\psi_{(3k+2)j}(z)\,dz + \int_0^L \phi''_{(3k+3)i}(z)\,\phi_{(3k+3)j}(z)\,dz\right],$$

$$g_{ij} = \sum_{k=0}^{3}\left[\int_0^L \psi''''_{(3k+1)i}(z)\,\psi_{(3k+1)j}(z)\,dz + \right.$$

$$\left. \int_0^L \phi''''_{(3k+2)i}(z)\,\phi_{(3k+2)j}(z)\,dz + \int_0^L \psi''''_{(3k+3)i}(z)\,\psi_{(3k+3)j}(z)\,dz\right],$$

in terms of which the POD-based reduced-order model assumes the form,

$$\ddot{a}_j(t) + \sum_{i=1}^{p}\left(-\frac{EA}{mc_j}e_{ij} + \frac{EI}{mc_j}g_{ij}\right)a_i(t) = \frac{1}{mc_j}[F_{1y}\psi_{1j}(0) + F_{1x}\phi_{1j}(0)],$$

$$j = 1,...,p, \qquad (3.37)$$

or

$$\{\ddot{a}(t))\} + [B]\{a(t))\} = \{F(t)\} \qquad (3.38)$$

218   3 Order Reduction by Proper Orthogonal Decomposition (POD) Analysis

where braces represent $(p \times 1)$ vectors, and brackets $(p \times p)$ matrices. For the numerical simulations a damping term $[D]\{\dot{a}(t)\}$ was added to simulate energy dissipation due to structural damping in the elastic members or at the joints. The damping matrix was estimated as $[D] = [T][D_d][T]^{-1}$; $[T]$ is the $(p \times p)$ matrix of eigenvectors (modal matrix) of $[B]$, and the $(p \times p)$ diagonal matrix $[D_d]$ has as elements the scalars $(D_d)_{ii} = E\eta/w_i$, $i = 1,\ldots,p$, where w_i is the i-th eigenvalue of $[B]$, E the (common) modulus of elasticity of the elastic members, and η a small damping coefficient.

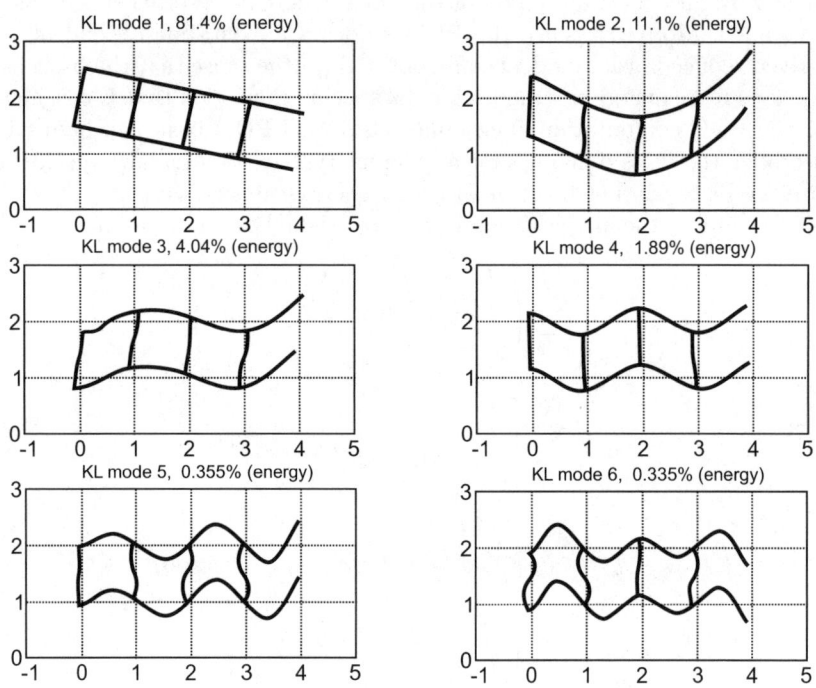

Figure 3.23. The six leading POD modes of the truss for a vertical impulsive excitation.

Numerical integration of (3.37), (3.38) evaluates the transient modal amplitudes $a_j(t)$ and provides estimates for the axial and transverse displacements of the structural members of the truss (through Equations (3.32)). These estimates can then be compared to direct numerical simulations of the truss dynamics in order to assess the accuracy of the POD-based low-order model. An example of the reduction of the truss dynamics is now provided. The geometrical and material properties of the four bay truss are as follows: $EA = 2.216 \times 10^6 Pa\,m^2$, $EI = 5.87 Pa\,m^4$, $m = 0.0555 kg/m$,

3.3 Coherent Spatial Structures in Extended Systems 219

$L_x = L_y = 0.903m$. It is assumed that all structural members have identical material properties, and no structural disorder exists.

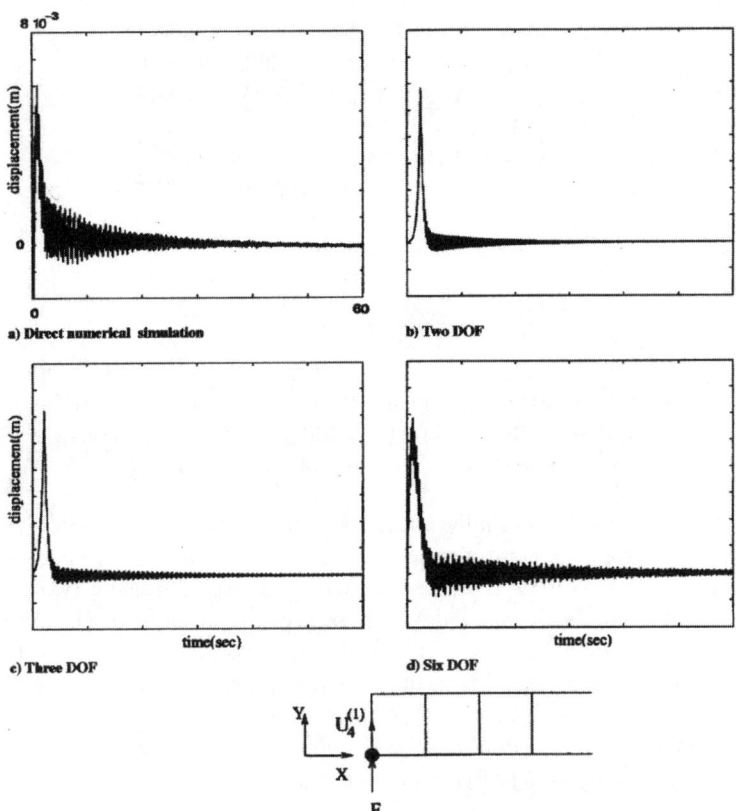

Figure 3.24. POD-based transient response reconstruction for the displacement $u_4^{(1)}$.

First, excitation of the truss by a vertical impulsive force $F_{1y}(t) = \delta(t)$ and $F_{1x}(t) = 0$ is considered (cf. Figure 3.22). In Figure 3.23 the six leading POD modes of the truss dynamics are depicted, together with the percentage of energy captured by each POD mode. Note that the five leading POD modes capture 99% of the energy of the time series of the truss. In additional numerical simulations with a finite-time sinusoidal pulse (not reported here, see [140]), it was shown that fewer POD modes are needed to capture the aforementioned percentage of energy. This indicates that when impulsive excitation is applied the energy of the transient dynamics is distributed over POD modes of higher order.

220 3 Order Reduction by Proper Orthogonal Decomposition (POD) Analysis

The POD-based reduced-order models of the truss dynamics are now discussed in more detail. For the six degrees-of-freedom (DOF) low-order model the matrix $[B]$ is computed as follows,

$$[B] = \begin{bmatrix} -2 & -30.6 & 26 & 102 & -4415.3 & -4544.2 \\ 17.1 & 440.1 & -125.3 & 232 & -393 & 3492.1 \\ -4 & -53.1 & 1211.7 & -442.4 & -3711 & -322.5 \\ 1.4 & -23 & 35.1 & 2159.2 & 575.1 & -754 \\ 16 & -16.2 & -131 & 44.2 & 20914.5 & 2502.3 \\ -2 & 32.1 & 32.5 & 233.8 & 1961.9 & 14073.5 \end{bmatrix}, \quad (3.39)$$

whereas the damping matrix $[D]$ is given by:

$$[D] = \begin{bmatrix} 1.344 & 0.067 & -0.001 & -0.111 & 0.252 & 0.363 \\ -0.054 & 0.069 & 0.0054 & 0.0014 & -0.0066 & -0.03 \\ -0.0016 & 0.0022 & 0.044 & 0.005 & 0.0059 & -0.001 \\ -0.001 & 0.0004 & -0.0004 & 0.032 & -0.001 & 0.001 \\ -0.0011 & 0.00002 & 0.00024 & 0.00011 & 0.01 & -0.0012 \\ 0.00049 & -0.00014 & -0.00011 & -0.00043 & -0.00058 & 0.0129 \end{bmatrix}. \quad (3.40)$$

Reduced-order models of smaller dimension $m < 6$ can be obtained by taking the $(m \times m)$ upper-left submatrices of the above matrices. Reconstructed transient responses of the truss are shown in Figure 3.24 using two-, three-, and six-DOF reduced-order models. These are compared to direct numerical simulations. As more DOF are added to the reduced models the reconstructed solutions converge to the direct numerical simulations. It is interesting to note that in this case, *a six DOF reduced model approximates the transient response of a set of 24 coupled partial differential equations of motion* (the original continuous model of the four bay truss).

In Figure 3.25 the receptance FRF $[\bar{u}_1(0,\omega)/\bar{F}_{1y}(\omega)]$ (where overbar denotes Fourier transform and ω the frequency variable) of the truss is considered. The FRF is approximated by means of the reduced-order model (3.37) and compared to the numerically simulated FRF. Two-, three- and six-DOF reduced-order models are considered for reconstructing the FRF.

As a general conclusion, the resonances of the reduced-order FRFs match the leading resonances of the numerical simulation. As more DOF are added to the reduced-order models, more resonances are captured; of particular interest of the plot of Figure 3.25e is that the reconstructed FRF appears to discriminate against the small second peak (mode) of the numerical simulation at approximately $4.5Hz$, and instead matches the higher frequency stronger resonances of the FRF. That the resonances of the reduced-order models and the numerical simulation match is due to the fact that the POD modes can be construed as trial functions for the Rayleigh quotient of the truss; hence, the extracted POD modes can be used to model the dynamics of the truss in the frequency domain, at least for the leading low-frequency

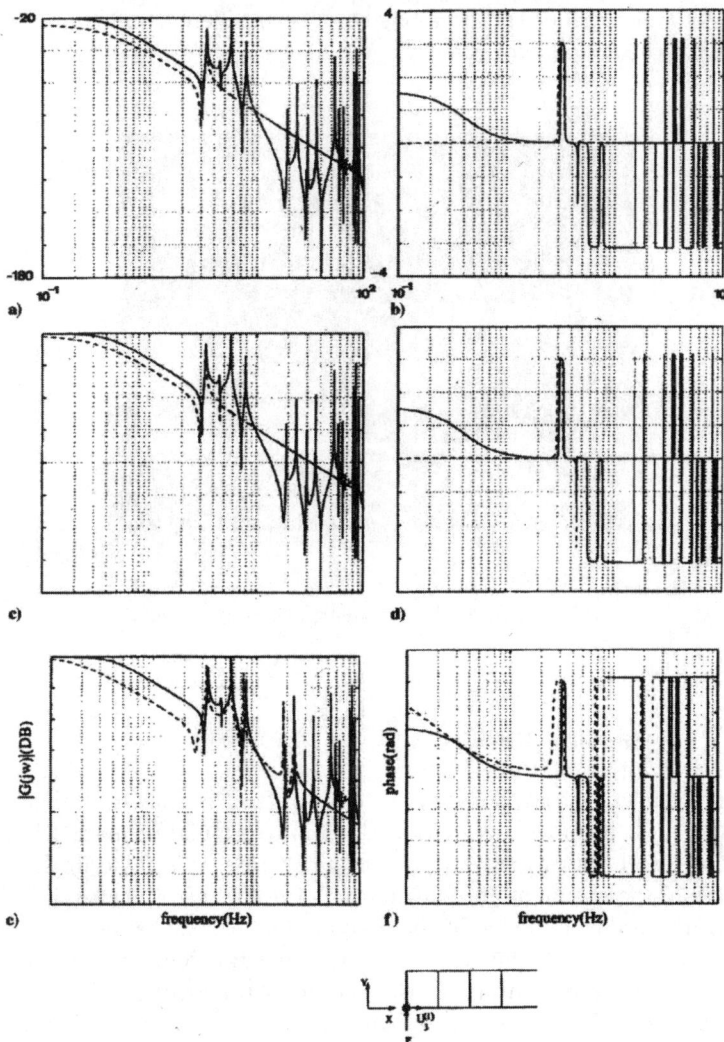

Figure 3.25. Reconstructions of the receptance FRF $[\bar{u}_1(0,\omega)/\bar{F}_{1y}(\omega)]$ using POD-based reduced-order models: (a) modulus and (b) phase of two-DOF model; (c) modulus and (d) phase of three-DOF model; (e) modulus and (f) phase of six-DOF model; —— numerical simulation, - - - - - reduced-order model.

modes. In many engineering applications it is the excitation of these low-frequency modes that are of primary interest in the study of their dynamic response.

Up to now only numerical results of application of the POD method were considered. In the next Section an application of POD analysis to the dynamics of an experimental three-bay truss is considered, and comparisons to theoretical predictions are given.

Figure 3.26. Experimental three-bay truss.

3.3.4 Experimental Results

The experimental truss possesses three bays and is shown in Figure 3.26. The structural members screw in the joints to simulate clamped boundary conditions, and in contrast to the theoretical truss considered in the preceding Section, there is an additional end structural member connecting the two joints of the right boundary of the system. The primary aims of the experimental study were, (a) to extract the experimental POD modes of the truss under near-impulsive excitation, (b) to compare the experimental POD mode shapes and energies to the theoretical ones, and (c) to prove the robustness of the POD mode computation and to show that it is feasible to implement a POD-based order reduction of the dynamics of practical large-scale flexible systems.

To extract the experimental POD modes it was necessary to perform simultaneous measurements of times series of the response of all eight joints of the truss. This was achieved by means of eight PCB 356A11 miniature triaxial accelerometers. In the experimental tests the out-of-plane accelerations of the truss were ignored, and only in-plane responses in the x- and y-directions at each joint were measured. In addition, the assumption of negligible joint rotations was made. The 16 signals for the in-plane accelerations at the eight joint locations were fed to a WCA Zonic eight-channel signal analyzer and to three Tektronics four-channel signal analyzers for post processing4. To trigger

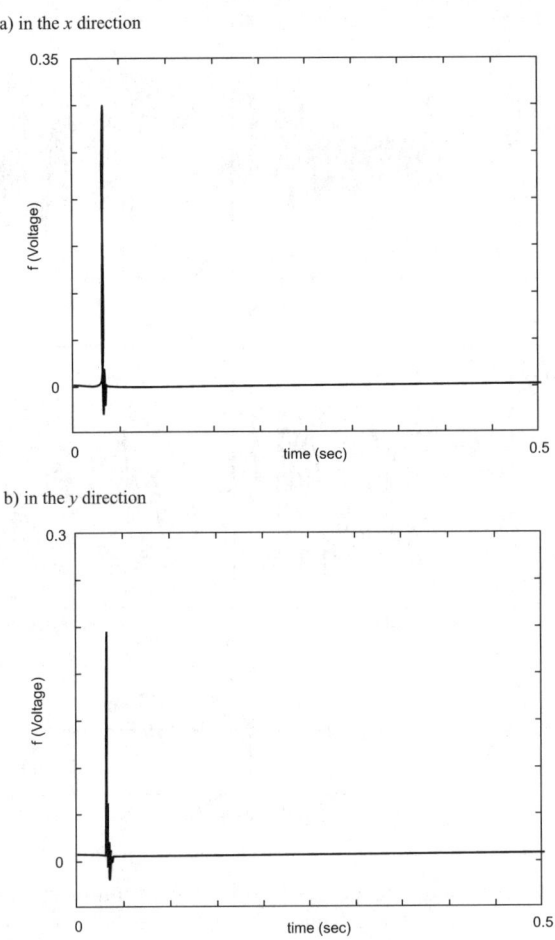

Figure 3.27. Near-impulsive excitations of the truss (sensitivity of the force transducer is $3.14 mV/N$).

and synchronize the data acquisition of the four signal analyzers, one channel from each was fed the signal from the force transducer of modal impact hammer used to excite the truss; this guaranteed the simultaneous data acquisition from all four analyzers. The truss was excited by two different types of near-impulsive excitations in the x- and y-directions; these are depicted in Figure 3.27.

In Figure 3.28 four acceleration frequency spectra of the truss are shown for near-impulsive excitation in the x-direction. For comparison purposes the experimental spectra (dashed lines) are superimposed to theoretically predicted ones (solid lines) that are generated by the DGM approach. The geometric and material properties of the experimental truss that are also used

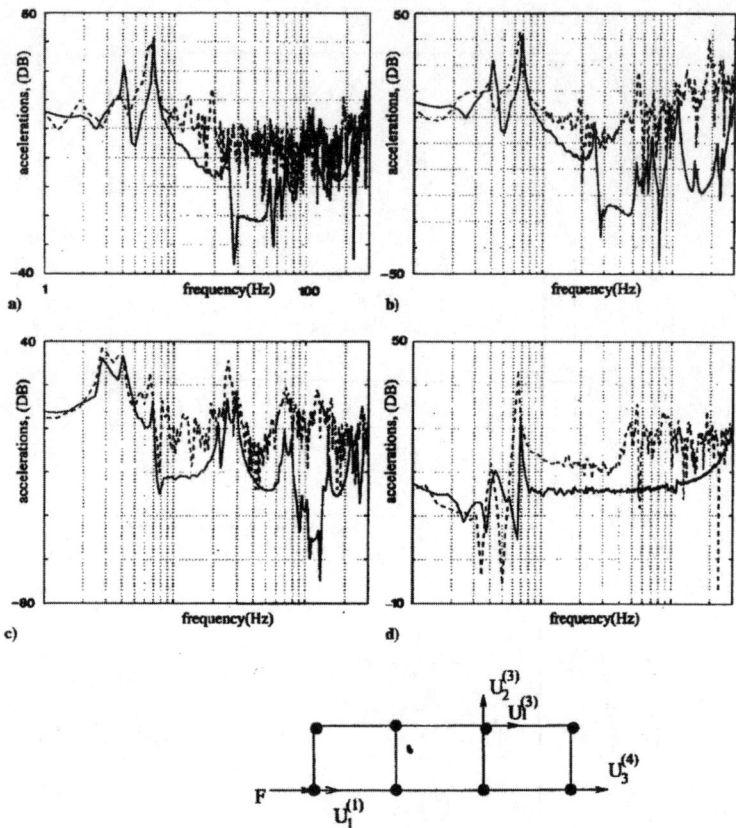

Figure 3.28. Acceleration frequency spectra for near-impulsive forcing in the x-direction: ———— theory, - - - - - experiment.

for the numerical simulations are, $EA = 2.216 \times 10^6 Pa\,m^2$, $EI = 5.587 Pa\,m^4$, $m = 0.0855 kg/m$, $L_x = 0.903m$, $L_y = 0.47065m$ and $M_0 = 0.0992kg$ (the mass of each joint). All structural members of the truss possess identical and uniform material properties and no other structural disorder exists.

The results indicate satisfactory agreement between theory and experiment at low frequencies, and divergence at higher ones. However, there is satisfactory agreement regarding the resonance locations of the theoretical and experimental systems. In Figures 3.29 and 3.30 the leading four theoretical and experimental POD modes are presented. There is satisfactory agreement between the dominant theoretical and experimental POD modes, although there is less energy captured by the experimental mode (35.24%) compared to the theoretical one (90.97%). When higher-order POD modes are considered there is more energy spreading in the experimental modes,

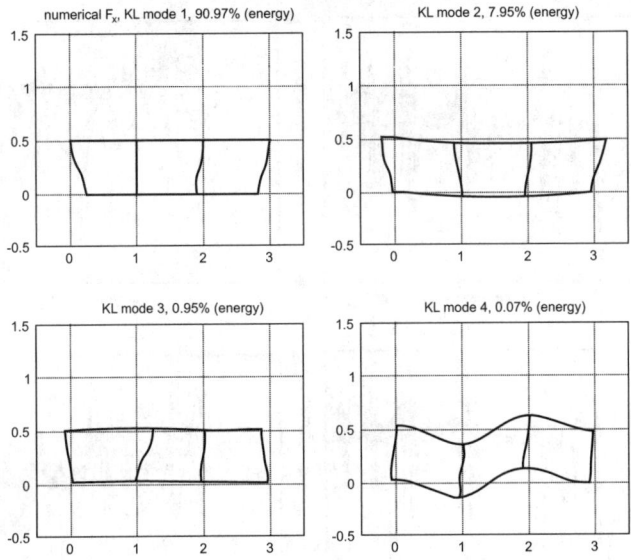

Figure 3.29. Theoretical POD modes for horizontal near-impulse excitation.

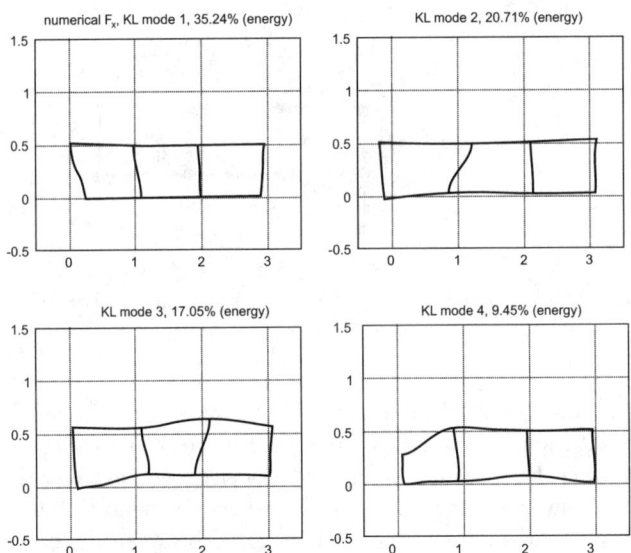

Figure 3.30. Experimental POD modes for horizontal near-impulse excitation.

and not so satisfactory agreement between the theoretical and experimental POD mode shapes.

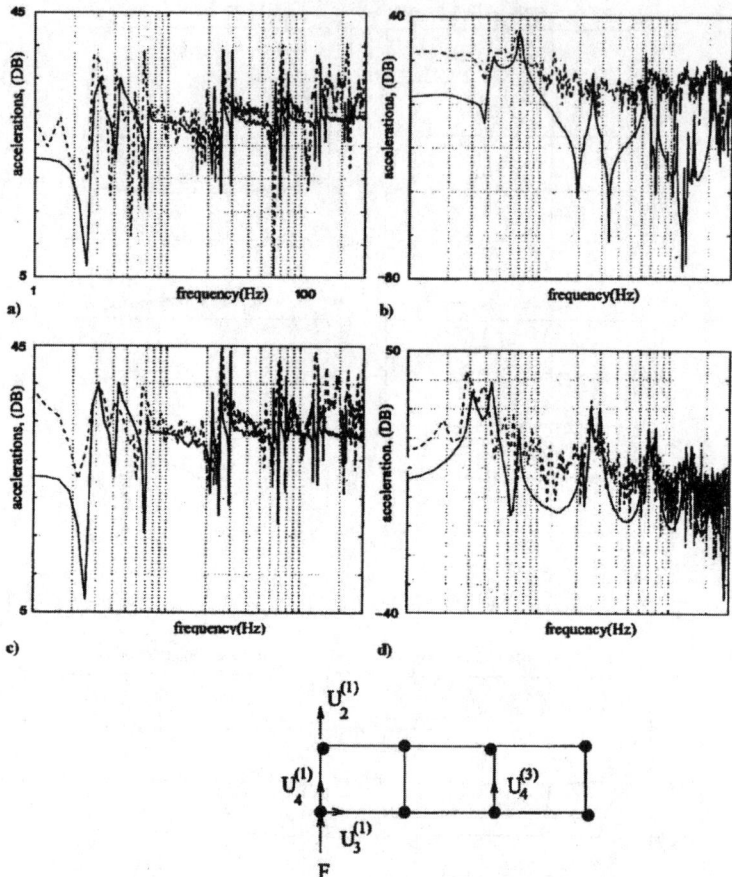

Figure 3.31. Acceleration frequency spectra for near-impulsive forcing in the y-direction: —— theory, - - - - - experiment.

Similar results were obtained when the truss was forced by a near-impulsive excitation in the y-direction. In Figure 3.31 four theoretical and experimental acceleration frequency spectra are compared, and in Figures 3.32 and 3.33 the resulting theoretical and experimental POD modes are depicted. Again, satisfactory agreement is noted between the dominant theoretical and experimental POD mode shapes, though the percentage of energy captured by the theoretical mode is higher than that captured by the experimental one (88.32% against 61.87%). Less agreement, however, is noted for the higher-order POD mode shapes and energies.

The differences observed between the higher-order theoretical and experimental POD modes can be partially attributed to the theoretical assumption

3.3 Coherent Spatial Structures in Extended Systems

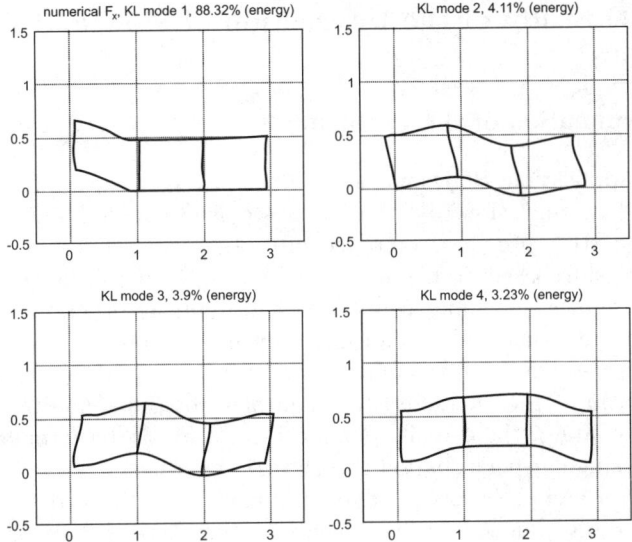

Figure 3.32. Theoretical POD modes for horizontal near-impulse excitation.

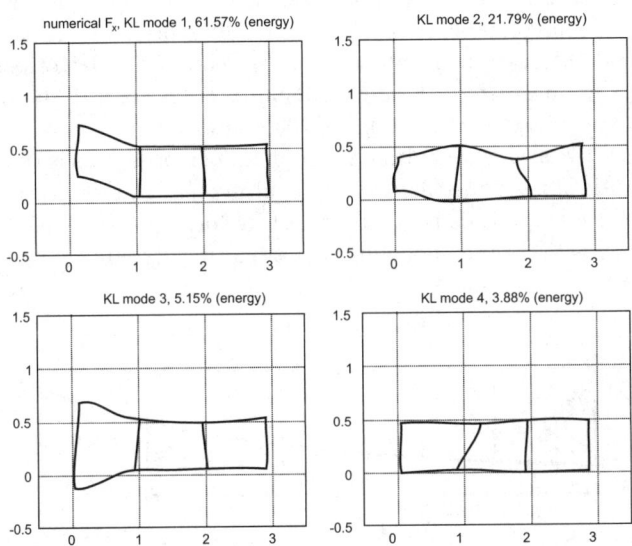

Figure 3.33. Experimental POD modes for horizontal near-impulse excitation.

that the joints undergo negligible rotations; this condition does not hold in the experiment, especially at higher frequencies.

3.4 POD Study of the Interaction of Slow and Fast Dynamics

3.4.1 Formulation of the Problem

In a final application, POD analysis will be applied to the study of the interaction of 'fast' and 'slow' dynamics in systems of coupled oscillators. Many engineering structures are composed of systems of coupled oscillators; examples are ship-based cranes with pendulating loads, airplane wings with attached engines, machines coupled to vibration absorbers, and extended spatial flexible structures with localized linear or nonlinear vibrating attachments. In cases where the coupling between the coupled oscillators is weak and the ratio of the characteristic frequencies of the connected oscillators is small (or large), the overall dynamical problem can be formulated in the context of singular perturbation theory [107, 77, 79, 81].

In such systems the overall dynamics can be partitioned into fast- and slow-oscillations ; indeed, for weak coupling there are oscillations of the soft and stiff subsystems that neighbor the degenerate motions of the corresponding uncoupled systems, which are slow and fast oscillations, respectively. As coupling increases, the interaction between the slow and fast dynamics increases, producing complex combined stiff-soft motions, and interesting instability and bifurcation phenomena. These complicated motions are associated with an increase of the dimensionality of the overall dynamics, since gradually more degrees of freedom of the coupled subsystems participate in the slow-fast dynamic interaction as coupling increases (this is especially true if one or both of the substructures are continuous). In this Section it will be shown that POD analysis is an effective tool for capturing the slow-fast dynamical interaction, and modelling the resulting dynamical phenomena in such systems.

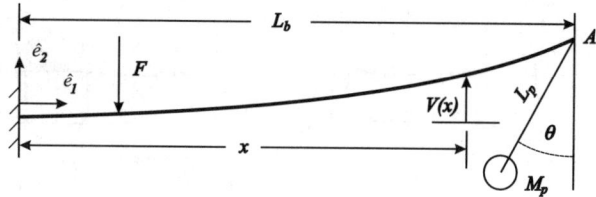

Figure 3.34. The beam-pendulum system under consideration.

The system considered for studying slow-fast dynamical interactions is depicted in Figure 3.34. It consists of a linear continuous beam with a pendulum attached at its right end. The displacement field $V(x,t)$ and the angle $\theta(t)$ are introduced to describe the oscillations of the two subsystems, and

3.4 POD Study of the Interaction of Slow and Fast Dynamics

only in-plane motions are considered. Introducing appropriate rescaling of the dependent and independent variables, and defining suitable non-dimensional parameters [80], the equations of motion of the system are expressed in the following normalized form:

$$\mu^2 \kappa_1^4 W_{\tau\tau}(\xi,\tau) + W_{\xi\xi\xi\xi}(\xi,\tau) + 2\zeta_b \mu W_{\xi\xi\xi\xi\tau} = \mu^2 \kappa_1^4 F(\xi,\tau),$$

$$\ddot{\theta}(\tau) + \left[1 + \mu^2 W_{\tau\tau}(1,\tau)\right] \sin\theta(\tau) + 2\zeta_p \dot{\theta}(\tau) = 0, \qquad (3.41)$$

where: $\xi \in [0,1]$ and τ are the normalized spatial and temporal independent variables; $W(\xi,\tau)$ is the displacement field of the beam from its position of static equilibrium; ζ_b, ζ_p are viscous damping coefficients; $F(\xi,\tau)$ is the normalized distributed external load; $\mu = \omega_p/\omega_1$, where ω_p is the linearized natural frequency of the (uncoupled) pendulum and the fundamental natural frequency of the (uncoupled) beam; and κ_1 is first root of the equation $\kappa_1 + \cos\kappa_1 = -1$. The boundary conditions of the problem are as follows:

$$W(0,\tau) = 0, \quad W_\xi(0,\tau) = 0, \quad W_{\xi\xi}(1,\tau) = 0,$$

$$W_{\xi\xi\xi}(1,\tau) + \beta\mu^2\kappa_1^4 \left[1 - T(\tau)\cos\theta(\tau)\right] = 0, \qquad (3.42)$$

where: $T(\tau) = \dot{\theta}^2(\tau) + [1 + W_{\tau\tau}(1,\tau)]\cos\theta(\tau)$ is the normalized tension at the support of the pendulum, and β represents a mass ratio parameter.

Examining the equations of motion it is realized that there are two sources of coupling between the subsystems: *direct coupling* is provided through a (nonlinear) parametric term in the second equation of motion; whereas *indirect coupling* is provided through the fourth (nonlinear) boundary condition which represents balance of forces at the right boundary of the beam. At this point the additional frequency normalization is introduced by setting ω_p and $\omega_1 = 1/\mu$.

For μ sufficiently small there exists weak coupling, and the frequency ratio parameter μ controls the interaction between the fast and slow dynamics of this system. This is concluded by examining the equations (3.41) and (3.42), and realizing that as $\mu \to 0$ the beam becomes increasingly stiff with respect to the pendulum which becomes increasingly soft; hence, as $\mu \to 0$ the beam generates high-frequency oscillations, whereas the pendulum generates slow-frequency ones. It is interesting to note that at the limit $\mu = 0$ the system becomes degenerate, since it reduces to the uncoupled pendulum,

$$W(\xi,\tau) = 0, \quad \ddot{\theta}(\tau) + \sin\theta(\tau) = 0, \quad (\mu = 0), \qquad (3.43)$$

e.g., to just one DOF.

From the above discussion one concludes that as μ increases from zero to small positive values the dimensionality of the dynamics should also increase, and, in addition, the interaction of the fast dynamics of the stiff beam with the slow dynamics of the soft pendulum should introduce complex nonlinear dynamical phenomena. This slow-fast interaction can be studied by asymptotically solving the system (3.41) and (3.42), using methods from singular perturbation theory (since the small parameter μ multiplies the highest derivative in the first equation in (3.41) and the dynamics of the system lose dimensionality, e.g., change qualitatively as $\mu \to 0$). Georgiou and co-workers [77, 78, 79, 80, 81] studied in detail this type of singularly perturbed problems encountered in stiff-soft coupled oscillators by defining and analytically studying low-order invariant manifolds of the motion (which they termed nonlinear normal mode - NNM invariant manifolds).

As pointed out by Georgiou et al. [81] for $0 < \mu \ll 1$ there exists a two-dimensional invariant manifold in the phase space of the weakly coupled, undamped and unforced system. This manifold contains predominantly slow-oscillations of the system. The pendulum slow dynamics is governed by a nonlinear oscillator which results as regular perturbation in μ of the uncoupled pendulum (3.43). In addition, there exist linear infinite-dimensional stable and unstable invariant manifolds that are defined by the relations $(\theta, \dot{\theta}) = (0, 0)$ (the stable one), and $(\theta, \dot{\theta}) = (\pm \pi, 0)$ (the unstable ones) which contain predominantly fast dynamics.

The interaction between the fast and slow dynamics as μ increases will be numerically and experimentally studied by POD analysis. This technique seems appropriate for the problem in hand since it can provide an accurate quantitative picture of dimensionality changes in the dynamics as the parameter μ varies; as pointed out earlier, such dimensionality changes characterize the slow-fast dynamical interaction as the small parameter tends to zero (and the dynamics from infinite-dimensional reduce to a one-dimensional). Numerical and experimental acceleration time series from different sensing locations along the beam will be analyzed by means of POD. Using this technique the optimal linear orthogonal basis of modes will be identified, by analyzing sets of spatio-temporal data over the space-time domain $[0, 1] \times [T_1, T_2]$; this domain is discretized into N and M points, respectively, with corresponding steps $\Delta \xi = 1/N$ and $\Delta \tau = (T_2 - T_1)/M$. In the following analysis the value of $N = 32$ was chosen.

The experimental fixture of the beam/pendulum system is shown in Figure 3.35. The beam substructure consists of two identical beams that are coupled at their tips by a rigid bar about which the attached pendulum can rotate freely. The un normalized length, width, and thickness of the beam are $L_b = 0.66m$, $D_d = 0.025m$, $H_b = 0.0032m$, respectively. In the experiment the strength of coupling between the beam and the pendulum are varied by varying the pendulum length from $0.0508m$ to $0.292m$; this results in a variation of the coupling parameter μ from 0.419 to 0.175, respectively. For fixed parameters β and μ the beam-pendulum fixture was forced by applying an un normalized external harmonic force $F = P \cos \hat{\Omega} t$, generated by an electromagnetic shaker (with sensitivity $54.174 mV/lbf$) applied at un normalized position $x = 0.025m$ from the clamped end of the beam. The un normalized forcing frequency $\hat{\Omega}$ is varied in the neighborhood of the fundamental

3.4 POD Study of the Interaction of Slow and Fast Dynamics

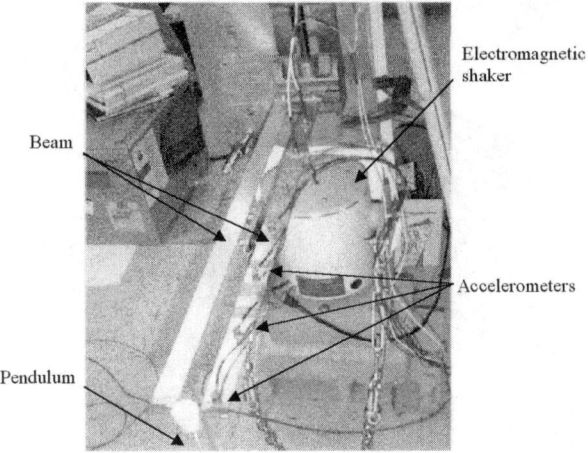

Figure 3.35. The experimental beam-pendulum system.

frequency of the beam $\hat{\omega}_1 = 28.525 rad/s$ (un normalized). Seven accelerometers are attached at the following (un normalized) sensing locations along the beam: $x_1 = 0.116m$, $x_2 = 0.205m$, $x_3 = 0.293m$, $x_4 = 0.382m$, $x_5 = 0.471m$, $x_6 = 0.559m$ and $x_7 = 0.645m$. Acceleration time series measurements at these sensing locations are used to generate the data for the ensuing POD analysis.

3.4.2 Regular Motions

Figure 3.36a depicts the experimental spatio-temporal behavior of the beam acceleration field during periodic oscillations of the coupled system. POD analysis reveals that the dynamics is captured by three POD modes. Figures 3.36b and 3.36c show qualitative agreement between the experimental and numerical leading POD modes.

In a series of experiments a 2:1 internal resonance (the fundamental frequency of the beam is twice the linearized natural frequency of the pendulum [165]) was imposed between the beam and pendulum dynamics by selecting the un normalized pendulum length as $L_p = 0.051m$. It is well known that such internal resonances give rise to essentially nonlinear dynamics, such as bifurcations, complicated regular or chaotic motions, and nonlinear exchanges of energy (nonlinear beat phenomena) between the modes in internal resonance. In Figure 3.37 the POD results of a quasi-periodic motion of the system in 2:1 internal resonance. During this motion the pendulum executes large-amplitude motions whereas the beam motion is reduced to small amplitudes. It was found that 4 POD modes are needed to capture the response; there is agreement between the experimental and numerical results (Figures 3.37b,c).

232   3 Order Reduction by Proper Orthogonal Decomposition (POD) Analysis

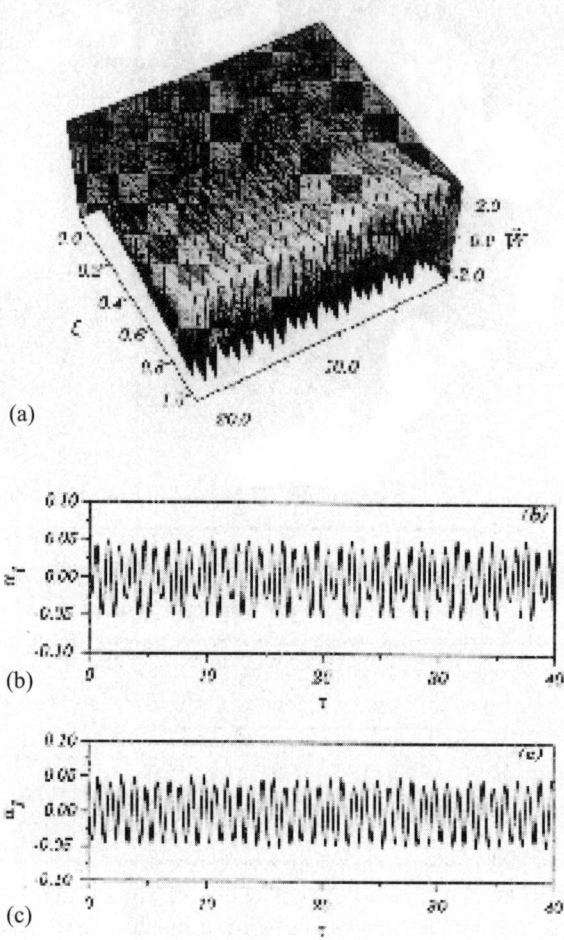

Figure 3.36. (a) Experimental acceleration field for periodic motion; amplitude of the leading POD mode: (b) numerical, (c) experimental.

For all cases of regular (periodic and quasi-periodic) motions it was found that four POD modes were needed to capture the dynamics, even for large coupling between the two subsystems. Figure 3.38a reveals that the experimental regular motions of the system are dominated by POD modes of nearly identical shapes. Similar agreement is revealed by examining the numerical POD modes of Figure 3.38b computed by analyzing numerical simulations. Figure 3.38c depicts the corresponding eigenvalues of the modes presented; the lower plateau of the eigenvalue spectrum is due to noise.

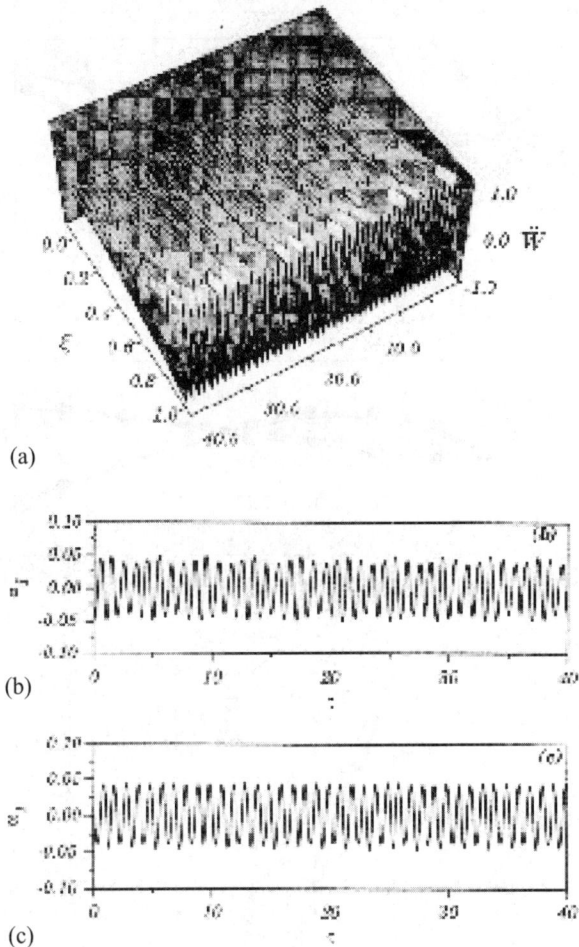

Figure 3.37. (a) Experimental acceleration field for quasi-periodic motion; amplitude of the leading POD mode: (b) numerical, (c) experimental.

3.4.3 Chaotic Motions

When forced close to the beam fundamental frequency the system can respond with irregular, chaotic motions which do not reside on the slow and fast invariant manifolds. If the system is forced with a frequency close to the linearized natural frequency of the pendulum and provided that the coupling is sufficiently weak, the system can undergo predominantly slow, chaotic motions (they reside on the slow invariant manifold); in that case, if the coupling is increased the system may undergo slow-fast chaotic motions that way from the neighborhood of the slow manifold.

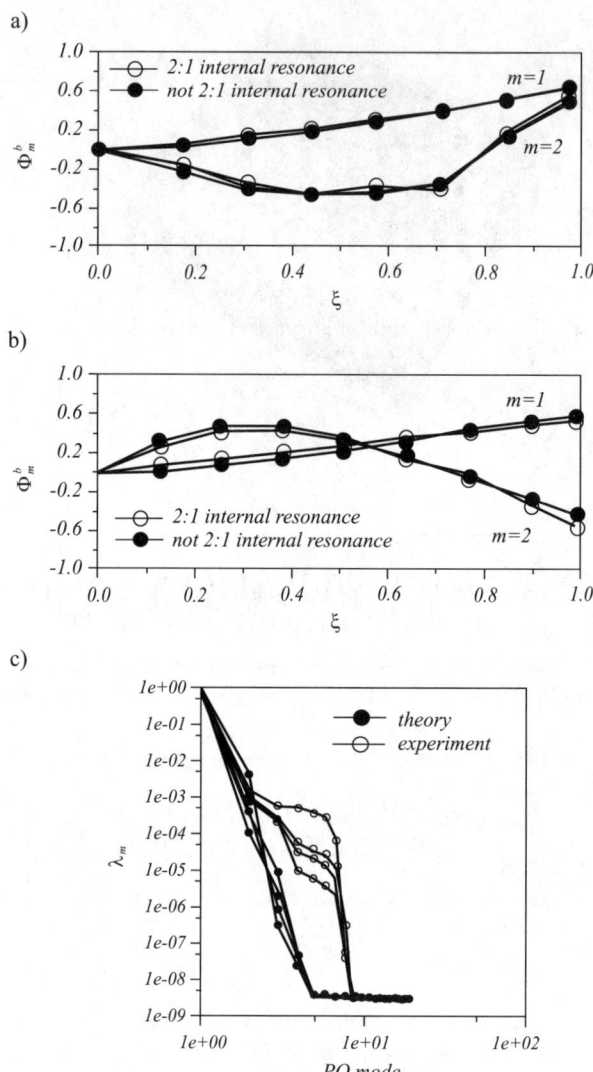

Figure 3.38. The two leading POD modes for various regular motions: (a) Experimental and (b) numerical mode shapes; (c) experimental and numerical POD eigenvalue spectra.

Of interest is to examine energy exchanges between the two subsystems as the forcing amplitude increases. Since detailed experimental results have been reported in [80], here only a summary of findings will be given. For weak coupling and forcing frequency close to the fundamental beam frequency, above a certain threshold of the forcing amplitude the pendulum undergoes chaotic vibrations whereas the beam experiences small amplitude modulations; this is termed *weak chaotic regime*. POD analysis of this response indicates that there are three dominant POD modes, with most of the energy of the leading POD mode being contained in the beam subsystem.

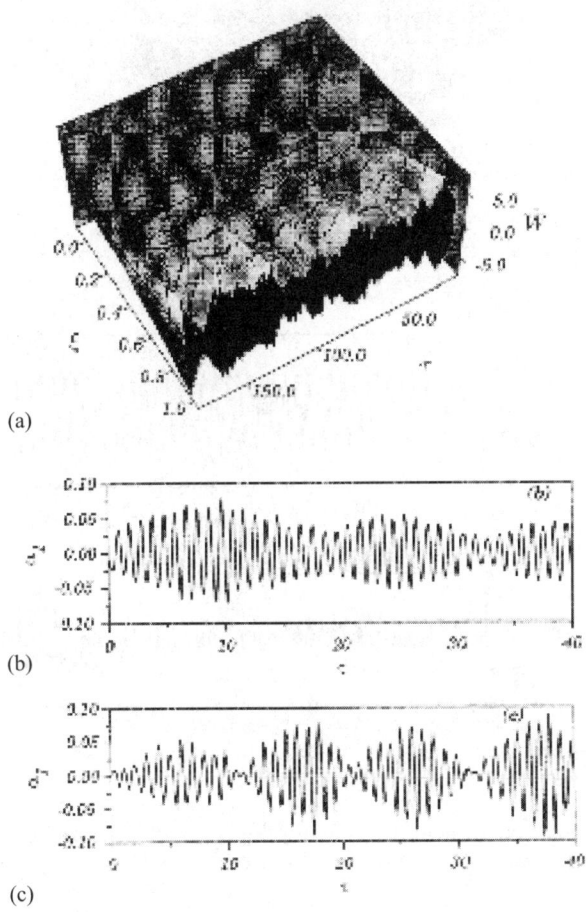

Figure 3.39. (a) Experimental acceleration field for chaotic motion; amplitude of the leading POD mode: (b) numerical, (c) experimental.

For increased forcing amplitude the modulations of the beam (fast) dynamics also increase indicating stronger interaction between the two subsystems (and also between the fast and slow dynamics). Indeed, above a certain forcing level there occurs a chaotic regime where complete energy transfer from the pendulum to the beam takes place; this is termed the *strong chaotic regime*. The corresponding chaotic motions possess six dominant POD modes,

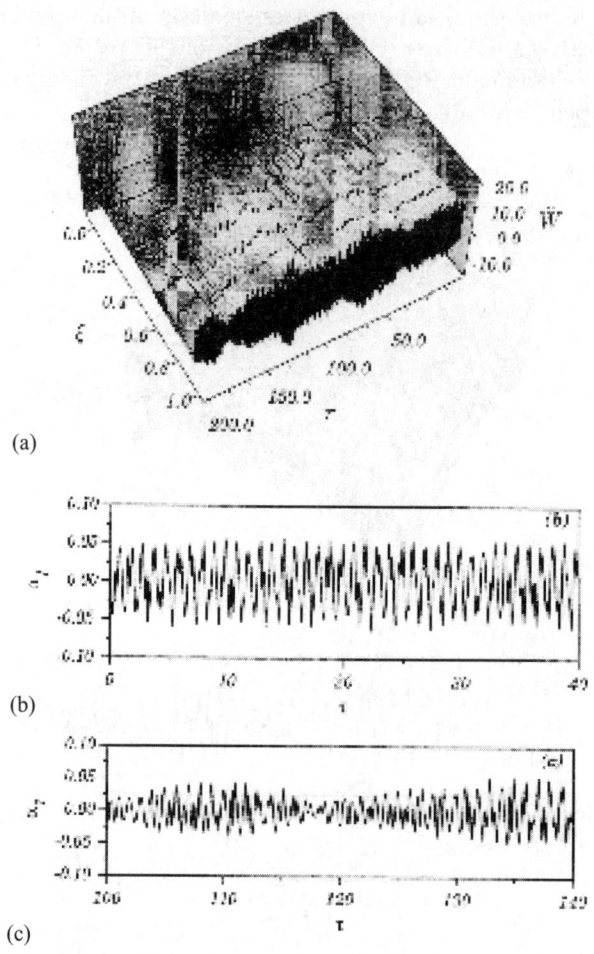

Figure 3.40. (a) Experimental acceleration field for chaotic motion of the system in 2:1 internal resonance; amplitude of the leading POD mode: (b) numerical, (c) experimental.

indicating an increase in the dimensionality of the dynamics. There is a critical value for which the chaotic regime becomes abruptly periodic, and the state of the system becomes similar to that discussed in the previous Section with the dynamics being captured by four dominant POD modes. Over a certain a certain forcing range the response remains periodic , with significant reduction of the beam amplitude. For certain forcing amplitudes this periodic regime co-exists with a chaotic regime where the pendulum rotates and vibrates irregularly with gradual energy transfer from the beam to the pendulum.

3.4 POD Study of the Interaction of Slow and Fast Dynamics

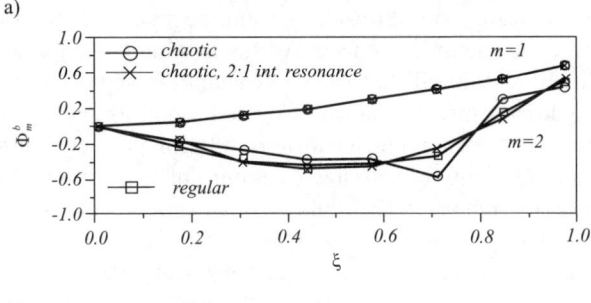

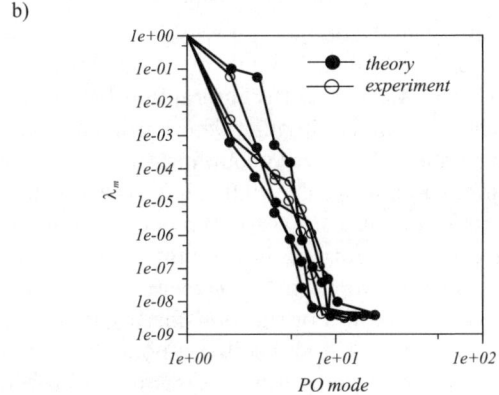

Figure 3.41. (a) Experimental mode shapes for regular and chaotic motions of the system; (b) Numerical and experimental POD eigenvalue spectra for chaotic motions.

In Figure 3.39 a chaotic motion of the system is depicted (actually, this is a transient chaotic response since after $20min$ of physical time the response settles into the periodic regime shown in Figure 3.36). The experimental and numerical amplitudes of the dominant POD mode show large amplitude modulations and are in quantitative agreement.

Similar phenomena are encountered in the system where a 2:1 internal resonance between the beam and the pendulum exists. The chaotic motions that co-exist with periodic ones are characterized by six dominant POD modes. In addition there occur abrupt transitions between regular (periodic or quasi-periodic) to chaotic motions resulting in irregular energy exchanges between the two subsystems. In Figure 3.40 the chaotic regime resulting from the abrupt loss of stability of the quasi-periodic orbit depicted in Figure 3.37 is shown. Note in this case the chaotic modulations in the amplitudes of the dominant experimental and numerical POD modes, in contrast to the smoothly modulated amplitudes of the chaotic regime of Figure 3.39 (where no 2:1 internal resonance exists).

In Figure 3.41 the mode shapes for various regular and chaotic motions of the system, and the eigenvalue spectra of the leading POD modes for chaotic motions depicted. It is noted that the mode shapes in the two cases are similar and consistent for all measurements. This led Georgiou et al. (1999) to the conclusion that the regular and chaotic oscillations of the system bear the spatial signature of the free fast-slow dynamics of the stiff-soft system.

The interaction between the infinite-dimensional fast dynamics of the beam and the two-dimensional slow-dynamics of the pendulum give rise to regular and chaotic motions and to interesting nonlinear dynamical phenomena. These include abrupt instabilities and transitions into chaotic regimes, irregular energy exchanges and complete energy transfers between subsystems, and fluctuations in the dimensionality of the dynamics as forcing or frequency parameters vary. Moreover, the interaction between the stiff and soft components of the system is singular, since from a point of view of singular perturbation in the limit of zero coupling the dimensionality of the system reduces to a single oscillator. The POD analysis is an effective tool for decomposing the dynamics of this complex dynamical system, and for studying quantitatively the dimensionality of the dynamics. Moreover, POD-based reduced-order models can be computed for reconstructing the response and better understanding the mechanics of the slow-fast dynamical interactions. Clearly, the interaction between stiff and soft components of structures is an interesting area of research with many applications to engineering practice.

4 Analytic Modelling of Discreteness Effects in Continuous Systems

4.1 The Method of Non-smooth Transformations

In this Chapter an analytical technique is presented for studying *discreteness effects* in continuous systems with microstructure or discrete loading effects. Discreteness effects are points of nonsmoothness (loss of some degree of differentiability) in the response of the dynamical system due to singularities in the governing equations of motion caused either by microstructure (for example, periodic inhomogeneities in the material or discrete supports) or by the forcing distribution (for example, non-smooth force distributions such as point loads).

The method relies on the use of non-smooth variable transformations to replace a small-scale independent variable of the problem with a pair of non-smooth variables. By doing so, the original *non-smooth* equations of motion are replaced by a smooth set of equations with twice the dimensions of the original problem. Even though of higher dimensions, there are two features of the transformed problem that makes it attractive from an analytical point of view:

a) Singular terms that appear in the original problem are eliminated exactly (e.g., with no approximation), thus rendering the transformed problem directly amenable to analysis by standard methods of (smooth) nonlinear dynamics.

b) In terms of the new variables, the transformed equations of motion are formulated as a (linear or nonlinear) boundary value problem (BVP) over a finite interval; this BVP is amenable to regular perturbation analysis in terms of the transformed variables, and enables analytic treatment in terms of known functions; this provides significant physical insight into the solution.

Once the smooth transformed problem is solved, the inverse transformations provide the discreteness effects in the original problem in analytical form. It is noted that the method enables the analytical treatment of certain (even strongly) nonlinear problems, where no other analytical methods are directly applicable. This is especially true in problems focusing on the analytic computations of nonlinear orbits in strongly nonlinear regimes, such as in neighborhoods of homoclinic or heteroclinic orbits [190], and in problems of elastic continua supported by discrete arrays of nonlinear resilient elements [189]. Moreover, viewed in the right context, the analytical approximations provided by the method of non-smooth transformations improve as the nonlinear effects become stronger; clearly, this feature is the opposite of what holds for standard asymptotic techniques which are based on linear generating functions and are applied under the assumption of weak nonlinearity [165].

The basic elements of the method were originally conceived by Pilipchuck [187, 188], who was influenced in the development of his theory by the earlier works on non-smooth transformations by Zhuravlev [253, 254, 255]. Some applications of this method to the dynamics of discrete nonlinear oscillators were given in [242]. In this work application of the method is made to certain representative problems involving continuous oscillators. The aim is to present the novel and unique aspects of the method, and its capacity to analytically solve challenging problems in the dynamics of these systems. First, a brief exposition of the method of non-smooth transformations is provided.

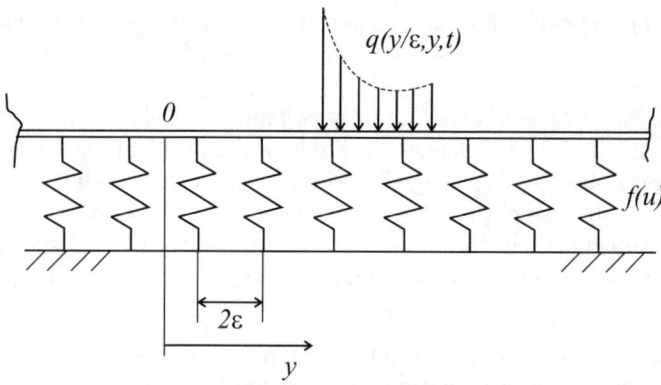

Figure 4.1. Forced string of infinite spatial extent resting on a periodic array of nonlinear springs.

The basic elements of the method will be presented using a concrete example, namely, the problem of the dynamical response of an elastic string supported by a discrete periodic array of nonlinear stiffnesses. To this end, one considers the system depicted in Figure 4.1, that is a linear string of infinite spatial extent, supported by an infinite set of periodically spaced nonlinear stiffness elements, and forced by distributed external excitation. The governing equation of motion of this system is,

$$\rho\frac{\partial^2 u}{\partial t^2} - T\frac{\partial^2 u}{\partial y^2} + 2f(u)\sum_{k=-\infty}^{\infty}\delta\left(\frac{y}{\varepsilon} - 1 - 2k\right) = q\left(\frac{y}{\varepsilon}, y, t\right), \quad -\infty < y < \infty$$

(4.1)

4.1 The Method of Non-smooth Transformations

where: $u = u\left(\frac{y}{\varepsilon} - 1 - 2k\right)$ denotes the transverse displacement of the string, the uniform density and internal tension of the string, the nonlinear supporting stiffness, and the transverse load. The infinite series of delta functions in the equation of motion are due to the discrete nature of the supports, and represent singularities that cannot be treated directly by traditional analytical methods of nonlinear dynamics. These singularities will be cause non-smooth effects in the response, e.g., discreteness effects. There is the need to eliminate these singularities in order to 'smoothen' the problem and render it amenable to analytical treatment; the method of non-smooth transformations addresses directly this task.

Assuming that $0 < \varepsilon \ll 1$, the problem (4.1) possesses two spatial scales, a *long scale* y, and a *short scale* y/ε. In the following perturbation analysis, these scales will be treated as independent from each other. From a physical point of view, the small parameter ε is the dimension that scales the discreteness effects in the dynamics due to the discrete nature of the stiffness supports. The fact that the analysis will be carried out using two independent spatial variables (a long and a short one) indicates that from an asymptotic point of view the response of the string will be decomposed into long- and short-dynamics, in similarity to the slow- and fast-dynamics (albeit in the time domain) that were discussed in Section 3.4. Hence, it is anticipated that the response of the string will consist of short-scale oscillations (with wavelengths comparable to the discreteness parameter ε), modulated by smooth envelopes governed by the long-scale. Note that for generality the external transverse load in (4.1) is also assumed to depend on the two scales y and y/ε, in order to be able to excite the string in both short and long spatial scales.

Due to the discrete nature of the supporting stiffnesses, the governing equation (4.1) possesses non-smooth (nonlinear) terms which must be eliminated before standard techniques from the theory of smooth nonlinear dynamics can be applied. This elimination will be performed by introducing a non-smooth transformation of the short-scale y/ε. First, this transformation will be introduced. To this end, the following pair of non-smooth variables is defined:

$$\tau(\phi) = \frac{2}{\pi} \arcsin\left[\sin\left(\frac{\pi\phi}{2}\right)\right], \quad e(\phi) = \tau'(\phi). \tag{4.2}$$

Then, the smooth variable $\phi \in R$ is replaced by the pair of non-smooth variables as follows:

$$\phi \to (\tau(\phi), e(\phi)). \tag{4.3}$$

By construction, the non-smooth variables (4.2) are periodic in ϕ, and $e(\phi)$ satisfies the relationship, $e^2(\phi) = 1$.

A general periodic function $g = g(\phi)$ with normalized period $\Pi = 4$ can be expressed in terms of the non-smooth variables as follows,

$$g(\phi) = X(\phi) + e(\phi) Y(\phi), \tag{4.4}$$

where:

$$X(\tau) = (1/2)\left[g(\tau) + g(2-\tau)\right], \quad Y(\tau) = (1/2)\left[g(\tau) - g(2-\tau)\right]. \quad (4.5)$$

Moreover, the derivative of $g(\phi)$ with respect to ϕ is expressed as,

$$g'(\phi) = Y'(\tau) + eX(\tau) + e'Y(\tau), \quad (4.6)$$

where primes denote differentiation of a function with respect to its argument. The last term in (4.6) is singular:

$$e'(\phi) = 2 \sum_{k=-\infty}^{\infty} \left[\delta(z+1-4k) - \delta(z-1-4k)\right]. \quad (4.7)$$

Assuming that the derivative $g'(\phi)$ is a continuous function, it is necessary to eliminate the singularities (4.7) that are localized at the set of points $\phi \in N = \{\phi/\tau(\phi) = \pm 1\}$, it is necessary to impose the following additional 'smoothness' condition:

$$Y|_{\phi \in N} = Y|_{\tau=\pm 1} = 0. \quad (4.8)$$

Higher derivatives can be similarly treated. For example, the second derivative is a smooth function given by,

$$g''(\phi) = X''(\tau) + eY''(\tau) \quad (4.9)$$

provided that the following additional 'smoothening' condition is imposed:

$$X'|_{\phi \in N} = X'|_{\tau=\pm 1} = 0. \quad (4.10)$$

In addition, it can be proven that any smooth function $w(g)$ of the aforementioned periodic function can be expressed in terms of the non-smooth variables (2) using the following relations:

$$w(g) = R_2 + eI_2,$$

$$R_w = (1/2)\left[w(X+Y) + w(X-Y)\right], \quad I_w = (1/2)\left[w(X+Y) - w(X-Y)\right]. \quad (4.11)$$

The terms R_w, I_w are termed the $R-$ and $I-$components of the function $w(g)$. We note the similarity of (4.11) with the classical complex representation in terms of real and imaginary parts, with playing the role of the imaginary constant.

Considering the governing equation (4.1), one introduces the pair of non-smooth transformations (4.2) in order to replace the short spatial scale of the problem, y/ε. Hence, the variables in (4.1) are expressed in the following form:

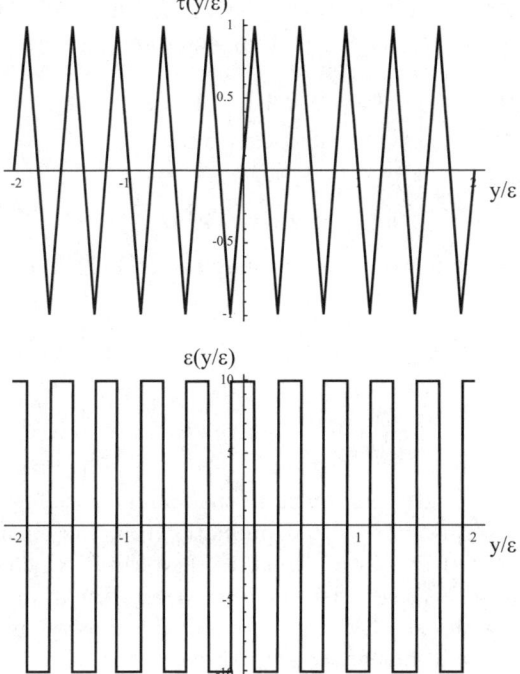

Figure 4.2. The non-smooth variables $\tau = \tau(y/\varepsilon)$ and $e = e(y/\varepsilon)$.

$$u = U(\tau, y, t) + eV(\tau, y, t), \quad q = Q(\tau, y, t) + eP(\tau, y, t), \quad \tau = \tau(y/\varepsilon), \quad e = e(y/\varepsilon), \tag{4.12}$$

where: $\tau = \tau(y/\varepsilon)$ and $e = e(y/\varepsilon)$ are the non-smooth functions of the short-scale variable defined by (4.2), and depicted in Figure 4.2. In expressing the solution in the form (4.12) it was assumed that the spatial variables y/ε and y are independent, with each being used to describe the micro- and macro-spatial dependence of the solution. An interesting feature of the non-smooth transformations used in (4.12) is that the non-smooth terms in the governing equation of motion can also be expressed in terms of the new non-smooth variables as,

$$2f(u) \sum_{k=-\infty}^{\infty} \delta\left(\frac{y}{\varepsilon} - 1 - 2k\right) = -f(u)\,\mathrm{sgn}(\tau)\,\tau''. \tag{4.13}$$

Note, however, that the second generalized derivate τ'' in the above equation possesses singularities that must be eliminated from the analysis; this issue is addressed below.

Substituting (4.12) and (4.13) into (4.1), taking into account that $\tau'\tau'' = 0$ [since in a generalized sense it can be viewed at the derivative of the constant function [153, 189], and setting equal to zero separately the $R-$ and

I−components of the resulting transformed expression [187, 188], one obtains the following two transformed governing equations for the variables U and V,

$$\rho\frac{\partial^2 U}{\partial t^2} - T\left[\frac{\partial^2 U}{\partial y^2} + \frac{2}{\varepsilon}\frac{\partial^2 V}{\partial y \partial \tau} + \frac{1}{\varepsilon^2}\frac{\partial^2 U}{\partial \tau^2}\right] = Q(\tau, y, t),$$

$$\rho\frac{\partial^2 V}{\partial t^2} - T\left[\frac{\partial^2 V}{\partial y^2} + \frac{2}{\varepsilon}\frac{\partial^2 U}{\partial y \partial \tau} + \frac{1}{\varepsilon^2}\frac{\partial^2 V}{\partial \tau^2}\right] = P(\tau, y, t), \qquad (4.14)$$

complemented by the following additional 'smoothening' conditions,

$$-\left(\frac{1}{\varepsilon}\frac{\partial V}{\partial y} + \frac{1}{\varepsilon^2}\frac{\partial U}{\partial \tau}\right)\bigg|_{\tau=\pm 1} = \mp\frac{1}{T}R_f,$$

$$\frac{\partial V}{\partial y}\bigg|_{\tau=\pm 1} = V|_{\tau=\pm 1} = 0, \qquad (4.15)$$

where: R_f denotes the R−component of the nonlinear stiffness characteristic $f(u)$. The first of equations (4.15) eliminates from the equation of motion all singular terms that are proportional to the term τ'' (including the corresponding singular terms that arise due to the transformation of the nonlinear stiffness support terms (4.13)). The second of the expressions (5.13b) render smooth the transformations of the spatial derivatives of u.

Comparing the system (4.14), (4.15) to the original governing equation (4.1) one notes that they completely equivalent. Although of double the dimension compared to the original problem, the transformed set of equations is smooth since it does not contain any generalized functions. Hence, *by replacing the short spatial variable by the pair of non-smooth variables one is able to 'smoothen' the dynamical problem exactly*, i.e., with no analytical approximation; however, in order to achieve one must also increase the dimensionality of the problem. Moreover, examining the transformed problem (4.14), (4.15) one concludes that it is in the form of a nonlinear boundary value problem (NLBVP) in terms of the variable τ, and its solution is carried only in the interval $\tau \in [-1, 1]$. This provides an additional simplification from an analytical point of view, since, it will be shown that the analytical solution can be expressed in terms of a regular perturbation expansion in τ without introducing secularities in terms of the original short-variable y/ε.

The plan is to analytically approximate the solution of the transformed problem using standard techniques from the theory of nonlinear dynamics. Then, the discreteness effects in the original problem due to the microstructure can then be analytically accounted for by following the inverse path and express the solution in terms of the original short-scale spatial variable. In the next Section this analysis is carried out in detail.

4.2 Elastic Continuum on Discrete Elastic Nonlinear Foundation

4.2.1 Asymptotic Analysis

The asymptotic solution of equations (4.14), (4.15) is expressed in the following series form of ascending powers of the small parameter:

$$U(\tau,y,t) = \sum_{k} \varepsilon^k U_k(\tau,y,t), \quad V(\tau,y,t) = \sum_{k} \varepsilon^k V_k(\tau,y,t). \tag{4.16}$$

Substituting (4.16) into (4.14), (4.15) one obtains an hierarchy of governing equations and accompanying smoothness conditions at various orders of approximation.

At $O(\varepsilon^0)$ one computes the following leading order approximations:

$$\frac{\partial^2 U_0}{\partial \tau^2} = 0 \Rightarrow U_0(\tau,y,t) = A_0(y,t)\tau + B_0(y,t),$$

$$\frac{\partial^2 V_0}{\partial \tau^2} = 0 \Rightarrow V_0(\tau,y,t) = C_0(y,t)\tau + D_0(y,t). \tag{4.17}$$

Note that since the variable τ is defined over the finite interval $[-1,1]$ it does not introduce any secular terms in the solutions. Applying the $O(\varepsilon^0)$ smoothness conditions,

$$\left.\frac{\partial U_0}{\partial \tau}\right|_{\tau=\pm 1} = 0, \quad \left.V_0\right|_{\tau=\pm 1} = 0 \tag{4.18}$$

one expresses the leading order solutions as,

$$U_0(\tau,y,t) = B_0(y,t), \quad V_0(\tau,y,t) = 0, \tag{4.19}$$

where the yet un dertermined function $B_0(y,t)$ (which depends on the long spatial variable and on time) will be evaluated by considering the next orders of approximation.

At $O(\varepsilon^1)$ one obtains a problem with trivial solutions:

$$\left.\begin{array}{l}\frac{\partial^2 U_1}{\partial \tau^2} = -2\frac{\partial^2 V_0}{\partial y \partial \tau}, \quad \frac{\partial^2 V_1}{\partial \tau^2} = -2\frac{\partial^2 U_0}{\partial y \partial \tau} \\ \left.\frac{\partial U_1}{\partial \tau}\right|_{\tau=\pm 1} = 0, \quad \left.V_1\right|_{\tau=\pm 1} = 0\end{array}\right\} \Rightarrow U_1(\tau,y,t) = 0, \quad V_1(\tau,y,t) = 0$$

$$\tag{4.20}$$

Proceeding to the problem at $O(\varepsilon^2)$ one obtains the following equations:

$$\frac{\partial^2 U_2}{\partial \tau^2} = -2\frac{\partial^2 V_1}{\partial y \partial \tau} + \frac{\rho}{T}\frac{\partial^2 U_0}{\partial t^2} - \frac{\partial^2 U_0}{\partial y^2} - \frac{Q(\tau,y,t)}{T},$$

$$\frac{\partial^2 V_2}{\partial \tau^2} = -2\frac{\partial^2 U_1}{\partial y \partial \tau} + \frac{\rho}{T}\frac{\partial^2 V_0}{\partial t^2} - \frac{\partial^2 V_0}{\partial y^2} - \frac{P(\tau,y,t)}{T}, \tag{4.21}$$

with the smoothness conditions,

$$\left.\frac{\partial U_2}{\partial \tau}\right|_{\tau=\pm 1} = \mp \frac{1}{2T}\left[f(U_0+V_0)+f(U_0-V_0)\right], \quad V_2|_{\tau=\pm 1}=0. \quad (4.22)$$

Solving this linear problem for U and V_2 using the method of variation of parameters one obtains:

$$U_2(\tau,y,t) = -\frac{\tau^2}{2T}f(B_0(y,t)) + \frac{\tau^2+2\tau}{4T}\int_{-1}^{1}Q(\xi,y,t)\,d\xi+$$

$$\frac{1}{T}\int_{-1}^{\tau}(\xi-\tau)Q(\xi,y,t)\,d\xi + B_2(y,t),$$

$$V_2(\tau,y,t) = \frac{1}{T}\int_{-1}^{\tau}(\xi-\tau)P(\xi,y,t)\,d\xi - \frac{\tau+1}{2T}\int_{-1}^{1}(\xi-1)P(\xi,y,t)\,d\xi, \quad (4.23)$$

where: $B_2(y,t)$ is an arbitrary function (resulting as a constant of integration in terms of the short-scale variable) that is determined at the next order of approximation. In addition to (4.23) one obtain the following equation [resulting from the imposition of the boundary conditions (4.22)] that evaluates the function $B_0(y,t)$ in (4.19):

$$\rho\frac{\partial^2 B_0(y,t)}{\partial t^2} - T\frac{\partial^2 B_0(y,t)}{\partial y^2} + f(B_0(y,t)) = \frac{1}{2}\int_{-1}^{1}Q(\xi,y,t)\,d\xi, \quad -\infty<y<\infty. \quad (4.24)$$

In order to determine the governing equation for the function $B_2(y,t)$ one must proceed to the problems at $O(\varepsilon^3)$ and $O(\varepsilon^4)$,

$$\frac{\partial^2 V_3}{\partial \tau^2} = -2\frac{\partial^2 U_2}{\partial y\partial \tau} + \frac{\rho}{T}\frac{\partial^2 V_1}{\partial t^2} - \frac{\partial^2 V_1}{\partial y^2}, \quad V_3|_{\tau=\pm 1}=0 \quad (4.25)$$

and

$$\frac{\partial^2 U_4}{\partial \tau^2} = -2\frac{\partial^2 V_3}{\partial y\partial \tau} + \frac{\rho}{T}\frac{\partial^2 U_2}{\partial t^2} - \frac{\partial^2 U_2}{\partial y^2},$$

$$\left.\frac{\partial U_4}{\partial \tau}\right|_{\tau=\pm 1} = \mp\frac{1}{2T}\left[\frac{(U_1+V_1)^2}{2}\frac{d^2f}{d(U_0+V_0)^2} + \frac{(U_1-V_1)^2}{2}\frac{d^2f}{d(U_0-V_0)^2} + \right.$$

$$\frac{(U_2+V_2)^2}{2}\frac{d^2f}{d(U_0+V_0)^2} + \frac{(U_2-V_2)^2}{2}\frac{d^2f}{d(U_0-V_0)^2} +$$

$$\left. (U_1+V_1)\frac{df}{d(U_1+V_1)} + (U_1-V_1)\frac{df}{d(U_1-V_1)}\right] = \mp\frac{1}{T}U_2\frac{df}{dB_0}, \quad (4.26)$$

4.2 Elastic Continuum on Discrete Elastic Nonlinear Foundation

where it is recognized that $f = f(U_0 + V_0) = f(B_0(y,t))$. The solution of the set (4.25) is expressed as,

$$V_3(\tau, y, t) = -\frac{2}{T} \int_{-1}^{\tau} (\alpha - \tau) \int_{-1}^{\alpha} \frac{\partial Q}{\partial y}(\xi, y, t)\, d\xi\, d\alpha +$$

$$\frac{(\tau+1)}{T} \int_{-1}^{1} (\alpha - 1) \int_{-1}^{\alpha} \frac{\partial Q}{\partial y}(\xi, y, t)\, d\xi\, d\alpha -$$

$$\frac{(\tau+3)(\tau^2-1)}{6T} \int_{-1}^{1} \frac{\partial Q}{\partial y}(\xi, y, t)\, d\xi + \frac{\tau(\tau^2-1)}{3T} \frac{\partial f(B_0(y,t))}{\partial y}. \quad (4.27)$$

Then, the equation governing $B_2(y,t)$ is obtained by substituting (4.23) and (4.27) into the first of equations (4.26), solving for $U_2(\tau, y, t)$ and substituting into the corresponding smoothness conditions [the rest of equations (4.26)]. After some algebraic manipulations one finds the following equation:

$$\rho \frac{\partial^2 B_2(y,t)}{\partial t^2} - T \frac{\partial^2 B_2(y,t)}{\partial y^2} + f'(B_0(y,t)) B_2(y,t) =$$

$$\frac{\rho}{6T} \left[\frac{\partial^2 f(B_0(y,t))}{\partial t^2} - \frac{1}{2} \int_{-1}^{1} \frac{\partial^2 Q}{\partial t^2}(\xi, y, t)\, d\xi - 3 \int_{-1}^{1} \int_{-1}^{\gamma} (\xi - \gamma) \frac{\partial^2 Q}{\partial t^2}(\xi, y, t)\, d\xi\, d\gamma \right] -$$

$$\frac{1}{6} \left[\frac{\partial^2 f(B_0(y,t))}{\partial t^2} - \frac{1}{2} \int_{-1}^{1} \frac{\partial^2 Q}{\partial y^2}(\xi, y, t)\, d\xi - 3 \int_{-1}^{1} \int_{-1}^{\gamma} (\xi - \gamma) \frac{\partial^2 Q}{\partial y^2}(\xi, y, t)\, d\xi\, d\gamma \right] +$$

$$\frac{f'(B_0(y,t))}{2T} \left[f(B_0(y,t)) - \int_{-1}^{1} \left(\xi - \frac{1}{2}\right) Q(\xi, y, t)\, d\xi \right], \quad -\infty < y < \infty.$$

(4.28)

Summarizing, the approximate solution for the response of the string on a discrete periodic array of nonlinear springs is given by,

$$u(y/\varepsilon, y, t) = \left\{ B_0(y, t) + \varepsilon^2 \left[-\frac{\tau^2}{2T} f(B_0(y,t)) \right. \right.$$

$$\left. \left. + \frac{\tau^2 + 2\tau}{4T} \int_{-1}^{1} Q(\xi, y, t)\, d\xi + \frac{1}{T} \int_{-1}^{\tau} (\xi - \tau) Q(\xi, y, t)\, d\xi + B_2(y, t) \right] \right\} +$$

$$e\varepsilon^2 \left\{ \frac{1}{T} \int_{-1}^{\tau} (\xi - \tau) P(\xi, y, t) \, d\xi - \frac{\tau + 1}{2\tau} \int_{-1}^{1} (\xi - 1) P(\xi, y, t) \, d\xi \right\} + O(\varepsilon^3),$$
(4.29)

where: $\tau(y/\varepsilon)$ and $e(y/\varepsilon)$ are non-smooth variables defined by (4.2), and $B_0(y,t)$ and $B_2(y,t)$ are computed by solving (4.24) and (4.28), respectively. By construction, the asymptotic approximation is periodic with respect to the short-scale y/ε, but not with respect to the long-scale variable y (as pointed out previously these two spatial variables are treated independently from each other). Moreover, the derived solution automatically incorporate discreteness effects (discontinuities) that arise due to the discrete nature of the elastic support. These non-smooth effects are accounted for in the analytic construction by the non-smooth variables $\tau(y/\varepsilon)$ and $e(y/\varepsilon)$, which as shown below produce discontinuities in the first spatial derivative of the envelope of the transverse oscillation. Hence, the solution is C^0 – but not C^1 – differentiable with respect to y.

To proceed with explicit expressions, it is assumed that the external load has a harmonic dependence with respect to time, possesses a localized distribution with respect to y, and is periodic with respect to the scale y/ε,

$$q(y/\varepsilon, y, t) = H(y) \left\{ m_0 + \sum_{k=1}^{\infty} \left[m_k \cos\left(\frac{k\pi y}{2\varepsilon}\right) + n_k \sin\left(\frac{k\pi y}{2\varepsilon}\right) \right] \right\} \cos \omega t,$$
(4.30)

where: m_i and n_i are Fourier coefficients. Using computer algebra, the corresponding expressions for the transformed forcing components are determined as follows:

$$Q(\tau, y, t) = \frac{1}{2} [q(\tau, y, t) + q(2 - \tau, y, t)] =$$

$$H(y) \left\{ m_0 + \sum_{j=1}^{\infty} \left[m_{2j} \cos(j\pi\tau) + n_{2j-1} \sin\left(\frac{2j-1}{2}\pi\tau\right) \right] \right\} \cos \omega t \quad (4.31)$$

and

$$P(\tau, y, t) = \frac{1}{2} [q(\tau, y, t) - q(2 - \tau, y, t)] =$$

$$H(y) \left\{ m_0 + \sum_{j=1}^{\infty} \left[m_{2j-1} \cos\left(\frac{2j-1}{2}\pi\tau\right) + n_{2j} \sin(j\pi\tau) \right] \right\} \cos \omega t. \quad (4.32)$$

The transverse response of the string is then computed as,

$$u(y/\varepsilon, y, t) = \left\{ B_0(y, t) + \varepsilon^2 \left[-\frac{\tau^2}{2T} f(B_0(y, t)) + \right. \right.$$

$$H(y) \cos \omega t \left[-\frac{m_0}{2T} + \frac{1}{T} \sum_{j=1}^{\infty} \left\{ m_{2j} \frac{\cos j\pi\tau - (-1)^j}{j^2 \pi^2} + \right. \right.$$

$$n_{2j-1}\frac{4\sin\left[(2j-1)\pi\tau/2\right]-4(-1)^j}{(2j-1)^2\pi^2}\Bigg\}\Bigg]+B_2(y,t)\Bigg]\Bigg\}+$$

$$e\varepsilon^2 H(y)\cos\omega t\left\{\frac{1}{\pi^2 T}\sum_{j=1}^{\infty}\left[m_{2j-1}\frac{4\cos\left[(2j-1)\pi\tau/2\right]}{(2j-1)^2}+n_{2j}\frac{\sin j\pi\tau}{j^2}\right]\right\}+O(\varepsilon^3) \tag{4.33}$$

and the equations (4.24) and (4.28) assume the forms,

$$\rho\frac{\partial^2 B_0(y,t)}{\partial t^2}-T\frac{\partial^2 B_0(y,t)}{\partial y^2}+f(B_0(y,t))=m_0 H(y)\cos\omega t \tag{4.34}$$

and

$$\rho\frac{\partial^2 B_2(y,t)}{\partial t^2}-T\frac{\partial^2 B_2(y,t)}{\partial y^2}+f'(B_0(y,t))B_2(y,t)=$$

$$\frac{1}{6T}\left[\rho\frac{\partial^2 f(B_0(y,t))}{\partial t^2}-T\frac{\partial^2 f(B_0(y,t))}{\partial y^2}+3f'(B_0(y,t))f(B_0(y,t))\right]-$$

$$\frac{1}{2T}\cos\omega t\left[m_0+\frac{8}{\pi^2}\sum_{j=1}^{\infty}\frac{(-1)^j n_{2j-1}}{(2j-1)^2}\right]\left\{\left[\rho\omega^2-f'(B_0(y,t))\right]H(y)+T\frac{\partial^2 H(y)}{\partial y^2}\right\} \tag{4.35}$$

with the domain of the solution being $-\infty<y<\infty$.

To derive closed form solutions one needs to specify a specific for the spring nonlinearity $f(\cdot)$. Two specific stiffness nonlinearities will be considered, namely, hardening cubic and clearance.

4.2.1.1 Springs with Hardening Cubic Nonlinearities

In the first application, hardening cubic springs are considered, with $f(u)=\alpha_1 u+\alpha_3 u^3$, with α_1,α_2 positive quantities. Moreover, the following distribution is assumed (cf. Figure 4.3):

$$q(y/\varepsilon,y,t)=10\sec hy\cos\left(\frac{\pi y}{2\varepsilon}\right)\cos\omega t\ . \tag{4.36}$$

This corresponds to a localized distribution at $y=0$ with large wavenumber content. Clearly, this excitation is the simplest possible of expressions (4.30), with $H(y)=10\sec hy$, $m_1=0$, and all other Fourier coefficients equal to zero.

The analytical expressions (4.33)–(4.35) can be directly applied to compute an approximation for the response of the string. To this end, one seeks time-periodic solutions of equations (4.34) and (4.35) with period $T=2\pi/\omega$, by expressing them in series forms using the method of harmonic balance:

$$B_0(y,t)=\sum_p \hat{A}_p(y)\cos p\omega t,\quad B_2(y,t)=\sum_p \hat{C}_p(y)\cos p\omega t\ . \tag{4.37}$$

Focusing only in the leading approximations of the series, and introducing the notations,

$$\sigma=\rho\omega^2-\alpha_1,\quad \mu=(3/4)\alpha_3/\sigma,\quad z=y\left(\sigma/T\right)^{1/2},$$

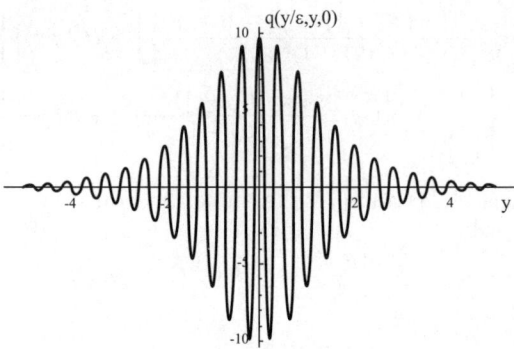

Figure 4.3. Spatial distribution of the external load at $t = 0$.

$$\hat{A}_1(y(z)) \equiv A_1(z), \quad \hat{C}_1(y(z)) \equiv C_1(z),$$

one derives the following set of ordinary differential equations that govern the normalized leading-order approximations in (4.37):

$$A_1''(z) + A_1(z) - \mu A_1^3(z) = 0, \tag{4.38}$$

$$C_1''(z) + C_1(z) - 3\mu A_1^2(z) C_1(z) = F[A_1(z)], \tag{4.39}$$

where:

$$F[A_1(z)] = \frac{1}{\sigma} \left\{ \frac{1}{6T} \alpha_1 \sigma A_1''(z) + \frac{1}{12T} \left[2\alpha_1 \sigma - 4\alpha_1^2 + 9\alpha_3 \sigma A_1'^2(z) \right] A_1(z) + \right.$$

$$\left. \frac{3}{8T} \alpha_3 \sigma A_1''(z) A_1^2(z) + \frac{1}{8T} \alpha_3 (\sigma - 11\alpha_1) A_1^3(z) - \frac{15}{16T} \alpha_3^2 A_1^5(z) \right\}.$$

In the above equations primes denote differentiation with respect to z, and it is assumed that $\omega^2 > \alpha_1/\rho$ so that the parameters μ and σ are non-negative.

Clearly, the dynamical systems (4.38), (4.39) possess qualitative different solution depending on the values of the parameters and the initial conditions. The solutions considered herein focus in solitary solutions with localized slope distributions (non-localized solutions also exist but will not be considered in this work, the method for analyzing them is similar to the one followed here). Equation (4.38) has the solitary solution:

$$A_1(z) = \mu^{-1/2} \tanh\left(2^{-1/2} z\right). \tag{4.40}$$

When substituting this expression in the (4.39) one finds that the resulting nonhomogeneous linear system with parametric excitation possesses bounded as well as unbounded solutions; since the later lead to physically unacceptable motions of the string, one focuses only on bounded solutions.

4.2 Elastic Continuum on Discrete Elastic Nonlinear Foundation

It can be shown [238, 241] that by carefully evaluating the initial conditions $C_1(0)$ and $C_1'(0)$ one can compute a special bounded, solitary solution of (4.39) that provides the leading higher order correction to the envelope (4.40). The general solution of (4.39) can be expressed in analytic form as follows,

$$C_1(z) = \left[d_1 - \int_0^z F[A_1(z)]\Phi_2(\xi)d\xi\right]\Phi_1(z) + \left[d_2 + \int_0^z F[A_1(z)]\Phi_1(\xi)d\xi\right]\Phi_2(z), \tag{4.41}$$

where the two linearly independent homogeneous solutions are given by,

$$\Phi_1(z) = \mathrm{sech}^2\left(\frac{z}{\sqrt{2}}\right),$$

$$\Phi_2(z) = \frac{1}{16\sqrt{2}}\left[6\sqrt{2}z + 8\sinh\left(\sqrt{2}z\right) + \sinh\left(2\sqrt{2}z\right)\right]\Phi_1(z)$$

and d_1, d_2 are constants of integration. The first homogeneous solution $\Phi_1(z)$ is a bounded function of z, but the second homogeneous solution $\Phi_2(z)$ becomes unbounded as $z \to \pm\infty$. Hence, for boundness of the second summation term in (4.41) as $z \to \pm\infty$, the second constant of integration must be evaluated as follows:

$$d_2 = -\int_0^\infty F[A_1(z)]\Phi_1(\xi)d\xi. \tag{4.42}$$

Performing the above integration using computer algebra, this constant is evaluated as $d_2 \simeq 1.07484$. In addition, the first constant of integration must be set equal to zero, $d_1 = 0$, in order to eliminate a shift in the argument of $C_1(z)$ [238]. With these evaluations of the constants, the explicit evaluation of (4.41) gives:

$$C_1(z) = \left\{\mathrm{sech}^6(2^{1/2}z)[z(5400\alpha_3^2 + 1584\mu\alpha_1\alpha_3 - 128\mu^2\alpha_1^2 - 3024\mu\sigma\alpha_3 - 192\mu^2\sigma\alpha_1 + \right.$$

$$(5400\alpha_3^2 + 3168\mu\alpha_1\alpha_3 - 384\mu^2\alpha_1^2 - 1584\mu\sigma\alpha_1^2 - 384\mu^2\sigma\alpha_1)\cosh(2^{1/2}z) +$$

$$(3240\alpha_3^2 + 1584\mu\alpha_1\alpha_3 - 384\mu^2\alpha_1^2 - 1008\mu\sigma\alpha_3 - 192\mu^2\sigma\alpha_1)\cosh(2\sqrt{2}z) +$$

$$(360\alpha_3^2 - 128\mu^2\alpha_1^2 - 144\mu\sigma\alpha_3)\cosh(3\sqrt{2}z)) + \sqrt{2}[(-1620\alpha_3^2 - 1188\mu\alpha_1\alpha_3 +$$

$$256\mu^2\alpha_1^2 + 144\mu\sigma\alpha_3 + 144\mu\sigma\alpha_1)\sinh(\sqrt{2}z) + (-2070\alpha_3^2 - 660\mu\alpha_1\alpha_3 + 160\mu^2\alpha_1^2 +$$

$$996\mu\sigma\alpha_3 + 80\mu^2\sigma\alpha_1)\sinh(2\sqrt{2}z) + (-420\alpha_3^2 - 132\mu\alpha_1\alpha_3 + 240\mu\sigma\alpha_3 +$$

$$16\mu^2\sigma\alpha_1)\sinh(3\sqrt{2}z) + (-45\alpha_3^2 - 66\mu\alpha_1\alpha_3 - 16\mu^2\alpha_1^2 + 6\mu\sigma\alpha_3 +$$

$$\left. 8\mu^2\sigma\alpha_1)\sinh(4\sqrt{2})]]\right\}\left\{(-\mathrm{sech}^2(z/\sqrt{2})/(24576\sqrt{2}\mu^{5/2}\sigma T)\right\} +$$

$$\{1.07484 + [(-45\alpha_3^2 + 33\mu\alpha_1\alpha_3 - 48\mu^2\alpha_1 + 114\mu\sigma\alpha_3 - 4\mu^2\sigma\alpha_1 + (60\alpha_3^2 - 64\mu^2\alpha_1^2 +$$

$$24\mu\sigma\alpha_3)\cosh(\sqrt{2}z) + (115\alpha_3^2 - 33\mu\alpha_1\alpha_3 - 16\mu^2\alpha_1^2\sigma\alpha_3 +$$

$$4\mu^2\sigma\alpha_1)\cosh(2\sqrt{2}z))\operatorname{sech}^4(2^{-1/2}z)\tanh^2(2^{-1/2}z)]/(384\sqrt{2}\mu^{5/2}\sigma T)\bigg\} \times$$
$$\left[6\sqrt{2}z + 8\sinh(\sqrt{2}z) + \sinh(2\sqrt{2}z)\right]\left[\operatorname{sech}^2(z/\sqrt{2})/16\sqrt{2}\right]. \qquad (4.43)$$

In Figure 4.4 $C_1(z)$ is depicted, resembling the tanh-dependence of the $O(1)$ approximation $A_1(z)$; due to the careful evaluation of the constants of integration this function remains bounded as $z \to \pm\infty$.

The discreteness effects in the envelope of the spatial displacement of the string are depicted in Figure 4.5, where $u(y/\varepsilon, y, 0)$ is presented. Note that in accordance with previous discussion the envelope of the oscillation possesses $C^)$ – but not C^1 – differentiability with respect to y.

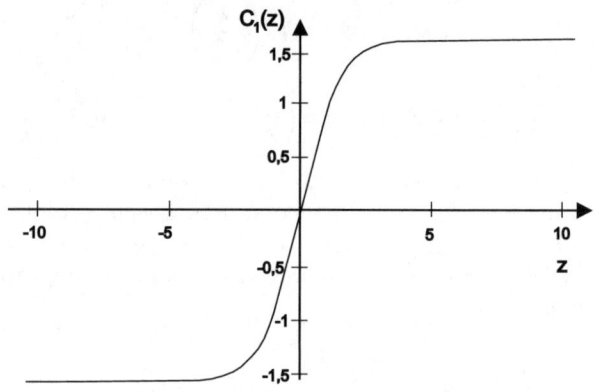

Figure 4.4. The solution $C_1(z)$ (expressions (4.41)–(4.43)).

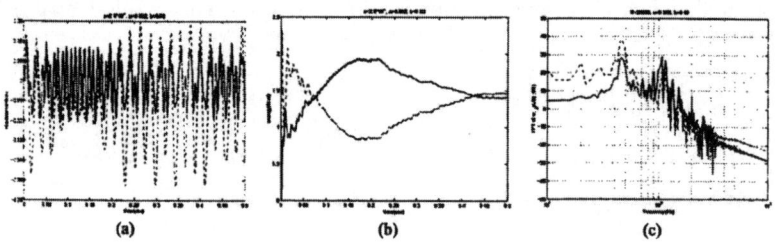

Figure 4.5. The envelope $u(y/\varepsilon, y, 0)$ of the standing solitary wave; note discreteness effects.

4.2.1.2 Springs with Clearance Nonlinearities
In the second application springs with clearance nonlinearities are considered, with stiffness characteristics given by:

4.2 Elastic Continuum on Discrete Elastic Nonlinear Foundation

$$f(u) = \begin{cases} a(u - c_0), & u \geq c_0 \\ 0, & |u| < c_0 \\ a(u + c_0), & u \leq -c_0 \end{cases} \quad (4.44)$$

The parameter a defines the strength of the restoring force, and c_0 the half-clearance of the supports; in the limit $\alpha \to \infty$ the supports become rigid barriers and the string suffers vibro-impacts at the points of the supports.

Assuming that an external excitation of the form (4.30), the leading order approximations of the response are computed by expressions (4.33)–(4.35). Again, solitary standing wave solutions with frequency ω are sought by the method of harmonic balance. Expressing these solutions in the series forms,

$$B_0(y,t) = \sum_p \hat{A}_p(y) \cos p\omega t, \quad B_2(y,t) = \sum_p \hat{C}_p(y) \cos p\omega t \quad (4.45)$$

and considering only the principle harmonic components, one multiplies through by $\cos \omega t$ and integrates from 0 to $2\pi/\omega$ with respect to t. Integral terms involving $f\left(\hat{A}_1(y) \cos \omega t\right)$ or $f'\left(\hat{A}_1(y) \cos \omega t\right)$ are summed over the intervals $[0, \hat{t}]$, $[(\pi/\omega) - \hat{t}, (\pi/\omega) + \hat{t}]$ and $[(2\pi/\omega) - \hat{t}, 2\pi/\omega]$, where $\hat{t}$ is the time instant when the string overcomes the half-clearance and reaches the spring; it is computed by the following relations:

$$\cos \omega \hat{t} = c_0/\hat{A}(y), \quad \hat{A}(y) \geq c_0$$
$$\cos \omega \hat{t} = -c_0/\hat{A}(y), \quad \hat{A}(y) \leq -c_0 \quad (4.46)$$
$$\hat{t} = 0, \quad \left|\hat{A}(y)\right| < c_0.$$

By completing the integration and substituting for $\hat{t}$, one determines the following equations governing the leading-order harmonic normalized components of (4.45):

$$A_1''(w) + A_1(w) - \frac{2\alpha\varphi}{\pi} \left[\cos^{-1}\left(\frac{c_0}{|A_1(w)|}\right) - \frac{c_0}{|A_1(w)|} \sqrt{1 - \left(\frac{c_0}{|A_1(w)|}\right)^2} \right] A_1(w) =$$

$$-\varphi m_0 \bar{H}(w), \quad |A_1(w)| \geq c_0,$$
$$A_1''(w) + A_1(w) = -\varphi m_0 \bar{H}(w), \quad |A_1(w)| < c_0 \quad (4.47)$$

and

$$C_1''(w) + C_1(w) - \frac{2\alpha\varphi}{\pi} \left[\cos^{-1}\left(\frac{c_0}{|A_1(w)|}\right) + \frac{c_0}{|A_1(w)|} \sqrt{1 - \left(\frac{c_0}{|A_1(w)|}\right)^2} \right] C_1(w) =$$

$$\frac{E}{2T} \left[\bar{H}(w) + \bar{H}''(w)\right] - \frac{2\alpha^2 \varphi}{3T\pi} \left[\cos^{-1}\left(\frac{c_0}{|A_1(w)|}\right) - \frac{c_0}{|A_1(w)|} \sqrt{1 - \left(\frac{c_0}{|A_1(w)|}\right)^2} \right] \times$$

$$A_1(w) - \frac{\alpha\varphi(m_0 + 3E)}{3T\pi}\left[\cos^{-1}\left(\frac{c_0}{|A_1(w)|}\right) + \frac{c_0}{|A_1(w)|}\sqrt{1 - \left(\frac{c_0}{|A_1(w)|}\right)^2}\right]\bar{H}(w),$$

$$|A_1(w)| \geq c_0,$$

$$C_1''(w) + C_1(w) = \frac{E}{2T}\left[\bar{H}(w) + \bar{H}''(w)\right], \quad |A_1(w)| < c_0. \qquad (4.48)$$

In the above equations the following normalizations were used,

$$\varphi = (\rho\omega^2)^{-1}, \quad E = m_0 + \frac{8}{\pi^2}\sum_{j=1}^{\infty}\frac{(-1)^j n_{2j-1}}{(2j-1)^2}, \quad w = y(\varphi T)^{-1/2},$$

$$\hat{A}_1(y(w)) \equiv A_1(w), \quad \hat{C}_1(y(w)) \equiv C_1(w), \quad \bar{H}(w) = H(y(w))$$

with primes denoting differentiations with respect to w.

In the numerical simulations the external excitation is set equal to zero, $\bar{H}(w) = 0$, in which case the dynamical system (4.47) becomes autonomous, with discontinuous trajectories. Due to the complex nature of the clearance nonlinearity resort to numerical simulations is made. In Figure 4.6 a series of phase plots of (4.47) is given, for varying clearance c_0 and stiffness characteristic α. For large values of α the vibro-impact limit is approached, and C^2- discontinuities are evident in the phase plots corresponding to $\alpha = 20\rho\omega^2$ and $500\rho\omega^2$. For $\alpha > \rho\omega^2$ and positive c_0 the phase plot of the system possesses a pair of heteroclinic orbits that correspond to standing solitary waves similar to the ones computed in the previous Section for the case of hardening cubic stiffnesses.

After computing $A_1(w)$, $C_1(w)$ is evaluated by solving numerically (4.48). In Figure 4.7 the solutions for $A_1(w)$ and $C_1(w)$ are presented for the localized standing wave with $\rho = 1$, $\omega = 1.3$, $T = 1$, $c_0 = 1.5$ and varying a. For comparison purposes, in the same plots the corresponding time-periodic solutions of the linear system with $a = 0$ are superimposed; clearly, linear systems cannot possess localized standing waves, but only spatially periodic ones.

In Figure 4.8 the envelopes $u(y/\varepsilon, y, 0)$ are depicted for solitary standing waves of systems with $\rho = 1$, $\omega = 1.3$, $T = 1$, $c_0 = 1.5$ and varying values of a. These envelopes were computed by evaluating $B_0(y,t)$ and $B_2(y,t)$ by means of (4.45) using the leading order numerical solutions (4.47) and (4.48), and then using the expressions (4.33)–(4.35), (4.16) and (4.12). In this case the discreteness effects are much smaller in magnitude but still exist. Near the vibro-impact limit ($\alpha \gg 1$) the envelopes 'saturate' to the value $u = c_0$, a fact that agrees to physical intuition.

In the next Section the method of non-smooth transformations will be applied to a static buckling problem under discrete periodic compressive forcing. An asymptotic analysis will be undertaken to determine the effects of the discrete static forcing on the critical buckling loads and on the buckled deformed states.

4.2 Elastic Continuum on Discrete Elastic Nonlinear Foundation 255

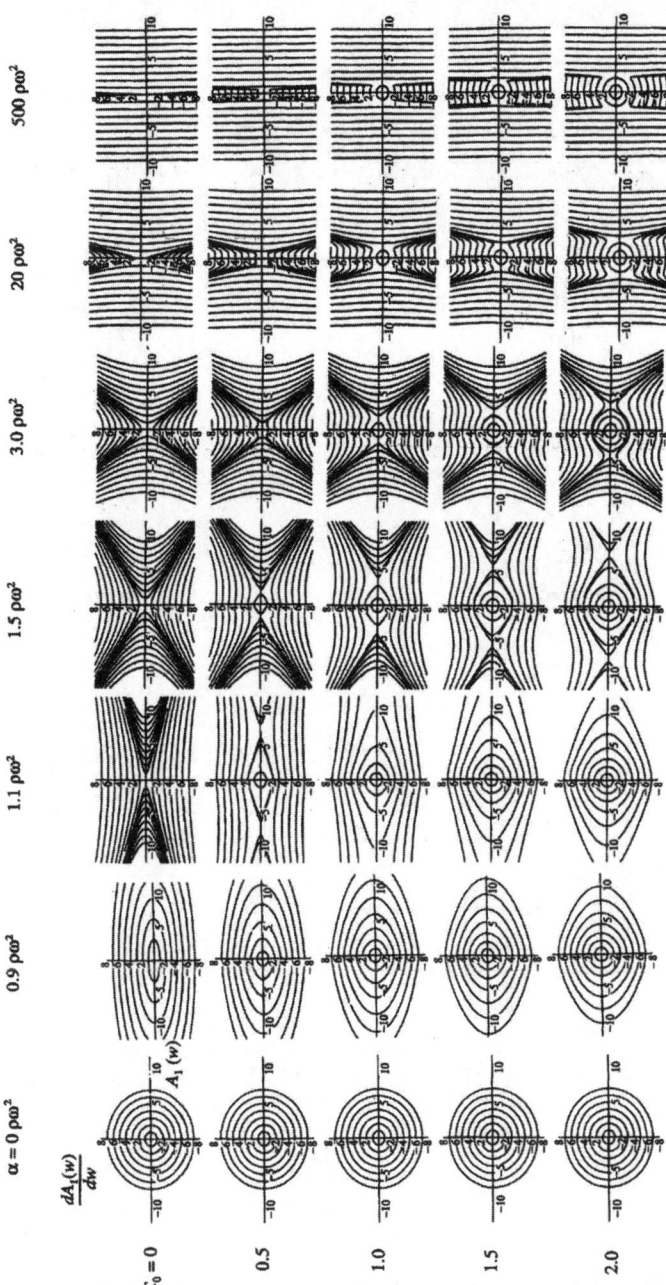

Figure 4.6. Phase plots of the unforced system with $\rho = 1$, $\omega = 1.3$, $T = 1$.

256   4 Analytic Modelling of Discreteness Effects in Continuous Systems

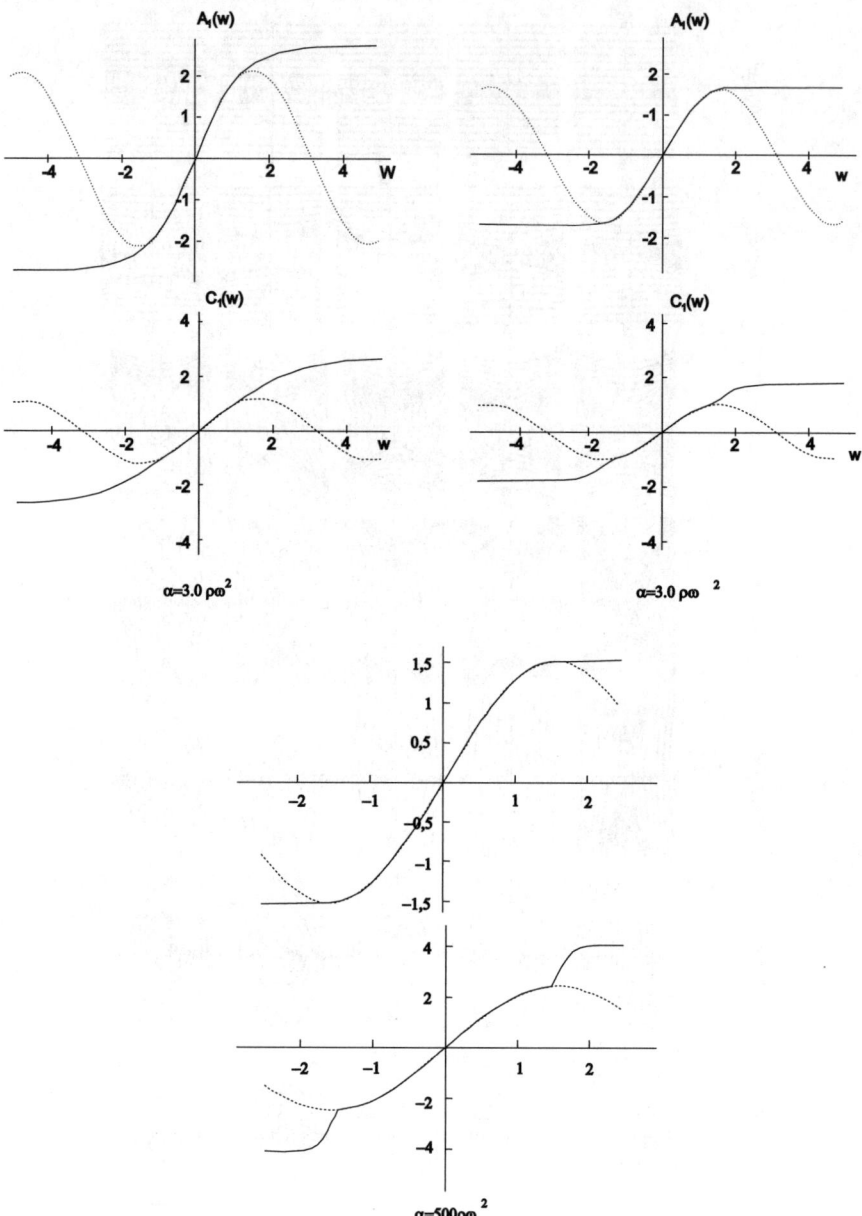

Figure 4.7. The solitary standing solutions (solid lines) for varying values of a superimposed to the corresponding linear system with $a = 0$ (dashed lines).

4.3 Static Buckling of a Circular Ring Compressive Loads

In this Section the method of non-smooth transformations is applied to the analytic study of the problem of static buckling of an elastic ring under

4.3 Static Buckling of a Circular Ring Compressive Loads 257

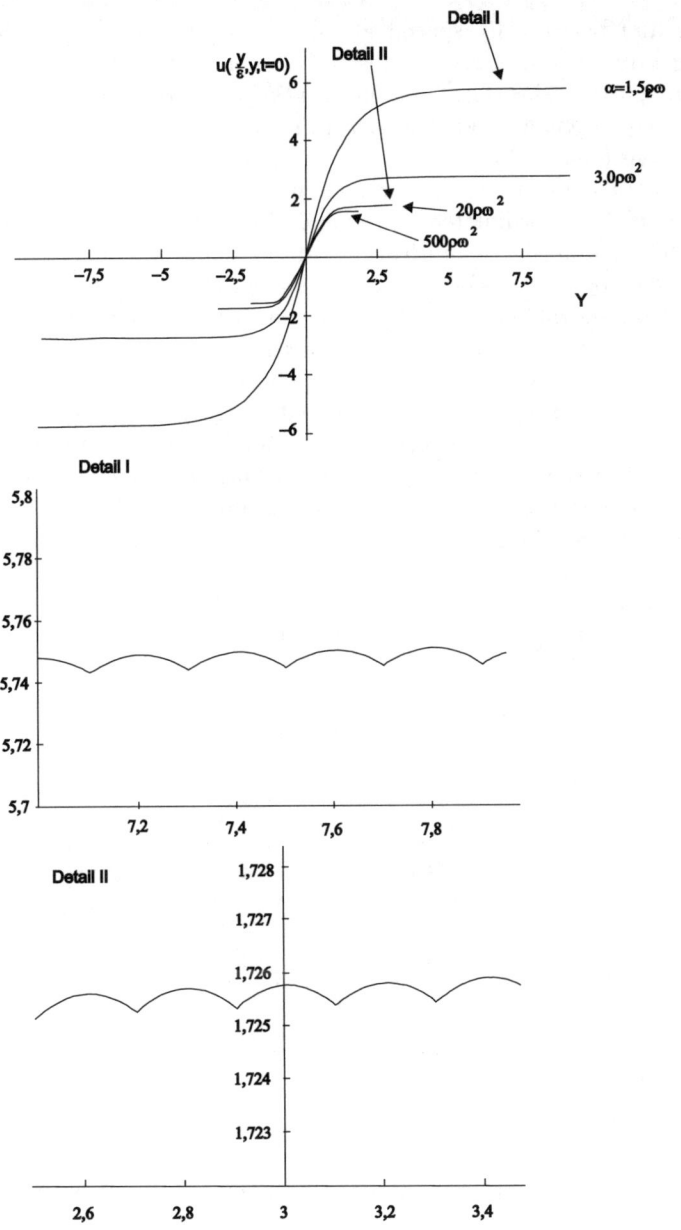

Figure 4.8. The envelopes $u(y/\varepsilon, y, 0)$ of standing solitary waves for systems with parameters $\rho = 1$, $\omega = 1.3$, $T = 1$, $c_0 = 1.5$ and varying values of a; note discreteness effects.

258   4 Analytic Modelling of Discreteness Effects in Continuous Systems

discrete and periodically spaced loads. The singularities in the governing equations due to the discrete loads will be eliminated exactly by replacing the short spatial scale of the problem with the set of non-smooth variables; the resulting equations can be analyzed using standard asymptotic techniques. The analysis follows closely [240].

The problem of the stability of a circular ring under uniform hydrostatic pressure has been formulated and studied in previous works. Seide and Albano [206] examined the bifurcations of circular rings under concentrated loads by employing transfer matrices and solving linear eigenvalue problems; Kabanov and Astrakharchuck [97] developed algorithms for the stability analysis of variable-thickness rings under nonuniform loads; Chaskalovic and Naili [45] studied rings obeying Bernulli-Euler theory; Atanackovic [10] examined rings with constitutive laws that accounted for axial compressibility, whereas Schmidt [205] on rings with shear deformation and axial compressibility; Fu and Waas [74] studied the effect of ring thickness on postbuckling behavior, whereas, a review of numerical methods focusing in buckling of thin shells is given by Riks and Rankin [198].

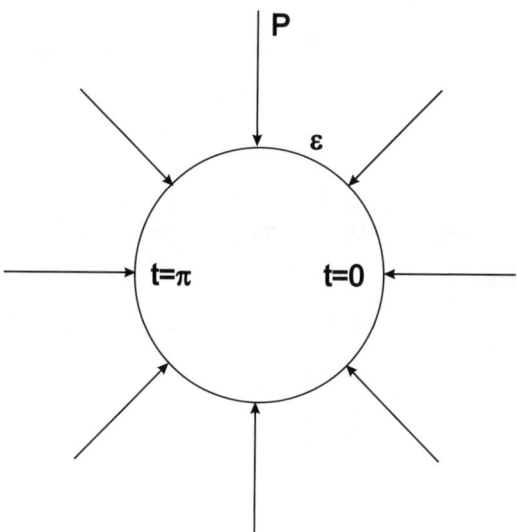

Figure 4.9. The circular ring under discrete, periodically spaced compressive loads.

The circular ring shown in Figure 4.9 is considered. The ring is linearly elastic, with material obeying the Euler-Bernulli theory, and is forced by a periodic array of N compressive static loads. It is assumed that all loads are identical and equally spaced, and that they are applied normally to the undeformed ring configuration. The normalized internal force, moment, displacement and rotation distributions are governed by the following set of

4.3 Static Buckling of a Circular Ring Compressive Loads

ordinary differential equations (Atanackovic, 1998),

$$\dot{q}(t) = -\frac{2\pi\tilde{\lambda}}{N}\sum_{k=0}^{N-1}\delta\left(2t - \frac{2\pi(2k+1)}{N}\right) - n(t)\left[1 + m(t)\right],$$

$$\dot{n}(t) = q(t)\left[1 + m(t)\right],$$
$$\dot{m}(t) = -q(t) \qquad (4.49)$$

and

$$\dot{u}(t) = \cos\theta(t),$$
$$\dot{v}(t) = \sin\theta(t),$$
$$\dot{\theta}(t) = 1 + m(t), \qquad (4.50)$$

where: $\delta(\cdot)$ denotes the Dirac function, and dot denotes differentiation with respect to t. Complementing these relations are the following periodicity conditions:

$$q(0) = q(2\pi), \quad n(0) = n(2\pi), \quad m(0) = m(2\pi),$$
$$u(0) = u(2\pi), \quad v(0) = v(2\pi), \quad \theta(0) = \theta(2\pi) + 2\pi. \qquad (4.51)$$

In the equations above, the normalized variables are defined as follows: $t = s/R$ where s denotes arc length of the undeformed ring axis, and R the radius of the undeformed ring; $q = QR^2/EI$ where q is the shear force, and EI the modulus of rigidity of the ring; $n = AR^2/EI$ where A is the axial force; $m = MR/EI$ where M is the bending moment; $u = x/R$ and $v = y/R$, where x and y are horizontal and vertical deformations, respectively, of an arbitrary point of the undeformed ring; $\tilde{\lambda} = [2P/(2\pi R/N)](R^3/EI) = NPR^2/(\pi EI)$, where P is the magnitude of the applied compressive concentrated load; and θ is the angle between the axial force N and the horizontal axis in the deformed ring configuration.

The set of singularities in the first of equations (4.49) prevents the direct application of asymptotic techniques to this problem. Hence, in similarity to the string problem discussed in the previous Section, one must first eliminate these singularities from the governing equations using the method of non-smooth transformations. To this end, one assumes that the concentrated loads are densely placed by requiring that the distance between adjacent loads is small, $\varepsilon = 2\pi/N \ll 1$. Then it is anticipated that the solution of the problem will possess two spatial scales, a long-scale t and a short one t/ε (recall, that this was also the case for the discretely supported string in the previous Section). It follows that the solution is sought in the two-scale form $w = w(t, t/\varepsilon)$, where $w = q, n, m, u, v$ or θ, and the short and long spatial scales will be treated as independent variables.

The short-scale variable of the problem is now replaced by the non-smooth variables,

$$t/\varepsilon \to \{\tau(t/\varepsilon), e(t/\varepsilon)\},$$

where the definitions (4.2) are used. The dependent variables of the problem are then expressed in the form:

$$q(t,t/\varepsilon) = Q_1(t,\tau) + eQ_2(t,\tau), \quad n(t,t/\varepsilon) = N_1(t,\tau) + eN_2(t,\tau),$$
$$m(t,t/\varepsilon) = M_1(t,\tau) + eM_2(t,\tau), \quad u(t,t/\varepsilon) = U_1(t,\tau) + eU_2(t,\tau),$$
$$v(t,t/\varepsilon) = V_1(t,\tau) + eV_2(t,\tau), \quad \theta(t,t/\varepsilon) = \Theta_1(t,\tau) + e\Theta_2(t,\tau). \qquad (4.52)$$

Substituting (4.52) into (4.49)–(4.51), using the chain rule to express differentiations with respect to t in terms of the new variables, and eliminating the resulting singular terms by setting their coefficients equal to zero, one obtains an alternative set of equations that do not contain any singular terms.

This procedure will be demonstrated by considering the first equation in (4.49), and expressing the derivative on the left hand side as follows,

$$\dot{q} = \frac{1}{\varepsilon}\frac{\partial Q_1}{\partial \tau}e + \frac{\partial Q_1}{\partial t} + \frac{1}{\varepsilon}Q_2\tau'' + \frac{1}{\varepsilon}\frac{\partial Q_2}{\partial \tau} + \frac{\partial Q_2}{\partial t}e \qquad (4.53)$$

where the relation $(\tau')^2 = e^2 = 1$ is used, and a prime denotes differentiation of a function with respect to its argument. The term τ'' is singular, as it represents an infinite series of Dirac functions:

$$\tau'' = -2\,\text{sgn}(\tau)\sum_{k=-\infty}^{\infty}\delta\left(\frac{2t}{\varepsilon} - 1 - 2k\right). \qquad (4.54)$$

This expression is similar to the singular term in the first of equations (4.49) in the domain $t \in [0, 2\pi)$, since:

$$\sum_{k=0}^{N-1}\delta\left(2t - \frac{2\pi(2k+1)}{N}\right) = \frac{1}{\varepsilon}\sum_{k=0}^{N-1}\delta\left(\frac{2t}{\varepsilon} - 1 - 2k\right) = \frac{1}{\varepsilon}\sum_{k=0}^{\infty}\delta\left(\frac{2t}{\varepsilon} - 1 - 2k\right),$$
$$t \in [0, 2\pi). \qquad (4.55)$$

It follows that the singularities in the equation can be cancelled by the singularities in the expression for the derivative (4.53). Indeed, expressing the singularities in the governing equation in terms of τ'', taking into account (4.53), setting to zero the coefficient multiplying τ'' in the resulting expression, and setting separately equal to zero the $R-$ and $I-$components, one obtains the following set of transformed equations,

$$\frac{\partial Q_1}{\partial \tau} = -\varepsilon\frac{\partial Q_2}{\partial t} - \varepsilon\left[N_1 M_2 + N_2(1 + M_1)\right],$$
$$Q_1(0,\tau) = Q_1(2\pi,\tau),$$
$$\frac{\partial Q_2}{\partial \tau} = -\varepsilon\frac{\partial Q_1}{\partial t} - \varepsilon\left[N_2 M_2 + N_1(1 + M_1)\right], \qquad (4.56)$$

4.3 Static Buckling of a Circular Ring Compressive Loads

$$Q_2(0,\tau) = Q_2(2\pi,\tau),$$
$$Q_2|_{\tau=\pm 1} = \mp\varepsilon\lambda,$$

where: $\lambda = -\tilde{\lambda}/2$. The last of the above relations is a smoothening condition that eliminates the singular terms (proportional to τ'') from the transformed governing equation. The above set of equations replaces the first of equations (4.49) and the first of the periodicity conditions (4.51). The transformed equations are smooth, but are of higher dimension compared to the original problem. Note also, that the smoothening condition appears to impose symmetry conditions on the I–part of the response, $Q_2(t,\tau)$; this, however, is not reflected in the physical buckling state of the ring, which as shown below may become unsymmetric.

Working in a similar fashion, one obtains the following set of transformed equations that complements relations (4.56)–(4.61), and fully replaces the original problem (4.49)–(4.51):

$$\frac{\partial N_1}{\partial \tau} = -\varepsilon\frac{\partial N_2}{\partial t} - \varepsilon\left[Q_1 M_2 + Q_2\left(1 + M_1\right)\right],$$

$$N_1(0,\tau) = N_1(2\pi,\tau),$$

$$\frac{\partial N_2}{\partial \tau} = -\varepsilon\frac{\partial N_1}{\partial t} - \varepsilon\left[Q_2 M_2 + Q_1\left(1 + M_1\right)\right], \quad (4.57)$$

$$N_2(0,\tau) = N_2(2\pi,\tau),$$

$$N_2|_{\tau=\pm 1} = 0,$$

$$\frac{\partial M_1}{\partial \tau} = -\varepsilon\frac{\partial M_2}{\partial t} - \varepsilon Q_2, \quad M_1(0,\tau) = M_1(2\pi,\tau),$$

$$\frac{\partial M_2}{\partial \tau} = -\varepsilon\frac{\partial M_1}{\partial t} - \varepsilon Q_1, \quad M_2(0,\tau) = M_2(2\pi,\tau), \quad (4.58)$$

$$M_2|_{\tau=\pm 1} = 0,$$

$$\frac{\partial U_1}{\partial \tau} = -\varepsilon\frac{\partial U_2}{\partial t} + \varepsilon I_c, \quad U_1(0,\tau) = U_1(2\pi,\tau),$$

$$\frac{\partial U_2}{\partial \tau} = -\varepsilon\frac{\partial U_1}{\partial t} + \varepsilon R_c, \quad U_2(0,\tau) = U_2(2\pi,\tau), \quad (4.59)$$

$$U_2|_{\tau=\pm 1} = 0,$$

$$\frac{\partial V_1}{\partial \tau} = -\varepsilon\frac{\partial V_2}{\partial t} + \varepsilon I_s, \quad V_1(0,\tau) = V_1(2\pi,\tau),$$

$$\frac{\partial V_2}{\partial \tau} = -\varepsilon\frac{\partial V_1}{\partial t} + \varepsilon R_s, \quad V_2(0,\tau) = V_2(2\pi,\tau), \quad (4.60)$$

$$V_2|_{\tau=\pm 1} = 0,$$

$$\frac{\partial \Theta_1}{\partial \tau} = -\varepsilon \frac{\partial \Theta_2}{\partial t} + \varepsilon M_2, \quad \Theta_1(0,\tau) = \Theta_1(2\pi,\tau) + 2\pi,$$

$$\frac{\partial \Theta_2}{\partial \tau} = -\varepsilon \frac{\partial \Theta_1}{\partial t} + \varepsilon(1+M_1), \quad \Theta_2(0,\tau) = \Theta_2(2\pi,\tau), \quad (4.61)$$

$$\Theta_2|_{\tau=\pm 1} = 0,$$

where:

$$R_c = (1/2)\left[\cos(\Theta_1 + \Theta_2) + \cos(\Theta_1 - \Theta_2)\right],$$
$$I_c = (1/2)\left[\cos(\Theta_1 + \Theta_2) - \cos(\Theta_1 - \Theta_2)\right],$$
$$R_s = (1/2)\left[\sin(\Theta_1 + \Theta_2) + \sin(\Theta_1 - \Theta_2)\right],$$
$$I_s = (1/2)\left[\sin(\Theta_1 + \Theta_2) - \sin(\Theta_1 - \Theta_2)\right].$$

The transformed equations (4.56)–(4.61) form 6 nonlinear boundary value problems (NLBVPs) in terms of the non-smooth variable $\tau \in [-1,1]$, with the normalized load λ playing the role of the (nonlinear) eigenvalue. Moreover, the solutions of the NLBVPs ((4.59)–(4.61)) can be derived by direct integrations once the solutions of NLBVPs ((4.56)–(4.58)) have been derived; hence, in what follows only these later set of problems will be analyzed. However, the NLBVP (4.61) must also be considered at certain stages of the analysis in order to provide necessary compatibility conditions for the overall solution.

Considering ε to be the perturbation parameter of the problem, one seeks the solutions of (4.56)–(4.58) and (4.61) in the following regular perturbation series form:

$$W_i(t,\tau) = W_i^{(0)}(t,\tau) + \varepsilon W_i^{(1)}(t,\tau) + \varepsilon^2 W_i^{(2)}(t,\tau) + \ldots,$$
$$\lambda = \lambda_0 + \varepsilon\lambda_1 + \varepsilon^2\lambda_2 + \ldots, \quad (4.62)$$

where: $W = Q, N, M, \Theta$ and $i = 1, 2$. Substituting (4.62) into the NLBVPs under consideration and matching the coefficients of equal power of ε, one obtains an hierarchy of subproblems at different orders of approximation of the solution.

$O(\varepsilon^0)$ System

Considering terms of $O(\varepsilon^0)$, one obtains the following leading-order approximations:

$$Q_1^{(0)}(t,\tau) = A_0(t), \quad A_0(0) = A_0(2\pi), \quad Q_2^{(0)}(t,\tau) = 0,$$
$$N_1^{(0)}(t,\tau) = B_0(t), \quad B_0(0) = B_0(2\pi), \quad N_2^{(0)}(t,\tau) = 0,$$
$$M_1^{(0)}(t,\tau) = C_0(t), \quad C_0(0) = C_0(2\pi), \quad M_2^{(0)}(t,\tau) = 0,$$
$$\Theta_1^{(0)}(t,\tau) = \gamma_0(t), \quad \gamma_0(0) = \gamma_0(2\pi) + 2\pi, \quad \Theta_2^{(0)}(t,\tau) = 0. \quad (4.63)$$

The yet undetermined, t–dependent functions in the above expressions result as constants of integration with respect to τ; these are evaluated by solving higher order problems. Note that at the leading order of approximation no information on the normalized load λ is extracted.

$O(\varepsilon^1)$ System

Solving the $O(\varepsilon)$ problem one obtains the following approximations,

$$Q_1^{(1)}(t,\tau) = A_1(t), \quad A_1(0) = A_1(2\pi), \quad Q_2^{(1)}(t,\tau) = -\lambda_0 \tau,$$

$$N_1^{(1)}(t,\tau) = B_1(t), \quad B_1(0) = B_1(2\pi), \quad N_2^{(1)}(t,\tau) = 0,$$

$$M_1^{(1)}(t,\tau) = C_1(t), \quad C_1(0) = C_1(2\pi), \quad M_2^{(1)}(t,\tau) = 0,$$

$$\Theta_1^{(1)}(t,\tau) = \gamma_1(t), \quad \gamma_1(0) = \gamma_1(2\pi), \quad \Theta_2^{(1)}(t,\tau) = 0, \qquad (4.64)$$

together, with the following expressions that govern the yet undetermined functions of the previous order of approximation [by enforcing the periodicity conditions in (4.63)]:

$$\dot{A}_0(t) = -B_0(t)\left[1 + C_0(t)\right] + \lambda_0,$$

$$\dot{B}_0(t) = -A_0(t)\left[1 + C_0(t)\right],$$

$$\dot{C}_0(t) = -A_0(t),$$

$$\dot{\gamma}_0(t) = 1 + C_0(t). \qquad (4.65)$$

The last of the equations above is decoupled from the first three, which constitute a NLBVP with λ_0 as the eigenvalue. It turns out that this is identical to the eigenvalue problem that results when the ring is forced by a uniform (hydrostatic) pressure distribution. In essence, this smooth problem governs the dependence of the solution on the long variable t; discreteness effects appear at higher orders and are manifested by the presence of the non-smooth variables $\tau(t/\varepsilon)$ and $e(t/\varepsilon)$ in the solution. The solution of the first-order problem (4.65) is know discussed.

The base (unbuckled) state of the ring is determined by setting the derivatives in the first three equations equal to zero, leading to the relation $\hat{B}_0\left[1 + \hat{C}_0\right] = \lambda_0$; solving the last of equations (4.65) one determines uniquely the base state as follows:

$$\hat{A}_0 = 0, \quad \hat{B}_0 = \lambda_0, \quad \hat{C}_0 = 0, \quad \hat{\gamma}_0(t) = t. \qquad (4.66)$$

To analytically study the bifurcation that leads to buckling on introduces the following perturbation of the base state,

$$\begin{Bmatrix} A_0(t) \\ B_0(t) \\ C_0(t) \\ \gamma_0(t) \end{Bmatrix} = \begin{Bmatrix} 0 \\ \lambda_0 \\ 0 \\ t \end{Bmatrix} + \mu \begin{Bmatrix} x_1(t) \\ x_2(t) \\ x_3(t) \\ x_4(t) \end{Bmatrix}, \quad 0 < \mu \ll 1, \qquad (4.67)$$

where: μ is a second perturbation parameter, independent from ε. This second parameter governs the proximity of the buckled state to the base state of the ring. Substituting (4.67) into (4.65) one obtains the following NLBVP in the neighborhood of the base state:

$$\dot{x}_1 = -\lambda_0(t)x_3 - x_2(1 + \mu x_3),$$

$$\dot{x}_2 = x_1(1 + \mu x_3),$$

$$\dot{x}_3 = -x_1,$$

$$\dot{x}_4 = x_3,$$

$$x_i(0) = x_i(2\pi), \quad i = 1, 2, 3, 4.$$
(4.68)

Seeking a solution in the form,

$$x_i(t) = x_i^{(0)}(t) + \mu x_i^{(1)}(t) + \mu^2 x_i^{(2)}(t) + ..., \quad i = 1, 2, 3, 4,$$

$$\lambda_0 = \lambda_0^{(0)} + \mu \lambda_0^{(1)} + ...$$
(4.69)

substituting into (4.68), and matching coefficients of equal powers of μ, one obtains a series of linear, constant parameter boundary value problems that can be solved explicitly. Omitting the details of this analysis, the results are summarized as follows,

$$\begin{Bmatrix} A_0(t) \\ B_0(t) \\ C_0(t) \\ \gamma_0(t) \end{Bmatrix} = \begin{Bmatrix} 0 \\ \lambda_0 \\ 0 \\ t \end{Bmatrix} + \delta \begin{Bmatrix} \sin jt \\ -(1/j)\cos jt \\ (1/j)\cos jt \\ -(1/j^2)\sin jt \end{Bmatrix} +$$

$$\delta^2 \begin{Bmatrix} (1/2j^3)\sin 2jt \\ (1/4j^4)\left[2j^2 - (1+j^2)\cos 2jt\right] \\ (1/4j^4)\cos 2jt \\ (1/8j^5)\sin 2jt \end{Bmatrix} + O(\delta^3),$$
(4.70)

$$\delta = \pm \left[\frac{8j^4(\lambda_0 - 1 + j^2)}{3(1 - j^2)}\right]^{1/2}, \quad j = 2, 3, ...,$$

where the small perturbation parameter $|\delta| \ll 1$ was obtained by expressing μ in terms of λ_0 from the last of relations (4.69), and substituting into the local perturbation solutions for $A_0(t)$, $B_0(t)$, $C_0(t)$ and $\gamma_0(t)$. The plus and minus signs in the definition of δ correspond to the two bifurcating branches of buckled states, which at this order are identical. As shown below, higher order corrections perturb the symmetry of the bifurcating branches. Also, it is noted that from a mathematical point of view the periodicity conditions in the last of the set of equations (4.68) provide compatibility relations that determine uniquely the solutions of the first three of equations of this set.

The critical buckling loads correspond to $\delta = 0$, and are given by $\lambda_{0j} = 1 - j^2$; from a physical point of view these are the critical buckling loads of the hydrostatically forced circular ring. By changing the index j one examines

4.3 Static Buckling of a Circular Ring Compressive Loads

solutions in the neighborhoods of higher order bifurcation points (buckling states). Note that by their construction the previous asymptotic analysis is valid only in the neighborhoods of the bifurcation points, e.g., for $0 < \lambda_0 - 1 + j^2 \ll 1$, $j = 2, 3, \ldots$. The parameter δ is a measure of the distance of the buckled state from the initiation of that state (e.g., the bifurcation point). The solution corresponding to $j = 2$ corresponds to the first buckling load, and will be examined closely in the next order of approximation.

$O(\varepsilon^2)$ System

Proceeding to the next order of approximation, one derives the following expressions:

$$Q_1^{(2)}(t,\tau) = A_2(t), \quad A_2(0) = A_2(2\pi), \quad Q_2^{(2)}(t,\tau) = -\lambda_1 \tau,$$

$$N_1^{(2)}(t,\tau) = -\lambda_0 \left[1 + C_0(t)\right] \frac{\tau^2}{2} + B_2(t), \quad B_2(0) = B_2(2\pi), \quad N_2^{(2)}(t,\tau) = 0,$$

$$M_1^{(2)}(t,\tau) = \lambda_0 \frac{\tau^2}{2} + C_2(t), \quad C_2(0) = C_2(2\pi), \quad M_2^{(2)}(t,\tau) = 0,$$

$$\Theta_1^{(2)}(t,\tau) = \gamma_2(t), \quad \gamma_2(0) = \gamma_2(2\pi), \quad \Theta_2^{(2)}(t,\tau) = 0 \quad (4.71)$$

and the complementary equations that evaluate the undetermined functions of the previous order of approximation,

$$\dot{A}_1(t) = -B_0(t)C_1(t) - B_1(t)\left[1 + C_0(t)\right] + \lambda_1,$$

$$\dot{B}_1(t) = -A_0(t)C_1(t) + A_1(t)\left[1 + C_0(t)\right],$$

$$\dot{C}_1(t) = -A_1(t),$$

$$\dot{\gamma}_1(t) = C_1(t), \quad (4.72)$$

with the periodicity conditions in (4.64) enforced. One notes the first appearance of discreteness effects in the terms Q_2^2, $N_1^{(2)}$ and $M_1^{(2)}$. Again, this represents a NLBVP with λ_1 as eigenvalue.

It is noted, however, that this problem admits the trivial solution,

$$A_1(t) = B_1(t) = C_1(t) = \gamma_1(t) = \lambda_1 = 0 \quad (4.73)$$

indicating absence of $O(\varepsilon)$ t–dependent terms in the solution.

$O(\varepsilon^3)$ System

The solutions at this order of approximation are given by:

$$Q_1^{(3)}(t,\tau) = A_3(t), \quad A_3(0) = A_3(2\pi),$$

$$Q_2^{(3)}(t,\tau) = \left\{ B_0(t)\frac{\lambda_0}{6} - \lambda_0\left[1 + C_0(t)\right]^2 - \lambda_2 \right\}\tau - \left\{ B_0(t) - \left[1 + C_0(t)\right]^2 \right\}\lambda_0 \frac{\tau^3}{6},$$

266   4 Analytic Modelling of Discreteness Effects in Continuous Systems

$$N_1^{(3)}(t,\tau) = B_3(t), \quad B_3(0) = B_3(2\pi), \quad N_2^{(3)}(t,\tau) = 0,$$
$$M_1^{(3)}(t,\tau) = C_3(t), \quad C_3(0) = C_3(2\pi), \quad M_2^{(3)}(t,\tau) = 0,$$
$$\Theta_1^{(3)}(t,\tau) = \gamma_3(t), \quad \gamma_3(0) = \gamma_3(2\pi), \quad \Theta_2^{(2)}(t,\tau) = -\frac{\lambda_0}{6}\left(\tau - \tau^3\right). \quad (4.74)$$

The undetermined functions of the previous order of approximation are governed by the following set:

$$\dot{A}_2(t) = -B_0(t)C_2(t) - B_2(t)\left[1 + C_0(t)\right] - \frac{\lambda_0}{6}B_0(t) + \lambda_0\left[1 + C_0(t)\right]^2 + \lambda_2,$$

$$\dot{B}_2(t) = -A_0(t)C_2(t) + A_2(t)\left[1 + C_0(t)\right],$$

$$\dot{C}_2(t) = -A_2(t),$$

$$\dot{\gamma}_2(t) = C_2(t) + \frac{\lambda_0}{6}. \quad (4.75)$$

The periodicity conditions in (4.71) were enforced.

The problem (4.75) does not admit a trivial solution. Taking into account the local solution (4.70), the above equations are written in the form of the following parametrically varying linear boundary value problem with eigenvalue λ_2:

$$\left\{\begin{array}{c}\dot{A}_2(t)\\ \dot{B}_2(t)\\ \dot{C}_2(t)\end{array}\right\} + \left[\begin{array}{ccc} 0 & 1+\delta(1/j)\cos jt & 1-j^2-\delta(1/j)\cos jt\\ -1-\delta(1/j)\cos jt & 0 & -\delta\sin jt\\ 1 & 0 & 0 \end{array}\right]\left\{\begin{array}{c}A_2(t)\\ B_2(t)\\ C_2(t)\end{array}\right\} =$$

$$\left\{\begin{array}{c}-\left[(1-j^2)/6\right]\left[1-j^2+\delta(1/j)\cos jt\right]+\left[(1-j^2)/6\right]\left[1+2\delta(1/j)\cos jt\right]+\lambda_2\\ 0\\ 0\end{array}\right\} + O(\delta^2),$$

$$\dot{\gamma}_2(t) = C_2(t) + \frac{\lambda_0}{6}. \quad (4.76)$$

This problem is solved by expressing the dependent variables and the eigenvalue in formal series in terms of δ, e.g., $A_2(t) = A_2^{(0)}(t) + \delta A_2^{(1)} + \ldots$, $\lambda_2 = \lambda_2^{(0)} + \delta\lambda_2^{(1)} + \ldots$; substituting into (4.76) one solves am hierarchy of linear boundary value problems with constant coefficients at successive orders of δ. The following asymptotic solution is then derived:

$$A_2(t) = 0 + O(\delta^2),$$

$$B_2(t) = \frac{1-j^2}{6} + \delta\frac{1-j^2}{6}\cos jt + O(\delta^2),$$

$$C_2(t) = -\frac{1-j^2}{6} + O(\delta^2),$$

$$\gamma_2(t) = 0 + O(\delta^2), \quad \lambda_2 = 0 + O(\delta^2), \quad j = 2, 3, \ldots. \quad (4.77)$$

4.3 Static Buckling of a Circular Ring Compressive Loads

These results indicate that the correction to the critical buckling load due to discreteness effects is at least of $O(\varepsilon^2\delta^2, \varepsilon^3)$, and thus small; this is in agreement with numerical results reported in [206, 97]. In addition, from the expression of $B_2(t)$ one concludes that the post-buckled discretely loaded ring depends on the sign of δ; as a result, the two bifurcating (buckling) states of the ring corresponding to the positive and negative values of δ are not identical (as in the case of hydrostatic loading), and asymmetries develop. Again, this analytical result is in agreement with the numerical results of [206].

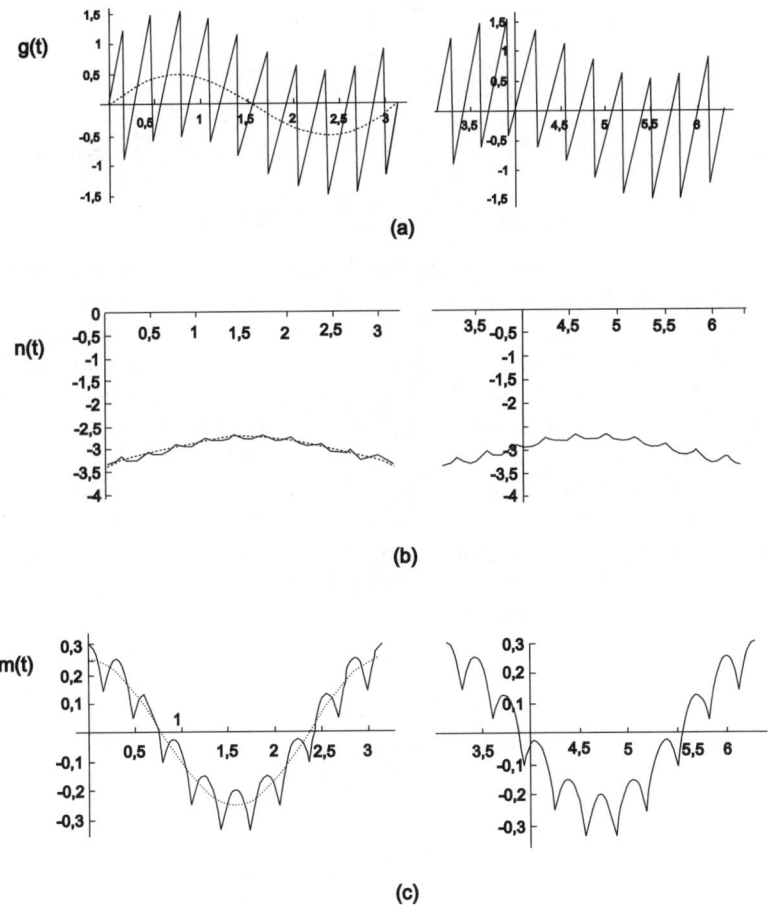

Figure 4.10. Numerical results for $0 \leq t < \pi$ and $\pi \leq t < 2\pi$: Normalized (a) shear, (b) axial, and (c) bending distribution; ——— discrete, and - - - - - uniform forcing distribution.

268   4 Analytic Modelling of Discreteness Effects in Continuous Systems

Summarizing all previous findings, the post-buckled state of the discretely loaded circular ring is approximated as follows:

$$q(t, t/\varepsilon) = \left[A_0(t) + O\left(\delta^3, \varepsilon^2\delta^2, \varepsilon^3\right)\right] +$$

$$e\left\{-\varepsilon\lambda_0\tau + \varepsilon^3\left\{\left[B_0(t)\frac{\lambda_0}{6} - \lambda_0\left[1 + C_0(t)\right]^2 - \lambda_2\right]\tau - \frac{\lambda_0\tau^3}{6}\left[B_0(t) - \left[1 + C_0(t)\right]^2\right]\right\} + O\left(\varepsilon^3\delta^3, \varepsilon^4\right)\right\},$$

$$n(t, t/\varepsilon) = \left\{B_0(t) + \varepsilon^2\left\{-\lambda_0\left[1 + C_0(t)\right]\frac{\tau^2}{2} + B_2(t)\right\} + O\left(\delta^3, \varepsilon^2\delta^2, \varepsilon^3\right)\right\} + e\left[O\left(\varepsilon^4\right)\right],$$

$$m(t, t/\varepsilon) = \left\{C_0(t) + \varepsilon^2\left\{\lambda_0\frac{\tau^2}{2} + C_2(t)\right\} + O\left(\delta^3, \varepsilon^2\delta^2, \varepsilon^3\right)\right\} + e\left[O\left(\varepsilon^4\right)\right],$$

$$\tilde{\lambda} = -2\lambda_0 + O\left(\varepsilon^2\delta^2, \varepsilon^3\right), \quad j = 2, 3, \dots . \quad (4.78)$$

Note that in contrast to the normalized shear force q, the normalized axial force n and bending moment m do not possess any e–dependent terms; this result indicates the presence of C^0–discontinuities in the spatial distribution of the shear force, but only C^p-discontinuities, $p \geq 1$, in the spatial distributions of the axial force and bending moment. The remaining variables in (4.50) are determined by integrating the corresponding governing equations.

The solutions (4.74) are depicted in Figure 4.10 for $j = 2$, $\varepsilon = 2\pi/20$, $\delta = 0.5$. These parameters correspond to the first buckling mode of a ring with 20 compressive normalized loads of magnitude $\lambda = -3.0176$. For comparison purposes, the results for the hydrostatically forced buckled ring with $\varepsilon \to 0$ are also shown. The discreteness effects in the spatial distributions of the depicted quantities are evident.

5 Continuous Systems with Non-smooth Nonlinearities

Non-smooth nonlinearities are rather common in engineering problems involving continuous systems. Examples are clearance nonlinearities due to loose joints that connect substructures, vibro-impact nonlinearities when vibrating media contact rigid boundaries, dry friction forces generated at the interface between sliding elastic members, and micro- and macro-slip effects at mechanical joint interfaces. Hence, the study of non-smooth nonlinearities is important for being able to understand and (more important) predict the dynamical phenomena caused by assemblies of continuous systems that exhibit this type of nonlinearities. The dynamics of systems with non-smooth elements is difficult to analyze since the motion of such systems is inherently multi-phase, that is, the dynamical response is governed by different linear or nonlinear differential equations at different phases of the motion, with continuity conditions being imposed at the intersections between phases. In this Chapter some specific continuous systems with non-smooth nonlinearities are examined in detail, with the aim to demonstrate the variety of complex phenomena that can be exhibited by such systems.

5.1 Transient Localization Due to Backlash Nonlinearities

In this first application, transient nonlinear localization and beat phenomena are studied in a system of two rods coupled with a nonlinear backlash spring. The method of proper orthogonal decomposition (POD) introduced in Chapter 3 is used to reduce the order of the dynamics, and to study nonlinear effects by considering energy transfers between leading POD modes. Moreover, the computed POD modes are used to discretize the governing partial differential equations, thus creating accurate and computationally efficient low-dimensional nonlinear models of the system. Reconstruction of transient nonlinear responses using these low dimensional models reveals the accuracy of the order reduction. Finally, Poincaré maps are utilized to study the nonlinear localization and beat phenomena caused by the clearance connecting the coupled rods.

5.1.1 Transient Localization and Beat Phenomena Due to Clearance

The system under consideration is depicted in Figure 5.1. It consists of two identical rods connected by a spring with clearance nonlinearity. The load displacement curve of the nonlinear spring is as depicted in that Figure. Rod 2 is connected to the ground by a linear spring. The transient force F(t) applied to rod 1 is also depicted in Figure 5.1, and local coordinate systems for the two rods are adopted. The governing equations of motion are given by,

$$m\frac{\partial^2 w_1(x,t)}{\partial t^2} - EA\frac{\partial^2 w_1(x,t)}{\partial x^2} = k_1\delta(x_1-L)(w_2(0,t)-w_1(L,t))+F(t)\delta(x_1-0),$$

$$m\frac{\partial^2 w_2(x,t)}{\partial t^2} - EA\frac{\partial^2 w_2(x,t)}{\partial x^2} =$$
$$-k_1\delta(x_2-0)(w_2(0,t)-w_1(L,t)) - k_2\delta(x_2-L)w_2(L,t), \quad (5.1)$$

where: k_1 and k_2 are the stiffnesses of the springs given by,

$$k_1 = \begin{cases} S, \Delta w < a, \Delta w > b \\ 0, a \leq \Delta w \leq b \end{cases}, \quad k_2 = S,$$

where: $\Delta w = w_1(0) - w_1(L)$, and S is a stiffness characteristic. The system is subject to the following initial conditions:

$$w_1(x,0) = \frac{\partial w_1(x,0)}{\partial t} = w_2(x,0) = \frac{\partial w_2(x,0)}{\partial t} = 0. \quad (5.2)$$

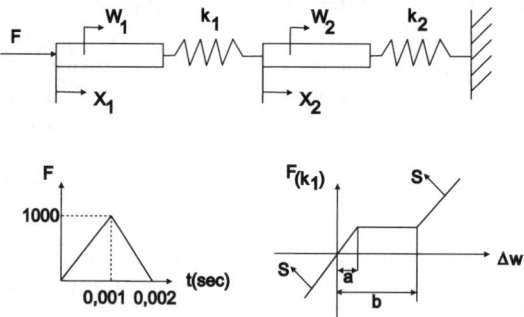

Figure 5.1. The continuous system with clearance nonlinearity.

5.1 Transient Localization Due to Backlash Nonlinearities

The numerical simulation of equations (5.1) and (5.2) is computed by Galerkin discretization, i.e., by expressing the longitudinal displacements in the following series forms:

$$w_1(x,t) = \sum_{i=1}^{\infty} A_{1,i}(t)\phi_i(x), \quad w_2(x,t) = \sum_{i=1}^{\infty} A_{2,i}(t)\phi_i(x), \quad (5.3)$$

where: $\phi_i(x)$ represents the ith normalized linear physical mode, and $A_{p,i}(t)$, $p=1,2$ the amplitude of the ith mode of rod p. For the simulations we employ the modes corresponding to free-free boundary conditions, $\phi_i(x) = \cos\frac{i\pi}{L}x$. Employing the well-known orthogonality properties of modes $\phi_i(x)$, the partial differential equations of motion (1) can be approximately replaced by the following set of ordinary differential equations governing the time evolution of the modal amplitudes:

$$\ddot{A}_{1,j}(t) + p_j^2 A_{1,j} = \frac{k_1\phi_j(L)}{mC_j}\left\{\sum_{i=1}^{5}[A_{2,i}(t)\phi_i(0) - A_{1,i}(t)\phi_i(L)]\right\} + \frac{F(t)}{mC_j}\phi_j(0),$$

$$\ddot{A}_{2,j}(t) + p_j^2 A_{2,j} = -\frac{k_1\phi_j(0)}{mC_j}\left\{\sum_{i=1}^{5}[A_{2,i}(t)\phi_i(0) - A_{1,i}(t)\phi_i(L)]\right\} -$$

$$\frac{k_2\phi_j(L)}{mC_j}\left\{\sum_{i=1}^{5}A_{2,i}(t)\phi_i(L)\right\}, \quad (5.4)$$

where the index j is truncated to 5 (only five physical modes are used), p_j is the natural frequency of the jth physical mode, and $C_j = \int_0^L \phi_j^2(x)dx$.

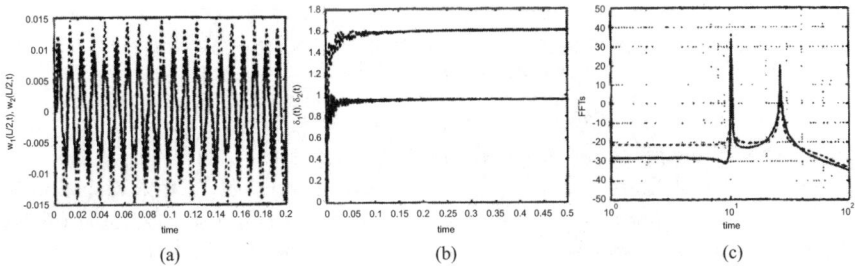

Figure 5.2. Middle-point responses of the linear system of two rods: (a) Time series, (b) energy distributions, (c) FFTs; - - - - - - rod 1, ——— rod 2.

The transient responses at the middle points of the two rods (cf. Fig. 5.1) are studied for the system with parameters $m = 0.0853 kg/m$ (mass per unit

272   5 Continuous Systems with Non-smooth Nonlinearities

length), $EA = 2.24 \cdot 10^6 Pa \cdot m^2$ (where E is the modulus of the elasticity and A the cross section of the rod), and $L = 1.0m$ (length of the rods). Figure 5.2a depicts the transient responses of the linear system (no clearance, $a = b = 0$) with $S = 9.0 \cdot 10^4 N/m$. In Figure 5.2b the time averaged energy distributions $\delta_1(t)$ and $\delta_2(t)$ of the left and right rods are plotted, respectively; these are defined by,

$$\delta_i(t) = \frac{1}{t}\int_0^t E_i(t)dt, \quad i = 1, 2 ,$$

where: $E_i(t)$ is the instantaneous total energy of rod i at time t.

In Figure 5.3a we depict the nonlinear transient responses at the middle points of the rods for clearance parameters $a = 0.002m$, $b = 0.03m$. The comparison between the time averaged energy distributions of the two cases shows that for the specific clearance used more transient energy localization is achieved in the directly excited rod. Moreover, we observe that the motion of the two rods resembles a subharmonic oscillation, with rod 2 (the unexcited one) oscillating with higher frequency than rod 1. Increased clearance, however, is not always associated with increased transient energy localization as the following set of simulations demonstrates.

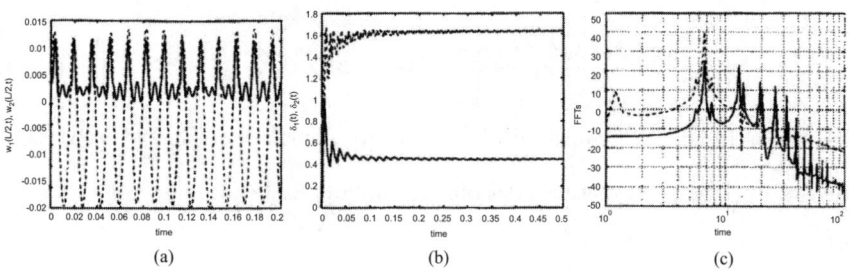

Figure 5.3. Middle-point responses of the system of two rods with backlash: (a) Time series, (b) energy distributions, (c) FFTs; - - - - - - rod 1, ——— rod 2.

In the following series of simulations the clearance of the spring was kept fixed, and the stiffness of the joint was varied. The effect of this variation on the nonlinear dynamics of the system is complex. In Figure 5.5 the transient middle-point rod responses are plotted for a joint with $S = 2.6 \cdot 10^N/m$, $a = 0.002m$, $b = 0.03m$. In this case a very clear transient beat phenomenon takes place as evidenced in the plots of the time-averaged energy distributions. The beat phenomenon eliminates the transient localization in the first rod. If one compares these results with the corresponding linear transient responses depicted in Figure 5.4, one concludes that the beat of Figure 5.5b is due solely to the nonlinear clearance, and eliminates the localization in the linear system. The same transient nonlinear beat phenomenon can be observed

5.1 Transient Localization Due to Backlash Nonlinearities 273

when the stiffness of the coupling element is set to $3.5 \cdot 10^4 N/m$ as shown in Figures 5.6 and 5.7. Again, the localization observed in the linear system is eliminated due to clearance.

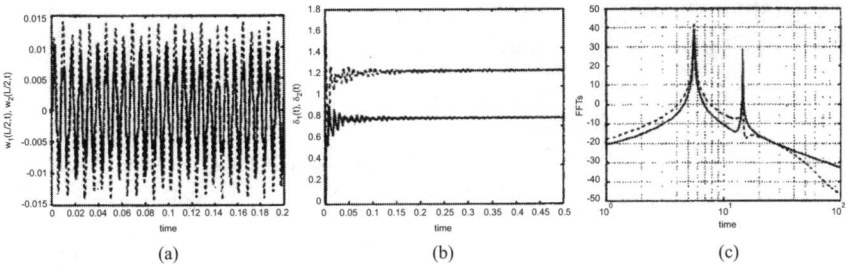

Figure 5.4. Middle-point responses of the system of the system of two linear rods: (a) Time series, (b) energy distributions, (c) FFTs; - - - - - - rod 1, ——— rod 2, $a = 0.0m$, $b = 0.0m$, $S = 2.6 \cdot 10^4 N/m$.

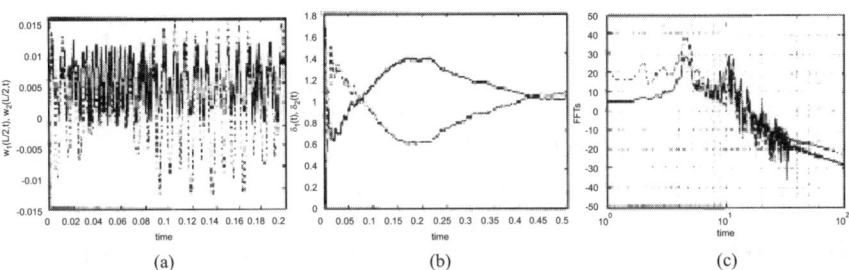

Figure 5.5. Middle-point responses of the system of the system of two rods with backlash: (a) Time series, (b) energy distributions, (c) FFTs; - - - - - - rod 1, ——— rod 2, $a = 0.002m$, $b = 0.03m$, $S = 2.6 \cdot 10^4 N/m$.

In Figures 5.2c and 5.3c the Fast Fourier Transforms (FFTs) of the transient responses at the middle points of the rods are also depicted when the stiffness of the coupling spring is equal to $9.0 \cdot 10^4 N/m$. From these plots, one observes that in the nonlinear case the energy spreads over a wide frequency range due to the clearance of the nonlinear coupling element. Moreover, the subharmonic localized motion of rod 2 is evidenced by the stronger higher frequency 'peaks' in the FFT of the response of that rod in the plot of Figure 5.3c. In Figures 5.4c, 5.5c, 5.6c, and 5.7c the corresponding FFTs of the transient responses are depicted for the cases when the coupling stiffness is $2.6 \cdot 10^4 N/m$ and $3.5 \cdot 10^4 N/m$. The nonlinear plots correspond to beat

274   5 Continuous Systems with Non-smooth Nonlinearities

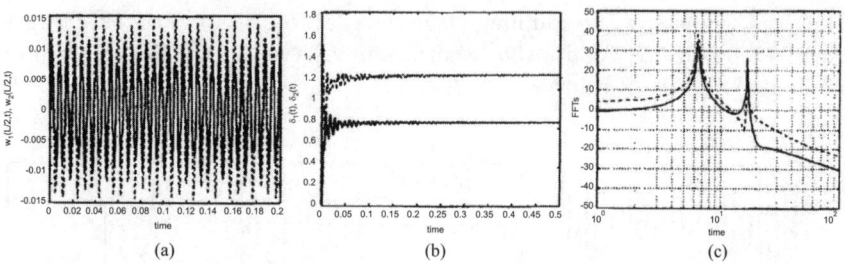

Figure 5.6. Middle-point responses of the system of the linear system of two rods: (a) Time series, (b) energy distributions, (c) FFTs; - - - - - - rod 1, ——— rod 2, $a = 0.0m$, $b = 0.0m$, $S = 3.5 \cdot 10^4 N/m$.

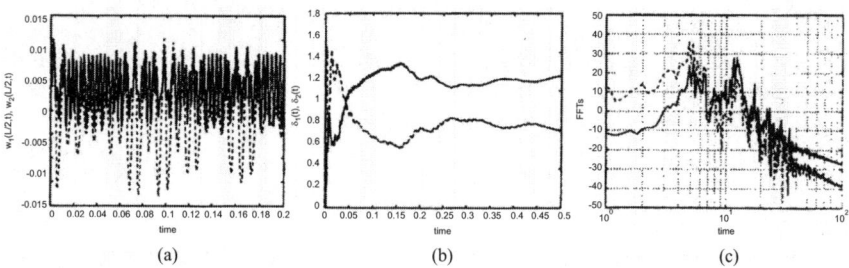

Figure 5.7. Middle-point responses of the system of the system of two rods with backlash: (a) Time series, (b) energy distributions, (c) FFTs; - - - - - - rod 1, ——— rod 2, $a = 0.002m$, $b = 0.03m$, $S = 3.5 \cdot 10^4 N/m$.

phenomena between the rods, as evidenced by the wider spread of energy in higher frequency ranges, and by the relative strength of the FFTs of the responses of rod 2 between 80 and $160Hz$.

5.1.2 Order Reduction and Response Reconstruction

As discussed in Chapter 3, the method of proper orthogonal decomposition (POD) enables one to extract spatial information from a set of time series data available on a spatio-temporal domain. The direct version of the method will be used to compute the POD modes of the system with clearance nonlinearity considered herein. To this end, the rod displacement at location x_j at time instant t_n, $n = 1, 2, \ldots, N$ is denoted by $v(x_j, t_n)$. The aim of the K-L decomposition is to find the optimal orthogonal eigenfunctions (POD modes) that best 'fit' the 'cloud of measurements' $v(x_j, t_n)$. To this end, a two-point correlation matrix is constructed, where the (i, j)-th element is computed as follows:

$$K_{i,j} = \frac{1}{N} \sum_{n=1}^{N} v(x_i, t_n) v(x_j, t_n) \,. \tag{5.5}$$

5.1 Transient Localization Due to Backlash Nonlinearities

By solving the linear eigenvalue problem,

$$K\phi = \lambda\phi \tag{5.6}$$

one computes the POD modes ϕ; the corresponding eigenvalues λ represent the amount of the energy of the signal captured by the corresponding mode. By construction, the resulting POD modes form an orthogonal basis. In Figure 5.8 the leading POD mode of the coupled nonlinear system is depicted for different system parameters. Clearly, the POD modes are different from the corresponding physical vibration modes, since they are in no way associated with natural frequencies; however, the POD modes provide the basic coherent spatial structures of the system, and their energies provide a measure of the dimensionality of the dynamics of the system. Indeed, if most of the energy of the dynamical response is captured by m K-L modes, it can be stated that the dimensionality of the system is approximately m.

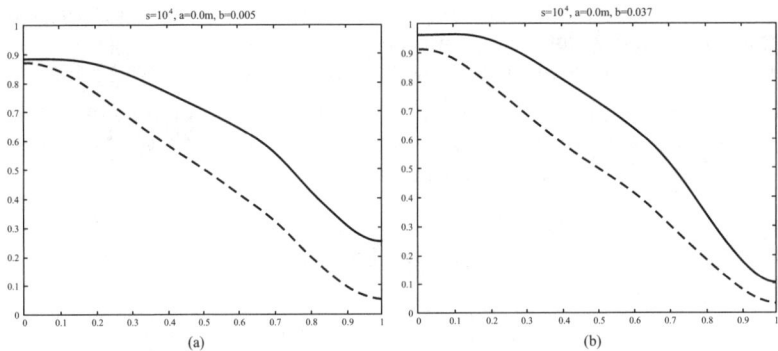

Figure 5.8. The leading POD mode of the system with (a) $a = 0.0m$, $b = 0.005m$, $S = 10^4 N/m$, (b) $a = 0.002m$, $b = 0.03m$, $S = 3.5 \cdot 10^4 N/m$; ------ rod 1, ——— rod 2.

When one considers systems composed of many elements, the way of constructing the correlation matrix is very important in order to have computationally efficient order reduction. Considering, for example the system of Figure 5.1, if the correlation matrix is constructed separately for each element of the system, the discretization of the governing partial differential equations (PDEs) is computationally inefficient. Instead, if one introduces *coupled POD modes*, i.e., POD modes defined over the entire spatial domain of the structure, one greatly reduces the order of the discretized system and makes the order reduction computationally accurate and efficient (a similar methodology was adopted for discretizing the coupled PDEs of the four-bay truss structure in Chapter 3, Section 3.3.3).

In the numerical simulations of the two-coupled rods discussed in the previous Section, a system of 10 coupled ODEs was considered since 5 physical modes were utilized to model the dynamics of each rod. In what follows it will be shown that a reduced order model composed of only two coupled ODEs is sufficient for reconstructing the transient nonlinear response when one bases the order reduction on coupled POD modes.

The correlation matrix K is computed in the spatial domain of the two rods, i.e., one selects M positions on the two rods to construct a $M \times M$ symmetric real matrix K. Hence, the i-th POD mode includes two components, $\phi_i^{(1)}$, $\phi_i^{(2)}$, where the superscripts stand for the components of the particular mode in rods 1 and 2, respectively; the corresponding eigenvalue λ_i indicates the relative importance of the POD mode. In this way, the displacements of the rods can be expressed as follows:

$$w_1(x,t) = \sum_{i=1}^{N} A_i(t)\phi_i^{(1)}(x), \quad w_2(x,t) = \sum_{i=1}^{N} A_i(t)\phi_i^{(2)}(x), \quad 0 \leq x \leq L \ . \quad (5.7)$$

Note that since one considers a single coupled POD mode having components in both rods 1 and 2, one needs only one modal amplitude $A_i(t)$ to describe the transient modal evolution in both rods. Substituting (5.7) into (5.1), multiplying each equation by the relevant POD mode component, $\phi_i^{(p)}$, $p = 1, 2$, integrating with respect to x from 0 to L, and adding the two resulting equations, the following low order model is obtained,

$$\ddot{A}_j(t) \left[\int_0^L \phi_j^{(1)^2}(x)dx + \int_0^L \phi_j^{(2)^2}(x)dx \right] -$$

$$\frac{EA}{m} \sum_{i=1}^{p} A_i(t) \left[\int_0^L \phi_i^{(1)} \phi_j^{(1)} dx + \int_0^L \phi_i^{(2)} \phi_j^{(2)} dx \right] =$$

$$\frac{k_1}{m} \left[\phi_j^{(1)}(L) - \phi_j^{(2)}(0) \right] \left\{ \sum_{i=1}^{p} [A_i(t)\phi_i^{(2)}(0) - A_i(t)\phi_i^{(1)}(L)] \right\} -$$

$$\frac{K_2 \phi_j^{(2)}(L)}{m} \sum_{i=1}^{p} A_i(t)\phi_i^{(2)}(L) - \frac{F(t)}{m} \phi_j^{(1)}(0), \quad (5.8)$$

where: p is the number of POD modes needed to reconstruct the response of the system. For accurate reconstructions, p is selected by the criterion:

$$\frac{\sum_{i=1}^{p} \lambda_i}{\sum_{i=1}^{M} \lambda_i} \geq 99\%,$$

that is, one considers sufficient POD modes to capture 99% of the energy of the signal.

5.1 Transient Localization Due to Backlash Nonlinearities 277

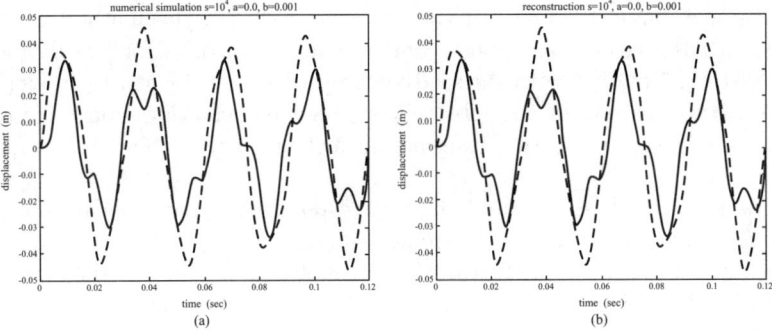

Figure 5.9. Transient response simulations at the middle points of the rods with $a = 0.0m$, $b = 0.01m$, $S = 10^4 N/m$, based on: (a) Five vibration modes per rod, (b) two POD modes; - - - - - - - rod 1, ———— rod 2.

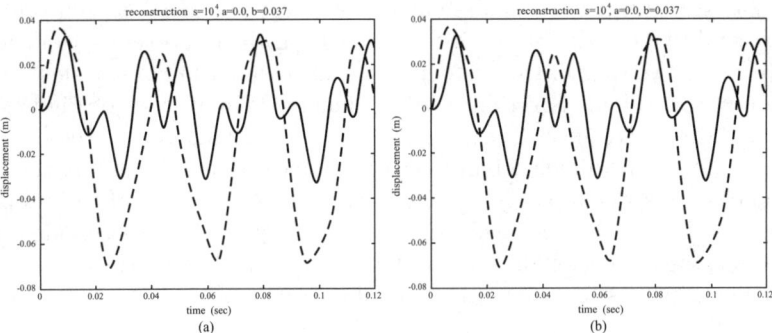

Figure 5.10. Transient response simulations at the middle points of the rods with $a = 0.0m$, $b = 0.037m$, $S = 10^4 N/m$, based on: (a) Five vibration modes per rod, (b) two POD modes; - - - - - - - rod 1, ———— rod 2.

The transient responses at the middle points of the two rods reconstructed by using only the two leading POD modes of the system are depicted in Figure 5.9. The numerical simulations for the same responses obtained by considering 5 vibration modes per rod are also provided for comparison purposes. It can be seen that by using two POD modes one reconstructs almost completely the transient responses obtained by 5 vibration (physical) modes. A similar reconstruction for stronger clearance nonlinearity is given in Figure 5.10. These results strengthen the conclusions of Section 3, and re-confirm that the PODs modes provide excellent orthogonal bases for decomposing the responses of flexible systems. The POD decomposition method can be applied to the analysis of systems with linear, nonlinear, smooth or non-smooth dynamics, since it does not distinguish between linear or nonlinear data; moreover, by directly considering the time series responses which contain the complete dynamics of the systems analyzed, the method of POD deals with the richest and most complete source of data for the dynamics of a system. Coupled with its relative simplicity, these features render the POD method as one of the most effective system identification techniques for analyzing systems with strong nonlinearities, such as the one considered herein.

The POD mode shapes and their corresponding eigenvalues are not only used for reconstructing the dynamic response of the system, but also for studying certain nonlinear features of the global dynamics. In the system of coupled rods studied herein, two factors affect the nonlinearity of the system, namely, the stiffness of the spring and the clearance of the joint. In Figure 5.11a the energies contained in the two leading POD modes are depicted for varying stiffnesses of the joint, keeping the clearance fixed to $a = 0.002m$, $b = 0.03m$. These energies are related to the eigenvalues of equation (5.6). It can be clearly seen that when the stiffness of the spring is nearly $2.6 \cdot 10^4 N/m$ and $3.5 \cdot 10^4 N/m$, there is more enhanced energy transfer between the first and second POD modes of the system, indicating vigorous exchange of energy between the two rods of the system, and thus, elimination of transient localization in rod 1. This is verified by the numerical simulations of Figures 5.5 and 5.7. Moreover, since only two POD modes are dominant at these values of the stiffness, the dimensionality of the nonlinear dynamics during the beat phenomena is approximately equal to two. For other values of the coupling stiffness (when nonlinear localization occurs) only the first K-L mode dominates and the dimensionality of the dynamics is nearly one. Hence, the beat phenomena increase the dimensionality of the dynamics, a feature compatible with intuition.

In Figure 5.11b the energy contained in the first two POD modes is depicted for varying clearance b (cf. Figure 5.1) while keeping the stiffness of the spring and parameter a fixed at and $0.002m$, respectively. For small values of b a single POD mode captures most of the energy and the dimensionality of the dynamics is nearly one. However, as larger values of b are realized, enhanced transfer of energy between the first and the second POD modes takes place, and the dimensionality of the system increases to two. This is also clear from the corresponding numerical simulations of Figure 5.5, where for high values of b a beat phenomenon exists in the energy plots, indicating participation of multiple modes in the response.

From the above analysis of the energy contained in the leading POD modes, one concludes that more enhanced energy transfer between the leading POD modes takes place when the beat phenomenon exists in the time-averaged energies of the two rods. Since this transient beat phenomenon appears to be the mechanism for eliminating the transient motion confinement in this system, it will be studied in more detail in the next Section, where the global dynamics of a reduced system will be examined by numerical Poincaré maps.

5.1 Transient Localization Due to Backlash Nonlinearities

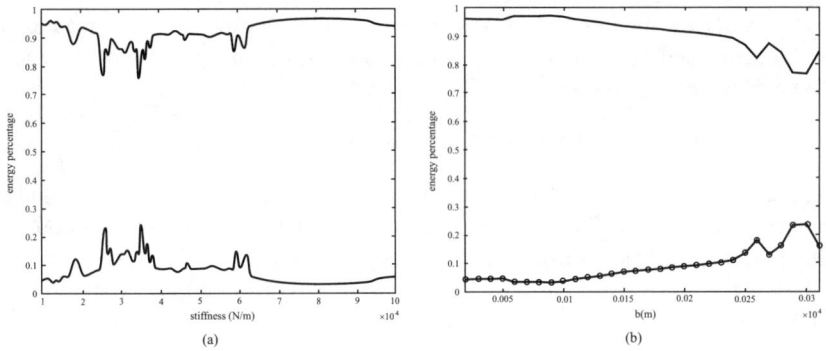

Figure 5.11. Energy distributions among the leading two POD modes for a system with: (a) $a = 0.002m$, $b = 0.03m$ and varying stiffness S, (b) $a = 0.002m$, $S = 2.6 \cdot 10^4 N/m$ and varying b.

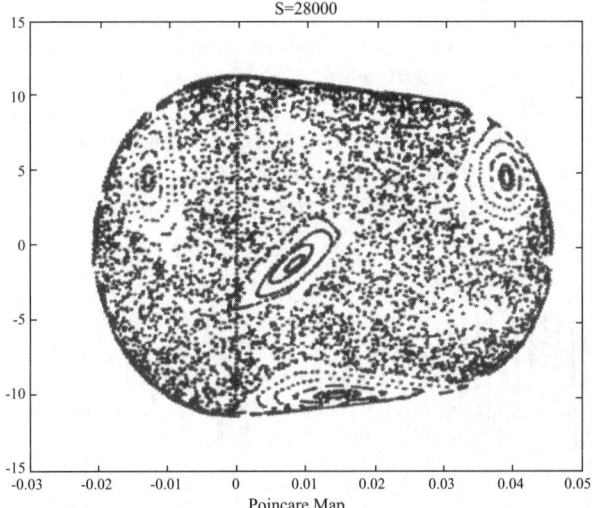

Figure 5.12. The Poincaré map of the system with $S = 2.6 \cdot 10^4 N/m$, $a = 0.002m$, $b = 0.03m$.

5.1.3 Nonlinear Analysis

From the FFT plots of Figures 5.3–5.6 it appears that to study the beat phenomenon it is sufficient to study a two degrees-of-freedom (DOF) reduced model based on the first two vibration modes of the linear system. Based on this two-DOF model Poincaré maps will be constructed in order to study the nonlinear localization and beat phenomena observed in the numerical simulations.

Considering the first (rigid body) mode of rod 1 and the first flexible mode of the grounded rod 2 [142], one expresses approximately the displacements of the two rods as follows:

$$w_1(x,t) = a_1(t)\cos(\beta_1 x), \quad w_2(x,t) = a_2(t)\cos(\beta_2 x), \quad (5.9)$$

where: $\beta_1 = 0$ and $\beta_2 \tan(\beta_2 L) = K_2/EA$ (the first root of this equation is only considered). The nonlinear spring characteristic is decomposed into linear and nonlinear components as shown in Figure 5.12. Discretizing the governing equations of motion by means of (8), one obtains the following ODEs that govern the modal amplitudes,

$$\begin{bmatrix}\ddot{a}_1(t)\\ \ddot{a}_2(t)\end{bmatrix} + \begin{bmatrix}\frac{S}{m} & -\frac{S}{m}\\ -\frac{S}{m} & (\beta_2)^2\frac{EA}{m} + \frac{K_2}{mc_1}(1+\cos^2(\beta_2 L))\end{bmatrix}\begin{bmatrix}a_1(t)\\ a_2(t)\end{bmatrix} = \begin{bmatrix}\frac{-f(a_1(t)-a_2(t))}{m}\\ \frac{f(a_1(t)-a_2(t))}{mc_1}\end{bmatrix}, \quad (5.10)$$

where: $c_1 = \int_0^L \cos^2(\beta_2 x)dx$.

The dynamical system (5.10) is of dimension 4. Fixing the energy at a certain level H, one reduce the flow of the dynamical system to a 3-dimensional isoenergetic manifold (then, one of the initial conditions becomes a depended quantity, say, $\dot{a}_2(0) = f(a_1(0), \dot{a}_1(0), a_2(0); H)$). One then 'cuts' the dynamics on the isoenergetic manifold with the 2-dimensional cross section: and constructs the 2-dimensional Poincaré map of (5.10).

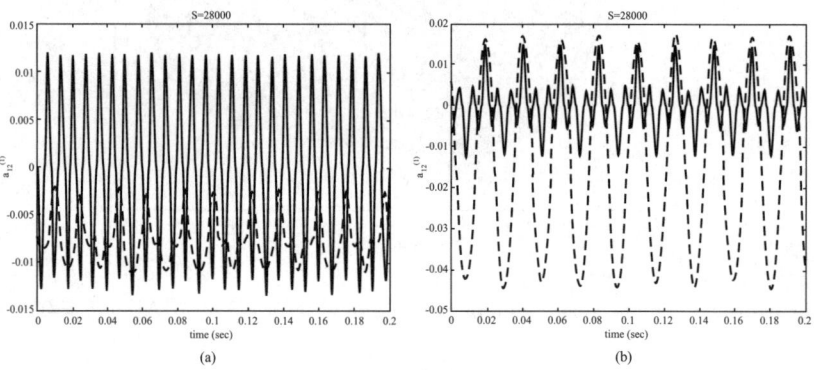

Figure 5.13. Time series of - - - - - - - - $a_1(t)$, ——— $a_2(t)$: (a) Periodic orbit, (b) subharmonic orbit.

In Figure 5.12 the Poincaré map for the system with $S = 2.6 \cdot 10^4 N/m$ is depicted. In this case there exist a periodic orbit and a subharmonic orbit in the dynamics of the system. Studying the time series of the modal amplitudes $a_1(t)$ and $a_2(t)$ with initial conditions corresponding to the periodic orbit (cf. Fig. 5.13a), or to the subharmonic orbit (cf. Fig. 5.13b), one concludes that

5.1 Transient Localization Due to Backlash Nonlinearities   281

the single periodic orbit is responsible for the beat phenomenon, whereas the subharmonic orbit for the nonlinear localization (motion confinement).

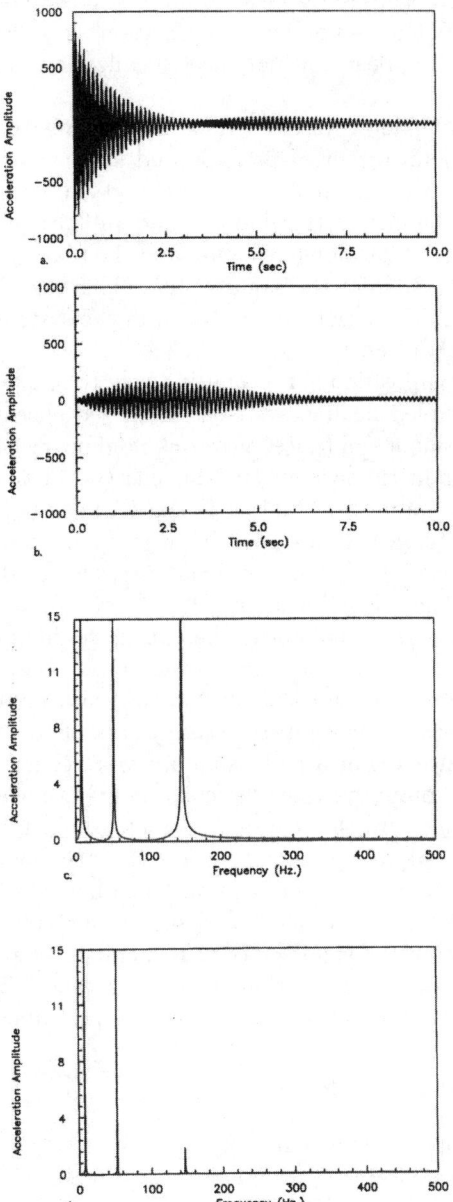

Figure 5.14. The Poincaré map of the system with $S = 3.5 \cdot 10^4 \, N/m$, $a = 0.002m$, $b = 0.03m$.

Considering the Poincaré map for $S = 3.5 \cdot 10^4 N/m$ (cf. Fig. 5.14) there are two subharmonic orbits and one periodic orbit. From the plots of $a_1(t)$ and $a_2(t)$ corresponding to these orbits (cf. Fig. 5.15), one notes that again the single periodic orbit is responsible for the beat phenomenon, and the subharmonic orbits for the nonlinear localization. For increased stiffness, the beat phenomena disappear and nonlinear localization in rod 1 exists in the system.

The Poincaré map for a system with stiffness $S = 9.0 \cdot 10^4 N/m$, is depicted in Figure 5.16. Again a pair of periodic and subharmonic orbits exists. In Figure 5.17 the transient responses for $a_1(t)$ and $a_2(t)$ are shown with initial conditions corresponding to the periodic and subharmonic orbits. Note the resemblance of the subharmonic response and the (exact) transient nonlinear responses of Figure 5.3a for the same system. This is a clear indication that nonlinear localization (motion confinement) in the system of coupled rods is caused by the subharmonic orbit.

The POD decomposition is a promising way to study complex nonlinear phenomena in coupled flexible systems. Increased energy transfer between the leading POD modes indicates vigorous nonlinear energy exchanges between components of the system, and an increase of the dimensionality of the dynamics. An interesting finding is that depending on the parameters of the clearance nonlinearity nonlinear transient motion confinement or beat phenomena may occur. A low-dimensional analysis based on Poincaré maps revealed that nonlinear localization resulted from the excitation of a subharmonic orbit, whereas, nonlinear beat phenomena resulted from the excitation of a periodic orbit. Hence, depending on the initial conditions of the system, localization or delocalization (beat phenomena) may occur.

A further general conclusion from the results of this Section is that the presence of clearance nonlinearities can induce interesting nonlinear phenomena that have no counterparts in the corresponding linear systems with no clearances. Examples are the transient motion confinement phenomena observed in the coupled system of rods with clearance nonlinearities. The fact that such a small structural modification (such as adding a small clearance at the joint between substructures) can cause such qualitatively new dynamics is an indication of the significance that ought to be paid to the study of non-smooth nonlinearities in continuous systems, especially those arising due to loose mechanical joints or to other types of joint imperfections.

5.1.4 Additional Examples

Additional examples of nonlinear motion confinement phenomena in elastic continua with backlash joints are given in the works by Nayfeh [166], and Nayfeh and Vakakis [167]. Transient wave propagation in linear waveguides coupled by means of backlash joints were numerically studied in these works, and it was shown that under certain conditions spatial entrapment of the propagating disturbance could be achieved.

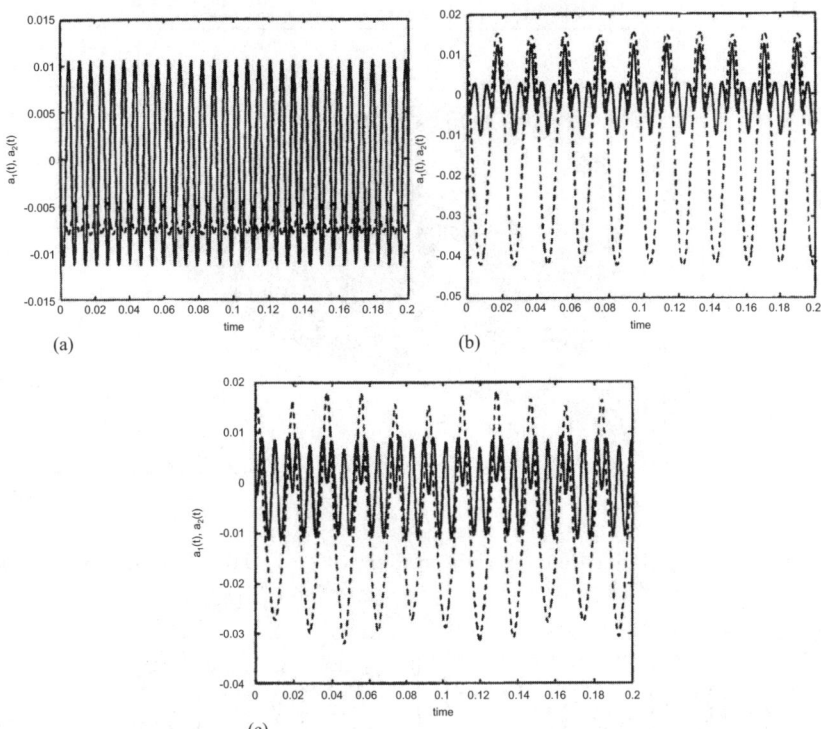

Figure 5.15. Time series of $- - - - - - - a_1(t)$, ——— $a_2(t)$: (a) Periodic orbit, (b) subharmonic orbit 1, (c) subharmonic orbit 2.

It is noted that the transient dynamics and the active or passive vibration suppression of spatially extended continuous systems was studied in numerous works [175, 93, 173, 137, 145]. A method to passively control the transmission of unwanted vibrations in periodic flexible structures is through the appropriate design of their joints [156, 71]. Onoda et al. [174] considered preloaded backlash joints to couple flexible elements of a truss, and showed that they cause transfer of vibrational energy to higher frequency modes, thereby facilitating energy dissipation; additional energy dissipation at the backlash joints due to hysteresis was also detected. These results, coupled with the nonlinear energy confinement findings of Section 5.1, indicate that backlash joints are good candidates for efficient passive energy confinement and dissipation of propagating disturbances in periodic elastic networks.

A possible configuration of the backlash joint appears in Figure 5.18, consisting of a pin that slides in a groove. A preloaded spring prevents the pin from slipping when the force across the joint is below a certain threshold; hence, for sufficiently low loads the joint behaves as a linear stiffness (a feature which provides static support to the truss), whereas, for high loads it behaves

284   5 Continuous Systems with Non-smooth Nonlinearities

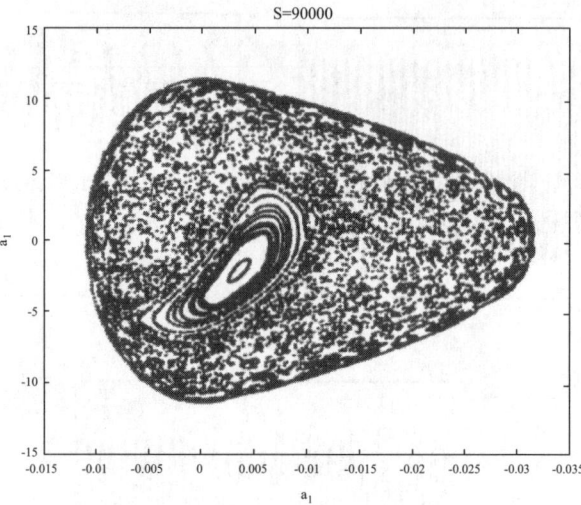

Figure 5.16. The Poincaré map of the system with $S = 9.0 \cdot 10^4 N/m$, $a = 0.002m$, $b = 0.03m$.

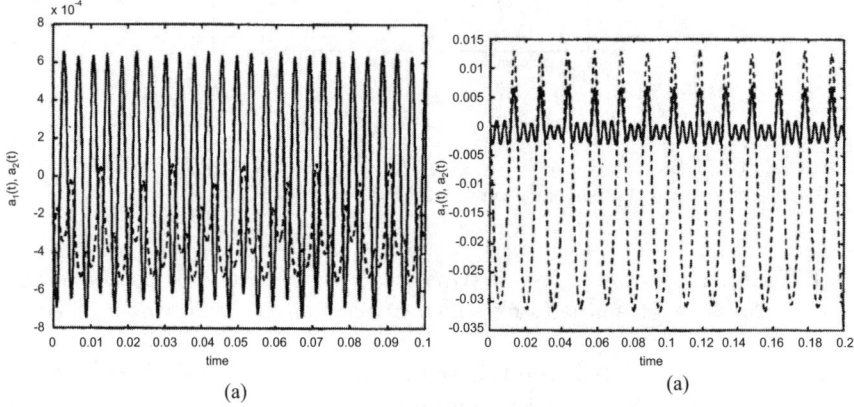

Figure 5.17. Time series of - - - - - - - $a_1(t)$, ——— $a_2(t)$:(a) Periodic and (b) subharmonic orbit.

as a nonlinear backlash element. Energy dissipation can also be modeled by assuming dry friction forces applied to the pin when it slides in the groove.

In Figure 5.19 the transient responses of the system of two longitudinal rods jointed by a backlash joint subject to a single-period sine wave forcing function are depicted [166, 167]. The force is applied on the free left boundary condition of the left rod, whereas the right boundary of the right rod is grounded. Comparison with the corresponding response of the linear system (with backlash eliminated) is also given in that Figure. The computational

5.1 Transient Localization Due to Backlash Nonlinearities   285

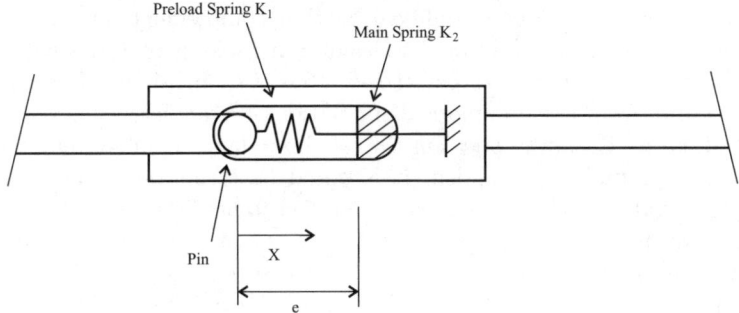

Figure 5.18. The clearance joint.

results were obtained using the finite-element method. Note the transient beat phenomena in the plots of the instantaneous energies of the rods of the linear system, and the enhanced motion confinement in the system with the backlash. These finite-element results are in qualitative agreement with the numerical results of Sections 5.1.1 and 5.1.2 obtained through discretization and low-order reduction of the equations of motion.

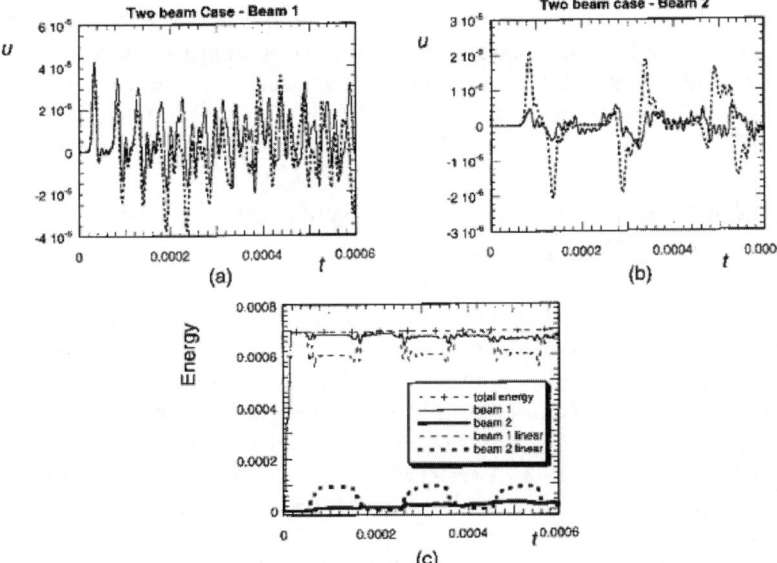

Figure 5.19. Transient response of a two-rod system subject to longitudinal external excitation: Displacement of the mid-point of (a) rod 1, and (b) rod 2 (- - - - - - linear, ——— nonlinear system); (c) instantaneous total energy in each rod.

To further demonstrate nonlinear motion confinement due to backlash joint nonlinearities, an additional series of numerical simulations with a system of ten coupled rods was performed [166, 167]. In Figure 5.20 representative results from these numerical simulations are depicted, for a transient longitudinal excitation applied on the left boundary of the system, whereas the right boundary is grounded. It is noted that the amplitude of the vibration of rod 1 of the system with clamped joints (the nonlinear system) remains consistently higher than that of the system with linear joints (compare Figures 5.20a and 5.20b); however, in the second and higher-order rods, the opposite occurs, with the amplitudes of the nonlinear vibration being higher than the corresponding linear responses. The transient motion confinement phenomenon in the nonlinear case is more clearly evidenced by considering the instantaneous energies of the rods in the linear and nonlinear cases. Considering the linear system (Figure 5.20c), as time progresses energy gets eventually transferred from the first to the second and higher-order rods, whose vibration levels eventually exceed that of the first rod; indeed, after sufficient time transient beat phenomena develop and energy exchanges between the first and higher-order rods take place. In contrast, the rate of energy 'leakage' from the first to the other rods decreases significantly, so that motion confinement develops; moreover, transient beat phenomena are absent in the nonlinear case.

In the next Section the study of non-smooth dynamics of continuous systems is extended to the case of vibro-impact nonlinearities, where it is shown that motion confinement phenomena take place, similar to the clearance case. This should be expected in view of the fact that vibro-impacts can be regarded as the limited case (for large stiffness) of the clearances.

5.2 Nonlinear Localization in Systems of Vibro-impacting Beams

Impacts between vibrating systems and rigid (immovable) barriers represent a very common source of structural nonlinearity in engineering applications; for example, vibro-impact nonlinearities in systems due to loose joint connections are often encountered in practice. Representative monographs and books edited in this area are the ones by Brogliato [39], Babitsky [26, 27], Pfeiffer and Glocker [183], Awrejcewicz and Lamarque [22]. The modelling and analysis of continuous structures exhibiting vibro-impacts is a challenging problem in dynamics [229, 160, 210, 211, 64, 40]. The first main reason is that the points of the structure and time instants where the impacts occur are determined by the solution itself, and are not a priori known. As a result, in theory the vibro-impact problem must be solved simultaneously with tests that detect the occurrence (or absence) of the impacts. The second source of difficulties concerns the mathematical modeling of the structure-barrier in-

5.2 Nonlinear Localization in Systems of Vibro-impacting Beams

teraction; depending on the specific model of impact adopted in the analysis, impacts can be modeled as purely elastic or inelastic.

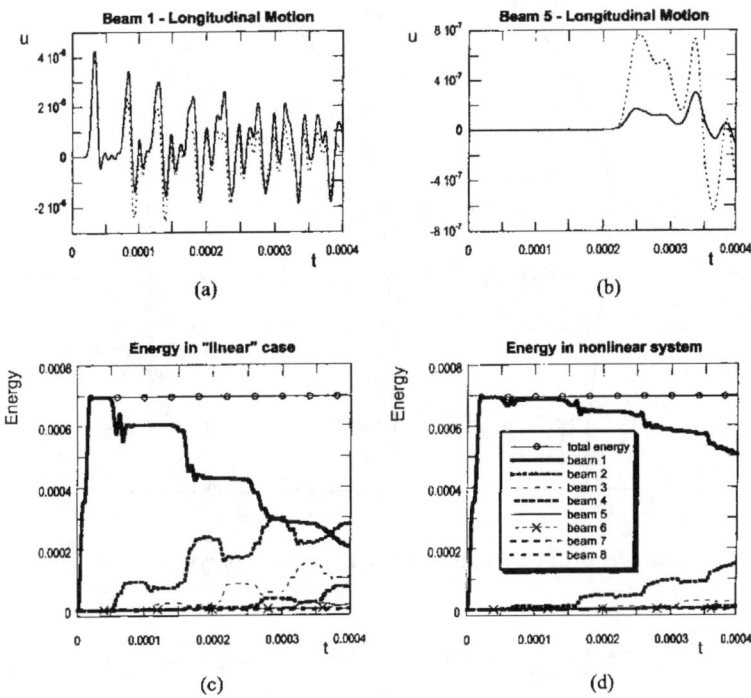

Figure 5.20. Transient response of a ten-rod system subject to longitudinal external excitation: Displacement of the mid-point of (a) rod 1, and (b) rod 5 (- - - - - - linear, ——— nonlinear system); instantaneous total energies in, (c) the linear, and (d) the nonlinear system (with backlash joints).

In this Section the dynamics of systems of coupled beams undergoing vibro-impacts is examined numerically and experimentally, with the aim to study localization and motion confinement phenomena that are absent in the corresponding linear systems. The basic configuration of the systems considered appear in Figure 5.21. It consists of two identical elastic beams coupled by a linear stiffness. Vibro-impact nonlinearities are induced by inelastic collisions of the beams with rigid constraints situated at positions $x = v$. Assuming linear Bernoulli theory to model the beam vibrations, the governing differential equations of motion are given by:

$$m\frac{\partial^2 u_1}{\partial t^2} + d\frac{\partial u_1}{\partial t} + EI\frac{\partial^4 u_1}{\partial x^4} = -k\delta(x-a)\left[u_1(a,t) - u_2(a,t)\right] + B_1 + I_1 + P(x,t),$$

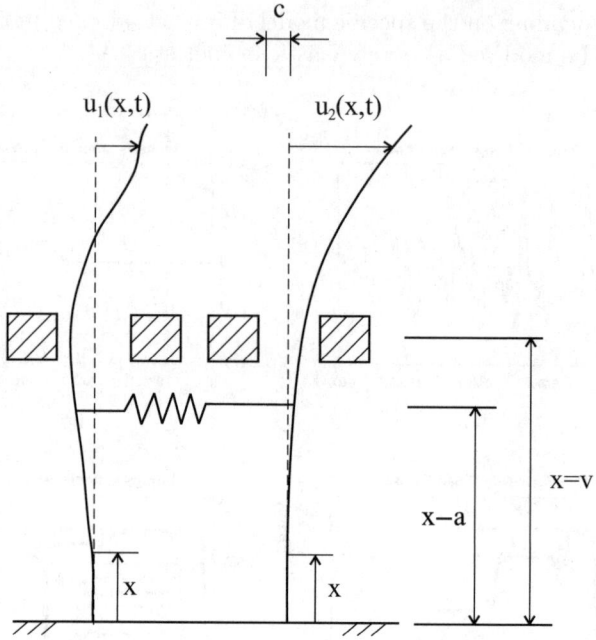

Figure 5.21. System of coupled vibro-impacting beams.

$$m\frac{\partial^2 u_2}{\partial t^2} + d\frac{\partial u_2}{\partial t} + EI\frac{\partial^4 u_2}{\partial x^4} = k\delta(x-a)\left[u_1(a,t) - u_2(a,t)\right] + B_2 + I_2, \quad (5.11)$$

where the geometric parameters are defined in Figure 5.21, m, d and EI are the uniform mass, viscous damping and elasticity distributions per unit length of the beams, $u_{1,2}(x,t)$ the transverse displacements fields of the two beams, $P(x,t)$ the distributed external excitation applied to beam 1, and $B_{1,2}, I_{1,2}$ the dissipation and nonlinear stiffness vibro-impact forces, respectively, exerted at beams 1 and 2 at the time instants $t = t_i$, $i = 1, 2, \ldots$ of collisions with the rigid constraints. The explicit expressions of these terms are discussed below.

Regarding the dissipative terms $B_{1,2}$, they model the energy dissipation due to inelastic impacts of the beams with the rigid constraints. They are applied only at the time instants $t = t_i$, $i = 1, 2, \ldots$ and are mathematically expressed as follows:

$$B_{1,2} = -\beta_{1,2}\left(u_{1,2}(v,t)\right)\delta(x-v)\frac{\partial u_{1,2}(v,t)}{\partial t}, \quad (5.12)$$

where the dissipation coefficients $\beta_{1,2}\left(u_{1,2}(v,t)\right)$ are defined as,

$$\beta_p = 0 \text{ if } |u_p(v,t)| < c, \text{ and } \beta_p = \beta_0 \neq 0 \text{ if } |u_p(v,t)| \geq c, \quad p = 1, 2. \quad (5.13)$$

5.2 Nonlinear Localization in Systems of Vibro-impacting Beams

The parameter c denotes the clearance between each of the beams and the rigid barriers (cf. Figure 5.21), and the parameter β_0 is chosen empirically to match experimental observations. Note that if one sets $\beta_0 = 0$, then one assumes purely elastic impacts between the beams and the rigid barriers.

The vibro-impact nonlinear effects are modeled by the forces $I_{1,2}$. Modeling of these forces is a challenging computational task since the time instants $t = t_i$, $i = 1, 2, \ldots$ where the impacts occur are determined by the solution itself. An alternative technique which circumvents this problem is to model the vibro-impact nonlinear interactions by clearance stiffnesses of very high stiffness characteristics. Then, one expresses the vibro-impact terms as follows,

$$I_{1,2} = \begin{cases} -a\delta(x-v)\left[u_{1,2}(v,t) - c\right] & \text{if } u_{1,2}(v,t) \geq c \\ 0 & \text{if } c \geq u_{1,2}(v,t) \geq -c \\ -a\delta(x-v)\left[u_{1,2}(v,t) + c\right] & \text{if } u_{1,2}(v,t) \leq -c \end{cases}, \qquad (5.14)$$

where the sign convention for the transverse displacements shown in Figure 5.21 is adopted, and a denotes the simulated large stiffness characteristic of the rigid barrier.

The nonlinear set of equations (5.11) is solved by discretization, The transverse displacements are expressed in the series forms,

$$u_1(x,t) = \sum_{i=1}^{\infty} A_{1,i}(t)\,\Phi_i(x), \quad u_2(x,t) = \sum_{i=1}^{\infty} A_{2,i}(t)\,\Phi_i(x), \qquad (5.15)$$

where: $\Phi_i(x)$ represents the i-th normalized linear cantilever mode, and $A_{p,i}(t)$, $p = 1, 2$ the amplitude of the i-th mode of beam p. Employing the orthogonality conditions satisfied by the cantilever modes, the original governing differential equations of motion are replaced by the following set of nonlinear ordinary differential equations that govern the time evolution of the modal amplitudes:

$$\ddot{A}_{1,j}(t) + d_{1,j}\dot{A}_{1,j}(t) + \omega_j^2 A_{1,j}(t) = -\frac{k\Phi_j(a)}{mC_j}\left\{\sum_{i=1}^{\infty}[A_{1,i}(t) - A_{2,i}(t)]\,\Phi_i(a)\right\} -$$

$$\frac{\beta_1\left(u_1(v,t)\right)\Phi_j(v)}{mC_j}\sum_{i=1}^{\infty}\dot{A}_{1,j}(t)\Phi_i(a) +$$

$$\frac{1}{mC_j}\int_0^L P(x,t)\Phi_j(x)\,dx - \frac{a\Phi_j(v)}{mC_j}\gamma_1\left(u_1(v,t)\right),$$

$$\ddot{A}_{2,j}(t) + d_{2,j}\dot{A}_{2,j}(t) + \omega_j^2 A_{2,j}(t) = \frac{k\Phi_j(a)}{mC_j}\left\{\sum_{i=1}^{\infty}[A_{1,i}(t) - A_{2,i}(t)]\,\Phi_i(a)\right\} -$$

$$\frac{\beta_2\left(u_2(v,t)\right)\Phi_j(v)}{mC_j}\sum_{i=1}^{\infty}\dot{A}_{2,j}(t)\Phi_i(a) - \frac{a\Phi_j(v)}{mC_j}\gamma_2\left(u_2(v,t)\right), \quad (5.16)$$

where: $j = 1, 2, \ldots, \omega_j^2$ is the natural frequency squared of the j-th linearized cantilever mode, $C_j = \int_0^L \Phi_j^2(x)dx$ is the normalization coefficient for the j-th mode, and the functions γ_p are defined as follows:

$$\gamma_{1,2}\left(u_{1,2}(v,t)\right) = \begin{cases} [u_{1,2}(v,t) - c] & \text{if } u_{1,2}(v,t) \geq c \\ 0 & \text{if } c \geq u_{1,2}(v,t) \geq -c \\ [u_{1,2}(v,t) + c] & \text{if } u_{1,2}(v,t) \leq -c \end{cases}, \quad (5.17)$$

For computational purposes the set (5.16) is truncated to a finite number of terms by considering only the leading N cantilever modes in the discretization. The time evolution of the modal amplitudes is then evaluated by integrating the truncated set using a 4-th order Runge-Kutta numerical algorithm. A convergence study was also performed in order to assess the validity of the truncated set, e.g., to determine the smallest possible integer N for which accuracy of the numerical simulations is ensured. Moreover, the initial conditions that complement equations (5.16) are determined by the formulas:

$$A_{p,j}(0) = C_j^{-1}\int_0^L u_p(x,0)\Phi_j(x)dx,$$

$$\dot{A}_{p,j}(0) = C_j^{-1}\int_0^L \frac{\partial u_p(x,0)}{\partial t}\Phi_j(x)dx, \quad p = 1, 2. \quad (5.18)$$

Numerical simulation results obtained by solving the truncated model are presented in the next Section.

5.2.1 Numerical Results - Nonlinear Localization Due to Vibro-impacts

In the first series of numerical simulations the external excitation acting on beam 1 is taken as zero, $P(x,t) = 0$ and only excitations due to nonzero initial conditions are considered. The aim of this first set of numerical simulations is to show that the coupled assembly of beams possesses transient motion confinement properties. Moreover, only one-sided impacts of the two beams with the rigid barriers on their right are considered (cf. Figure 5.21). The values of the various coefficients in the discretized equations of motion (5.16) are listed at Table 5.1; these values were determined by performing modal tests with two aluminum beams that formed an experimental fixture with configuration similar to Figure 5.21. The functions $\gamma_{1,2}$ in (5.16) are,

5.2 Nonlinear Localization in Systems of Vibro-impacting Beams

$$\gamma_{1,2}(u_{1,2}(v,t)) = \begin{cases} 0 & \text{if } c \geq u_{1,2}(v,t) \geq -c \\ [u_{1,2}(v,t) + c] & \text{if } u_{1,2}(v,t) \leq -c \end{cases}, \quad (5.19)$$

whereas, the stiffness characteristic a in the vibro-impact terms is assigned the large value, $a = 2.5 \times 10^4$ to simulate the large impact forces exerted by the barriers at the time instants when the vibro-impacts with the beams occur. The clearance was taken $c = 2 \times 10^{-3}m$, whereas two different choices regarding the dissipation coefficient β_0 are considered, namely, $\beta_0 = 0$ (for purely elastic impacts), and $\beta_0 = 35.0$ (for inelastic impacts). Thus, he effect of energy dissipation due to inelastic impacts on the vibro-impact dynamics will be examined.

The initial conditions for the numerical simulations are given by (5.18) with,

$$u_1(x,0) = \frac{Fx^2}{6EI}(3L - x), \quad u_2(x,0) = \frac{\partial u_1(x,0)}{\partial t} = \frac{\partial u_2(x,0)}{\partial t} = 0.$$

These initial conditions correspond to initial deflection of beam 1 with a deformation identical to its first linearized cantilever mode, with the other beam initially at rest. The parameter F is chosen so that the initial tip displacement of beam 1 is equal to $u_1(L,0) = 0.067m$.

Table 5.1. Coefficients of the discretized model (5.16).

$C_1 = 1.421$	$a = 0.306m$
$C_2 = 0.738$	$f = 0.091m$
$C_3 = 0.770$	$L = 0.766m$
$d_{1,1} = 0.475$	$v = 0.50m$
$d_{1,2} = 0.750$	$\Phi_1(v) = -1.440$
$d_{1,3} = 3.0$	$\Phi_2(v) = -0.909$
$d_{2,1} = 0.375$	$\Phi_3(v) = 1.245$
$d_{2,2} = 0.650$	$\omega_1 = 52.449 rad/s$
$d_{2,3} = 5.40$	$\omega_2 = 328.718 rad/s$
$k = 117.0N/m$	$\omega_3 = 920.513 rad/s$
$m = 0.753kg/m$	

An initial convergence study was performed to determine the lowest number of linearized cantilever modes needed for accurate discretization of the governing partial differential equations of motion. It was found that $N = 3$ modes are sufficient to model the dynamics, and that increasing the dimensionality of the discretized model produces negligible improvement in the results. It is noted that this result is tied to the specific initial conditions used, which produced an oscillation of the beams close to the frequency of the first linearized cantilever mode; for other types of initial excitation, or for nonzero external forcing distributions the number of modes N that is sufficient for the accurate discretization might change.

292   5 Continuous Systems with Non-smooth Nonlinearities

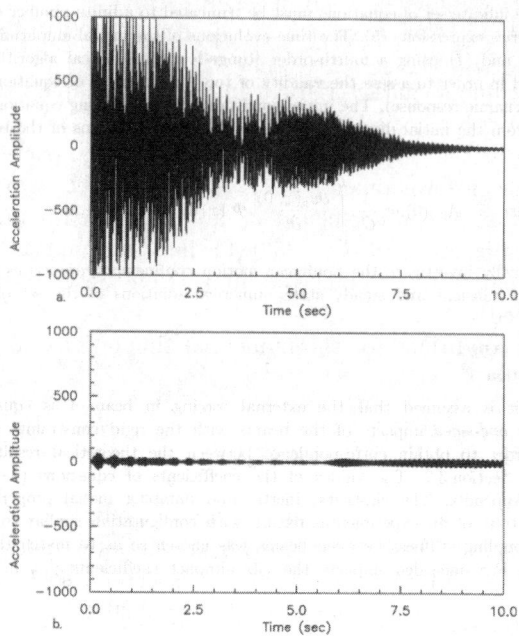

Figure 5.22. Transient motion confinement due to purely elastic vibro-impacts: Tip acceleration of, (a) beam 1, (b) beam 2.

In Figures 5.22 and 5.23 the simulated transient accelerations at the tips the two vibro-impacting beams are depicted for purely elastic and inelastic impacts, respectively. For comparison, in Figure 5.24 the numerical accelerations of the corresponding linear system (with no impacts with the rigid barriers) are also shown. By studying the results of Figures 5.22 and 5.23 one notes that in the vibro-impact case motion confinement of the transient response in the directly excited beam 1 takes place, with small vibrational energy being 'spread' in beam 2. No such confinement is observed in the linear case (with no vibro-impacts, Figure 5.24), where the well-known beat phenomenon takes place and energy is exchanged back and forward between the two linear beams; this beat is due to fact that the two beams are nearly identical, and, hence, they oscillate with nearly identical frequencies. Considering the vibro-impact responses one notes that although the energy dissipation (or absence of) due to inelastic vibro-impacts has a profound effect on the response of the directly excited beam, it does not significantly affect the 'spreading' of the motion to the unexcited one. It is concluded that passive motion confinement occurs in the system with purely elastic or inelastic vibro-impacts, a phenomenon which can be attributed to the excitation by the initial conditions of a localized nonlinear normal mode (NNM) of the cou-

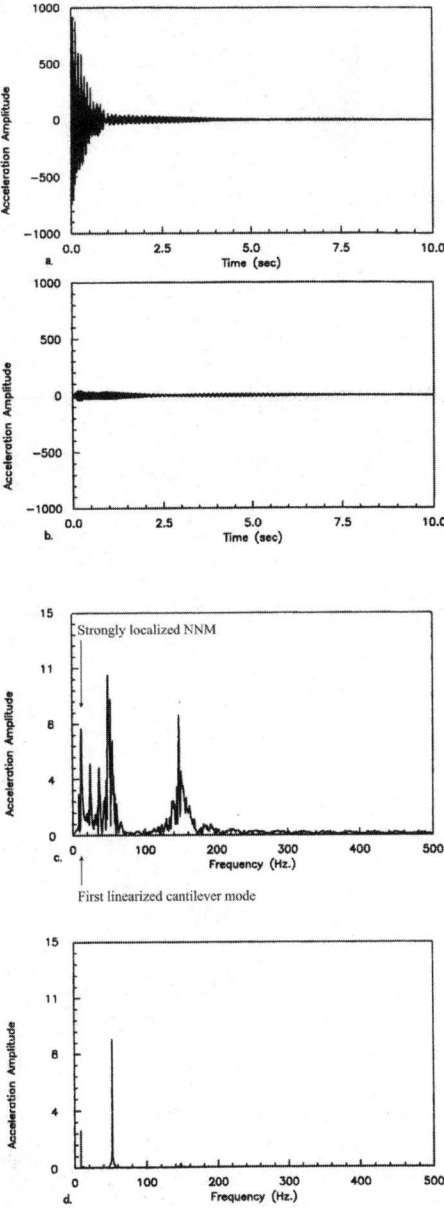

Figure 5.23. Transient motion confinement due to inelastic vibro-impacts: tip acceleration time series of, (a) beam 1, (b) beam 2; acceleration frequency spectra of, (c) beam 1, (d) beam 2.

294   5 Continuous Systems with Non-smooth Nonlinearities

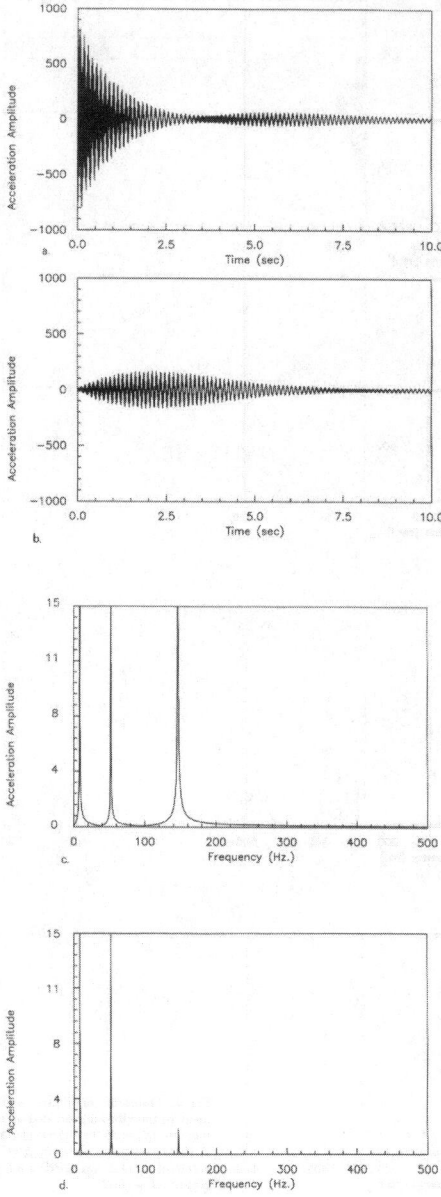

Figure 5.24. Transient accelerations of the linear system with no vibro-impacts: Tip acceleration time series of, (a) beam 1, (b) beam 2; Acceleration frequency spectra of, (c) beam 1, (d) beam 2.

5.2 Nonlinear Localization in Systems of Vibro-impacting Beams 295

pled system, whose energy is spatially confined in the directly excited beam. This localized mode is generated through a NNM bifurcation that occurs at a critical coupling to nonlinearity ratio [238, 242], and can only exist in the nonlinear system (e.g., it has no counterpart in linear theory). Indeed, the nonlinear motion confinement responses of Figures 5.22 and 5.23 are in qualitative agreement with previous results reported in [11] where a similar continuous system but with smooth grounding stiffness nonlinearities was considered.

Comparing the Fast Fourier Transforms (FFTs) of Figures 5.23 and 5.24 one obtains a clear picture of the dynamics of passive motion confinement. Indeed, the frequency spectra of the linear accelerations possess only three peaks, which is consistent with the fact that only three linearized cantilever modes are used in the discretization process; moreover, the amplitudes of the leading modes of beams 1 and 2 are comparable in magnitude, a feature which is consistent with the occurrence of the linear beat phenomenon. By contrast, the frequency spectra of the vibro-impact responses possess numerous peaks due to the nonlinearity induced by the vibro-impacts. Considering the spectrum of Figure 5.23c (directly excited beam 1), one observes a pair of closely spaced modes close to the first linearized eigenfrequency: one of these modes is the linearized cantilever mode, whereas the other (higher one) is a strong new mode which has no counterpart in linear theory; this is the strongly localized NNM referred to earlier, which is generated at a NNM bifurcation for sufficiently low coupling stiffness to nonlinearity ratios. It is the excitation of precisely this mode that leads to the transient motion confinement phenomena observed in the vibro-impact case. Indeed, one notes that the amplitude of this mode in the corresponding spectrum of the unexcited beam 2 is small, indicating motion confinement in the directly excited beam.

A final point concerning the transient simulations centers on the importance of including dissipation terms to model energy loss during vibro-impacts. As shown, inelastic impacts can significantly reduce the energy of the directly excited beam, and, constitute the main mechanism for energy dissipation in the vibro-impacting system. By ignoring this dissipation and assuming purely elastic impacts one eliminates one of the main sources of damping in the system, and (as shown below) produces results that are in disagreement with experimental measurements.

In the second series of numerical simulations the steady state response of the vibro-impacting beams is examined by considering a distributed force applied at beam 1 with harmonic time dependence:

$$P(x,t) = -P\delta(x-f)\sin\omega t \ . \tag{5.20}$$

Hence, it is assumed that a concentrated harmonic force is acting at position $x = f$ of beam 1. The initial conditions of the problem are assumed to be zero (actually the specific initial conditions are of no importance when considering the steady state solution of the system, since they represent decaying transients), and double-impacts between each beam with its right and left rigid barriers are considered. In the following numerical simulations the clearance is taken equal to $c = 0.008m$, the structural parameters of the problem are listed in Table 5.1, and the other parameters are assigned the values, $a = 25 \times 10^3$, $\beta_0 = 35.0$, $p = 13.4N$, $f = 0.091m$. As in the first set of simulations, three linearized cantilever modes per beam are considered for the discretization of the governing equations.

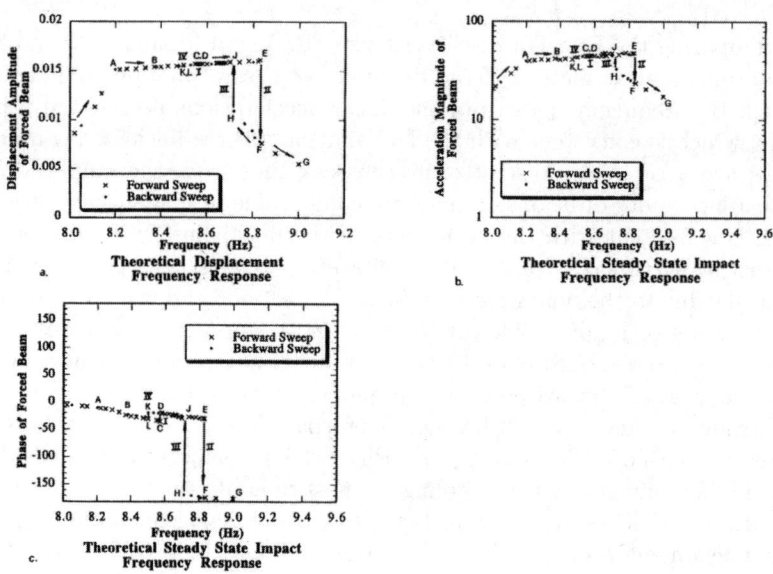

Figure 5.25. Numerical FRCs for the directly forced beam 1: (a) displacement, (b) acceleration, (c) phase; both forward and backward sweeps are considered.

In Figures 5.25 and 5.26 the frequency response curves (FRCs) for the steady state displacements and accelerations of the two beams are depicted. The FRCs are constructed by keeping the forcing amplitude fixed and varying the frequency ω; at each frequency the system was allowed to reach steady-state before its response was recorded. In the steady state numerical simulations depicted, the frequency was varied in a neighborhood of the first linearized cantilever natural frequency.

Both forward and backward frequency sweeps were considered. It is noted that in frequent occasions the steady state reached did not represent a periodic motion, but rather a quasi-periodic one; this was consistent with the experimental results reported later. The displacement FRCs depict the magnitudes of the zero-to-peak amplitudes of the steady state displacements, whereas, the acceleration FRCs depict the magnitudes of the fundamental harmonics (at the values of the excitation frequency) of the steady state accelerations. Because the time series of the steady state accelerations pos-

5.2 Nonlinear Localization in Systems of Vibro-impacting Beams 297

sessed nonsmooth effects due to the vibro-impacts, depicting zero-to-peak amplitudes for the accelerations was of limited use.

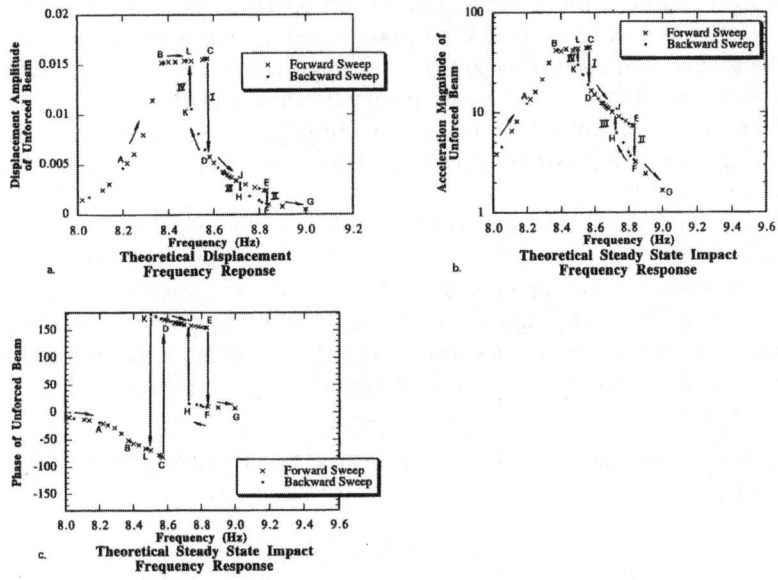

Figure 5.26. Numerical FRCs for the unforced beam 2: (a) displacement, (b) acceleration, (c) phase; both forward and backward sweeps are considered.

Considering the numerical FRCs it is observed that for frequencies sufficiently lower than the first linearized cantilever natural frequency the amplitudes of vibration of the two beams are too small for vibro-impacts to occur; as a result, the response of the system is linear. At point A of the plots the directly forced beam starts impacting with its rigid constraints and the FRCs become nonlinear. For small frequency increases above point A the steady state response of the first mode of beam 1 gets saturated, and the vibrational energy is transferred to its two higher modes and to the unforced beam 2. At certain frequencies in that saturated region the response of beam 1 is quasi-periodic and more than one frequency components dominate the steady state response. At point B the response of beam 2 gets sufficiently large for vibro-impacts to occur, and for small frequency increases above this point both beams oscillate with relatively large (saturated) amplitudes in nearly in-phase fashion; both periodic and quasi-periodic responses exist in that region.

A jump takes place at point C (at $8.58Hz$), labeled as jump I in Figures 5.25 and 5.26, and the system settles into a nonlinear localized steady state motion where most of the energy of vibration is spatially confined to the directly forced beam 1. On the localized branch KE the two beams vibrate in near anti-phase fashion, in full agreement with theoretical and experimental in previous works. For further increases of frequency the nonlinear steady state localization is eliminated by jump II (at $8.84Hz$), and the system settles into a linear, out-of-phase periodic response with no vibro-impacts occurring. For backward frequency sweeps two additional jumps III and IV are identified, occurring between the high-frequency linear and localized branches, and the localized and nonlocalized (saturated) branches, respectively. Hence, two nonlinear hysteresis loops are detected, defined between jumps IV and I, and III and I, respectively.

These numerical results prove that the vibro-impact flexible assembly under consideration possesses nonlinear motion confinement properties. The localized phenomena observed both in the transient and steady state regimes, are caused by the vibro-impact nonlinearities, and provide a way to passively confine vibrational energy in a substructure of the system. In the next Section experimental results are provided that confirm the theoretical findings.

5.2.2 Experimental Results - Nonlinear Localization Due to Vibro-impacts

The experimental setup consisted of two coupled cantilever aluminum beams of length $L = 0.766m$ with clamped boundary conditions imposed by vices. The coupling between the two beams was by means of a linear spring composed of surgical plastic. The dimensions of the beams were identical to those of the theoretical system (Table 5.1), and their elastic and damping properties were identified by means of modal testing (sine-sweeps). The average material properties of the beams are also equal to those listed in Table 5.1. A modal shaker was connected to beam 1 at a distance equal to $f = 0.091m$ from its clamped end. To create symmetry within the system, a bungie and an adjustable tension cable were connected to the unforced beam 2 at a distance equal to f from its clamped end. At a distance equal to $v = 0.50m$ from the clamped end of each beam, rigid constraints (aluminum blocks) were placed with varying impacting clearances c. An accelerometer was placed close to the free end of each beam to measure its transient or steady state response. All experimental measurements presented in this Section are beam accelerations.

Preliminary modal tests were performed to verify that the uncoupled beams possessed nearly identical dynamical properties, thus ensuring symmetry. This was necessary to ensure that the detected localization phenomena were due to the nonlinearity and not to structural disorder. Indeed the primary flexural modes of the beams had a 0.209% difference in their natural frequencies, the second ones a 0.141% difference, and the third ones a less than 0.1% difference. Similar negligible differences were detected for the other modal properties of the two beams, and it was concluded that the two (uncoupled) beams possessed nearly identical dynamics. Hence, the motion

5.2 Nonlinear Localization in Systems of Vibro-impacting Beams

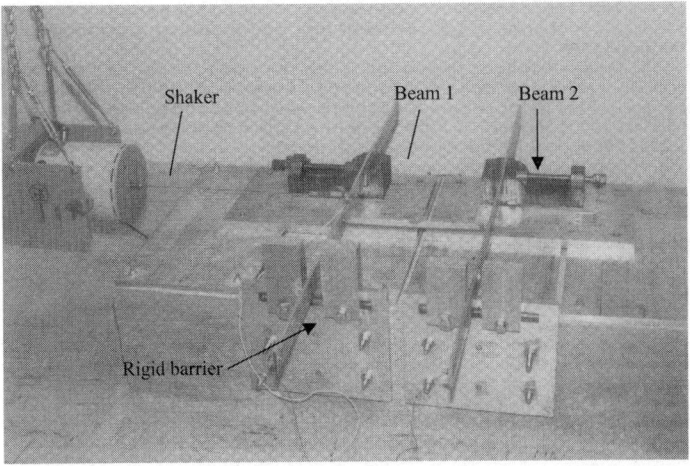

Figure 5.27. The experimental fixture.

confinement phenomena reported here can be only attributed to the vibro-impact nonlinearity.

In the first series of experimental tests transient nonlinear motion confinement was studied in the system. The use of one-sided impacts for the transient tests was dictated by the necessity to achieve sufficiently large initial displacements in the directly excited beam 1 in order to get sufficiently strong nonlinear effects. For the transient tests the clearances between the beams and the rigid constraints were fixed to $c = 0.002m$, e.g., identical to the value used in the numerical simulations of the previous Section. The initial tip displacement of beam 1 was set to $u_1(Lm0) = 0.067m$, with beam 2 being initially at rest, $u_2(x,0) = 0$; again, these initial conditions were identical to the ones used in the theoretical simulations. In Figure 5.28 typical transient responses of the two beams are depicted, from were it is observed that the transient motion is mainly confined to the directly excited beam 1, in full agreement with the numerical simulations. It is noted that the experimental results are in better agreement to the numerical results corresponding to inelastic vibro-impacts (cf. Figure 5.23), a result that shows that the energy dissipation due to inelastic collisions at the time instants of impact is significant and plays an important role in the dynamics. Comparing the experimental and theoretical FFTs of Figures 5.23c,d and 5.28c,d, one notes that there is qualitative agreement at least up to $150Hz$. Above that frequency there is no significant theoretical response due to the fact that only three modes were included in the theoretical model (the natural frequency of the third mode was close to $150Hz$).

300   5 Continuous Systems with Non-smooth Nonlinearities

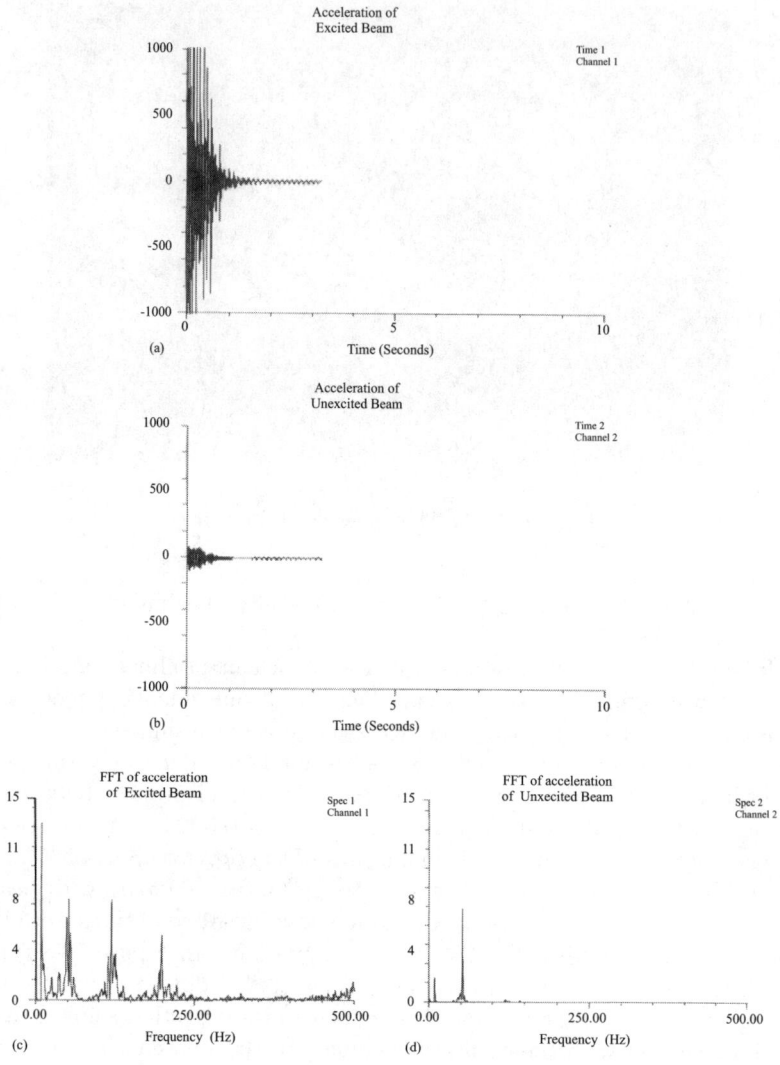

Figure 5.28. Experimental transient motion confinement in the vibro-impact fixture: acceleration of (a) beam 1, (b) beam 2, FFT of acceleration of (c) beam 1, (d) beam 2.

When the rigid barriers are removed the system becomes linear and the transient dynamics exhibits a beat phenomenon, with energy being exchanged between the primary cantilever modes of the two beams (cf. Figure 5.29).

An interesting observation can be made regarding the third 'peaks' in the acceleration of the directly excited beam 1 (cf. Figures 5.23c and 5.28c). In particular, the third experimental 'peak' is shifted to lower frequency com-

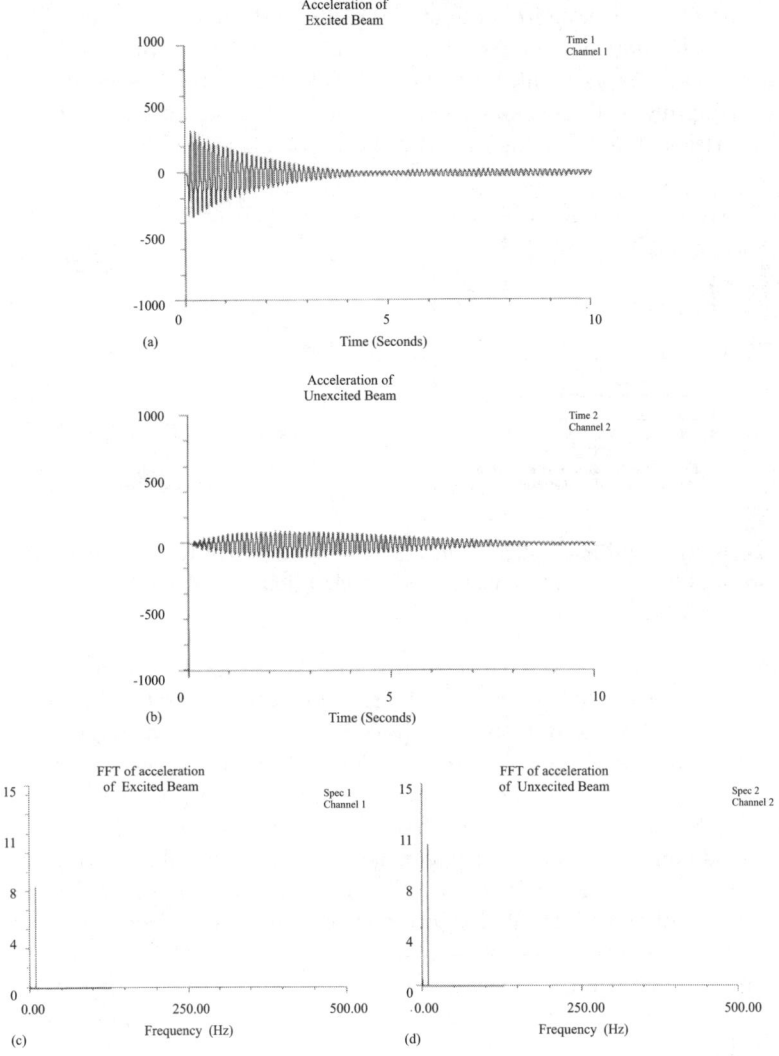

Figure 5.29. Experimental transient motion confinement in the linear fixture: acceleration of (a) beam 1, (b) beam 2, FFT of acceleration of (c) beam 1, (d) beam 2.

pared to the theoretical one. It was experimentally verified that the third experimental 'peak' is close to the third natural frequency of the cantilever beam with a grounding stiffness at the grounding position; this should not be surprising since, when the motion is localized in beam 1, the unexcited beam 2 oscillates with small amplitude, and the system resembles a single beam with a grounding stiffness at the coupling position. In essence, the localization phenomenon alters the modal configuration of the system. The fact that no such frequency 'shifts' were observed for the first two 'peaks' of the theoretical and experimental FFTs may be attributed to the specific position of the coupling stiffness, and to the hardening effects of the vibro-impact nonlinearities on each of the first two modes of the assembly.

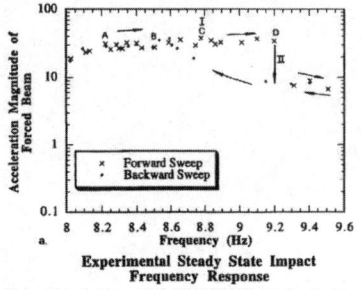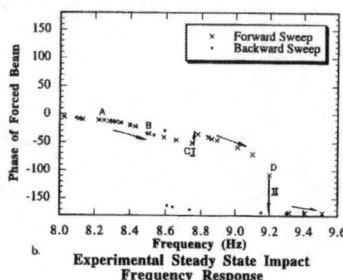

Figure 5.30. Experimental nonlinear FRCs for the directly excited beam 1: (a) fundamental harmonic component of the acceleration, (b) phase.

In the second set of experimental tests, the assembly with double-sided rigid barriers was used, in analogy to the theoretical model of the previous Section. Harmonic excitation was applied to beam 1 by means of a modal shaker, for clearance $c = 0.008m$. The level of the force chosen for the experiments was kept constant for forward and backward frequency sweeps. In Figures 5.30 and 5.31 depict the frequency response curves (FRCs) of the flexible assembly for forcing frequencies close to the natural frequency of the primary cantilever mode of each beam. Only the fundamental harmonic components of the FFTs of the accelerations are depicted in these Figures; hence, these experimental results should be compared to the theoretical FRCs of Figures 5.25b,c and 5.26b,c. At each frequency the system was allowed to reach a steady state (periodic or quasi-periodic) motion before recording it for the plots.

In general, the experimental results confirm the features predicted by the corresponding numerical simulations. In the experimental plots, the notation used for the theoretical bifurcation points and jumps was preserved. The only feature not observed experimentally, was the jump from the branch of low-amplitude anti-phase solutions to the branch of localized motions when during the backward frequency sweep. The lack of this jump could be attributed to the small fluctuations in the forcing amplitude applied from the shaker to beam 1, and to a difference in the damping distribution between theory and experiment. It is well known that in close-to-resonance modal tests the level of actual force applied by the shaker to the structure is very sensitive to frequency fluctuations. In the experimental tests the level of force

5.3 Vibro-impact Motions of a Rotary System

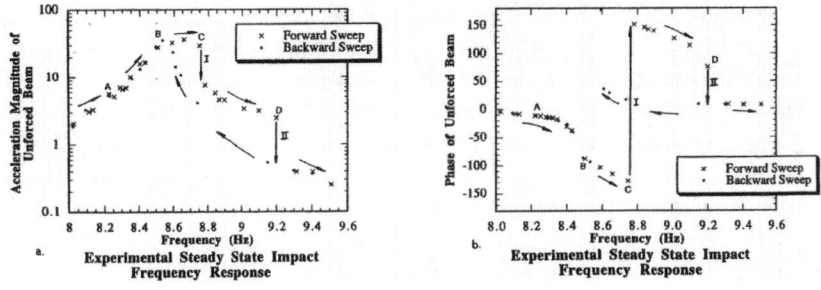

Figure 5.31. Experimental nonlinear FRCs for the unexcited beam 2: (a) fundamental harmonic component of the acceleration, (b) phase.

applied by the shaker was monitored by an oscilloscope and its fundamental harmonic component was kept to a constant level. Due to the highly nonlinear behavior of the system, the shaker-structure interaction occurring close to resonance results in fluctuations in the actual force applied; this could have contributed in the elimination of the backward jumps. However, the existence of the branch of nonlinear localized steady state motions was clearly verified by the experimental results.

These results demonstrate that the introduction of even small clearances in an otherwise linear continuous system, can result in strong vibro-impact nonlinearities which can cause new types of nonlinear dynamical phenomena that have no counterparts in linear theory. In particular, small clearances can lead to motion confinement phenomena which destroy the symmetry in response (of a perfectly symmetric system). Since clearances constitute a very common form of strong nonlinearities that are encountered often in engineering applications (loose joints is an example), attention should be given to understand their effects in the dynamic response. In the next Section a final example is given of complex dynamics due to vibro-impact nonlinearities in a rotary system.

5.3 Vibro-impact Motions of a Rotary System

Rotary flexible systems have been analyzed extensively [65, 208, 94, 193, 70, 125]. In this Section the effects of vibro-impacts on the dynamics of a rotor-flywheel system are examined numerically and experimentally. It is shown that even small clearances give rise to complex vibro-impact dynamics, such as sudden transitions to chaotic motions, bifurcations and jump phenomena. This finding is consistent with the other vibro-impact studies presented in this Chapter. For a more detailed exposition on the dynamics of the vibro-impacting rotor the reader is referred to [23].

The schematic of the continuous rotor-flywheel system considered is depicted in Figure 3.8, and the values of its geometric and material parameters are identical to the ones listed in Table 3.4, except for $a = 0.4064m$ and $e = 0.0254m$. Instead of incorporating the effects of the flywheel in the boundary conditions, they are added in the governing partial differential equations of motion as a localized terms at the right end of the rotor. The basic equations for horizontal and vertical deflections of the rotor are similar to the classic (Bernoulli) beam equation; however, coupling between these two displacements is provided through gyroscopic terms. It is noted that rotary inertia effects are important factors in high-speed machinery; however, for the system under consideration the speeds are not high enough since interest is focused only at speed ranges up to a range between the first and second critical speeds. Hence, in the model considered the rotary inertia effects are dropped from the governing equations.

The vibro-impact effects of the rotor with a rigid circular ring located at position $x = a$, are modeled by a high-stiffness spring with a clearance from the rotor. In addition, the flywheel is assumed to possess a mass imbalance m_e positioned at a distance e from the axis of the rotor. Assuming that the rotor rotates with speed Ω, the governing equations of motion of the rotor assume the form,

$$EI\, u_{1xxxx}(x,t) + m u_{1tt}(x,t) + d_x u_{1t}(x,t) + M_D \delta(x-L) u_{1tt}(x,t) +$$
$$2\Omega I_D u_{2xt}(x,t)\delta'(x-L) = \delta(x-L)Q_1(t) + \delta(x-a)P_1(u_1(a,t)),$$
$$EI\, u_{2xxxx}(x,t) + m u_{2tt}(x,t) + d_y u_{2t}(x,t) + M_D \delta(x-L) u_{2tt}(x,t) +$$
$$2\Omega I_D u_{1xt}(x,t)\delta'(x-L) = \delta(x-L)Q_2(t) + \delta(x-a)P_2(u_2(a,t)), \quad (5.21)$$

with boundary conditions:

$$u_{1,2}(0,t) = u_{1,2x}(0,t) = 0,$$
$$u_{1,2xx}(L,t) = u_{1,2xxx}(L,t) = 0. \quad (5.22)$$

In the above equations $u_1(x,t)$ and $u_2(x,t)$ are the rotor displacements in the vertical and horizontal directions, respectively; EI and m are the flexural rigidity and mass per unit length of the rotor, M_D and I_D the mass and moment of inertia of the cross section of the attached flywheel, $d_x = d_y$ and L the viscous damping coefficients per unit length and the length of the rotor; P_1 and P_2 represent the vertical and horizontal forces, respectively, due to the vibro-impacts; and Q_1, Q_2 are the external forces acting on the rotor due to the mass imbalance of the flywheel. As in Section 3.2.2, these forces are defined by the following relations:

$$Q_1(t) = m_e \Omega^2 e \cos \Omega t, \quad Q_2(t) = m_e \Omega^2 e \sin \Omega t - M_D g,$$
$$P_1(u_1(a,t)) = -\sigma \left(K_i \frac{(r-c)u_1}{r} + d_r \frac{(\dot{u}_1 u_1 + \dot{u}_2 u_2) u_1}{r^2} \right)\bigg|_{x=a},$$

5.3 Vibro-impact Motions of a Rotary System

$$P_2(u_2(a,t)) = -\sigma \left(K_i \frac{(r-c)u_2}{r} + d_r \frac{(\dot{u}_1 u_1 + \dot{u}_2 u_2) u_2}{r^2} \right)\Bigg|_{x=a},$$

$$\sigma = \begin{cases} 1, & r = \sqrt{u_1^2 + u_2^2} \geq c \\ 0, & r < c \end{cases}. \tag{5.23}$$

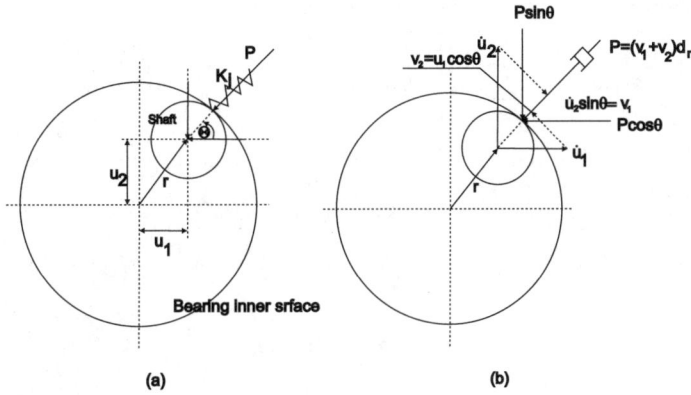

Figure 5.32. Elastic and damping forces generated due to vibro-impacts.

In Figure 5.32 the way of computing the elastic and damping components of the vibro-impact forces is depicted. The impacting forces P_1 and P_2 contain terms due to the impacts of the rotor with the distributed stiffness (that models the compliance of the circular ring), and damping terms that model energy dissipation due to inelastic impacts. In the numerical simulations elastic impacts were assumed by setting $d_r = 0$.

The governing equations (5.21) were discretized using as projection basis the eigenfunctions of the following simplified linear rotor connected to a flywheel at its right boundary and being clamped at its left end,

$$EI\, u_{xxxx}(x,t) + m u_{tt}(x,t) + M_D \delta(x-L) u_{tt}(x,t) = 0, \tag{5.24}$$

with boundary conditions:

$$u(0,t) = u_x(0,t) = 0, \quad u_{xx}(L,t) = u_{xxx}(L,t) = 0. \tag{5.25}$$

The eigensolution of this problem is presented in [23]; the derived eigenfunctions satisfy the orthogonality conditions,

$$m \int_0^L \phi_i(x) \phi_j(x) dx + M_D \phi_i(L) \phi_j(L) = \delta_{ij},$$

$$EI \int_0^L \phi_i''(x)\phi_j''(x)dx = \omega_i^2 \delta_{ij}, \qquad (5.26)$$

where the material properties of the rotor were assumed to be uniform.

The projection of the dynamics of the vibro-impacting rotary system to an N-dimensional space spanned by the first N modes is performed by expressing the horizontal and vertical deflections as follows:

$$u_1(x,t) = \sum_{i=1}^{N} a_i(t)\phi_i(x), \quad u_2(x,t) = \sum_{i=1}^{N} b_i(t)\phi_i(x). \qquad (5.27)$$

Then, the following reduced-order system of $2N$ governing differential equations results:

$$\ddot{a}_j + \omega_j^2 a_j + 2\Omega I_D \sum_{i=1}^{N} \phi_i'(L)\phi_j'(L)\dot{b}_i + (d_x/m)\sum_{i=1}^{N}[\delta_{ij} - M_D\phi_i(L)\phi_j(L)]\dot{a}_i =$$

$$P_1\phi_j(a) + Q_1\phi_j(L),$$

$$\ddot{b}_j + \omega_j^2 b_j + 2\Omega I_D \sum_{i=1}^{N} \phi_i'(L)\phi_j'(L)\dot{a}_i + (d_y/m)\sum_{i=1}^{N}[\delta_{ij} - M_D\phi_i(L)\phi_j(L)]\dot{b}_i =$$

$$= P_2\phi_j(a) + Q_2\phi_j(L),$$

$$j = 1,\ldots,N. \qquad (5.28)$$

The next step is to numerically solve this set of coupled differential equations (the coupling is provided through the gyroscopic terms), and to perform a convergence study in order to determine the smallest number of modes N (e.g., the truncation of the series (5.27)) that leads to accurate numerical results. Details of the convergence study can be found in [23]; it was determined that $N = 5$ modes and a time step of $\Delta t = 0.001s$ are sufficient for performing accurate numerical simulations in the ranges of speed Ω considered.

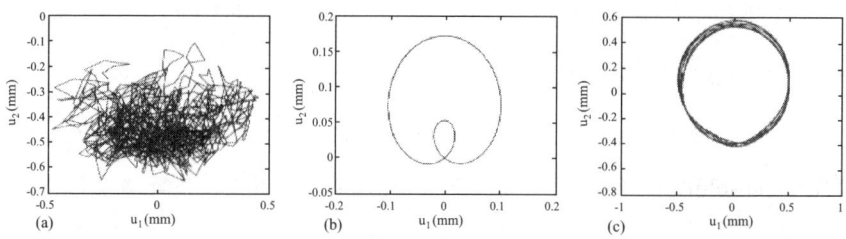

Figure 5.33. Orbits of the rotor at $x = 2034m$ for clearance $c = 0.002m$: (a) chaotic, $\Omega = 400rpm$, (b) periodic, $\Omega = 820rpm$, (c) quasi-periodic, $\Omega = 1200rpm$.

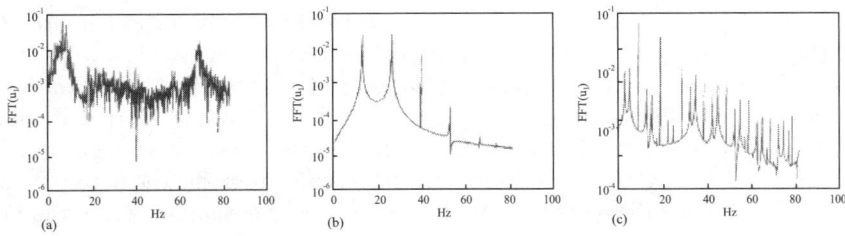

Figure 5.34. FFTs of the time series of the horizontal deflections of the rotor corresponding to the orbits of Figure 5.33.

Representative results of the numerical simulations are presented below. For the numerical simulations shown below the parameters used were the ones defined previously, with the exception of $c = 0.002m$, $d_x = d_y = 1.0 Ns/m$, $r_e m_e = 7.5 \times 10^{-3} Kgrm$. The orbits of the rotor at $x = 0.2034m$ (almost the half point between the right clamped boundary and the point of impacts) are depicted in Figure 5.33 for different speeds of rotation Ω. The FFTs of the corresponding horizontal deflections of the rotor are presented in Figure 5.34. Three types of motion encountered for the system under consideration are shown in these Figures, namely, periodic, quasi-periodic and chaotic motions. The presence of 1/2 subharmonic motion is evident from the plot of Figure 5.34b corresponding to the rotational speed $\Omega = 820 rpm$. The chaotic motion at $\Omega = 400 rpm$ shows the characteristic broadband frequency spectrum. The motion at $\Omega = 820 rpm$ is periodic with dominant peaks at $1\times, 2\times, \ldots$ the forcing frequency Ω. The motion at $\Omega = 1200 rpm$ is quasi-periodic, as there are many spontaneous side bands occurring close to integral multiples of the forcing frequency. In the particular period-2 quasi-periodic motion depicted in Figures 5.33 and 5.34 there are side bands close to $(1/2)\times, (3/2)\times, \ldots$ the forcing frequency.

Numerical simulations were performed for speeds in the range $\Omega \in [300 \div 2000 rpm]$. For each simulation sufficient time was given for the initial transients to decay, and then the steady-state motions were recorded. To distinguish between co-existing steady state motions an impulsive force was applied to the flywheel, that initiated a transition (jump) to a different solution than the one assumed by the system before the impulse was applied. The bifurcation diagrams depicting the steady state horizontal and vertical deflection amplitudes of the rotor at $x = 0.635m$ for varying speed of rotation, are depicted in Figure 5.35. The bifurcation diagrams were constructed as follows. For fixed speed Ω the steady state time series of the horizontal and vertical rotor response were sampled initially for a fixed number of cycles, and the maxima of these responses were determined. Then the steady state responses of the rotor were sampled in time starting from the time instants of maximum amplitudes, at periods of the rotor rotation $T = 2\pi/\Omega$. The resulting discrete data points were then plotted along the vertical line corresponding to the fixed value of Ω. If the (horizontal and vertical) motions of the rotor are periodic at the rotation frequency all collected data points coincide, and one obtains a single data point at that value of Ω in the bifurcation diagram. If the motion is periodic with k times the period T, the bifurcation diagram contains k different data points at the particular speed of rotation. Finally, if the motion is chaotic or quasi-periodic the data points in the bifurcation diagram fill densely one or more vertical line segments at the speed of rotation.

Examining the bifurcation diagrams and considering forward frequency sweeps, one notes that the motion remains periodic in the speed range 200 to 320 rpm. In this range the impact forces are relatively weak. However, as the speed of rotation increases the motion suffers chaotic explosions, e.g., it becomes suddenly chaotic for small variations in the speed of rotation. These chaotic motions exist up to 600 rpm. At this value of speed of rotation the motion attains large amplitudes and the rotor rides on the periphery of the rigid disk (where the vibro-impacts occur). From 600 to 1078 rpm the motion is periodic at the speed of rotation. Just after 1078 rpm a sudden jump to a chaotic regime occurs and the motion remains chaotic up to 1185 rpm. At that speed the rotor settles into a quasi-periodic orbit filling two disjoint line segments in the bifurcation diagrams. As frequency increases this orbit eventually bifurcates to a period-2 (non chaotic) orbit until a speed of 1420 rpm. Then the motion becomes again chaotic until the speed of 1500 rpm. A period-2 appears suddenly in the range 1500 to 1520 rpm, and then becomes chaotic in a sudden explosion as speed increases. It is interesting to note that the transitions to chaos occur suddenly, without any precursors (such as, period doubling routes to chaos). Such chaotic explosions are typical in multi-dimensional nonlinear dynamical systems, whereas in lower dimensional systems more regular transitions to chaos occur (for example, through period doublings or breakup of invariant tori). The observed chaotic explosions can be attributed to the complex dynamic interactions between various nonlinear modes, leading to sudden energy transfers and instabilities.

For a backward frequency sweeps, there occurs a jump from a low-amplitude chaotic motion to a high-amplitude periodic motion at approximately 900 rpm. The fact that the speed of rotation for this jump is smaller than the corresponding speed for the reverse jump mentioned earlier (at 1078 rpm) gives rise to a nonlinear hysteresis loop; this feature is typical in this type of bifurcation diagrams.

An experimental fixture was constructed to verify the occurrence of the complex dynamics predicted by the numerical simulations [23]. The rotor consisted of a brass shaft of diameter $(1/2)''$ supported by two pillow block bearings. The bearings were selected so to impose nearly clamped boundary conditions near the driven end of the shaft. The bearings were then locked onto the shaft by means of a cam locking collar in order to avoid axial slippage of the shaft at higher speeds of rotation. The shaft was machined at

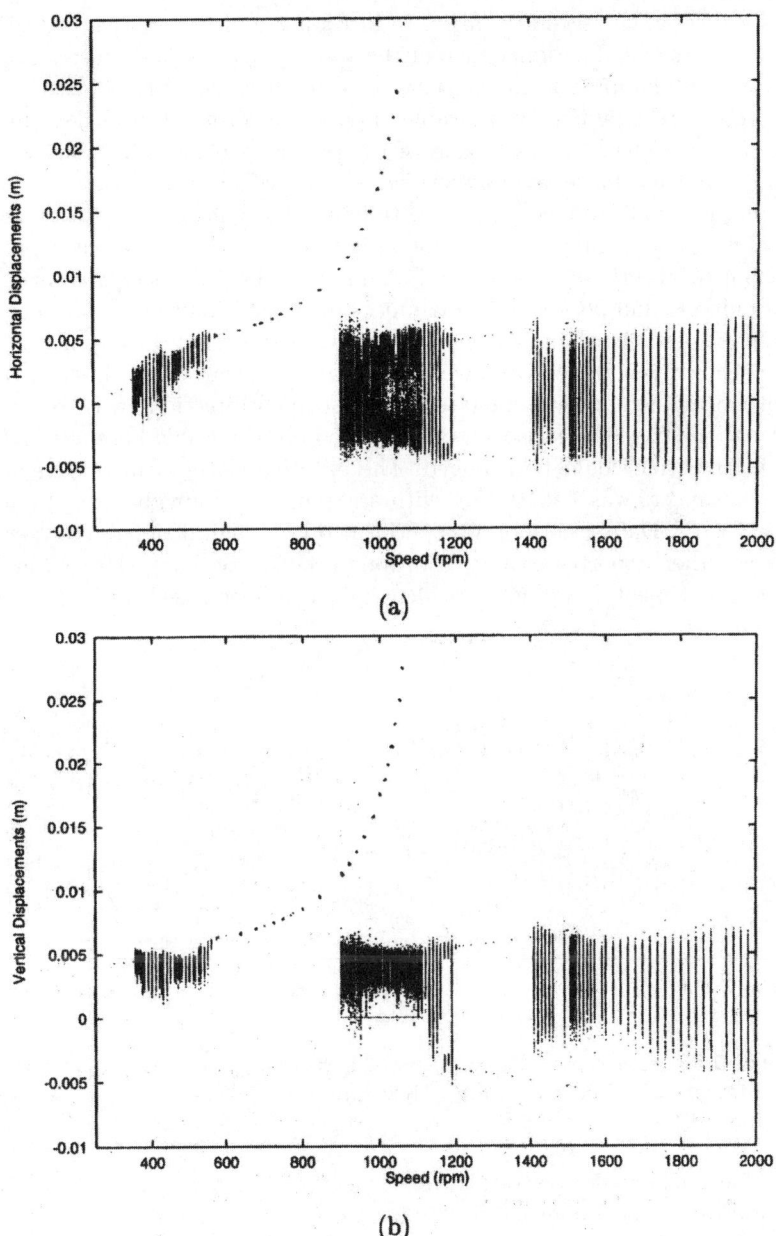

Figure 5.35. Bifurcation diagrams for varying Ω at the position $x = 0.635m$ of the rotor: (a) horizontal, and (b) vertical rotor deflections.

310   5 Continuous Systems with Non-smooth Nonlinearities

the location of the second bearing to introduce a clearance that would lead to vibro-impacts. The inner race of the second bearing was constrained not to move by clamping it to the outer race of the pillow block bearing. The bearings were mounted on movable supports in order to add flexibility in choosing the length of the length of the overhang rotor. The shaft was rotated by means of a servo motor which was controlled by a Smart Drive unit; the motor could produce speeds up to $5000 rpm$. The shaft was attached to the motor by a flexible jaw type coupling that possessed sufficient degrees of freedom to absorb shaft/motor misalignments. The flywheel was made out of aluminum, and possessed a small mass imbalance in order to produce harmonic loads to the shaft. The horizontal and vertical deflections of the shaft at the chosen sensing position were measured by means of two laser displacement sensors which were mounted in orthogonal directions with respect to each other. The laser sensors were attached on a movable base, so that the sensing position could be changed. The effective range of measurement for the transducers was $1 \times 10^{-2} m$, with a maximum bandwidth of data acquisition $3 kHz$. Data recording was performed by a digital signal analyzer; the data was then exported to a workstation for post processing. The parameters of the experimental system were identical to the ones used in the numerical simulations.

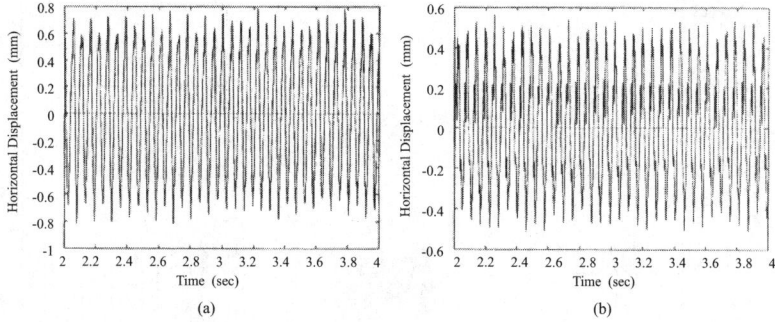

Figure 5.36. Experimental time series of horizontal deflections of the shaft: (a) $\Omega = 1078 rpm$, $c = 0.0015m$, (b) $\Omega = 1020 rpm$, $c = 0.002m$.

Experiments were performed for clearances of 1.5 and $2 mm$, for speeds of rotation in the range 300 to $1800 rpm$. Confirming the numerical simulations, two different coexisting steady state solutions were detected for a certain range of rotor speeds. Both solutions were stable, and their realization depended on the initial conditions. Transitions between solutions were induced by providing impulsive forces to the flywheel. The horizontal and vertical deflections of the shaft at the sensing position were recorded for a period of 8s with a bandwidth of $500 Hz$. This bandwidth was sufficient for accurate measurements, since the maximum speed of the rotor was only 30 Hz (1800 rpm).

To obtain the frequency-amplitude plots a forward frequency sweep was first performed, followed by a backward frequency sweep; these tests fully documented the two steady state solutions, and the nonlinear hysteresis loop of the system. Representative experimental time series of horizontal deflections of the shaft are depicted in Figure 5.36, and the corresponding orbits in a projection of the phase space are shown in Figure 5.37.

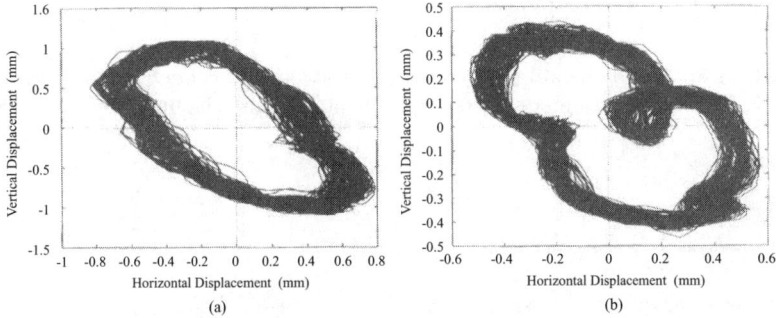

Figure 5.37. Orbits in a projection of the phase space of the time series of Figure 5.36.

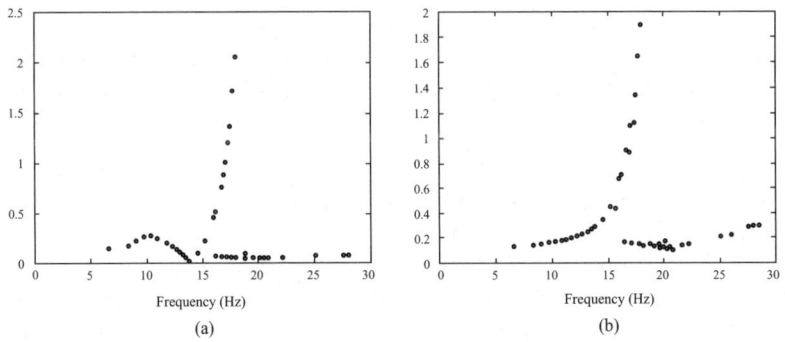

Figure 5.38. Fundamental FFT components of the horizontal deflection of the rotor for $c = 0.0015m$: (a) numerical, (b) experimental result (the units are in $10^{-3}m$).

Fixing the clearance to $c = 0.0015\,m$, for low speeds up to 475 rpm the motion of the shaft is periodic. In the range 475–768 rpm the motion becomes chaotic, but as the first critical speed of the shaft is reached the motion

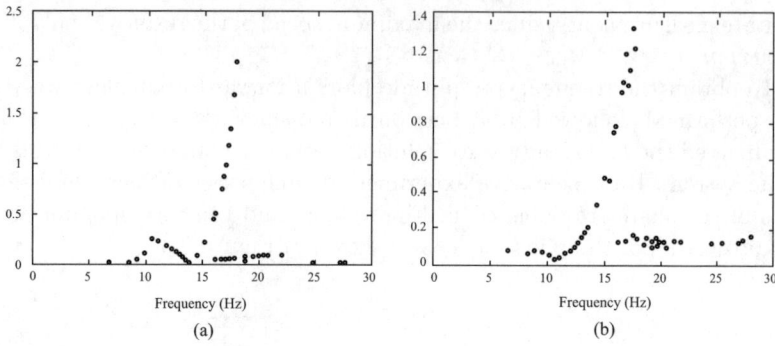

Figure 5.39. Fundamental FFT components of the vertical deflection of the rotor for $c = 0.0015m$: (a) numerical, (b) experimental result (the units are in $10^{-3}m$).

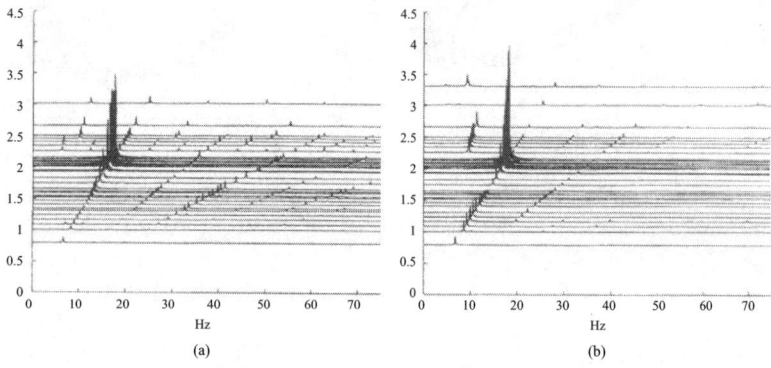

Figure 5.40. Waterfall plots of the horizontal deflection of the rotor at $x = 0.1936m$ obtained during forward frequency sweep for $c = 0.0015m$: (a) experiment, (b) simulation.

becomes quasi-periodic. After the first critical speed is passed the motion becomes again chaotic at $1176\,rpm$ until $1300\,rpm$ where a period-2 motion appears. Two jumps (one for forward and the other for backward frequency sweeps) are detected, in agreement to numerical predictions. When the clearance is increased to $c = 0.002\,m$ similar steady state responses are detected; there is a region from 870 to $940\,rpm$ where the motion is chaotic, and there appears to be a transition to a period-2 motion close to the region of the jump from high- to low-amplitude steady state motion.

To compare numerical and experimental results in Figures 5.38 and 5.39 the fundamental harmonic components (at the frequency of the rotation speed Ω) of the horizontal and vertical rotor deflections are plotted versus Ω. These results correspond to parameters $c = 0.0015\,m$ and $d_x = d_y = 2.0\,Ns/m$, and the measurement location is taken as $x = 0.193\,m$ along the length of the ro-

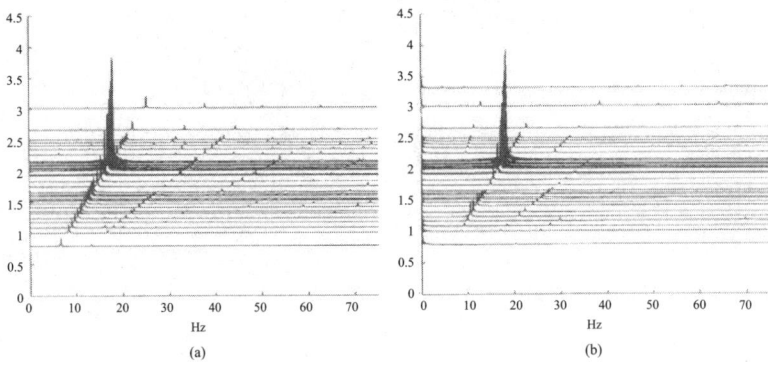

Figure 5.41. Waterfall plots of the vertical deflection of the rotor at $x = 0.1936m$ obtained during forward frequency sweep for $c = 0.0015m$: (a) experiment, (b) simulation.

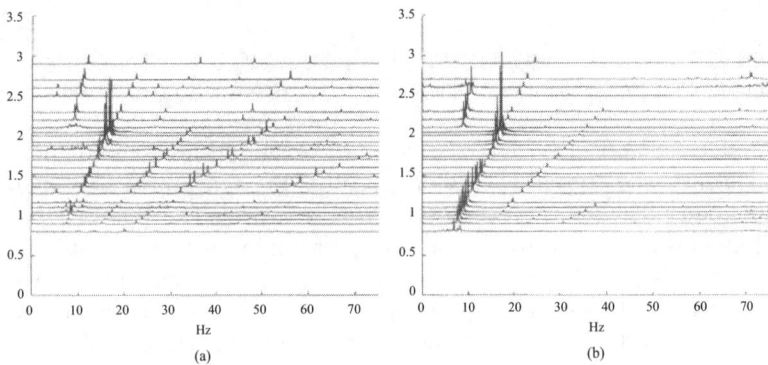

Figure 5.42. Waterfall plots of the horizontal deflection of the rotor at $x = 0.1936m$ obtained during forward frequency sweep for $c = 0.002m$: (a) experiment, (b) simulation.

tor. In both numerical and experimental plots there occur two jumps: one for forward sweep from smaller to larger amplitudes at $1078\,rpm$, and the reverse jump for backward sweep at $991\,rpm$. Hence, the experimental measurements verify the occurrence of the jumps between the steady state solutions, close to the speeds predicted by the numerical simulations. Moreover, the steady state deflection amplitudes are also comparable. Similar results are reported for the case of increased clearance $c = 0.002\,m$ [23].

In Figures 5.40–5.43 numerical and experimental Waterfall plots for systems with clearances $c = 0.0015\,m$ and $0.002\,m$ are presented. These plots were constructed by plotting the Fourier spectra of the rotor displacement for varying rotation speeds. The range of speeds considered in the plots was 300–$1800\,rpm$. For a linear rotor (with no clearances) the spectra contain

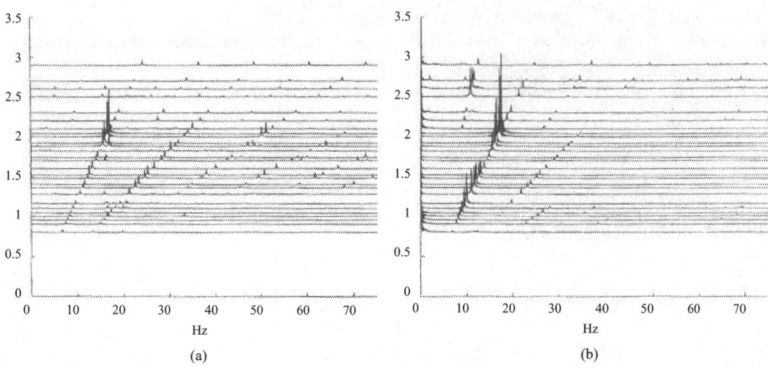

Figure 5.43. Waterfall plots of the vertical deflection of the rotor at $x = 0.1936m$ obtained during forward frequency sweep for $c = 0.002m$: (a) experiment, (b) simulation.

only Fourier components at integral multiples of the running speed Ω. In the nonlinear rotor system under consideration, however, subsynchronous ripples begin to develop as Ω increases. Indeed, there are cases where these subsynchronous components exceed in magnitude the synchronous components, although near the critical speeds of the rotor the synchronous components are dominant. The appearance of strong subsynchronous ripples in the Waterfall plots is attributed to the vibro-impact nonlinearities exhibited by the rotor. In terms of hydrodynamics this phenomenon is commonly referred to as oil whirl or oil whip, and it can be quite destructive regarding the operation of the rotor. Indeed, if the rotor is running between its first and second critical speeds one would assume that its operation is safe; however, if a subsynchronous harmonic component with frequency equal to the first critical speed develops, then it can directly excite the leading mode of the rotor, leading to large deflections. Hence, in t nonlinear rotor there exists the possibility of large-amplitude resonance motions even if the speed of rotation is away from the neighborhoods of its critical speeds. The subsynchronous components are harmless if the running speed is less than the first critical speed of the rotor, but they should be of concern at speeds above the first critical speed. In the depicted Waterfall plots there occur $(1/2)$, $(3/2)$ and $(5/2)$ subsynchronous components, with the $(1/2)$ component being the dominant one. This component becomes even more evident when the clearance is increased to $c = 0.002m$. In general there is agreement between the numerical and experimental Waterfall plots, which confirms the capacity of the numerical model to capture the complex dynamics of the vibro-impacting rotary system.

References

1. Algazi V.R., Brown K.L., Reddy M.J., Transform Representation of the Spectra of Acoustic Segments with Applications, *IEEE Trans. Speech Audio Proc.*, 1, 180-195, 1993.
2. Amba-Rao C.L., On the vibration of a rectangular plate carrying a concentrated mass. *Trans. ASME*, 1964, E31, No 3, pp. 550-551.
3. Ambarcumian S.A., *General theory of anisotropic plates*. Moscow, Nauka, 1974, pp. 448, in Russian.
4. Ambarcumian S.A., *Theory of anisotropic plates*. Moscow, Nauka, 1967, pp. 268, in Russian.
5. Amiro I.J., Palamarčuk V.G., Nosačenko A.M., Free vibrations of a rib cylindrical shell with a stiffly joint absolutely hard body. *Probl. Pročnosti*, 1978, No 2, pp. 56-60, in Russian.
6. Amiro I.J., Palamarčuk V.G., Free vibrations of a rib cylindrical shell with the masses on springs. *Probl. Pročnosti*, 1979, No 5, pp. 83-85, in Russian.
7. Andriejev L.V., Dyszko A.L., Pavlienko I.D., Fundamental free frequency estimation of plates and shells with an amortised mass. *Izv. VUZ Aviac. techn.*, 1980, No 2, pp. 89-93, in Russian.
8. Arman J.-L. P., *Priloženia tieorii optimalnowo uprawlienia sistiemami*. Mir, Moskva, 1977, in Russian.
9. Arutiunian N.H., Abramian B.L., *Torsion of Elastic Bodies*. Fizmatgiz, Moscow, 1963, in Russian.
10. Atanackovic T.M., Buckling of a Compressive Elastic Ring, *Acta Mechanica*, 127, 121-134, 1998.
11. Aubrecht J., Vakakis A., Tsao T.-C., Bentsman J., 1996, Experimental Study of Nonlinear Transient Motion Confinement in a System of Coupled Beams, *J. Sound Vib.*, 195 (4), 629-648.
12. Aubry N., *A Dynamical System / Coherent Structure Approach to the Fully Developed Turbulent Wall Layer*, Ph.D. Thesis, Cornell University, Ithaca, NY, 1987.
13. Aubry N., Holmes P., Lumley J.L., Stone E., The Dynamics of Coherent Structures in the Wall Region of the Wall Boundary Layer, *J. Fluid Mech.*, 192, 115, 1998.
14. Aubry N., Lian W.-Y., Titi E.S., Preserving Symmetries in the Proper Orthogonal Decomposition, *SIAM J. Sc. Comp.*, 14, 483, 1993.
15. Awrejcewicz J., Krysko V.A., Kutsemako N., Free Vibrations of Doubly Curved In-plane Non-Homogeneous Shells. *Journal of Sound and Vibration*, 1999, 225 (4), pp. 701-722.
16. Awrejcewicz J., Krysko V.A., 3D Theory Versus 2D Approximate Theory of the Free Orthotropic (Isotropic) Plates and Shells Vibrations. Part I, Deriva-

tion of Governing Equations. *Journal of Sound and Vibration*, 1999, 226 (5), pp. 807-829.
17. Awrejcewicz J., Krysko V.A., 3D Theory Versus 2D Approximate Theory of the Free Orthotropic (Isotropic) Plates and Shells Vibrations. Part II, Numerical Algorithms and Analysis. *Journal of Sound and Vibration*, 1999, 226 (5), pp. 831-871.
18. Awrejcewicz J., Krysko V.A., *Dynamics of Continuous Systems*. Warsaw, Wydawnictwa Naukowo-Techniczne, 1999, in Polish.
19. Awrejcewicz J., Krysko V.A., *Optimization of Plate and Shell Surface*. Proceedings of the Tenth World Congress on the Theory of Machines and Mechanisms, Oulu, Finland, June 20-24, 1999, 2128-2133.
20. Awrejcewicz J., Krysko V.A., *Optimization of Plate and Shell Surfaces*. Proceedings of the ASME design Engineering Technical Conferences, Las Vegas, Nevada, September 12-15, 1999, DETC99/VIB-8139, published on CD-Rom.
21. Awrejcewicz J., Krysko V.A., Vibration analysis of the plates and shells of moderate thickness. *Journal of Technical Physics*, 1999, 40, 3, pp. 277-305.
22. Awrejcewicz J., Lamarque C.-H., *Bifurcation and Chaos in Nonsmooth Mechanical Systems*, World Scientific, Singapore 2003 (to appear).
23. Azeez M.A.F., Vakakis A.F., Numerical and Experimental Analysis of a Continuously Overhung Rotor Undergoing Vibro-impacts, *Int.J. Nonlinear Mech.*, 34, 415-435, 1999.
24. Azeez M.F.A., *Theoretical and Experimental Studies of the Nonlinear Dynamics of a Class of Vibro-impact Systems*, Ph.D. Thesis, University of Illinois at Urbana-Champaign, Urbana, IL, 1998.
25. Azeez M.F.A., Vakakis A., Proper Orthogonal Decomposition of a Class of Vibro-impact Oscillations, *J. Sound Vib.*, 240(5), 859-889, 2001.
26. Babitsky V.I. (Ed.), *Dynamics of Vibro-impact Systems*, Springer Verlag, Berlin and New York, 2001.
27. Babitsky V.I., *Theory of Vibro-impact Systems and Their Applications*, Springer Verlag, Berlin and New York, 1998.
28. Bani P., On the vibration of a rectangular plate with a concentrated mass placed on it in a transverse magnetic field. *Intern. J. Pure and Appl. Math*, 1973, 4, No 4, pp. 461-467.
29. Baničuk N.V., *On one of variational problems with unknown limit and on optimal configurations of elastic bodies*. Applied Mathematics and Mechanics (PMM), 1975, 39, 6, pp. 1082 - 1092, in Russian.
30. Baničuk N.V., *Optimal conditions for drilling forms in elastic bodies*. PMM, 1977, 41, 5, pp. 920 - 925, in Russian.
31. Baničuk N.V., *Optimization of Elastic Bodies*. Science, Moscow 1980, in Russian.
32. Barg A.J., Kowalskij B.S., *On calculations of concentration coefficients during torsions*. Prikladnaja miechanika, 1971, 7, No 11, pp. 13 - 18, in Russian.
33. Barsi S.F., Kahyai K., Steady - state motion of a plate with a discontinuous mass. *Int. J. Nonlinear Mech*, 1974, 9, No 6, pp. 451-462.
34. Bayly P.V., Johnson E.E., Wolf P.D., Smith W.M., Ideker R.E., Measuring Changing Spatial Complexity in Ventricular Fibrillation Using the Karhunen-Loeve Decomposition of 506-Channel Epicardial Data, *Proceedings, Computers in Cardiology*, IEEE Computer Society Press, Los Alamitos, CA, 1993.
35. Bayly P.V., Virgin, L.N., An Empirical Study of the Stability of Periodic Motion in the Forced Spring-Pendulum, *Proc. Roy. Soc. London A*, 443, 391-408, 1993.

36. Benguedouar A., *Proper Orthogonal Decomposition in Dynamical Modeling: A Qualitative Dynamic Approach*, Ph.D. Thesis, Boston University, Boston, MA, 1995.
37. Bot Kin M.E., *Shape optimization of plate and shell structures*. AIAA Journal, 1982, 20, N 2, pp. 268 - 273.
38. Breuer K.S., Sirovich L., The Use of Karhunen-Loeve Procedure for the Calculation of Linear Eigenfunctions, *J. Comp. Physics*, 96, 277-296, 1991.
39. Brogliato B., *Nonsmooth Mechanics*, Springer Verlag, Berlin and New York, 1999.
40. Brogliato B., ten Dam A.A., Paoli L., Genot F., Abadie M., Numerical Simulation of Finite Dimensional Multibody Non-smooth Mechanical Systems, *Appl. Mech. Rev.*, 55 (2), 107-148, 2002.
41. Bublik B.N., *Numerical solutions to Dynamical Problems of Plates and Shells*. Naukowa dumka, Kiev, 1976, in Russian.
42. Burl J.B., Application of Karhunen-Loeve Expansion to the Reduced-Order Control of Distributed Parameter Systems, *Control and Computers*, 16 (1), 12-15, 1998.
43. Cetinkaya C., *Axisymmetric Wave Propagation in Weakly Coupled Layered Structures: Analytical and Computational Results*, Ph.D. Thesis, University of Illinois, Urbana, IL, 1994.
44. Chaftka R.T., Prasad B., *Review of optimal design of bending plates*. Rockets Technique and Cosmonautics, 1981, 19, No 6, pp. 104 - 113, in Russian.
45. Chaskalovic J., Naili S., Bifurcation Theory Applied to Buckling States of a Cylindrical Shell, *ZAMP*, 46, 149-155, 1995.
46. Chen P.Y., *Vibration Suppression of Flexible Structures Using Collocated Velocity Feedback and Nonlocal Actuator Control*, Ph.D Thesis, Cornell Univ., Ithaca, NY, 1990.
47. Chen R., Vibration of cylindrical panels carring a concentrated mass. *Trans. ASME*, 1970, E37, No 3, pp. 874-875.
48. Chen W.J., Pierre C., Exact Linear Dynamics of Periodic and Disordered Truss Beams: Localization of Normal Modes and Harmonic Waves, AIAA Paper 91-1114, *Proceedings AIAA/ASME/ASCE/AHS/ASC 32nd Structures, Structural Dynamics and Materials Conference*, Washington, D.C, 1991.
49. Chen W.J., Pierre C., Vibration Localization and Wave Conversion in Large Disordered Beam-Like Lattice Trusses, *AIAA J.*, 32 (8), 1724-1732, 1992.
50. Chomčenko A.N., Christienko A.S., On a problem of plates with concentrated masses vibrations. *Prikl. Miechanika,* 1972, 8, No 1, pp. 22-26, in Russian.
51. Christienko A.S., Kalko A.M., *An influence of rotational inertia of the added masses on shells vibrations for different boundary conditions*. Nikolajev Ship Construction Institute Press, 1973, 76, pp. 63-65, in Russian.
52. Christienko A.S., Tsybin E.N., *Vibration of an ortotropic closed cylindrical shell with an rotational inertia of jointed mass*. Nikolajev Ship Construction Institute Press, 1976, 117, pp. 3-8, in Russian.
53. Christienko A.S., *Free vibrations of nonsloped cylindrical shells with discrete additives*. Sudostr. i Morskie Sooruženia, 1973, 20, pp. 56-63, in Russian.
54. Christienko A.S., *Vibrations of cylindrical shells loaded by distributed and concentrated masses*. Izv. AN SSSR. Miechanika Tvierdovo Tiela, 1972, No 4, pp. 116-122, in Russian.
55. Chun Y.W., Hang E.Y., *Two-dimensional Shape Optimal Design*. Int. J. Numer. Meth. Eng., 1978, 13, No 2, pp. 311 - 336.

56. Cirkov V.P., *A change of stochastic vibration field caused by vibrational pick-ups.* In the book: *Theory of shells and plates.* Nauka, 1973, pp. 579-584, in Russian.
57. Conner M.D., Tang D.M., Dowell E.H., Virgin L.N., *J. Fluids Structures*, 11, 89-109, 1997.
58. Cusumano J.P., Sharkady M.T., Kimble B.W., Dynamics of a Flexible Beam Impact Oscillator, *Phil. Trans. Royal Soc. London*, 347, 421-438, 1994.
59. Demidovicz B.P., Maron I.A., *Basis of numerical mathematics.* Moscow, Nauka, 1966, in Russian.
60. Diemianulienko I.V., Koroliewa E.F., *Optimal design of turbomachine disks.* Izv. AN SSSR, ser. MTT, 1972, No 2, pp. 56 - 60, in Russian.
61. Djuvo G., Lions J.-L., *Inequalities in Mechanics and Physics.* Science, Moscow 1980, in Russian.
62. Emaci E., Azeez M.A.F., Vakakis A., Numerical and Experimental Results on the Dynamics of Flexible Periodic Trusses, *University of Illinois Report UIUC-ENG-97-4041*, Urbana, IL, 1997.
63. Emaci E., Azeez M.A.F., Vakakis A., Dynamics of Trusses: Numerical and Experimental Results, *J. Sound Vib.*, 214 (5), 953-964, 1998.
64. Emaci E., Nayfeh T.A., Vakakis A.F., Numerical and Experimental Study of Nonlinear Localization in a Flexible Structure with Vibro-impacts, *ZAMM*, 77 (7), 527-541, 1997.
65. Eshleman R.L., Eubanks R.A., On the Critical Speeds of a Continuous Rotor, *J. Eng. Industry*, 1180-1188, 1969
66. Feeny B.F., On Proper Orthogonal Co-ordinates as Indicators of Modal Activity, *J. Sound Vib.*, 255(5), 805-817, 2002.
67. Feeny B.F., Kappagantu R.V., On the Physical Interpretation of Proper Orthogonal Modes in Vibrations, *J. Sound Vib.*, 211(4), 607-616, 1998.
68. Feldmann U., Kreuzer E., Pinto F., Dynamic Diagnosis of Railway Tracks by Means of the Karhunen-Loeve Transformation, *Nonlinear Dynamics*, 22, 193-203, 2000.
69. Finlayson B.A., *The Method of Weighted Residuals and Variational Principles.* Series in Mathematics in Science and Engineering, 87, Academic Press, NY, 1972.
70. Flowers G.T., Wu T., Disk/Shaft Vibration Induced by Bearing Clearance Effects: Analysis and Experiments, *J. Vib. Acoust.*, 118, 204-208, 1996.
71. Folkman S.L., Rowsell E.A., Ferney G.D., Influence of Pinned Joints on Damping and Dynamic Behavior of a Truss, *J. Guid. Control Dyn.*, 18, 1398-1403, 1995.
72. Fomina A.A., Christienko A.S., *Universal program of free vibration frequency determination of a system: plate - mass - spring.* Nikolajev Ship Construction Institute Press, 1971, 50, pp. 71-75, in Russian.
73. Friedman P., Lust S., Bendiksen O., Free and Forced Response of Nearly Periodic Multispan Beams and Multibay Trusses, *Proceedings of the AIAA/ASME/ASCE/AHS/ASC 32nd Structuress, Structural Dynamics and Materials Conference*, Washington, D.C, 1991.
74. Fu L., Waas A.M., Initial Post-buckling Behavior of Thick Rings Under Uniform External Hydrostatic Pressure, *J. Appl. Mech.*, 62, 338-345, 1995.
75. Gajender N., Free vibration of a circular plate with a concentrated mass, spring and dashpot. *Acta Mech*, 1967, 4, No 3, pp. 273-287.
76. Gajender N., Vibration of a circular plate with a concentrated mass, spring system. *J. Sci and Eng. Res.*, 1965, 9, No 2, pp. 303-313.

77. Georgiou I.T., *Nonlinear Dynamics and Chaotic Motions of a Singularly Perturbed Nonlinear Viscoelastic Beam*, Ph.D. Thesis, Purdue University, West Lafayette, IN, 1993.
78. Georgiou I.T., Bajaj A.K., Corless M., Slow and Fast Invariant Manifolds and Normal Modes in a Two Degree-of-Freedom Structural Dynamical System with Multiple Equilibrium States, *Int. J. Nonlinear Mech.*, 33 (2), 275-300, 1998.
79. Georgiou I.T., Schwartz I., Slaving the In-Plane Motions of a Nonlinear Plate to its Flexible Motions: An Invariant Manifold Approach, *J. Appl. Mech.*, 64, 175-182, 1997.
80. Georgiou I.T., Schwartz I., Dynamics of Large Scale Coupled Mechanical Systems: A Singular Perturbation Proper Orthogonal Decomposition Approach, *SIAM J. Appl. Math.*, 59, 1178-1207, 1999.
81. Georgiou I.T., Schwartz I., Emaci E., Vakakis A., Interaction Between Slow and Fast Oscillations in an Infinite Degree-of-Freedom Linear System Coupled to a Nonlinear Subsystem: Theory and Experiment, *J. Appl. Mech.*, 66, 448-459, 1999.
82. Giershgorin S.A., Vibration of plates with the concentrated masses. *Prikl. Mat. i Mech.*, 1933, 1, No 1, pp. 25-37, in Russian.
83. Glovinski R., Lions J.-L., Triemolier R., *Cisliennoje issliedowanie wariacjonnych nierawienstw.* Mir, Moscow, 1979, in Russian.
84. Goldenvejzer A.L., Lidskij W.B., Towstik P.E., *Free vibrations of thin elastic shells.* Science, Moscow, 1979, in Russian.
85. Graham M.D., Kevrekedis I.G., Alternative Approaches to the Karhunen-Loeve Decomposition for Model Reduction and Data Analysis, *Comp. Chem. Eng.*, 20, 495-506, 1996.
86. Grigoliuk E.I., Sieliezov I.T., *Nonclassical theories of beams, plates and shells.* Moscow, 1973 (Achievements of Science and Technics. WINITI. *Mechanics of hard deformable bodies*), in Russian.
87. Griniev V.B., Filippov A.P., *An optimal circled plates.* Izv. AN SSSR, ser. MTT, 1977, 12, No 1, pp. 131 - 137, in Russian.
88. Guliajev V.I., Baženov V.A., Lizunov P.P., *Nonclassical plate theory and its application to solutions of the engineering problems.* Lvov, High school, 1978, in Russian.
89. Gutierrez R.H., Laura P.A., A note on the determination of the fundamental frequency of vibration of rectangular and circular plates supporting masses distributed over a finite area. *Appl. Acoust,* 1977, 10, No 4, pp. 303-313.
90. Hänel P.A., *Axialsymmetrische Eigenschwingunge der eingespannten Kreisplatte mit Finzel Masse.* Schiffbanforschung, 1969, 8, No 3-4, pp. 185-190.
91. Henon M., On the Numerical Computation of Poincaré Maps, *Physica D*, 5, 412-414, 1982.
92. Holmes P., Lumley J.L., Berkooz G., *Turbulence, Coherent Structures, Dynamical Systems and Symmetry,* Cambridge Univ. Press, Cambridge, UK, 1996.
93. Hyland D.C., Junkins J.L., Longman R.W., Active Control Technology for Large Space Structures, AIAA *J. Guid., Control, Dyn.*, 16 (5), 801-821, 1993.
94. Ishida Y., Nonlinear Vibrations and Chaos in Rotor Dynamics, *J. Japan Soc. Mech. Eng.*, 37 (2), 237-245, 1994.
95. Ivanov V.K., On incorrectly stated problems. *Mathematical Journal,* 1963, 61, No 2, pp. 61 - 73, in Russian.

96. Jacobsin M.J., Vibration of a rectangular sandwich plates with mass attachments. *J. Acoust. Soc. Amer.*, 1966, 40, No 3, pp. 677-683.
97. Kabalov V.V., Astrakharchuck S.V., Algorithm of Stability Analysis for Variable-thickness Rings Under Nonuniform Loads, *Prikl. Mekh.*, 19 (6), 75-79, 1983.
98. Kappagantu R.V., Feeny B.F., An Optimal Modal Reduction of a System with Frictional Excitation, *J. Sound Vib.*, 224(5), 863-877, 2000.
99. Karhunen K., Uber Lineare Methoden in der Wahrscheinlichkeitsrechnung, *Annals of Academic Science Fennicae Series A1, Math. Physics* 37, 1946.
100. Karmaker B.M., Amplitude frecuency characteristics of non linear vibrations of a clamped elliptic plates carrying a concentrated mass. *Int. J. Non-linear Mech*, 1978, 13, No 5-6, pp. 451-459.
101. Kasprzyk S., Vibration of a rectangular plates carrying a concentrated masses connected to it by elastic elements. *Strojnicky Časopis*, 1979, 30, No 3, pp. 309-323.
102. Kerschen G., *On the Model Validation in Non-linear Structures*, Ph.D. Thesis, University of Liege, Liege, Belgium, 2003.
103. Kerschen G., Golinval J.C., Physical Interpretation of the Proper Orthogonal Modes Using Singular Value Decomposition, *J. Sound Vib.*, 249(5), 849-865, 2002.
104. Kiecz V., Tieodoriescu P., *Introduction to generalized functions theory with applications in technique.* Moscow, Mir, 1978, pp. 520, in Russian.
105. Kilčevskij I.A., *Fundamentals of shells analytical mechanics.* Kiev, 1963, pp. 354, in Russian.
106. Kiričenko V.F., Krysko V.A., Method of variational iterations in theory of plates and its foundations. *Prikladnaja miechanika*, 1981, 17, No 4, pp. 71 - 76, in Russian.
107. Knobloch H.W., Aulbach B., Singular Perturbations and Integral Manifolds, *J. Math. Phys. Sci.*, 18 (5), 415-423, 1984.
108. Kollatz L., *Funkcjonalnyj analiz i wycislitielnaja matiematika.* Mir, Moskva, 1969.
109. Kollatz L., *Zadaci na sobstwiennyje znacienia.* Science, Moscow, 1968, in Russian.
110. Kolmogorov A.N., Fomin S.V., *Eliemienty tieorii funkcij i funkcjonalnowo analiza.* Nauka, Moskva, 1972, in Russian.
111. Komkov V., *Theory of optimal control using damping of simple elastic systems.* Mir, Moscow 1975, in Russian.
112. Korniszin M.S., *Non-linear Problems of Plates and Shallow Shells and Methods of Their Solutions.* Science, Moscow, 1964, in Russian.
113. Koroliev V.I., *Ringed anisotropic plates and shells made from armir composits.* Moscow, Maszgiz, 1965, in Russian.
114. Kovalczuk P.S., Kozlov S.V., Nosačenko A.N., On the problem of a cylindrical shell with an added mass dynamical instability. *Dokl. AN USSR. Ser. A*, 1980, No 1, pp. 43-47, in Russian.
115. Kozlov S.V., On determination of free vibration frequencies and the small vibration modes of an ortotropic cylindrical shell with the added masses. *Prikl. Miechanika*, 1981, 17, No 2, pp. 46-51, in Russian.
116. Kozlov S.V., On the parametric instability zones of an ortotropic cylindrical shell with the added masses. *Dokl. AN USSR.* Ser. A, 1981, No 2, pp. 47-51, in Russian.

117. Krysko V.A., *Nonlinear statics and dynamics of nonhomogeneous plates and shells*. Saratow, SGU, 1976, in Russian.
118. Krysko V.A., Červotkin B.V., Analysis of three dimensional dynamical problem for a rectangular plated with attached mass. *Izv. WUZ. Design and Architecture*, 1982, No 1, pp. 33 - 36, in Russian.
119. Krysko V.A., *Non-linear Statics and Dynamics of Non-homogeneous Shells*. Saratov State University Press, Saratov 1976, in Russian.
120. Kulikov E.L., Kiričenko V.F., Pavlov S.P., On solution to synthesis problems using stationary functionals. *Differential Equations*, 1979, 15, No 4, pp. 738 - 739, in Russian.
121. Kurant R., Gilbiert D., *Mietody matiematicieskoj fiziki, t. 1, 2*. Gostiechizdat, Moscow, 1951, in Russian.
122. Laura P.A., Gutierrez P.A., Transverse vibration of thin elastic plates with concentrated masses and internal elastic supports. *J. Sound and Vibr.*, 1981, 75, No 1, pp. 135-143.
123. Laura P.A., Gutierrez R.H., Vibration of simply-supported plates of arbitrary shape carrying concentrated masses and subjected to a hydrastatic state of in plane stresses. *J. Sound and Vibr.*, 1977, 55, No 1, pp. 49-53.
124. Laura P.A., Smith G.A., Vibration of rib-stiffened thin elastic plates carrying concentrated masses. *J. Acoust. Soc. Amer.*, 1968, 43, No 2, pp. 332-335.
125. Lee C.-W., Yun J.-S., Dynamic Analysis of Flexible Rotors Subjected to Torque and Force,' *J. Sound Vib.*, 192(2), 439-452, 1996.
126. Lee T.H., Vibrations of shallow spherical shells with concentrated mass. *Appl. Mech. Trans. ASME*, 1966, 33, No 4, pp. 938-939.
127. Lichodied A.I., Malinin A.A., Free vibrations of shape changeable homogeneous shells and with the concentrated masses. *Izv. AN SSSR. Miechanika Tvierdovo Tiela*, 1972, No 2, pp. 167-170, in Russian.
128. Lichodied A.I., Malinin A.A., On approximate formulas for vibration frequencies of smooth and with the concentrated additives shells. *Prikl. Miechanika*, 1973, 9, No 2, pp. 25-31, in Russian.
129. Lichodied A.I., Malinin A.A., Vibrations of the supported rotational shells with the concentrated masses and oscillators. *Izv. AN SSSR. Miechanika Tvierdovo Tiela*, 1971, No 1, pp. 42-47, in Russian.
130. Lichodied A.I., Dynamical influence of a mass distributed on a surface part of a shell. *Izv. AN SSSR. Miechanika Tvierdovo Tiela*, 1973, No 1, pp. 163-166, in Russian.
131. Lichodied A.I., Nonlinear vibrations of shells with the concentrated masses. *Izv. AN SSSR. Miechanika Tvierdovo Tiela*, 1973, No 1, pp. 141-146, in Russian.
132. Lin W.Z., Lee K.H., Lu P., Lim S.P., Liang Y.C., The Relationship Between Eigenfunctions of Karhunen-Loeve Decomposition and the Modes of Distributed Parameter Vibration System, *J. Sound Vib.*, 256(4), 791-799, 2002.
133. Lions J.-L., *Niekotoryje mietody rieszenia nieliniejnych krajewych zadać*. Mir, Moscow, 1972, in Russian.
134. Lions J.-L., *Optimalnoje uprawlienie sistiemami, opisywajemymi diffieriencjalnymi uprawnieniami w ciastnych proizwodnych*. Mir, Moscow, 1972, in Russian.
135. Litvinov V.G., On the problem of optimal control of eigenfrequencies with a changeable thickness. *J. of Computational Mathematics and Mathematical Physics*, 1979, 19, No 4, pp. 866 - 877, in Russian.
136. Loeve M.M., *Probability Theory*, Von Nostrand, Princeton, NJ, 1955.

137. Lu L.-Y., Utka S., Wada B.K., On the Placement of Active Members in Adaptive Truss Structures for Vibration Control, *Smart Mat. Str.*, 1, 8-23, 1992.
138. Lurie K.A., Čerkaev A.V., *On application of the Prager's theorem to design of thin shells*. Izv. AN SSSR, ser. MTT, No 6, pp. 157 - 159, in Russian.
139. Lurie K.A., Optimal control of a fluid in a channel with magnetic field. *Applied Mathematics and Mechanics* (PMM), 1964, 21, 1, pp. 45 - 49, in Russian.
140. Ma X., *Order Reduction, Identification and Localization Studies of Dynamical Systems*, Ph.D. Thesis, University of Illinois at Urbana-Champaign, Urbana, IL, 2000.
141. Ma X., Vakakis A., Karhunen-Loeve Decomposition of the Transient Dynamics of a Multibay Truss, *AIAA J.*, 37(8), 939-946, 1999.
142. Ma X., Vakakis A., Nonlinear Transient Localization and Beat Phenomena Due to Backlash in a Coupled Flexible System, *J. Vib. Acoustics*, 123, 36-44, 2001.
143. Ma X., Vakakis A., Bergman L., Karhunen-Loeve Modes of a Truss: Transient Response Reconstruction and Experimental Verification, *AIAA J.*, 39(4), 687-696, 2001.
144. Ma X., Vakakis A.F., Nonlinear Transient Localization and Beat Phenomena Due to Backlash in a Coupled Flexible System, *J. Vib. Acoust.*, 123 (1), 36-44, 2001.
145. Machens K.U., Dyer I., Energy Partitioning in a Truss Structure, *Acoustica*, 81, 587-595, 1990.
146. Magrab E.B., *Vibrations of a rectangular plate carrying a concentrated mass*. Trans. ASME, 1968, 35, No 2, pp. 411-412.
147. Makarovskij E.L., *On location load optimization on the thin plates using a fundamental frequency criterion. Mathematical methods of cybernetics*. Kiev, 1978, pp. 35-39, in Russian.
148. Malinin A.A., Vibration of rotational shells with the added masses and internal elastic joints. *Prikl. miechanika,* 1975, 11, No 2, pp. 29-33, in Russian.
149. Malinin A.A., Vibrations of rotational shells with concentrated and distributed elastic joints and masses. *Izv. AN SSSR. Miechanika Tvierdovo Tiela*, 1974, No 4, pp. 154-157, in Russian.
150. Maliutin I.S., Stability and vibrations of cylindrical shell with points support. *Izv. AN SSSR. Miechanika Tvierdovo Tiela*, 1976, No 2, pp. 186-189, in Russian.
151. Mari J.L., Glangeaud, Wave Separation Application, *Revue De LInstitut Francis Du Petrole*, 49, 21-75, 1994.
152. Martin A.V., Optimal design of elastic structures for multi-purpose loading. *J. of Optimiz. Theory and Applic.,* 1970, 6, No 1, pp. 22 - 40.
153. Maslov V.P., Omelianov G.A., Asymptotic Soliton-like Solutions of Equations with Small Dispersion, *Advances in Mathematical Sciences - Usp. Mat. Nauk.*, 36(3), 63-126, 1981.
154. Mead D.J., Wave Propagation and Natural Modes in Periodic Systems: II. Multi-coupled Systems With and Without Damping, *J. Sound Vib.*, 40(1), 19-39, 1975.
155. Michlin S.G., *Variational Methods in Mathematical Physics*. 2-e izv., pierierab. i dop. Science, Moscow, 1970, in Russian.
156. Miller D.W., Hall S.R., Von Flotow A., Optimal Control of Power Flow at Structural Junctions, *J. Sound Vib.*, 140, 475-497, 1990.
157. Mindlin R.D., Influence of rotatory enertia and shear on flexual motions of istropic elasticplates. *J. Appl, Mech*, 1951.

158. Mizochata S., *Tieoria urawnienij s ciastnymi proizwodnymi*. Mir, Moscow, 1977, in Russian.
159. Moehlis J., Smith T.R., Holmes P., Faisst H., 2002, Models for Turbulent Plane Couette Flow Using Proper Orthogonal Decomposition, *Physics of Fluids*, 14(7), 2493-2507.
160. Moon F., Shaw S., Chaotic Vibrations of a Beam with Nonlinear Boundary Conditions, *Int. J. Nonlinear Mech.*, 6, 465-477, 1993.
161. Morozow N.F., *On non-linear vibrations of thin plates taking into account rotational inertia*. DAN SSSR, 1967, 176, No 3, pp. 522 - 525, in Russian.
162. Muchojed A.P., Muchojed N.I., *On excited vibrations of ortotropic circled plate with a stiff additive. Calculation and Applied Mathematics*. Kiev, 1981, No 44, pp. 62-65, in Russian.
163. Naghdi P.M., On the theory of thin elastic shells. *Quart. Appl. Math.*, 1957, 14, pp. 365-380.
164. Navratna D.R., *Vibration of a rectangular plate with a concentrated mass, spring and dashpot*. Trans. ASME, 1963, E30, No 1, pp. 31-36.
165. Nayfeh A.H., Mook D., *Nonlinear Oscillations*, Wiley Interscience, New York, 1985.
166. Nayfeh T.A., *Application of Nonlinear Localization to Engineering Structures*, Ph.D. Thesis, University of Illinois, Urbana, IL, 1998.
167. Nayfeh T.A., Vakakis A.F., Passive Transient Wave Confinement Due to Nonlinear Joints in Coupled Flexible Systems, *Nonlinear Dynamics*, 25, 333-354, 2001.
168. Nigul U.K., On application of the Lurie symbolic method in a three dimensional plate dynamical theory. *Izv. AN Est. SSR. Ser. fiz.-mat. i tiechn. n.*, 1963, No 2, pp. 146-155, in Russian.
169. Novožilov V.V., *Theory of elasticity*. Leningrad, Sudpromgiz, 1953, in Russian.
170. Oden Ž.T., *Koniecznyje eliemienty w nieliniejnoj miechanikie splosznych sried*. Mir, Moscow, 1976, in Russian.
171. Okamoto M., Sekiya T., On the method for calculation of the natural frequenies of a plate with a rigid body. *Proc. 21st Jap. Nat. Cong. Appl. Mech.* Tokyo, 1971, 21, Tokyo, 1973, pp. 299-307.
172. Olhoff N., *Optimalnoje projektirowanie konstrukcij*. Mir, Moscow, 1981, in Russian.
173. Onoda J., Minegusi K., Semi-active Vibration Suppression of Truss Structures by Coulomb Friction, *J. Spacecrafts Rockets*, 31, 67-74, 1994.
174. Onoda J., Sano T., Minegusi K., Passive Damping of Truss Vibration Using Preloaded Joint Backlash, *AIAA J.*, 33, 1335-1341, 1995.
175. Onoda J., Watanabe N., Vibration Suppression by Variable-stiffness Member, *AIAA J.*, 29, 977-983, 1991.
176. Palamarčuk V.G., Dynamical instability of a ribbed cylindrical shell with a mass jointed by springs. *Prikl. Miechanika*, 1980, 16, No 6, pp. 22-27, in Russian.
177. Palamarčuk V.G., Dynamical instability of a system consisted of ribbed cylindrical shell and absolutely hard body. *Prikl. Miechanika*, 1978, 14, No 5, pp. 45-51, in Russian.
178. Palamarčuk V.G., Free vibrations of a ribbed cylindrical shell with a beam. *Prikl. Miechanika*, 1980, 16, No 3, pp. 45-49, in Russian.
179. Palamarčuk V.G., Free vibrations of a system consisted of a ribbed cylindrical shell and absolutely hard body. *Prikl. Miechanika*, 1978, 14, No 4, pp. 56-62, in Russian.

180. Palamarčuk V.G., On a problem of free vibrations of a ribbed shell with an absolutely hard body. *Prikl. Miechanika*, 1979, 15, No 9, pp. 37-41, in Russian.
181. Panteleev A.D., Medvedev N.G., *On coercivity of the three composite plate operator*. (In: *Matiematicieskaja fizika*. Kiev, Naukova Dumka, 1979, wyp. 26, pp. 121 -124) , in Russian.
182. Park H.M., Cho D.H., The Use of Karhunen-Loeve Decomposition for the Modeling of Distributed Parameter Systems, *Comp. Chem. Eng.*, 51, 81-98, 1996.
183. Pfeiffer F., Glocker C., *Multibody Dynamics with Unilateral Constraints*, Wiley Interscience, New York, 1996.
184. Pieliech B.L., Some problems of development theories and the calculation methods of anisotropic shells and plates with a finite tangential stiffness. *Mechanics of Polymers*, 1975, No 2, pp. 269-284, in Russian.
185. Pieliech B.L., *Theory of shells with a finite tangential stiffness*. Kiev, Naukowa dumka, 1973, in Russian.
186. Pierson B.L., An optimal control approach to minimum weight vibrating beam design. *J. Struct. Mech.*, 1977, t. 5, pp. 147 - 148.
187. Pilipchuck V.N., The Calculation of Strongly Nonlinear Systems Close to Vibration-impact Systems, *Prikl. Mat. Meck. (PMM)* 49 (6), 572-578, 1985.
188. Pilipchuck V.N., A Transformation for Vibrating Systems Based on a Non-Smooth Periodic Pair of Functions, *Doklady AN Ukr. SSR*, Ser. A, 4, 37-40, 1998 (in Russian).
189. Pilipchuck V.N., Vakakis A.F., Study of the Oscillations of a Nonlinearly Supported String Using Non-smooth Transformations, *J. Vib. Acoustics* 120, 434-440, 1998.
190. Pilipchuck V.N., Vakakis A.F., Azeez M.A.F., Study of a Class of Subharmonic Motions Using Non-smooth Temporal Transformations, *Physica D*, 100, 145-164, 1997.
191. Polia G., Siegio G., *Izoparametric Unenequalities in Mathematical Physics*. Fizmatgiz, Moscow, 1962, in Russian.
192. Prager V., *Osnowy tieorii optimalnowo projektirowania konstrukcij*. Mir, Moscow 1977, in Russian.
193. Raffa F.A., Vatta F., The Dynamic Stiffness Method for Linear Rotor-bearing Systems, *J. Vib. Acoust.*, 118, 332-339, 1996.
194. Rajaee M., Karlson S.K.F., Sirovich L., Low-dimensional Description of Free-Shear-Flow Coherent Structures and Their Dynamical Behavior, *J. Fluid Mech.*, 258, 1-29, 1994.
195. Rastrigin L.A., *Statical Search Methods*. Science, Moscow, 1968, in Russian.
196. Raszevskij P.K., *Lectures on differential algebra*. Moscow, Gostiechizdat, 1950, in Russian.
197. Ricards R.B., Tietiers G.A., *Stability of plates with composite materials*. Ryga, Zinatnie, 1973, in Russian.
198. Riks R., Rankin C.C., Computer Simulation of the Buckling Behavior of Thin Shells Under Quasi-static Loads, *Arch. Comp. Meth. Eng.*, 4 (4), 325-351, 1997.
199. Rodriguez J.D., Sirovich L., Low-dimensional Dynamics for the Complex Landau Equation, *Physica D*, 43, 77-86, 1990.
200. Sacienkov A.V., Konopliev J.G., Šiškin A.G., Free plates vibrations on the rolling learnings. *Prikl. Miechanika*, 1975, 11, No 5, pp. 62-68, in Russian.
201. Salenger G., Vakakis A.F., Discreteness Effects in the Forced Dynamics of a String on a Periodic Array of Nonlinear Supports, *Int. J. Nonlinear Mech.*, 33 (4), 659-673, 1998.

202. Salenger G., Vakakis A.F., Localized and Periodic Waves with Discreteness Effects, *Mech. Res. Comm.*, 25 (1), 97-104, 1989.
203. Sathyamoonthy M., Shear and rotatory inertia effects on plates. *J. Eng. Mech. Div. Proc. Amer. Soc. Civ. Eng.*, 1978, t. 104, No 5, pp. 1288 - 1293.
204. Schmidt H., Jensen F., A Full Wave Solution for Propagation in Multi-layered Viscoelastic Media with application to Gaussian Beam Reflection at Fluid-Solid Interfaces, *J. Acoust. Soc. Am.*, 77 (3), 813-825, 1985.
205. Schmidt R., A Critical Study of Postbuckling Behavior of Thick Rings Under Uniform External Hydrostatic Pressure, *ZAMM*, 59, 581-582, 1979.
206. Seide P., Albano E.D., Bifurcation of Circular Rings Under Normal Concentrated Loads, *J. Appl. Mech.*, 40, 233-238. 1973.
207. Shan A.H., Datta S.K., *Normal vibration of a rectangular plate with a ttached masses*. Trans. ASME, 1969, E36, No 1, pp. 130-132.
208. Shaw J., Shaw S., Instabilities and Bifurcations in a Rotating Shaft, *J. Sound Vib.*, 132 (2), 227-244, 1989.
209. Shaw J., Shaw S., Nonlinear Resonance of an Umbalanced Rotating Shaft with Internal Damping, *J. Sound Vib.*, 147 (3), 435-451, 1991.
210. Shaw S., Holmes P., A Periodically Forced Linear Oscillator with Impacts: Chaos and Long Period Motions, *Phys. Rev. Letters*, 51, 623-626, 1983.
211. Shaw S., Rand R.H., The Transition to Chaos in a Simple Mechanical System, *Int. J. Nonlinear Mech.*, 24 (1), 41-56, 1989.
212. Sheu C.Y., Elastic minimum-weight design for specified fundamental frequency. *Int. J. Solids and Struct.*, 1968, t. 4, N 10, pp. 953 - 958.
213. Shvartsman S.Y., Kevrekidis I.G., Low-dimensional Approximation and Control of Periodic Solutions in Spatially Extended Systems, *Phys. Rev. E*, 58 (1), 361-368, 1998.
214. Siarlie F., *Mietod koniecznych eliemientow dlia ellipticieskich zadać*. Mir, Moscow 1980, in Russian.
215. Siedov L.I., *Mechanics of a continuous medium*. Moscow, Nauka, 1973, in Russian.
216. Signorelli J., VonFlotow A., Wave Propagation, Power Flow and Resonance in a Truss Beam, *J. Sound Vib.*, 126 (1), 127-144, 1988.
217. Sirovich L., Turbulence and the Dynamics of Coherent Structures. Part I: Coherent Structures, *Quart. Appl. Math.*, 45, 561-571, 1987.
218. Sirovich L., Kirby M., An Eigenfunction Approach to Large Scale Transitional Structures in Jet Flow, *Physics of Fluids*, A2, 127-136, 1989.
219. Sirovich L., Knight B.W., Rodriguez J.D., Optimal Low-dimensional Dynamic Approximation, *Quart. Appl. Math.*, XLVIII, 535-548, 1990.
220. Solecki R., *Vibrations of plates with concentrated masses*. Bull Acad Polon. Sci Ser. Tech, 1961, No 4, pp. 209-215.
221. Srinivas S., Joga Rao C.V., Rao A.K., An exact analysis for vibration of simply supported homogeneous and laminated thick rectangular plates. *J. Sound and Vibr.*, 1970, 12, No 12, pp. 187-199.
222. Srinivas S., Rao A.K., A three-dimensional solution for plates and laminates. *J. Franklin Inst.*, 1971, 291, No 6, pp. 469-481.
223. Srinivas S., Rao C.V.J., Rao A.K., *Some results from an exact analysis of thick laminates in vibrations and buckling*. Trans. ASME, 1970, E37, No 3, pp. 868-870.
224. Stojan J.G., Makarovskij E.L., On a frequency rotational distribution of concentrated masses on thin plates. *Prikl. Miechanika*, 1980, 16, No 3, pp. 70-74, in Russian.

225. Stokey V.F., Zorovski C.F., Normal vibrations of uniform plate carrying any number of finite masses. *J. Appl. Mech.* Trans. ASME, 1959, 26, No 2, pp. 210-216.
226. Strieng G., Fiks Dž: *Tieoria mietoda koniecznych eliemientow.* Moscow, Mir, 1980, in Russian.
227. Tarman H., A Karhunen-Loeve Analysis of Turbulent Thermal Convection, *Int. J. Num. Meth. Fluids*, 22, 67-79, 1966.
228. Tarnopolskij J.M., Today Trends of Development of fiber composites. *Mechanics of Polymers*, 1972, No 3, pp. 541 - 552, in Russian.
229. Thompson J.M.T., Gaffari R., Chaos After Period-doubling Bifurcations in the Resonance of an Impact Oscillator, *Physics Letters* 91 A, 5-8, 1982.
230. Tichonov A.N., Arsenin V.J., *Methods of Solutions of Uncorrect Problems.* Science, Moscow, 1974, in Russian.
231. Tietiers G.A., Ricards P.B., Narusbierg V.L., *Optimizations of shells made from ringed composites.* Ryga, Zinatnie, 1978, in Russian.
232. Tietiers G.A., *Complex load and stability of shells made from polymer materials.* Ryga, Zinatnie, 1969, in Russian.
233. Timoshenko S.P., *On the correction for shear of the differential equation for transverse vibration of prismatic bar.* Philosophical Magazin, 1921, 41, No 6, pp. 744-746.
234. Timoshenko S.P., *Procznost i koliebania eliemientow konstrukcij.* Science, Moscow 1975, in Russian.
235. Triandaf I., Schwartz I.B., 1997, Karhunen-Loeve Mode Control of Chaos in a Reaction-Diffusion Process, *Phys. Rev. E*, 56 (1), 204-212.
236. Troickij V.A., Pietuchov L.V., *Form Optimizations of Elastic Bodies.* Science, Moscow, 1982, in Russian.
237. Vajnikko G.M., *On convergence of Galerkin type methods in eigenvalue problems.* Izv. WUZ, Mathematics, 1966, 51, No 2, pp. 37 - 45, in Russian.
238. Vakakis A., Passive Spatial Confinement of Impulsive Excitations in Coupled Nonlinear Beams, *AIAA J.*, 32(9), 1902-1910, 1994.
239. Vakakis A.F., Exponentially Small Splittings of Manifolds in a Rapidly Forced Duffing System, *J. Sound Vib.*, 170 (1), 119-129, 1994.
240. Vakakis A.F., Atanackovic T.M., Buckling of an Elastic Ring Forced by a Periodic Array of Compressive Loads, *J.Appl. Mech.*, 66(2), 361-367, 1999.
241. Vakakis A.F., Cetinkaya C., Analytic Evaluation of Periodic Responses of a Forced Nonlinear Oscillator, *Nonlinear Dynamics*, 7, 37-51, 1995.
242. Vakakis A.F., Manevitch L.I., Mikhlin Yu.V., Pilipchuk V.N., Zevin A.A., *Normal Modes and Localization in Nonlinear Systems*, Wiley Interscience, New York, 1996.
243. Vlasov V.Z., *General shells theory and its application in technique.* Moscow-Leningrad, Gostiechizdat, 1949, in Russian.
244. Volmir A.S., *Non-linear Dynamics of Plates and Shells.* Science, Moscow, 1972, in Russian.
245. Volmir A.S., *Stability of elastic systems.* Moscow, Nauka, 1967, in Russian.
246. VonFlotow A., Distrurbance Propagation in Structural Networks, *J. Sound Vib.*, 106, 433-450, 1986.
247. VonFlotow A., Traveling Wave Control for Large Spacecraft Structures, *AIAA J. Guid., Control, Dyn.*, 9 (4), 462-468, 1986.
248. Vorowič I.I., *On solution existence in non-linear theory of shells.* Izv. AN SSSR, series of Mathematics, 1955, 19, No 4, pp. 203 - 206, in Russian.
249. Vorowič I.I., *On solution existence in non-linear theory of shells.* DAN SSSR, 1957, 117, No 2, pp. 203 - 206, in Russian.

250. Whitney J.M., Beending Extensional Compling in Laminated Plates Under Transverse Louding. *Journal of Composite Materials*, 1969, t. 3, Inn, pp. 20 - 28.
251. Zahorian S.A., Rothenberg M., Principal Component Analysis for Low-order Redundancy Encoding of Speech Spectra, *J. Acoust. Soc. Am.*, 69, 519-524, 1981.
252. Zarutskij V.A., Forced vibrations of a longitudinally supported cylindrical shell with locally jointed mass. *Priklad. miechanika*, 1982, 18 (28), No 1, pp. 50-56, in Russian.
253. Zhuravlev V.F., A Method for Analyzing Vibration-impact Systems by Means of Special Functions, *Izv. AN SSSR, Mekhanica Tvergogo Tela*, 11 (2), 30-34, 1976.
254. Zhuravlev V.F., Investigation of Certain Vibration-impact Systems by the Method of Non-smooth Transformations, *Izv. AN SSSR, Mekhanica Tvergogo Tela*, 12 (6), 24-28, 1977.
255. Zhuravlev V.F., Equations of Motion of Mechanical Systems with Ideal Unilateral Constraints, *Prikl. Mat. Mekh.*, 42, 781-788, 1978.
256. Zukas J.A., Effects of transverse normal and shear strains in orthotropic shells. *AIAA Journal*, 1974, t. 12, pp. 1753 - 1755.
257. Žilin P.A., Ilicieva T.P., *Vibration spectra and modes of a rectangular cubicoid on a basis of a three dimensional elastic and plates theories.* Izv. AN SSSR. Miechanika Tvierdovo Tiela, 1980, No 2, pp. 94-103, in Russian.
258. Žilin P.A., *A new approach for development of a new linear theory of thin elastic shells.* In the book: *Calculations of space constructions.* Moscow, 1979, 19, in Russian.
259. Žilin P.A., *Linear theory of ribbed shells.* Izv. AN SSSR. Miechanika tvierdovo tiela, 1970, No 4, pp. 150-162, in Russian.

Index

N-dimensional space, 306
ε optimization, 124, 126
18-bay truss, 202–204, 210, 211

abrupt transition, 237
abruptly periodic chaotic regime, 236
accuracy, 3, 13, 52, 56, 69, 71–73, 77, 124, 130, 132, 135, 150, 151, 154, 155, 158, 218, 269, 290
acoustical, 178
aeroelastic problem, 178
Airy's function, 31
algebraic
– equation, 41, 52, 57, 58, 61–63, 66, 102, 104, 151
– system, 39, 47, 61
algorithm(s), V, VI, 3, 51, 52, 69, 72, 83, 84, 104, 107, 119, 120, 122, 135, 147, 148, 150, 151, 154, 155, 173, 175, 190, 201, 258
aluminum beam(s), 290, 298
anisotropic, 140
– body, 4, 7
anti-phase, 298, 302
– fashion, 298
antisymmetric, 14, 56, 64, 69, 74
approximate
– dimensionality, 188
artificial damping, 204
asymptotic
– analysis, 245, 254, 265
– approximation, 248
– point, 241
– technique(s), 240, 258, 259
axial compressibility, 258
axisymmetric wave propagation, 201

backlash
– joint, 282–285, 287
– nonlinearities, VII, 269, 285
backward
– frequency sweep(s), 296–298, 302, 308, 311–313
– jump, 303
bandwidth, 310
– maximum, 310
barrier, 253, 286, 289–292, 296, 300, 302
base space, 106
basis system, 106
bay, 201–205, 210, 211, 213, 215, 219, 220, 222, 275
– boundaries, 203
beam
– equation, 304
– substructure, 230
beam-pendulum, 228, 230, 231
bearing, 193, 308, 310
beat phenomena, 231, 269, 270, 272, 273, 278, 279, 281, 282, 285, 292, 295, 300
bending
– distribution, 267
– moment, 141–143, 161, 202, 213, 214, 259, 268
– motion, 205, 213
– vibration, 151, 201, 214
Bernulli-Euler theory, 258
bifurcating branches, 264
bifurcation
– diagram, 307–309
– point, 265, 302
bilinear, 98, 198
– form, 97, 99, 100, 120
borplastic, 122

boundary, VII, 95, 123, 143, 148, 149, 156, 158, 161, 174, 183, 202, 203, 205–210, 214, 222, 229, 284, 285, 305, 307
- conditions, VII, 10, 20, 22, 25, 27–29, 31, 33, 38, 39, 47, 48, 57, 60, 65, 95, 96, 102, 104, 108, 111, 120, 123, 126, 134, 141, 152, 153, 157, 162, 184, 189, 190, 201, 203, 213, 215, 229, 246, 271, 284, 298, 304, 305
- external, 158
- value problem, 160, 239, 262, 264, 266

bounded, 10, 22, 63, 105, 126, 127, 131, 146, 147, 166, 167, 170, 175, 176, 251, 252
- solution, 250

broadband frequency spectrum, 307

Bubnov-Galerkin method, VI, 102–104, 133, 164

buckled
- ring, 268
- state, 263–265

buckling, VII, 254, 256, 258, 263, 265, 267
- mode, 268
- problem, 254, 256

cam locking collar, 308

cantilever
- beam, 185, 193, 194, 301, 302
- mode, 200, 289–291, 295, 296, 298, 300, 302
- rotor, 196

centre, 2, 10, 13–15, 17, 25, 29, 32, 33, 38, 51, 53, 54, 56, 102, 133, 149, 156, 158

chain rule, 260

chaos, 194, 308

chaotic, 177, 191, 196, 198, 234, 236, 306, 308, 312
- explosion(s), 197, 308
- explosion, 194
- motion(s), 183, 196, 197, 231, 233, 235–238, 303, 307, 308

characteristic structure, 179

circle
- plate, 1, 133
- shell, 1

circular, 2, 5, 154, 155
- communication, 200
- ring, 258, 264, 268, 304, 305

clamped
- boundary conditions, 222, 298, 308
- joint, 201, 213, 285

clamped-simply supported beam, 185

clamping, 3, 22, 29, 33, 310

clearance, VI, VII, 183, 185, 188, 191, 193–196, 249, 253, 269, 272, 273, 278, 285, 289, 291, 296, 299, 302–304, 306, 310–314
- index, 188
- joint, 278, 285
- nonlinearities, VII, 252, 254, 269, 270, 272, 274, 277, 282

close-to-resonance, 302

closed set, 146

closure, 104

cluster, 200, 205, 211

co-existing, 201, 203, 236, 237, 307

coarse frequency sampling, 204

coherent spatial structures, 200, 210, 211, 275

compact, 131, 153, 167, 176, 180

complex
- dynamics, 303, 308, 314
- nonlinear dynamical phenomena, 229

composite(s), 1, 4, 86, 89, 93, 122

compressive
- load, 155, 256, 258, 259, 268
- static load, 258

constraint, VI, VII, 10, 124, 126, 130–133, 141, 144–146, 148, 150, 151, 155, 156, 158, 171, 173, 174, 287, 288, 297–299

construction(s), 145

continua, 5, 240, 282

continuity, 43, 96, 98, 131, 176, 269
- deformation, 31

continuous, V, VII, VIII, 1–3, 12, 13, 31, 68, 78, 81, 96, 100, 127, 129, 146, 147, 174, 179, 181–183, 191, 220, 228, 240, 242, 286
- system, VI, VII, 3, 4, 183, 185, 190, 191, 198, 200, 201, 213, 239, 269, 270, 282, 283, 285, 295, 303, 304
- system(s), 211

control
- parameter, 86–88, 175, 176
- strategies, 200
convergence, 72, 78, 102–104, 107, 120–122, 129–133, 185, 191, 199, 290, 291, 306
convergent, 107, 129, 130, 167, 169, 175
convexity, 147
corollary, 107, 120, 169
correlation matrix, 274–276
coupled
- nonlinear system, 275
- ordinary differential equations, 216
- oscillator(s), 228
- oscillators, 228, 230
- partial differential equations, 210, 220
- POD mode, 275, 276
criterion, 73, 159, 182, 276
critical
- buckling load, 264, 267
- clearance value, 195, 196
- load, 175
- speed, 304, 312, 314
cross
- section, 143, 154, 183, 191, 204, 272, 280, 304
cubicoid, 64, 73, 133
curvatures, 2, 3, 5, 9, 10, 13, 45, 53, 55, 94, 96, 102, 151, 163, 170
curvilinear
- orthogonal coordinate(s), 4, 5, 9
cusp(s), 156

damping, 1
- force, 305
- matrix, 218, 220
- properties, 298
data acquisition, 223, 310
defects monitoring, 195
deflection, 62, 68, 140, 143, 152, 170, 175, 304, 307, 310–314
- amplitude, 307, 313
deformation, 4–7, 10, 11, 13, 15, 17, 20, 23, 24, 31, 33, 40, 48, 53, 57, 99, 146–149, 154, 161, 170, 186, 187, 189, 210, 211, 258, 259, 291
degrees-of-freedom, 28, 40, 73, 78, 94, 179, 200, 211, 228, 279, 310

delocalization, 282
densely
- packed mode, 200
- packed resonance, 211
determination, 20, 38, 40, 42, 47–49, 53, 54, 56, 69, 71, 72, 81, 88, 110, 131, 133, 151
diagnosis of defects, 183, 195
differential
- equation(s), 3, 5, 10, 20, 25, 27, 29, 32, 38, 51, 60–62, 93, 101, 103, 104, 109, 144, 151, 152, 161, 164, 165, 170, 182, 183, 185, 191–193, 198, 201, 210, 213, 215, 216, 220, 250, 259, 269, 271, 275, 287, 289, 291, 304, 306
- operator, 96, 97, 144, 161, 164, 170
digital signal analyzer, 310
dimension, 15, 34, 96, 152, 156, 203, 220, 241, 244, 261, 280
dimensional subspaces
- finite, 129, 130
dimensionless, 21, 38, 48, 58, 74, 107, 134
Dirac's function, 3, 214, 259, 260
direct
- coupling, 229
- method, 181
directly excited beam, 292, 295, 299, 300, 302
disappear, 282
discrete
- additive, VI, 1
- array, 240
- loading effect, 239
- static forcing, 254
discreteness, 100
- effect(s), VII, 239, 241, 244, 248, 252, 254, 257, 263, 265, 267, 268
discretization, 101, 193, 215, 216, 275, 285, 289–291, 295, 296
discretized
- equation(s), 216, 290
-- ordinary differential, 190
dissipation
- coefficient, 288, 291
- term, 295
distributed dynamical system(s), 177

Index

distribution, VI, 4, 13, 41, 44, 65, 73, 78, 94, 119, 120, 126, 131, 133, 143, 144, 170, 173, 175, 183, 187, 188, 193, 195–198, 239, 248–250, 258, 263, 268, 272–274, 288, 291, 302
disturbance, 145, 200, 201, 282, 283
dominant mode(s), 189, 194, 196, 205, 210, 212, 213, 224, 226, 235–237, 278
double-impact, 295
double-sided rigid barrier, 302
dynamic(s)
- 'fast', 228, 238, 241
- 'slow', 228, 238, 241
- response, 221, 240, 278, 303
dynamical
- characteristics, 3, 4, 38, 83, 141
- phenomena, 1, 228, 229, 238, 269, 303

eigenfrequency, 3, 82, 295
eigenfunction(s), 3, 100, 104, 105, 112, 172, 177, 182, 274, 305
eigensolution, 101, 305
eigenvalue, 51, 64, 100, 101, 120, 177, 182, 201, 203, 218, 232, 262, 263, 266, 275, 276, 278
- lowest, 101
- problem, 51, 144, 146, 179, 181, 182, 211, 258, 263, 275
-- finite-dimensional, 177
-- reduced finite-dimensional, 181
- spectrum, 232, 234, 237, 238
eigenvector, 62, 182, 218
elastic
- continuum, 240, 245, 282
- impact, 289, 291, 292, 295, 305
- nonlinear foundation, 245
- parameter, 204
- string, 240
- support, 1
elasticity, VI, 1, 4, 7, 9, 24, 59, 82, 101, 183, 204, 218, 271, 288
energetic space, 96, 105
energy
- 'leakage', 285
- 'transfer', 187, 188
- deformation, 120
- dissipation, 218, 283, 284, 288, 291, 292, 305

- distribution, 188, 194–196, 198, 271–274, 279
-- plot, 198
- loss, 295
- plot, 195, 278
- spreading, 225
- surface, 189, 190
- transfer, 178, 235, 238, 269, 278, 282, 308
energy-capture, 217
engineering applications, V, VII, 182, 221, 286, 303
equation
- algebraic, 52, 102, 104, 151
- characteristic, 40, 44–47, 49–52, 56, 59, 64, 65, 69, 72
- ordinary homogeneous, 103
- transcendental, 64, 68, 69, 83
- variational, 25, 28
error, V, VI, 4, 54, 55, 59, 74, 77, 81, 88, 120–122, 132, 141, 155
error-prone stage, 204
essential boundary conditions, 189
Euclidean space, 146
excitation-dependent mode, 217
excited beam, 292, 295, 299, 300, 302
experimental
- acceleration field, 232, 233, 235, 236
- FFT, 299, 302
- fixture, 299, 308
- modal analysis, 183
- results, 222, 298, 299, 302, 303, 311, 312
- spatio-temporal behavior, 231
- spectrum, 223
- time series, 179, 310, 311
- truss, 222, 224
explicit solution, 69
explicitly satisfied, 190
exponential dichotomy, 201, 203
external
- edges, 148, 160
- forcing, 11, 171–173, 192, 204–206, 208, 216, 291, 304

fast
- dynamics, 179, 228–230, 235, 238
- invariant manifold, 233
- oscillations, 198, 228

Fast Fourier Transform, 204, 273, 295, 302, 307, 311, 312
fast-slow dynamics, 238
FFT, 204, 273, 295, 302, 307, 311, 312
- component(s), 311, 312
finite
- dimensional, 130, 131, 147
-- approximation, VI, 101, 129, 147, 151
-- basis, 182, 215
- elements, 115, 119, 120, 122, 148, 149, 151, 153–157, 200, 201, 285
finite-element, 135
- method, 102, 112, 120, 131, 133, 151, 285
- model, 189
finite-time sinusoidal pulse, 219
first
- critical speed, 311, 312, 314
- flexible mode, 280
- linearized
-- cantilever mode, 291
- vibration modes, 193, 194
first-order orthonormality condition, 182
flat vibration, 65, 69
flexible
- structure, 228, 283
- system(s), 179, 222, 277, 282, 303
flexural
- mode, 298
- rigidity, 304
fluid flow, 178
flywheel, 191–193, 304, 305, 307, 310
force transducer, 223
forcing frequency, 185–188, 230, 234, 302, 307
formal series, 266
forward sweep, 296, 308, 311–314
four-bay truss structure, 275
Fourier
- coefficient(s), 248, 249
- spectrum, 313
- transform, 204, 220
fourth boundary condition, 229
fractional-polynomial structure, 51
FRC, 105, 296, 297, 302, 303
free edge, 22

free-free boundary condition(s), 271
frequency
- 'peaks', 273
- 'shifts', 302
- acceleration spectrum, 223, 224, 226, 293, 294
- domain, 191
- fluctuations, 302
- fundamental, 53–56, 77, 81, 82, 85–88, 105, 120, 121, 123, 126, 130, 132, 134, 135, 139, 141, 155, 156, 158, 233
-- tone, 130
- ratio parameter, 229
- response curve, 296, 302
- spectrum, VI, 40, 45, 47, 73, 74, 77, 78, 84, 88, 90, 91, 94, 100, 101, 133, 145, 226, 295, 307
- sweep, 298, 302, 308, 312–314
frequency-amplitude plot, 311
functional, VII, 96, 126, 127, 129–132, 145, 155, 163, 171, 175, 176, 180, 181
- series, 130
functional-vectoral space, 96
fundamental
- beam frequency, 234
- frequency, 53–56, 77, 81, 82, 85–88, 105, 106, 120, 121, 123, 126, 130, 132, 134, 135, 139, 141, 155, 156, 158, 233
- solution, 61, 62, 64

Galerkin discretization, 271
Gauss formula, 5
generalized
- displacement, 202–204
- excitation vector, 203
- force, 203
geometric parameters, 65, 74, 84, 87, 91, 204, 288
global dynamics, 191, 278
graphitoplastic, 122
gyroscopic
- interaction, 194
- term, 304, 306

half-clearance, 253
Hamilton
- equation, 37
- principle, 10, 11

– rule, 25
Hamiltonian system, 178
hardening
– cubic, 249
– – stiffnesses, 254
– effect, 302
harmonic
– balance, 249, 253
– components, 253, 302, 303, 312, 314
– dependence, 248
– excitation, 170, 171, 173, 186, 302
– force, 183, 193, 230, 295
– load, VI, 170, 310
– vibrations, 1, 94, 202
heteroclinic orbit, 239, 254
high
– amplitude, 308
– frequency oscillations, 229
– modal density, 200, 211
– speed machinery, 304
– stiffness spring, 304
Hilbert space, 163, 175
homeomorphical, 96
homogeneous, 3, 22, 28, 29, 33, 38, 39, 43, 49, 63, 103, 104
– material, 10
– solution, 251
Hook's
– law, 8, 17
– principle, 5, 7
horizontal
– deflection, 306, 307, 310–313
– rotor deflection, 309, 312
hybrid form, 161
hydrostatic, 263, 267
– pressure, 258
hydrostatically forced, 264, 268
hyperbolic function, 68
hysteresis, 283, 298, 308, 311

identification tool, 183
impacting
– beam, 178, 183
– clearance, 298
impulsive
– excitation, 205, 218, 220
– force, 209, 219, 307, 310
– response, 201
– trapezoidal force, 210

in-phase fashion, 297
in-plane
– acceleration, 222
– response, 222
– vibration, 202, 213
indirect coupling, 229
inelastic, 287
– collision, 287, 299
– impact, 184, 288, 291, 292, 295, 305
– vibro-impact, 292, 293, 299
inertance, 204
inertia
– effect, 3, 29, 51, 57, 74, 77, 78, 89, 123, 160, 304
– rotation, 4, 40, 91
inertial moment, 16, 29, 38, 56
infinite
– degrees-of-freedom, 182
– series, 49, 241, 260
– spatial extent, 202, 205, 240
initial
– conditions, 10, 20, 23, 25, 27, 29, 162, 165, 170, 178, 184, 191, 250, 251, 270, 280, 282, 290–292, 295, 299, 310
– deflection, 291
– excitation, 291
instabilities, 145, 201, 203, 204, 238, 308
integral
– characteristics, 25, 37
– eigenvalue problem, 181, 182
– multiples, 307, 314
integrating by parts, 17, 111, 190
integro-differential equation, 111
interior, 129
internal, 158
– boundary, 202
– edge, 148
– resonance, 231, 236, 237
invariant
– manifold, 230, 233
– tori breakdown, 194
inverse operator, 96
inverse-Fourier function, 205
isoenergetic manifold, 280
isomorphism, 167
isotropic material, 8, 64, 81, 82, 86, 88, 122, 136, 140, 143

Index   335

joint imperfection, 282
jump, 90, 297, 298, 302, 303, 307, 308, 312, 313

K-L
– decomposition, 274
– mode, 275, 278
Kantorovitch method, 102
Karhunen-Loeve decomposition, 177
kinetic energy, 15, 25, 99
Kirchhoff-Love, 27, 28, 32, 33
– equation, 101, 160
– model, 33, 82, 101, 129, 131, 133, 151, 162
– theory, 31
Kronecker's symbol, 182

Lagrange's principle, 3
Lamé coefficients, 5
large deflections, 314
large-amplitude resonance, 314
large-scale structure, 189
laser displacement sensor, 310
leading resonance, 220
leading-order approximation, 250, 262
limit, VI, VII, 16, 34, 52, 53, 68, 69, 78, 96, 128, 129, 131, 133, 135, 146, 147, 158, 162, 167–170, 177, 178, 185, 188, 193, 196, 201, 203, 211, 229, 238, 253, 254, 286, 297
linear
– beam, 185, 292
– behavior, 188
– boundary, 266
– eigenvalue problem, 258, 275
– orthogonal basis, 177, 230
– rotor, 305, 313
– theory, 1, 4, 295, 303
– vibration theory, 213
linearized
– cantilever mode, 290, 291, 295, 296
– eigenfrequency, 295
local matrix, 119, 154
local perturbation solution, 264
localized standing wave, 254
long-dynamics, 241
long-scale, 241, 259
– variable, 248
loose joint, 269, 286, 303

loss of stability, 237
low-amplitude, 302, 308, 312
low-dimensional model, 177, 189, 269
low-frequency mode, 221
low-order
– invariant manifold, 230
– models, 178, 189, 193, 198, 201, 211, 219, 220
– reduction, 285
lower-upper factorization, 204

macro-slip effects, 269
mass imbalance, 192, 304, 310
material properties, 78, 141, 183, 219, 224, 298, 306
mathematical mode, 211
matrix
– coefficient, 66
– of eigenvectors, 218
maximum bandwidth, 310
mean surface, 5, 9, 24, 33, 65
mechanical joint(s), 269, 282
medium, 20, 40, 78
membrane
– analogy, 151, 152, 155
– force, 95
mesh, 9, 131, 148
micro-slip effects, 269
microstructure, 239, 244
middle-point response, 271–274
miniature triaxial accelerometer, 222
minimisation, 132
minimising series, 129, 130
minor, 59, 61, 62, 83
mixed-mode resonance, 205
modal
– amplitude, 215, 216, 218, 271, 276, 280, 289
– damping, 183
– impact hammer, 223
– matrix, 218
– shaker, 298, 302
– test(s), 290, 298, 302
movable support, 108, 310
multi-modal response, 194
multiply connected, 146, 148
mutually orthogonal, 100

natural

- boundary conditions, 189
- frequency, 183, 188, 200, 229, 231, 233, 271, 275, 290, 296–299, 301, 302

near-impulse excitation, 222, 223, 225–227

NLBVP, 244, 262–265

NNM, 230, 292
- bifurcation, 295

noise, 232

non-propagating, 203

non-smooth
- effect, 241, 248
- force distribution, 239
- transformation, VI, VII, 239–243, 254, 256, 259
- variable, 239, 241–244, 248, 258, 259, 262, 263

nondimensional set, 107

nonhomogeneous linear system, 250

nonlinear, VII, 83, 93, 111, 130, 131, 160, 168, 170, 175, 177, 179, 182–185, 187–189, 195, 196, 228, 229, 239–241, 244, 269, 272, 273, 276–279, 282–284, 289, 290, 295, 297–299, 302, 303, 308
- boundary value problem, 244, 262
- hysteresis, 298, 308, 311
- normal mode, 230, 292
- springs, 240, 247, 269, 270, 280
- transition, 185

nonlinearity ratio, 295

nonlocalized branches, 298

nonsmooth effect, 297

nonuniquess, 147

normal mode, 64, 230

normalized
- internal force, 258
- variable, 259

numerical
- acceleration, 208, 209, 292
- algorithm, VI, 154, 290
- instabilities, 201, 203, 204
- integration, 184, 191, 218
- inversion, 201, 203–205
- mode shape, 234
- receptance, 205–207
- simulations, 183–185, 193, 200, 218–221, 224, 232, 254, 271, 275, 277–279, 285, 290, 291, 295, 296, 299, 302, 305–308, 310, 313
- – 'exact', 192

numerically stable, 201, 203–205

numerous peaks, 295

objective function, 175, 179

oil
- whip, 314
- whirl, 314

operator, 96–98, 104, 105, 126, 162–165, 169, 170, 175, 176, 180, 334

optimization, VI, 83, 93, 124, 125, 127, 135, 137, 138, 140, 143–148, 150, 151, 155–160, 173, 175
- problem, VI, 93, 126, 133, 145, 147, 151, 176, 181

orbit, 191, 196, 197, 237, 239, 254, 280–284, 306–308, 311, 334

order reduction, VI, 177–179, 182, 198, 200, 222, 269, 274–276

ordinary differential equations, 60, 103, 104, 112, 164, 165, 190, 215, 216, 250, 259, 271, 289

orthogonal, 5
- base, VII, 177, 199, 211, 215, 217, 230, 275, 277
- condition(s), 189
- mode, 177, 212, 271
- – decomposition, 177

orthogonality, 100, 101, 215
- condition(s), 190, 289, 305

orthotropic, 4, 24, 33, 53, 61, 73, 94
- plate, 56–60, 65, 84
- shell, VI, 2, 3, 5, 9, 20, 29, 35, 38, 39, 45, 47, 49, 56, 58, 94

oscilloscope, 303

out-of-phase periodic response, 298

overhung rotor, 191–194

parameters
- dimensionless, 21, 58, 229
- geometrical, 41, 60, 65, 72, 74, 84, 86–88, 91, 134, 135, 140–142, 204, 288

parametric excitation, 250

partial differential equations, 93, 182, 183, 185, 191–193, 198, 201, 210, 213, 215, 220, 269, 271, 275, 291, 304

passively control, 283
penalty, 126, 132, 145
- multiplier, 171
period doubling, 194, 308
- route, 308
period-2, 307, 308
- motion, 312
periodic
- bay, 201
- motion, 197, 232, 296, 308
- orbit, 280–283, 308
- oscillations, 231
- regime, 236, 237
- set, 201, 202, 211
periodically spaced load, 258
periodicity conditions, 259
perturbation
- analysis, 239, 241
- series, 262
phase plot(s), 254, 255
physical buckling state, 261
pillow block bearing, 308, 310
plastic-glass material, 122
plate(s), V, VI, 1, 3, 4, 30, 51–53, 56, 57, 60, 62, 64, 65, 69, 71–78, 81–91, 93, 94, 100–102, 104, 120–126, 129, 130, 132–145, 147, 148, 151, 152, 155–158, 160, 161, 173
- circle, 133
- rectangular, 2, 93, 102, 133, 134, 141
- square, 155
- transversal-isotropic, 72, 73, 79–82, 86–88, 90, 91, 122, 123, 134, 140, 141
- triangle, 1
POD, VI, 177–179, 182, 183, 189, 191, 200, 201, 210, 215, 217, 230, 274, 278
- analysis, VII, 177, 185, 210, 211, 228, 230, 231, 235, 238
- decomposition method, 277
- energy distribution, 189, 190, 193–196
- mode(s), VII, 177, 178, 181–183, 185–191, 193–196, 198, 200, 210–213, 215, 217–220, 222, 224–227, 231–238, 274–279, 282
Poincaré map, 269, 278–282, 284
point load, 239
pointwise convergence, 191

Poisson's coefficient, 8, 33
polymers, 93
polynomial, 47, 51, 52, 157
positively definite, 96
post processing, 177, 185, 223, 310
post-buckled, 267, 268
postbuckling behavior, 258
potential energy, 2, 11, 17, 25, 33, 35, 36
pressure wave, 211
projection, 6, 106, 144, 152, 158, 161, 170, 171, 173, 305, 306, 311
proof, 96, 98, 100, 105–107, 127, 129–131, 147, 162, 163, 168, 174, 176
proper orthogonal decomposition, VI, 269, 274
purely elastic impact(s), 287, 291, 292, 295

quasi-ε-optimal, 125
quasi-periodic
- motion, 197, 231, 233, 307
- orbit, 196, 237, 308
- response, 297

radar antennas, 200
random
- process, 178
- signal decomposition, 178
rank, 61, 62, 65
Rayleigh
- formula, 127, 129
- principle, 2
- quotient, 101, 120, 147, 220
receptance, 204–207, 220, 221
reconstruction, 179, 189, 191, 198, 199, 219, 221, 269, 274, 276, 277
reduced-order
- model, 200, 210, 211, 213, 217, 220, 221, 238, 306
regular, 146, 177, 191, 194, 198, 230–232, 234, 237–239, 244, 262, 308
resilient elements, 240
resonance point, 204
response, V, VII, 173, 177, 179, 181, 188, 189, 191, 193, 194, 196, 198–201, 203–205, 210, 219–222, 231, 235–239, 241, 247–249, 253, 271–278, 282,

284–287, 292, 295–299, 303, 307, 312, 335, 336
restoring force, 253
reverse jump, 308, 313
rib, 1, 2, 12
rigid
– barrier, 286, 289
– barriers, 253, 289, 290, 292, 296, 300, 302
– clamping, 22
– constraints, 287, 288, 297–299
rigid-body deformation, 211
ring
– plate, 1
– shell, 1
rolling support, 102, 108, 120, 123, 124, 134, 141, 151, 152, 155
root(s), 40, 45, 52, 61, 62, 64, 65, 69, 72, 84, 229, 280
– multiple, 61
– simple, 61, 65
– triple, 62
rotary
– inertia effect(s), 304
– system, 303, 306, 314
rotation inertia, 4, 40, 56, 74, 77, 78, 81, 89, 91, 134
rotor, 191–196, 198, 200, 304–314
– dynamic system(s), 178
rotor-flywheel
– system, 303
rotor-flywheel system, 304

Schwartz iterational method, 120, 143, 154
self-conjugated, 164
semicontinuous, 175
sensing location, 177, 179, 181, 230, 231
shaft, 154, 156, 191, 308, 310, 311
shaft-flywheel interaction(s), 194
shaft/motor misalignment, 310
shaker, 230, 298, 302, 303
shaker-structure, 303
shear, 8, 10, 15, 33, 40, 143, 198, 201, 205, 258, 259, 267, 268
shearing force(s), 95
shell, V, VI, 1–6, 9–13, 16–18, 20, 22, 23, 25–35, 37, 38, 40–43, 45, 47–49, 51–56, 72, 82, 83, 91, 93, 94, 99–103, 105, 126, 129, 130, 132, 133, 144–148, 150–152, 155, 156, 158–161, 163, 165, 170–173, 175, 258
– spherical, 53, 55, 56, 151, 155
– surface, 2, 5, 9, 10, 18, 23, 32, 33, 132, 144, 151
ship-based crane(s), 228
short
– dynamics, 241
– scale, 241, 248, 259
– – oscillations, 241
– – spatial, 241, 242
– – variable, 243, 246
signal
– analyzer, 223, 310
– dependent, 178
simplified model, 29
simply supported rotor, 196
Simpson's rule, 119
sine-sweep(s), 298
singular, 41, 66, 238, 239, 242, 244, 260, 261
– perturbation, 238
– – theory, 228, 230
singularities, VII, 4, 48, 133, 239, 241–243, 258–260
singularly perturbed problems, 230
six degrees-of-freedom, 220, 221
sliding cover, 22
slope, 10, 20, 53, 170
– distribution(s), 250
slow
– manifold, 233
– oscillations, 228, 230
slow-fast
– chaotic motion, 233
– dynamic, 228
– dynamical interaction, 228, 230, 238
– interaction, 229
smoothening, 242, 244
– condition, 261
smoothly modulated amplitude, 237
snapshot, 179–182, 185
– method of, 181, 183
solitary
– solution(s), 250, 251
– standing solution(s), 256
– wave, 252–254, 257

space, VII, 17, 96, 105, 106, 119, 126, 145, 146, 162, 163, 167, 172, 174, 175, 179–182, 201, 203, 211, 230, 240, 258, 295, 306, 311
– conjugate, 175
– station, 200
spatially
– confined, 295, 297
– periodic, VII, 254
spatio-temporal
– behaviour, 231
– data, 230
– domain, 274
spectral line(s), VI, 144, 145, 147, 153, 173, 175
spectrum, VI, 40, 45, 47, 73, 74, 77, 78, 81, 84, 88–91, 94, 100, 101, 133, 223, 224, 232, 234, 237, 238, 293–295, 307
– optimization, 160, 173
spherical shell, 3, 53, 55, 56, 151, 155
spread, 273, 292
spreading, 225, 292
spring characteristic, 280
stable
– invariant manifold, 230
– solution, 146
standard asymptotic technique(s), 258
standing solitary wave, 252–254, 256, 257
static
– buckling, VII, 256
– – problem, 254, 256
– equilibrium, 229
stationary state, 171
stead state
– motion, 302, 307, 312
– response, 295, 297, 312
steady state
– localization, 298
– motion, 297, 303
– response, 298
steady-state, 202, 203, 296, 307
stiff
– spring-damper element, 184
– support, 102, 120, 121, 124, 143, 144
stiff-soft
– coupled oscillator(s), 230
– motion, 228

– system, 238
stiffness, 2, 4, 26, 33, 36, 48, 72, 81, 82, 86, 89, 94, 119, 122–125, 139, 141, 154, 198, 240, 241, 249, 270, 272, 273, 278, 279, 282, 283, 286–289, 295, 301, 302, 305
– characteristic(s), 33, 48, 58, 244, 252, 254, 270, 289, 291
– coefficient(s), 8, 58, 95, 107, 198
– cylindrical, 152
– finite transversal, 125
straightforward approach, 203
stress, 4, 22, 69, 77, 146, 148, 151, 154, 170
stressed component(s), 22
stretching-compression, 81
strong, 83, 128, 131, 150, 167, 169, 176, 188, 194, 196, 278, 295, 314
– chaotic regime, 235
– nonlinear effect(s), 188, 299
stronger
– clearance nonlinearity, 277
– resonance, 220
strongly localized NNM, 295
structural
– damping, 214, 218
– modification, 189, 201, 282
structure-barrier interaction, 287
subharmonic, 273
– orbit, 280, 282–284
– oscillations, 272
– response, 282
substructure, 201, 228, 230, 269, 282, 298
subsynchronous, 314
– ripple, 314
sudden
– jump, 308
– transition(s), 303
suddenly chaotic, 308
superimposed, 223, 254, 256
surface
– area, 145
– control algorithm, 147
– minimal projection, 158
– optimized, 157
– relaxed, 146
– shape, 145

surgical plastic, 298
sweep, 296–298, 302, 308, 311–314
switching point(s), 184
symmetric, VII, 4, 7, 14, 56, 64, 69, 78, 90, 96, 98–102, 104, 181, 182, 201, 276, 303
– forcing, 211
symmetry, 7, 10, 29, 96, 99, 100, 264, 298, 303
– condition(s), 261
synchronous, 314
synthesis, 144–147, 201
system(s)
– high modal density, 182, 200
– identification, VII, 177–179, 182, 183, 200, 201, 211, 278
– spatially extended, 182, 200, 201

target function, 93, 135, 142, 156, 158
ten-rod system, 287
tension, 4, 5, 7, 8, 20, 101, 229, 241, 298
tensor, 4, 15
– antisymmetric, 14
theoretical
– bifurcation point, 302
– FRC, 302
– model, 299
– simulation, 299
theories non-classical, VI, 4
theory
– elasticity, 4, 6, 9, 82
– Kirchhoff-Love, 31
– three dimensional, VI, 4, 59, 65, 73, 74, 77, 78, 81–84, 86, 88–91
– two dimensional, 23, 40, 59, 73, 77, 78, 82, 90
thermal behavior, 178
thin shell, 258
time series, 177, 179, 181, 182, 185, 187–189, 196, 205, 208–212, 219, 230, 231, 271–274, 277, 280, 283, 284, 293, 294, 296, 307, 310, 311
time-averaged
– energy, 278
– – distribution, 272
time-delay techniques, 178
time-periodic solution, 249, 254

Timoshenko's kinematic model, 4, 23, 27, 28, 51, 56, 74, 77, 81, 91, 94, 101, 123, 133, 134, 151, 160, 161
Timoshenko-like
– conical shell, 107
– equations, 102
tip acceleration, 292, 294
torsion, 154, 156
transcendental equation(s), 64, 68, 69, 83
transfer matrix, 201–203
transformation, VI, VII, 32, 39, 52, 56, 57, 147, 169, 239–244, 254, 256, 259
– weakly continuous, 129, 131
transient
– dynamics, 201, 205, 283
– modal evolution, 276
– motion, 285, 292, 293
– response(s), 220, 277, 286, 287, 299
transversal, 2–4, 7, 8, 48, 53, 56–58, 72, 74, 77, 81
– stiffness, 53, 58, 81, 82, 86, 94, 122–124, 139, 141
trapezoidal vertical impulsive excitation, 205
trial functions, 193, 198, 199, 220
triangle(s), 148–150, 153–156
– plate, 1
– shell, 1
truss, VII, 200–211, 215–220, 222–224, 226, 283, 284
– 18-bay, 202–204, 210, 211
– configuration, 202
– dynamics, 200, 201, 210–212, 217–221
– four-bay, 213–215, 220
– – structure, 275
– impulsively loaded, 205
– resonance(s), 204, 205
– three-bay, 222
– time series, 210
turbulent thermal convective system, 178
two-dimensional
– invariant manifold, 230
– model, 40
two-point correlation function, 180
two-scale form, 259

unbounded solution, 250

uncoupled
- beam, 201, 298
- pendulum, 229, 230
unexcited beam, 295, 301, 303
unforced
- beam, 297, 298
- system, 230, 255
uniform forcing distribution, 267
unique solution, 147
unstable invariant manifold, 230
unsymmetric, 261

variational
- equation, 10, 25, 28
- iteration, VI, 102, 104, 133
- step, 120
- vector, 6, 10, 14, 15, 17, 22, 25, 42, 43, 45, 48, 61, 62, 67, 68, 101, 102, 115, 134, 146, 154, 166, 170, 174, 175, 181, 202, 203, 218
vertical
- deflection, 306, 307, 310, 312–314
- rotor deflection, 304, 309, 312
vibrating structural assembly, 183
vibration
- absorber, 228
- mode, 56, 60, 69, 71, 73, 77, 78, 81, 90, 101, 103, 140, 143, 172, 177, 178, 186, 193, 194, 198, 200, 205, 211, 212, 215, 217, 275, 277, 279
vibrational energy, 283, 297, 298
vibro-impact
- 'weak', 198
- beam, 189, 292
- dynamics, 291, 303
- effects, 184
- fixture, 300
- limit, 254

- nonlinear effects, 289
- nonlinearities, VII, 269, 285–287, 298, 299, 302, 303, 314
- oscillations, 183
- problem, 286
- response, 292, 295
- term(s), 184, 289, 291
vibro-impacting
- beam, 183–185, 187, 196, 288, 292, 295
- beams, 286
- rotor, 303
- system(s), 178, 295
vibroisolation, 159, 160, 170, 171, 173–175
vigorous exchange, 278
viscous
- damping, 288, 304
-- coefficient(s), 191, 229, 304

waterfall
- plot, 312–314
wave
- propagation, 201, 282
- solution, 253
wavemode, 201, 203, 205
- conversion, 200
weak, 128, 163, 167–170, 176
- chaotic regime, 235
- coupling, 228, 229, 234
- nonlinearity, 240
Weierstrass theorem, 131, 147
workstation, 310

Young's modulus, 8

zero-to-peak amplitude, 296, 297

Printing: Strauss GmbH, Mörlenbach
Binding: Schäffer, Grünstadt